The Journal of Biological Chemistry

428 East Preston St., Baltimore, MD 21202 U.S.A.

1991 Minireview Compendium

CONTENTS*

*Page numbers refer to the original page numbers as printed in the 1991 issues.

Author Index

Minireview

THE JOURNAL OF BIOLOGICAL CHEMISTRY
Vol. 266, No. 1, Issue of January 5, pp. 1-4, 1991

Printed in U.S.A.

Actin: Protein Structure and Filament Dynamics

Marie-France Carlier

From the Laboratoire d'Enzymologie, Centre National de la Recherche Scientifique, 91198 Gif-sur-Yvette, Cedex, France

The property of monomeric globular actin (G-actin) to polymerize into noncovalent helical filaments (F-actin) is fundamental to its biological activity in all eukaryotic cells (1). In nonmuscle cells actin filaments are dynamic and undergo self-assembly and disassembly to extents and at turnover rates that are finely regulated by associated proteins and that vary from one place to another in the cell according to the different motile functions in which actin is involved. The changes in the average length of filaments will cause changes in the viscoelastic properties of the cytoplasm. It is therefore important to understand the unique structural properties of the actin molecule that are at the origin of its ability to polymerize and to elucidate the mechanism of polymerization itself in order to anticipate possible modes of regulation. The hydrolysis of actin-bound ATP, which is linked to filament assembly, was early shown to complicate the classical thermodynamics of actin polymerization developed by Oosawa (2) and to introduce the possibility of subunit flux in the filament, due to different critical concentrations at the two ends (3). In the past decade, more progress has been accomplished. We now understand better how ATP hydrolysis affects the dynamics of actin filaments; we also know more about the binding of ATP, the structure of the ATP binding site, and the mechanism of ATP hydrolysis. Very recently, the structure of the actin monomer has been solved at atomic resolution which provides new bases for the approach of structure-function relationship in actin and for analyzing its interaction with actin binding proteins. The present paper will focus on these new points and update previous reviews on the same subject (4–7). Other reviews cover the related fields of actin binding proteins (5, 8), actomyosin interaction in muscle contraction (9, 10) and in nonmuscle cells (11, 12), actin ADP-ribosylation by *Clostridium* toxins (13), and the cellular control of cortical actin assembly by external stimuli (14).

Actin Filament Structure

The actin filament can be described by either a one-start left-handed genetic helix of 5.9 nm pitch or by a two-start right-handed helix of 72 nm pitch (15). Torsional motion of actin subunits in the filament has been reported (16, 17). Accordingly disorder appears on images of actin filaments in the electron microscope. The variable twist can be accounted for by a model of cumulative angular disorder (18) that appears to be modulated by actin binding proteins (19). Reconstructions of actin filaments from images of negatively stained or frozen hydrated specimens lead to a structure in which the strongest actin-actin contacts are along the small pitch genetic helix (18, 19). Other works, however, offer the contrasting view that bonds along the long pitch two-strand helix are the strongest, as suggested by the occasional separation of the two strands (20).

ATP and Divalent Cation Binding to Actin

G-actin binds tightly one ATP and one divalent metal ion that can be Ca^{2+} or Mg^{2+} and is thought to be Mg^{2+} in the cell. The affinity of this metal ion was recently shown to be in the 10^9 M^{-1} range (21–23). The dissociation of tightly bound Ca^{2+} is slow (24) and rate-limiting in the overall process of nucleotide exchange (25). Actin also binds mono- or divalent cations (Mg^{2+}, Ca^{2+}, K^+ . . .) to a series of lower affinity sites with affinities in the 10^5–10^4 M^{-1} range (23, 26). How binding of cations to these different sites affects the conformation of actin (27), the binding of ATP, and the polymerization properties has been a subject of intense debate in recent years. Both kinetic and physical measurements of metal ion and nucleotide interactions with G-actin failed to provide a consensus answer to the issue of the proximity of tightly bound metal ion and ATP on actin (see Refs. 25 and 30 for review). The picture emerging now is that rapid binding of either Mg^{2+}, Ca^{2+}, or K^+ to low affinity sites represents the "monomer activation step," *i.e.* the preliminary step in actin polymerization (23, 28, 29); on the other hand, slow Mg^{2+} exchange for Ca^{2+} at the tight binding site is associated with a different conformational change of the protein (23). Recently use of $\beta\gamma$ Cr-ATP, an exchange inert analog of Mg-ATP, led to the conclusion that the tightly bound divalent metal ion directly interacts with the β- and γ-phosphates of ATP on actin and that metal-ATP is bound in the Λ configuration (30). These last two results are now confirmed by the recent three-dimensional structure of actin (31). The actin ligand therefore is the metal-ATP complex as described by the following scheme,

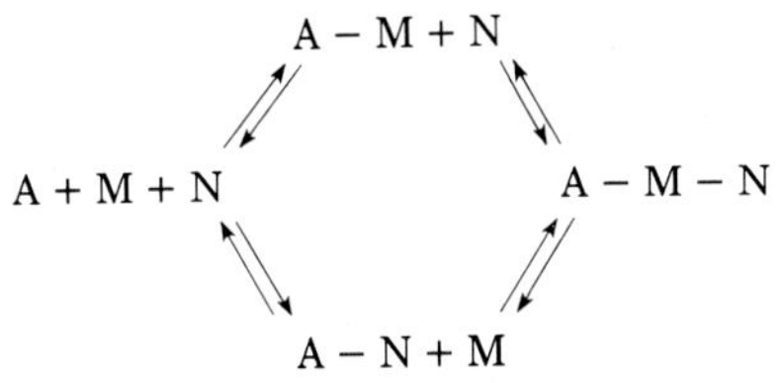

where A is G-actin, M the tightly bound metal ion (Ca^{2+} or Mg^{2+}), and N the nucleotide. The reported 3–5-fold higher affinity of Ca^{2+} *versus* Mg^{2+} for actin means that Ca-ATP binds 3–5-fold more tightly than Mg-ATP. Examination of the rate parameters for the binding of ATP, metal ion, and metal-ATP (Table I) shows that nucleotide exchange follows different routes according to whether Ca-ATP or Mg-ATP is bound. Ca-ATP dissociates at a much slower rate than via the consecutive dissociation of Ca^{2+} ion and nucleotide, which accounts for the dependence of the rate of ATP dissociation on Ca^{2+} ion (25); in contrast, the rate constant for dissociation of Mg-ATP is practically as slow as Mg^{2+} dissociation. Therefore, the fast nucleotide exchange observed on Mg-ATP-actin implies that ATP can dissociate from actin with Mg^{2+} remaining bound to actin. In conclusion actin can be either a Ca- or a Mg-ATPase, which we will see affects in turn the dynamics of actin filaments.

Three-dimensional Structure of Actin

Crystals were obtained from the 1:1 complexes of actin with profilin (32) and DNase I (31), two relatively small proteins that prevent actin from polymerizing and whose three-dimensional structures are already known. Crystals of profilin-actin (obtained in the absence of divalent metal ion) show a flat zigzagging ribbon structure similar to a flattened filament

TABLE I

Comparison of Ca-ATP-actin and Mg-ATP actin

ATP, ADP, and divalent metal ion binding parameters

		Ref.
K_D (ATP)[a]	10^{-10} M	68
k_+ (Ca-ATP)	7.10^6 M^{-1} s^{-1}	25
K_D (Ca)[b]	$1.9.10^{-9}$ M	22, 24
k_+ (Mg-ATP)	2.10^6 M^{-1} s^{-1}	25
K_D (Mg)[b]	10^{-8} M	22, 24
k_{-Ca} (ATP-G-actin)	0.06–0.18 s^{-1} (pH 8)–0.02–0.04 s^{-1} (pH 7)	24, 25
k_{-Ca} (ADP-G-actin)	2.8 s^{-1}	58
k_{-Mg} (ATP-G-actin)	0.013 s^{-1}–0.007 s^{-1}	22, 25
k_{-Mg} (ADP-G-actin)	0.04 s^{-1}	58

Polymerization parameters at barbed (B) and pointed (P) end of actin filament

Bound ligand	Conditions	C_c^B	C_c^P	k_+^B	k_+^P	$k_{hydrolysis}$	Ref.
		μM	μM	μM^{-1} s^{-1}	μM^{-1} s^{-1}	s^{-1}	
Ca-ATP	20 mM KCl, 0.1 mM $CaCl_2$	1	0.9	5.9	0.8		58, 63
	75 mM KCl, 0.2 mM $CaCl_2$	0.4	0.4				45
	100 mM KCl, 0.2 mM $CaCl_2$	0.44	0.73		1.5	0.01	60
Mg-ATP	150 mM KCl, 5 mM $MgSO_4$	0.1	1.2	3.0	1.2		61
	100 mM KCl, 1 mM $MgCl_2$	0.07		5.2			62
	1 mM $MgCl_2$	0.14	4	1.7	0.1	13	46, 47, 58, 60

[a] This equilibrium dissociation constant for ATP binding to G-actin was measured in the presence of Ca^{2+} ions and therefore can be attributed to the equilibrium dissociation constant of Ca-ATP.

[b] This equilibrium dissociation constant refers to binding of the metal ion to ATP-G-actin. The relative affinities of Ca^{2+} and Mg^{2+} for ADP-G-actin are reversed (59).

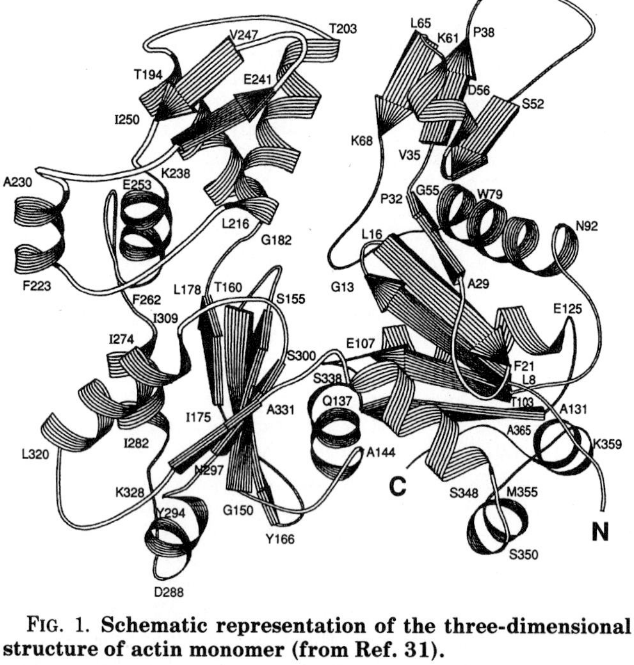

FIG. 1. **Schematic representation of the three-dimensional structure of actin monomer (from Ref. 31).**

with profilin molecules intercalating the F-actin subunits. Crystals of DNase I-actin (obtained by the addition of millimolar amounts of Mg^{2+} ions to the Ca-ATP-G-actin-DNase I 1:1 complex) have led to the three-dimensional structure of the actin monomer at 2.8-Å resolution shown in Fig. 1. Actin consists of a small and a large domain, each divided into two subdomains. The nucleotide is bound in a cleft between the two domains. The very high affinity of ATP is accounted for by the very tight arrangement of residues around the triphosphate moiety of the molecule. A single tightly bound Ca^{2+} interacts with the β- and γ-phosphates of ATP in the Λ conformation and also with residues D154, Q137, and D11. N-terminal residues 14–16, known also to interact with myosin (33), are in close vicinity to the γ-phosphate of ATP. The solved structure corresponds to Ca-ATP-actin and may be different from Mg-ATP-actin. An important issue is whether the solved structure is that of G-actin or of F-actin or yet a different one. Several points favor second interpretation including the facts that the crystallization buffer actually is a polymerization buffer and that DNase I binds to the pointed end of the actin molecule, thus mimicking a neighboring subunit in the filament. In the derived atomic model of the actin filament (34) which is consistent with previously determined radial positions of selected residues (35), the C terminus of actin is at the barbed end of the filament, as anticipated from its cross-linking to profilin and gelsolin (36, 37), and the strongest actin-actin bonds are along the two long pitch helices of the filament.

Mechanism of Polymerization and Involvement of ATP Hydrolysis

Many independent observations pointed to a rapid change in G-actin conformation upon addition of salts prior to nucleus formation or monomer incorporation into filaments. Recent findings that (i) the tight binding of either Mg-ATP or Ca-ATP determines actin conformation (23, 24, 27, 38) and nucleation rate (38–43), (ii) cation binding to low affinity sites allows both Ca-ATP-actin and Mg-ATP-actin to elongate filaments (41, 44); and (iii) the rate of elongation is similar regardless of the nature of the cations bound to the low affinity sites (28, 40) rather encourage the view that binding of mono- or divalent cations to multiple sites activates Ca-ATP-G-actin or Mg-ATP-G-actin.

It is well established that filaments are formed by the helical, as opposed to isodesmic, polymerization of actin. This mechanism involves an energetically unfavorable nucleation step followed by elongation off the nuclei by endwise addition of subunits. Kinetic analyses of polymerization curves indicate that the nucleus is a trimer (4, 5), in agreement with the helical structure of the filament (34).

The first steps of the polymerization process can be described as follows.

$$\begin{array}{llll}
A + nM & \underset{}{\overset{K}{\rightleftharpoons}} & AM_n = A & \text{monomer activation} \\
A' + A' & \overset{K_1}{\rightleftharpoons} & A'_2 & \\
A'_2 + A' & \overset{K_2}{\rightleftharpoons} & A'_3 & \text{nucleation} \\
A'_3 & \overset{K_{is}}{\rightleftharpoons} & F_3 & \\
F_i + A' & \underset{k_{-e}}{\overset{k_{+e}}{\rightleftharpoons}} & F_{i+1}\ (i > 3) & \text{elongation}
\end{array}$$

A' represents the activated monomer (either Ca-ATP-G-actin or Mg-ATP-G-actin) and M the cations binding to n low affinity sites. This scheme is valid for polymerization of both ATP- and ADP-actin, with different thermodynamic parameters. ATP hydrolysis accompanies the elongation process.

Conventional reversible helical polymerization models applied to actin involve the existence of a critical concentration, which actually is the elongation equilibrium constant $K_e = k_{-e}/k_{+e}$ (2). While this concept well describes the polymerization of ADP-actin, which is truly reversible, it does not apply to the polymerization of ATP-actin that is accompanied by the irreversible hydrolysis of ATP. In the presence of ATP, a steady-state monomer concentration can be measured at which net filament growth is 0, with one end undergoing net positive growth and the other end depolymerizing at the same rate (3, 45–48). This apparent critical concentration does not have the physical significance of a thermodynamic equilibrium constant. It results from a complex combination of the critical concentrations at the two ends in which the energetic balance of ATP hydrolysis must be included. A complete description of the polymerization mechanism and adequate fitting of a mathematical model to the polymerization curves therefore requires prior understanding of the involvement of ATP hydrolysis in the polymerization reaction.

The hydrolysis of ATP associated with actin polymerization consists of two temporally distinct steps (49) that follow the incorporation of the ATP monomer in the filament: chemical cleavage of ATP followed by the slower release of P_i in the medium. Cleavage of ATP is irreversible, while P_i release is reversible (50).

On Mg-ATP-F-actin, cleavage of ATP occurs essentially vectorially during the polymerization process (60), meaning that a clear-cut boundary exists between F-ADP-P_i and F-ATP subunits in the filament. The change in the composition of actin filament during polymerization is illustrated in Fig. 2. The net result of this mechanism is that filaments have terminal F-ATP subunits at fast growth rates, terminal F-ADP-P_i subunits at moderate growth rates, and F-ADP subunits in a regime of depolymerization. Evidence that the nature of the terminal subunits affects the dynamics of actin filaments is provided by the nonlinear dependence of the rate of elongation on monomer concentration, from which the rate constants for monomer association to and dissociation from filament ends can be derived (51). Data show that when filaments are growing, newly incorporated terminal subunits dissociate at a slower rate than F-ADP subunits, *i.e.* form a stabilizing cap.

Actin filaments entirely made of F-ADP-P_i subunits can be obtained by the binding of P_i (52) or its high affinity analogs AlF_4^- and BeF_3^- (53) to F-ADP-actin subunits. In contrast neither P_i (54) nor its analogs bind to G-ADP-actin. The reconstituted F-ADP-P_i filaments have the same dynamic properties as the transient F-ADP-P_i polymer, *i.e.* they lose subunits at a 5–10-fold slower rate than the F-ADP polymer (52), which explains the decrease in critical concentration of ADP-actin upon addition of P_i (55). At steady state F-ADP-P_i subunits are present at the fast growing barbed ends only, because polymerization is too slow at the pointed ends for any F-ADP-P_i subunits to accumulate. A large free energy change is linked to P_i release; indeed in the presence of saturating amounts of P_i, the nonlinear monomer concentration dependence of the rate of growth vanishes as does the difference in critical concentration between the two ends (52, 55). Finally, the stabilizing effect of P_i binding, which scales as kT (where k is the Boltzmann constant and T the absolute temperature), develops at physiological concentrations; hence P_i might regulate structural and dynamic properties of actin filaments *in vivo*.

ATP hydrolysis evidently regulates polymerization kinetics since the same set of rate constants does not describe the whole time course of polymerization, and elongation curves are not true exponentials as for a reversible polymerization mechanism; upon approaching steady state, the probability increases for terminal F-ADP subunits that dissociate at a fast rate (51). The involvement of ATP hydrolysis also complicates the analysis of monomer-polymer exchange reactions at steady state, because it provides a more dynamic pattern of filament turnover (via extensive depolymerization of a few individual filaments, a process also called dynamic instability (56)) than either the classical model based on reversible association-dissociation of subunits or the treadmilling model (3), which both fail to describe the kinetics of monomer-

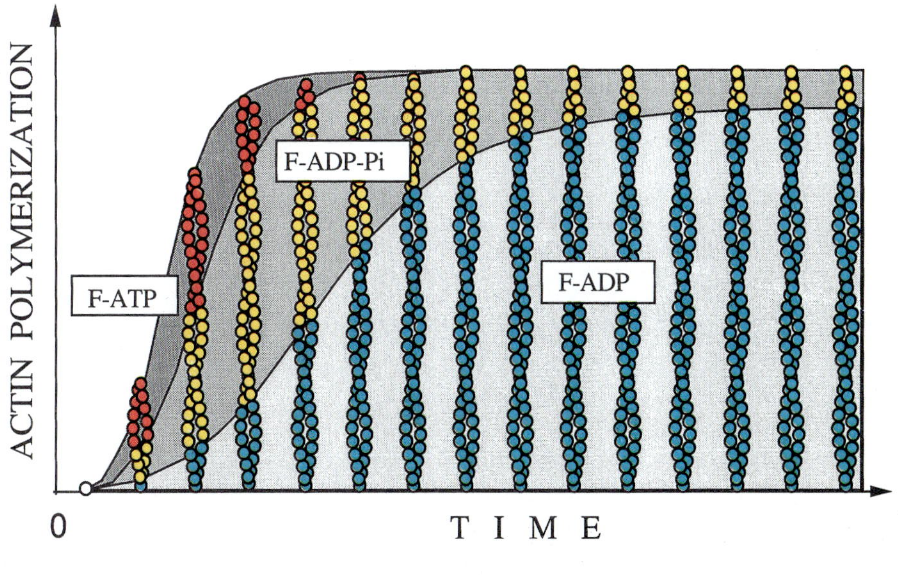

FIG. 2. **Schematic evolution of the actin filament during the time course of polymerization.** Filaments grow initially at a fast rate with terminal F-ATP subunits (in *red*); at later stages F-ADP-P_i subunits (in *yellow*) transiently accumulate; at steady state, the filament is made of F-ADP subunits (in *blue*) except for terminal F-ADP-P_i subunits at the barbed end.

polymer exchange adequately (57). In the cell, maintenance of this dynamic steady state costs a lot of ATP. When a dynamic state is not needed, blockage of the barbed ends by capping proteins provides stability of filaments and saves energy.

The main consequences of ATP hydrolysis on F-actin is therefore to destabilize the actin-actin interactions in the polymer and facilitate depolymerization. The liberation of P_i is the elementary step linked to this destabilization, *i.e.* to a structural change of the filament.

Comparison of the Polymerization Properties of Ca-ATP-Actin and Mg-ATP-Actin

Table I lists the polymerization, nucleotide binding, and ATP hydrolysis parameters measured for pure Ca-ATP-actin or Mg-ATP-actin under defined ionic conditions. Mg-ATP-actin exhibits a higher tendency to form nuclei (38–43), a faster elongation rate, and faster hydrolysis than Ca-ATP-actin. Mg-F-actin filaments are more dynamic than Ca-F-actin filaments because they display a larger difference between the dissociation rate constants of ADP and ADP-P_i (or ATP) subunits (41). In other words, ATP hydrolysis on Mg-F-actin causes a larger structural change of the filament than when it occurs on Ca-F-actin. Accordingly, the energetic difference between the two ends is large for Mg-actin while the critical concentrations (C_c) are very similar at the two ends of Ca-F-actin filaments. These differences should appear more clearly when the three-dimensional structure of Mg-actin is available.

Similarities to Other Systems, Conclusion and Perspectives

The regulation of actin filament dynamics by ATP hydrolysis is a general feature common to many other biological systems. The closely related microtubule system, the other main component of the cytoskeleton, exhibits the same regulation via GTP hydrolysis that accompanies tubulin polymerization (64). Other systems also use the chemical energy of ATP or GTP hydrolysis to regulate the strength of interaction between two components; G-proteins involved in signal transduction (65), RecA protein, which is involved in genetic recombination (66), and the microtubule-based motor kinesin (67) are other examples.

Knowledge of the three-dimensional structure of actin in different conformational states and subsequent site-directed mutagenesis experiments will certainly help, in the future, to elucidate the structure of the filament, which has important implications in the mechanism of actomyosin interaction, but also will enable an understanding of the mechanism of ATP hydrolysis and the related conformational switch in greater detail. The current structural and mechanistic data suggest that actin filaments can exist in several states. Whether these states have a biological significance and how the transitions between them may be used in the motile functions of actin is a challenging issue for future research.

REFERENCES

1. Korn, E. D. (1982) *Physiol. Rev.* **62,** 672–737
2. Oosawa, F. (1975) *Thermodynamics of the Polymerization of Protein,* Academic Press, London
3. Wegner, A. (1976) *J. Mol. Biol.* **108,** 139–150
4. Frieden, C. (1985) *Annu. Rev. Biophys. Biophys. Chem.* **14,** 189–210
5. Pollard, T. D., and Cooper, J. A. (1986) *Annu. Rev. Biochem.* **55,** 987–1035
6. Korn, E. D., Carlier, M.-F., and Pantaloni, D. (1987) *Science* **238,** 638–644
7. Pollard, T. D. (1990) *Curr. Op. Cell Biol.* **2,** 33–40
8. Vandekerckhove, J. (1990) *Curr. Op. Cell Biol.* **2,** 41–50
9. Hibberd, M. G., and Trentham, D. R. (1986) *Annu. Rev. Biophys. Biophys. Chem.* **15,** 119–161
10. Huxley, H. E. (1990) *J. Biol. Chem.* **265,** 8347–8350
11. Korn, E. D., and Hammer, J. A. (1988) *Annu. Rev. Biophys. Biophys. Chem.* **17,** 23–45
12. Spudich, J. A. (1989) *Cell Regul.* **1,** 1–11
13. Aktories, K., and Wegner, A. (1989) *J. Cell Biol.* **109,** 1385–1387
14. Stossel, T. P. (1989) *J. Biol. Chem.* **264,** 18261–18264
15. Amos, L. A. (1985) *Annu. Rev. Biophys. Biophys. Chem.* **14,** 189–210
16. Thomas, D. D., Seidel, J. C., and Gergely, J. (1979) *J. Mol. Biol.* **132,** 257–273
17. Yoshimura, H., Nishio, T., and Mihashi, K. (1984) *J. Mol. Biol.* **179,** 453–467
18. Egelman, E. H. (1985) *J. Muscle Res. Cell Motil.* **6,** 129–151
19. Stokes, D. L., and De Rosier, D. J. (1987) *J. Cell Biol.* **104,** 1005–1017
20. Aebi, U., Millonig, R., Salvo, H., and Engel, A. (1987) *Ann. N. Y. Acad. Sci.* **483,** 100–119
21. Konno, K., and Morales, M. F. (1985) *Proc. Natl. Acad. Sci. U. S. A.* **82,** 7904–7908
22. Gershman, L. C., Selden, L. A., and Estes, J. E. (1986) *Biochem. Biophys. Res. Commun.* **135,** 607–614
23. Carlier, M.-F., Pantaloni, D., and Korn, E. D. (1986) *J. Biol. Chem.* **261,** 10778–10784
24. Estes, J. E., Selden, L. A., and Gershman, L. C. (1987) *J. Biol. Chem.* **262,** 4952–4957
25. Nowak, E., Strzelecka-Golaszewska, H., and Goody, R. S. (1988) *Biochemistry* **27,** 1785–1792
26. Zimmerle, C. T., Patane, K., and Frieden, C. (1987) *Biochemistry* **26,** 6545–6552
27. Frieden, C., Lieberman, D., and Gilbert, H. R. (1980) *J. Biol. Chem.* **255,** 8991–8993
28. Zimmerle, C. T., and Frieden, C. (1988) *Biochemistry* **27,** 7766–7772
29. Selden, L. A., Estes, J. E., and Gershman, L. C. (1989) *J. Biol. Chem.* **264,** 9271–9277
30. Valentin, C., and Carlier, M.-F. (1989) *J. Biol. Chem.* **264,** 20871–20880
31. Kabsch, W., Mannherz, H. G., Suck, D., Pai, E. F., and Holmes, K. C. (1990) *Nature* **347,** 37–44
32. Schutt, C. E., Lindberg, U., Myslik, J., and Strauss, N. (1989) *J. Mol. Biol.* **209,** 735–746
33. Sutoh, K. (1982) *Biochemistry* **21,** 3654–3661
34. Holmes, K. C., Popp, D., Gebhard, W., and Kabsch, W. (1990) *Nature* **347,** 44–49
35. Kasprzak, A. A., Takashi, R., and Morales, M. F. (1988) *Biochemistry* **27,** 4512–4522
36. Vandekerkhove, J. S., Kaiser, D. A., and Pollard, T. D. (1989) *J. Cell Biol.* **109,** 619–626
37. Boyer, M., Feinberg, J., Hue, H.-K., Capony, J.-P., Benjamin, Y., and Roustan, C. (1987) *Biochem. J.* **248,** 359–364
38. Cooper, J. A., Buhle, E. L., Walker, S. B., Tsong, T. Y., and Pollard, T. D. (1983) *Biochemistry* **22,** 2193–2202
39. Tobacman, L. S., and Korn, E. D. (1983) *J. Biol. Chem.* **258,** 3207–3214
40. Gershman, L. C., Newman, J., Selden, L. A., and Estes, J. E. (1984) *Biochemistry* **23,** 2199–2203
41. Carlier, M.-F., Pantaloni, D., and Korn, E. D. (1986) *J. Biol. Chem.* **261,** 10785–10792
42. Newman, J., Estes, J. E., Selden, L. A., and Gershman, L. C. (1985) *Biochemistry* **24,** 1538–1544
43. Mozo-Villarias, A., and Ware, B. R. (1985) *Biochemistry* **24,** 1544–1548
44. Zimmerle, C. T., and Frieden, C. (1988) *Biochemistry* **27,** 7759–7765
45. Bonder, E. M., Fishkind, D. J., and Mooseker, M. S. (1983) *Cell* **34,** 491–501
46. Coué, M., and Korn, E. D. (1985) *J. Biol. Chem.* **260,** 15033–15041
47. Carlier, M.-F., Criquet, P., Pantaloni, D., and Korn, E. D. (1986) *J. Biol. Chem.* **261,** 2041–2050
48. Wegner, A., and Isenberg, G. (1983) *Proc. Natl. Acad. Sci. U. S. A.* **80,** 4922–4925
49. Carlier, M.-F., and Pantaloni, D. (1986) *Biochemistry* **25,** 7789–7792
50. Carlier, M.-F., Pantaloni, D., Evans, J. A., Lambooy, P. K., Korn, E. D., and Webb, M. R. (1988) *FEBS Lett.* **235,** 211–214
51. Carlier, M.-F., Pantaloni, D., and Korn, E. D. (1984) *J. Biol. Chem.* **259,** 9983–9986
52. Carlier, M.-F., and Pantaloni, D. (1988) *J. Biol. Chem.* **264,** 817–825
53. Combeau, C., and Carlier, M.-F. (1988) *J. Biol. Chem.* **264,** 17429–17436
54. Wanger, M., and Wegner, A. (1987) *Biochim. Biophys. Acta* **914,** 105–113
55. Rickard, J., and Sheterline, P. (1986) *J. Mol. Biol.* **191,** 273–280
56. Mitchison, T., and Kirschner, M. W. (1984) *Nature* **312,** 237–242
57. Brenner, S. L., and Korn, E. D. (1983) *J. Biol. Chem.* **258,** 5013–5020
58. Selden, L. A., Gershman, L. C., and Estes, J. E. (1986) *J. Muscle Res. Cell Motil.* **7,** 215–224
59. Selden, L. A., Gershman, L. C., Kinosian, H. J., and Estes, J. E. (1987) *FEBS Lett.* **217,** 89–93
60. Carlier, M.-F., Pantaloni, D., and Korn, E. D. (1987) *J. Biol. Chem.* **262,** 3052–3059
61. Tsukita, S., Tsukita, S., and Ishiwata, H. (1984) *J. Cell Biol.* **98,** 1102–1110
62. Lal, A. A., Korn, E. D., and Brenner, S. L. (1984) *J. Biol. Chem.* **259,** 8794–8800
63. Pollard, T. D., and Mooseker, M. S. (1981) *J. Cell Biol.* **88,** 654–659
64. Carlier, M.-F. (1989) *Int. Rev. Cytol.* **115,** 139–170
65. Casey, P. J., and Gilman, A. G. (1988) *J. Biol. Chem.* **263,** 2577–2580
66. Menetski, J. P., Varghese, A., and Kowalczykowski, S. C. (1988) *Biochemistry* **27,** 1205–1212
67. McIntosh, J. R., and Porter, M. E. (1989) *J. Biol. Chem.* **264,** 6001–6004
68. Engel, J., Fasold, H., Hulla, F. W., Waechter, F., and Wegner, A. (1977) *Mol. Cell. Biochem.* **18,** 3–13

Minireview

THE JOURNAL OF BIOLOGICAL CHEMISTRY
Vol. 266, No. 2, Issue of January 15, pp. 677–680, 1991

HNF-1, a Member of a Novel Class of Dimerizing Homeodomain Proteins

Dirk B. Mendel and Gerald R. Crabtree

From the Beckman Center for Molecular and Genetic Medicine, Howard Hughes Institute at Stanford University, Stanford University Medical School, Stanford, California 94305

Most hierarchical models that attempt to account for patterns of expression of genes in the developing organism postulate the existence of small groups of master regulators operating in a contingent temporal series. Genes expressed in a tissue-specific manner have been extensively studied with the expectation that these genes are likely to be under the control of a more general regulatory pathway. One fruitful approach has been to identify factors that induce expression of a family of tissue-specific genes. Consistent with the demonstration that tissue-specific gene expression is largely controlled at the transcriptional level (1), the majority of these factors identified to date, including MyoD (2), Pit-1 (3) or GHF-1 (4, 5), and Oct-2 (6, 7), have been DNA-binding transcription factors. However, less biased genetic approaches in invertebrates have also identified cytokines and membrane proteins, including adhesive molecules and receptors for cytokines, indicating the importance of the cellular environment in development.

The liver, perhaps because of its simple virtues of size, accessibility, the fact that it is composed primarily of a single cell type, and its relatively simple developmental origin, has been studied extensively in an attempt to understand the mechanisms underlying tissue-specific gene expression. To date, numerous liver-specific genes have been studied, and many of the transcription factors that regulate their expression, both those ubiquitously expressed and those found primarily in hepatocytes, have been identified. Growing evidence indicates that several transcription factors including HNF-1 (8, 9), C/EBP (10, 11), DBP (12), HNF-3 (13), HNF-4 (13), and LF-A1 (14) act together to regulate the development of the hepatocyte phenotype. Remarkably none of these factors seems to be restricted to the liver, an observation that suggests that more subtle mechanisms underlie tissue-specific expression than simply the restricted expression of a transcription factor. In this review we will focus on the role of HNF-1, also called APF, LFB1, ABF, HS, and AFP, and emphasize its potential to participate in regulator networks during development.

HNF-1 as a "Hepatocyte-specific" Transcription Factor

HNF-1 was initially identified in a search for molecules that coordinate the expression of a tissue-specific family of genes. The three fibrinogen genes were chosen among the many candidate gene families for several reasons. Fibrinogen is a disulfide-linked dimer in which each monomer is composed of three polypeptide chains, α, β, and γ, that are assembled in equimolar portions (15), suggesting that the fibrinogen genes are highly coordinately regulated (16). The work of Doolittle *et al.* (17) suggested that these three genes arose by triplication about 500 million years ago. This is sufficient time to randomize all sequences that do not have a common function, making possible the identification of regulatory sequences by simply searching for homology. Finally, these genes are closely linked genetically (18) indicating that chromosomal influences might not significantly perturb the regulation exerted by common transcription factors.

Transcription directed by either the α or β fibrinogen promoters is largely restricted to hepatocyte cell lines although only a limited number of cell types has been investigated (8, 19). The tissue-specific nature of both promoters appeared to be attributed to a single *cis*-acting element, since deletion or mutation of this element reduced promoter activity by ~95% in hepatocyte cell lines to a level comparable with that of the promoter in nonhepatocyte cell lines (8, 19). DNase I footprint analysis of the promoters demonstrated that a protein(s) present in nuclear extract of hepatocytes, but not of other cell types, bound to this *cis*-acting element in each promoter. The same protein(s) appeared to bind to the two elements since DNA-protein complexes formed with the elements comigrated in nondenaturing gels and the *cis*-acting elements competed with one another for binding of the protein(s). Furthermore, the promoter of the α_1-antitrypsin gene contained a site that bound to this protein. While early studies suggested that the α_1-antitrypsin site was not involved in tissue-specific expression (20), later studies have indicated that this site is necessary for tissue-specific expression (21, 22). These results suggested that the protein(s) which bound to these elements might be able to regulate the activity of other hepatocyte-specific genes. Due to the *apparent* restriction of this protein to hepatocytes it was named hepatocyte nuclear factor 1 (HNF-1) (23).

HNF-1 was purified approximately 200,000-fold from rat liver nuclear extracts by DNA affinity chromatography (9, 24) using the β fibrinogen promoter element (8). The purified protein, which migrated as a single ~88-kDa protein on sodium dodecyl sulfate-polyacrylamide gel electrophoresis gels under reducing conditions, was able to produce a footprint in the β fibrinogen promoter that was indistinguishable from that produced by the crude nuclear extract, indicating that no other protein in the extract was directly binding to this site. The purified protein was also tested for its ability to interact with other promoters. While HNF-1 was able to bind to the promoters of a variety of "hepatocyte-specific" genes including those of the α fibrinogen chain, albumin, α-fetoprotein, α_1-antitrypsin, and transthyretin, it did not interact with ubiquitously expressed promoters, including the adenovirus major late, SV-40 early, and β globin promoters (24). To date, HNF-1 has been shown to interact with defined functional sequences in 11 "hepatocyte-specific" genes in species as diverse as *Xenopus* and the human (Table I), providing strong support for the initial hypothesis that HNF-1 might be a factor that regulates a larger family of genes. From these sequences the HNF-1 consensus site appears to be the inverted palindrome GTTAATNATTAAC (24) to which HNF-1 can bind as a homodimer (25).

In most of the genes tested the HNF-1 site plays a prominent role in regulating transcription both *in vitro* and *in vivo*; however, other factors are also required for efficient transcrip-

TABLE I

Regulatory sequences interacting with HNF-1

Promoter	Species	Sequence	Position	Ref.
Fibrinogen α	Rat	GGTGATGATTAAC	−47	8, 24
Fibrinogen β	Rat	GTCAAATATTAAC	−84	8, 19, 24
Fibrinogen β	Human	ATTAAATATTAAC	−77	[a]
Albumin	Mouse	GTTAATGATCTAC	−52	9, 26, 27
Albumin	Rat	GTTAATGATCTAC	−53	24, 27–29
Albumin	Human	GTTAATAATCTAC	−51	30
Albumin	*Xenopus*	GTTAATAATTTTC	−53	31, 32
α-Fetoprotein	Mouse	GTTACTAGTTAAC	−50	24, 33, 34
α-Fetoprotein	Mouse	GTTAATTATTGGC	−116	33, 34
α-Fetoprotein	Rat	GTTACTAGTTAAC	−49	29, 35
α-Fetoprotein	Rat	GTTAATTATTGGC	−115	29, 35
α-Fetoprotein	Human	GTTACTAGTTAAC	−47	30
α-Fetoprotein	Human	GATTAATAATTAC	−3400	30
α-Fetoprotein	Human	GTTAATTATTGGC	−118	30
Transthyretin	Mouse	GTTACTTATTCTC	−118	36
Transthyretin	Rat	GTTACTTATTCTC	−116	13
Transthyretin	Human	GTTACTTATTCTC	−116	
α_1-Antitrypsin	Mouse	GTTAAT-ATTCAT	−63	8, 24
α_1-Antitrypsin	Human	GTTAAT-ATTCAC	−63	22, 37
Aldolase B	Human	GTGTTGAATAAAC	−74	38
Aldolase B	Chicken	AGGGAGAATAAAC	−71	
Pyruvate kinase	Rat	GTTATACTTTAAC	−79	39, 40
Hepatitis B virus pre-S	Human	GTTAATCATTACT	−75	24, 41, 42
Phosphoenolpyruvate carboxykinase	Rat	AACATTCATTAAC	−182	43, 44
α2,6-*S*-Transferase	Rat	GTTAATGTTTAAC	−66	[b]
CYP2E1	Rat	GCTAATAATAAAC	−95	45
HNF-1 consensus (26 sequences)		$G_{23}T_{21}T_{22}A_{22}A_{18}T_{20}NA_{22}T_{25}T_{20}A_{11}A_{17}C_{24}$		

[a] G. Courtois, unpublished data.
[b] E. Svensson, unpublished data.

tional activity. As an example, the albumin promoter has six *cis* elements, A–F, which serve as binding sites for transcription factors (26, 27, 46, 47). Basal activity of the promoter depends on the integrity of the C element which is a binding site for the ubiquitous factor NF-1. Optimal liver-specific transcriptional activity depends on the integrity of the B and D elements which bind the proteins HNF-1 and C/EBP (or the related factor, DBP), respectively. The other three elements are occupied by an additional ubiquitous factor (element E) and two more sites for C/EBP (elements A and F). Elimination of any of these elements has a measurable effect on the transcriptional activity of this promoter, though the B, C, and D elements are clearly the most critical since deletion of any of these sites results in a substantial reduction in promoter activity (26, 27). In addition, enhancer elements are also required for optimal transcription of this gene since deletion of these sequences reduces gene transcription by at least 1 order of magnitude (48).

What Is the Logic Underlying the Pattern of HNF-1 Expression?

Perhaps surprising in light of the proposed role of HNF-1 in inducing the hepatocyte phenotype (see below) is the recent demonstration that the message for HNF-1 is found in several nonhepatocyte tissues including the kidney, stomach, intestine, and at low levels in the thymus and spleen (49, 50). These observations conflict with a report (25) which described HNF-1 mRNA as being liver-specific. This discrepancy may be in part related to the fact that Northern blotting, which is relatively insensitive, was used for the later study, while ribonuclease protection, which is both sensitive and quantitative, was used for the former. The broader pattern of expression (which we believe to be correct) seems to cross conventional embryologic boundaries and raises questions as to the meaning underlying this distribution.

The mRNA present in the kidney appears to give rise to functional HNF-1 protein since it can be detected both in footprinting assays and it activates transcription of the α fibrinogen gene (49). In addition, a number of genes expressed selectively in the liver are also active in the kidney. For example, the gene for phosphoenolpyruvate carboxykinase is expressed in the kidney cortex and an element present in the phosphoenolpyruvate carboxykinase promoter binds HNF-1 with high affinity. This same region of the phosphoenolpyruvate carboxykinase promoter also binds a nuclear protein from rat kidney.[1] The α_1-antitrypsin promoter is active in the liver, kidney, and intestine and in macrophages such as those in the spleen. Finally, the α-fetoprotein promoter is active in the liver, embryonic yolk sac, and the intestine (51). Thus, many genes expressed in the liver and studied as examples of liver-specific gene expression are actually more widely distributed than commonly believed and their pattern of expression parallels that of HNF-1.

Does HNF-1 Play a Role in the Development of the Liver?

While classical embryological approaches have indicated that extracellular factors play a prominent role in the development of the liver (52), recent studies in other cell types indicate that a limited number of transcription factors are able to substantially effect the differentiation of particular tissues. Most notably, constitutive expression of the helix-loop-helix myogenic regulators myo D, myogenin, myf-5, or Herculin in fibroblasts induces these cells to activate genes characteristic of muscle cells (2, 53–56). However, the observation that most of the promoters that bind HNF-1 require other factors for activity suggests that HNF-1 will not behave like the helix-loop-helix myogenic regulators unless HNF-1 also regulates the expression of other transcription factors. The observation that HNF-1 plays a prominent role in the regulation of a wide variety of hepatocyte-specific genes has suggested that HNF-1 expression may be an early signal in the development of the liver and other endodermally derived organs. Although no direct link has been established to morphogenesis, indirect evidence suggests that HNF-1 does play a role in regulating the hepatocyte phenotype.

Using morphological criteria Deschatrette *et al.* (57) have isolated dedifferentiated variants (C2 cells) of the well differentiated Fao hepatocyte cell line. Biochemical characterization of the C2 cells demonstrated that C2 cells do not express the majority of liver-specific genes, including albumin and fibrinogen, but they do express some liver-specific genes such as tyrosine and alanine aminotransferases. The C2 cells do not express HNF-1, though they do express a ~70-kDa protein which has a similar DNA binding specificity as HNF-1 (58, 59). However, this protein, which has been referred to as a variant form of HNF-1 (vHNF or vAPF), does not appear to be an alternatively spliced form of HNF-1 (49) and, though it can bind to the HNF-1 site *in vitro*, does not appear to be able to induce transcription of HNF-1-dependent genes (58). The switch to the variant form is likely to be directed by an event higher in a developmental pathway since both transcriptional and posttranscriptional events underlie this switch (58).

Differentiated hepatocytes are one of the few cell types capable of performing gluconeogenesis. Deschatrette *et al.* (57) used this fact to select very rare (10^{-9}) spontaneous revertants of C2 cells capable of performing gluconeogenesis by selection in media lacking glucose. Remarkably, the revertants (Rev7 cells) regained the morphological characteristics of differentiated hepatocytes as well as the ability to express

[1] Richard Hanson, personal communication.

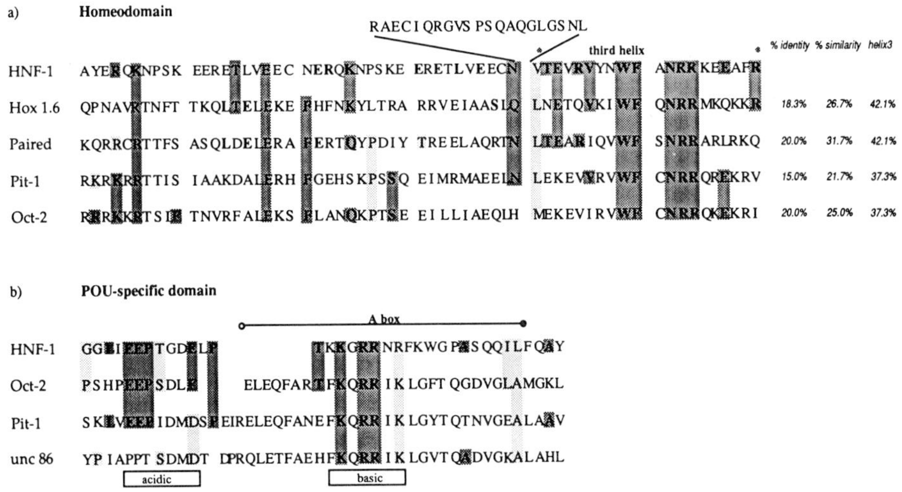

FIG. 1. **Alignment of the homeodomains of the proteins most closely related to HNF-1.** The *darker shading* indicates complete identity to at least one of the other proteins, while the *lighter shading* indicates similarity. The data labeled *helix 3* on the *right* refer to comparisons made only within this helix. (From Ref. 49.)

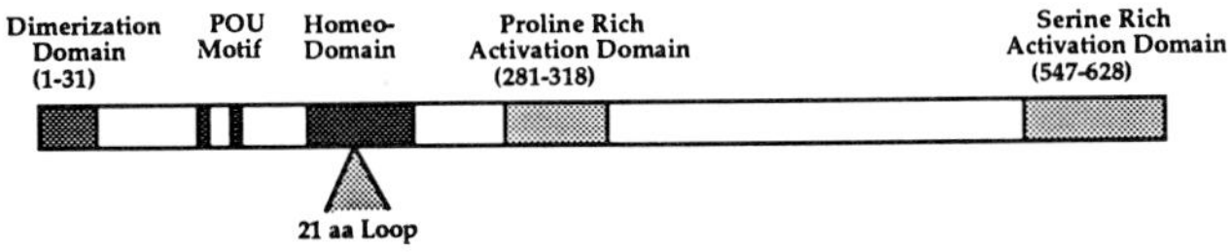

MVSKLTSLQQELLSALLSSGVTKEVLIQALEE **HNF-1β**
MVSKLSQLQTELLAALLESGLSKEALIQALGE **HNF-1α**
MHADLSQLQTEVEEAVQECRNAEEKAKKAITD **RAT MYOSIN CHAIN 2**

FIG. 2. **Functional domains within the HNF-1 molecule.** The *upper illustration* shows DNA-binding and activation domains as well as the 21-amino acid insertion between helix II and helix III of the homeodomain. A fourth helix is not apparent from the analysis of the sequence of HNF-1. In the *lower half* of the figure a sequence similar to the myosin heavy chain is shown that is essential for dimerization of the intact molecule and by itself is able to dimerize with the native protein (70). A recently discovered homologue of HNF-1, HNF-1β, shows near perfect conservation of the dimerization domain as well as helix III of the homeodomain (D. Mendel and G. Crabtree, unpublished observations). In this figure we have designed the form originally cloned (8, 25, 49) as HNF-1α while the newly cloned homologue is HNF-1β (D. Mendel, M. Graves and G. Crabtree, manuscript in preparation).

hepatocyte-specific genes. The Rev7 cells, like the Fao parent cell line, express HNF-1 (58, 59) but have posttranscriptional activation of many genes suggesting that reversion is complex (58, 60) and may not be fully explained by the reexpression of HNF-1.

A similar correlation between the presence of HNF-1 and the expression of the differentiated hepatocyte phenotype has also been observed in somatic hybrids generated by fusing hepatocytes and fibroblasts. As is generally the case for hybrids formed between two different differentiated cell types (61, 62), these cells do not express the majority of hepatocyte-specific genes (63) and likewise do not express HNF-1. Again, as is the case with the C2 cells, these hybrids appear to express the variant form of HNF-1 (58).

Structure and Expression of HNF-1 Suggest a Diverse Role in Development

Additional evidence that HNF-1 is a developmentally important protein is based on the presence of a homeodomain in the HNF-1 sequence. DNA-binding proteins have generally been limited to only a few DNA-binding motifs, with the zinc finger (64) and the helix-turn-helix (65) motifs being the most common. A specialized type of the helix-turn-helix DNA-binding motifs that is the subject of a great deal of interest is the homeodomain, a highly conserved DNA sequence encoding about 60 amino acids, characteristic of proteins that have important developmental roles (66, 67). In a few cases the homeodomain is coupled with a POU motif made up of about 100 conserved amino acids (68). Proteins containing these two DNA-binding motifs have generally been found to be developmental regulators, and screening for the presence of these structures has recently been employed as an approach to identify new proteins involved in development (69). The HNF-1 DNA-binding domain is a distant member of these families (Fig. 1) and contains both a divergent homeodomain and sequence motifs similar to the POU proteins that are essential for DNA binding (49, 70). This observation suggests that HNF-1 might have a role in producing the hepatocyte phenotype.

Though HNF-1 is a relative of the homeodomain proteins it has two characteristics which distinguish it from the other members of this family. First, its DNA-binding domain contains a 21-amino acid insertion (Fig. 1) which is not found in any of the other homeodomains (49, 71). Though the function of this structure is not known, its presence in the homeodomain suggests that it is likely to be involved in DNA binding or in protein-protein interactions once HNF-1 occupies its binding site. Second, as inferred by the original recognition that the binding sequence was a center of dyad symmetry (24), HNF-1 binds to its recognition sequence as a dimer (25, 70). The domain which mediates dimerization is similar to cardiac myosin but is much shorter than the region involved in myosin dimerization (Fig. 2) (70). Nevertheless this region forms an α helical structure in solution that could be part of a coiled-coil as a homodimer or a heterodimer. Recently we have found that HNF-1 exists as a dimer in the absence of its recognition sequence.[2]

Some other transcription factors, such as MyoD and members of the Fos/Jun family, which form dimers prior to binding to their recognition sequences form heterodimers in order to alter their DNA binding specificity. HNF-1 may likewise be able to form heterodimers with members of a similar family in order to alter its DNA binding specificity or transcriptional activity. Indeed, while most of the HNF-1 binding sites contain a perfect or near-perfect half-binding site, no sequence contains a perfect inverted repeat (Table I). This could be interpreted to suggest that a heterodimer containing one molecule of HNF-1 and a second HNF-1-like protein could select its binding site based on the sequence specificity of the

[2] D. Mendel, unpublished data.

second protein. Though no other members of an HNF-1 family have yet been verified, Nakao *et al.* (72) have reported that a protein related to HNF-1 interacts with the HNF-1 binding site in the α-fetoprotein promoter. Likewise, we have found a gene (HNF-1β) expressed in liver cells that has nearly identical dimerization and homeodomains as HNF-1 but widely differing activation domains[3] suggesting that HNF-1 may have broader functions than was initially anticipated.

One clear implication of the finding of HNF-1 in nonhepatic tissues is that HNF-1 cannot act alone to produce the hepatic phenotype. The vertebrate liver selectively expresses over 1000 different genes at levels ranging from 10% of the total mRNA for albumin to less than 0.001% for certain enzymes. To date, only about six distinctly different transcriptional control proteins selectively expressed in the liver have been described. Controlling the broadly varying activity of 1000 genes seems a formidable task for only six transcriptional control proteins. Not surprisingly, none of them have characteristics that suggest they alone could account for the hepatic phenotype or the full range of expression of genes in a single organ. The recent discovery of a molecule similar in structure to HNF-1 (HNF-1β)[3] with highly conserved dimerization and homeodomains suggests that HNF-1 may expand the range of its transcriptional control by heterodimer formation.

Acknowledgments—We wish to thank Matt Scott for his thoughtful reading of the manuscript and Jean Oberlindacher for help with its preparation.

REFERENCES

1. Derman, E., Krauter, K., Walling, L., Weinberger, C., Ray, M., and Darnell, J. E., Jr. (1981) *Cell* **23,** 731–739
2. Davis, R. L., Weintraub, H., and Lassar, A. B. (1987) *Cell* **51,** 987–1000
3. Ingraham, H. A., Chen, R., Mangalam, H. J., Elsholtz, H. P., Flynn, S. E., Lin, C. R., Simmons, D. M., Swanson, L., and Rosenfeld, M. G. (1988) *Cell* **55,** 519–529
4. Bodner, M., Castrillo, J.-L., Theill, L. E., Deerinck, T., Ellisman, M., and Karin, M. (1988) *Cell* **55,** 505–518
5. Lefevre, C., Imagawa, M., Dana, S., Grindlay, J., Bodner, M., and Karin, M. (1987) *EMBO J.* **6,** 971–981
6. Landolfi, N. F., Capra, J. D., and Tucker, P. W. (1986) *Nature* **323,** 548–551
7. Staudt, L. M., Singh, H., Sen, R., Wirth, T., Sharp, P. A., and Baltimore, D. (1986) *Nature* **323,** 640–643
8. Courtois, G., Morgan, J. G., Campbell, L. A., Fourel, G., and Crabtree, G. R. (1987) *Science* **238,** 688–692
9. Lichtsteiner, S., and Schibler, U. (1989) *Cell* **57,** 1179–1187
10. Johnson, P. F., Landschulz, W. H., Graves, B. J., and McKnight, S. L. (1987) *Genes & Dev.* **1,** 133–146
11. Landschulz, W. H., Johnson, P. F., Adashi, E. Y., Graves, B. J., and McKnight, S. L. (1988) *Genes & Dev.* **2,** 786–800
12. Mueller, C. R., Maire, P., and Schibler, U. (1990) *Cell* **61,** 279–291
13. Costa, R. H., Grayson, D. R., and Darnell, J. E., Jr. (1989) *Mol. Cell. Biol.* **9,** 1415–1425
14. Hardon, E. M., Frain, M., Paonessa, G., and Cortese, R. (1988) *EMBO J.* **7,** 1711–1719
15. Doolittle, R. F. (1981) *Sci. Am.* **245,** 126–135
16. Crabtree, G. R., and Kant, J. A. (1982) *J. Biol. Chem.* **257,** 7277–7279
17. Doolittle, R. F., Watt, K. W. K., Cottrell, B. A., Strong, D. D., and Riley, M. (1979) *Nature* **280,** 464–470
18. Kant, J. A., Fornace, A. J., Jr., Saxe, D., Simon, M. I., McBride, O. W., and Crabtree, G. R. (1985) *Proc. Natl. Acad. Sci. U. S. A.* **82,** 2344–2348
19. Huber, P., Laurent, M., and Dalmon, J. (1990) *J. Biol. Chem.* **265,** 5695–5701
20. Ciliberto, G., Dente, L., and Cortese, R. (1986) *Cell* **41,** 531–540
21. Shen, R. F., Li, Y., Sifers, R. N., Wang, H., Hardick, C., Tsai, S. Y., and Woo, S. L. C. (1987) *Nucleic Acids Res.* **15,** 8399–8415
22. Li, Y., Shen, R-F., Tsai, S. Y., and Woo, S. L. C. (1988) *Mol. Cell. Biol.* **8,** 4362–4369
23. Baltimore, D., Grosschedl, R., Weaver, D., Costantini, F., and Imanishi-Kari, T. (1985) *Cold Spring Harbor Symp. Quant. Biol.* **50,** 417–420
24. Courtois, G., Baumhueter, S., and Crabtree, G. R. (1988) *Proc. Natl. Acad. Sci. U. S. A.* **85,** 7937–7941
25. Frain, M., Swart, G., Monaci, P., Nicosia, A., Stämpfli, S., Frank, R., and Cortese, R. (1989) *Cell* **59,** 145–157
26. Lichtsteiner, S., Wuarin, J., and Schibler, U. (1987) *Cell* **51,** 963–973
27. Maire, P., Wuarin, J., and Schibler, U. (1989) *Science* **244,** 343–346
28. Crowley, C., Liu, C. C., and Levinson, A. D. (1983) *Mol. Cell. Biol.* **3,** 44–55
29. Jose-Estanyol, M., Poliard, A., Foiret, D., and Danan, J.-L. (1989) *Eur. J. Biochem.* **181,** 761–766
30. Sawadaishi, K., Morinaga, T., and Tamaoki, T. (1988) *Mol. Cell. Biol.* **8,** 5179–5187
31. Schorpp, M., Kugler, W., Wagner, U., and Ryffel, G. U. (1988) *J. Mol. Biol.* **202,** 307–320
32. Pierce, J. H., and Aaronson, S. A. (1982) *J. Exp. Med.* **156,** 873–887
33. Godbout, R., Ingram, R., and Tilghman, S. M. (1986) *Mol. Cell. Biol.* **6,** 477–487
34. Feuerman, M. H., Godbout, R., Ingram, R. S., and Tilghman, S. M. (1989) *Mol. Cell. Biol.* **9,** 4204–4212
35. Poliard, A., Bakkali, L., Poiret, M., Foiret, D., and Danan, J.-L. (1990) *J. Biol. Chem.* **265,** 2137–2141
36. Costa, R. H., Lai, E., and Darnell, J. E. (1986) *Mol. Cell. Biol.* **6,** 4697–4708
37. Monaci, P., Nicosia, A., and Cortese, R. (1988) *EMBO J.* **7,** 2075–2087
38. Izzo, P., Costanzo, P., Lupo, A., Rippa, E., and Salvatore, F. (1989) *FEBS Lett.* **257,** 75–80
39. Vaulont, S., Puzenat, N., Levrat, F., Cognet, M., Kahn, A., and Raymondjean, M. (1989) *J. Mol. Biol.* **20,** 205–219
40. Vaulont, S., Puzenat, N., Kahn, A., and Raymondjean, M. (1989) *Mol. Cell. Biol.* **9,** 4409–4415
41. Chang, H.-K., Wang, B.-Y., Yuh, C.-H., Wei, C.-L., and Ting, L.-P. (1989) *Mol. Cell. Biol.* **9,** 5189–5197
42. Nakao, K., Miyao, Y., Ohe, Y., and Tamaoki, T. (1989) *Nucleic Acids Res.* **17,** 9833–9842
43. Roesler, W. J., Vandenbark, G. R., and Hanson, R. W. (1989) *J. Biol. Chem.* **264,** 9657–9664
44. Trus, M., Benvenisty, N., Cohen, H., and Reshef, L. (1990) *Mol. Cell. Biol.* **10,** 2418–2422
45. Ueno, T., and Gonzalez, F. J. (1990) *Mol. Cell. Biol.* **9,** 4495–4505
46. Cereghini, S., Raymondjean, M., Carranca, A. G., Herbomel, P., and Yaniv, M. (1987) *Cell* **50,** 627–638
47. Herbomel, P., Rollier, A., Tronche, F., Ott, M.-O., Yaniv, M., and Weiss, M. C. (1989) *Mol. Cell. Biol.* **9,** 4750–4758
48. Pinkert, C. A., Ornitz, D. M., Brinster, R. L., and Palmiter, R. D. (1987) *Genes & Dev.* **1,** 268–276
49. Baumhueter, S., Mendel, D. B., Conley, P. B., Kuo, C. J., Turk, C., Graves, M. K., Edwards, C. A., Courtois, G., and Crabtree, G. R. (1990) *Genes & Dev.* **49,** 372–379
50. Kuo, C. J., Conley, P. B., Hsieh, C., Franke, U., and Crabtree, G. R. (1990) *Proc. Natl. Acad. Sci. U. S. A.* **87,** 9838–9842
51. Tilghman, S. M., and Belayew, A. (1982) *Proc. Natl. Acad. Sci. U. S. A.* **79,** 5254–5257
52. Houssiant, E. (1980) *Cell Differ.* **9,** 269–280
53. Wright, W. E., Sassoon, D. A., and Lin, V. K. (1989) *Cell* **56,** 607–617
54. Miner, J. H., and Wold, B. (1990) *Proc. Natl. Acad. Sci. U. S. A.* **87,** 1089–1093
55. Pinney, D. F., Pearson-White, S. H., Konieczny, S. F., Latham, K. E., and Emerson, C. P., Jr. (1988) *Cell* **53,** 781–793
56. Braun, T., Buschhausen-Denker, G., Bober, E., Tannich, E., and Arnold, H. H. (1989) *EMBO J.* **8,** 701–709
57. Deschatrette, J., Moore, E. E., Dubois, M., and Weiss, M. (1980) *Cell* **19,** 1043–1051
58. Baumhueter, S., Courtois, G., and Crabtree, G. R. (1988) *EMBO J.* **7,** 2485–2493
59. Cereghini, S., Blumenfeld, M., and Yaniv, M. (1988) *Genes & Dev.* **8,** 957–974
60. Friedman, J. M., Babiss, L. E., Weiss, M., and Darnell, J. E., Jr. (1987) *EMBO J.* **6,** 1727–1731
61. Thompson, E. B., and Gelehrter, T. D. (1971) *Proc. Natl. Acad. Sci. U. S. A.* **68,** 2589–2593
62. Weiss, M. C., and Chaplain, M. (1971) *Proc. Natl. Acad. Sci. U. S. A.* **68,** 3026–3030
63. Hillary, A. M., and Fournier, R. E. (1984) *Cell* **38,** 523–534
64. Miller, J., McLachlan, A. D., and Klug, A. (1985) *EMBO J.* **4,** 1609–1614
65. Sehgal, P. B., May, L. T., Tamm, I., and Vilcek, J. (1987) *Science* **235,** 731–732
66. Laughon, A., and Scott, M. P. (1984) *Nature* **310,** 25–31
67. Scott, M. P., Tamkun, J. W., and Hartzell, G. W. (1989) *Biochim. Biophys. Acta* **989,** 25–48
68. Herr, W., Sturm, R. A., Clerc, R. G., Corcoran, L. M., Baltimore, D., Sharp, P. A., Ingraham, H. A., Rosenfeld, M. G., Finney, M., Ruvkun, G., and Horvitz, H. R. (1988) *Genes & Dev.* **2,** 1513–1516
69. He, X., Treacy, M. N., Simmons, D. M., Ingraham, H. A., Swanson, L. W., and Rosenfeld, M. G. (1989) *Nature* **340,** 35–41
70. Nicosia, A., Monaci, P., Tomei, L., De Francesco, R., Nuzzo, M., Stunnenberg, H., and Cortese, R. (1990) *Cell* **61,** 1225–1236
71. Finney, M. (1990) *Cell* **60,** 5–6
72. Nakao, K., Lawless, D., Ohe, Y., Miyao, Y., Nakabayashi, H., Kamiya, H., Miura, K., Ohtsuka, E., and Tamaoki, T. (1990) *Mol. Cell. Biol.* **10,** 1461–1469

[3] D. Mendel, M. Graves, and G. R. Crabtree, unpublished data.

Minireview

THE JOURNAL OF BIOLOGICAL CHEMISTRY
Vol. 266, No. 3, Issue of January 25, pp. 1355–1358, 1991
Printed in U. S. A.

Cytoplasmic Transcription System Encoded by Vaccinia Virus

Bernard Moss, Byung-Yoon Ahn, Bernard Amegadzie, Paul D. Gershon, and James G. Keck

From the Laboratory of Viral Diseases, National Institute of Allergy and Infectious Diseases, National Institutes of Health, Bethesda, Maryland 20892

The poxviruses comprise the only known family of DNA viruses that propagate entirely within the cytoplasm of eukaryotic cells and that encode most, if not all, of the specific enzymes and factors needed for transcription and replication (reviewed in Refs. 1 and 2). The "life cycle" of vaccinia virus, the prototypal member of the family, is surprisingly complex with at least three temporally regulated classes of gene products (Fig. 1). The 200,000-base pair (bp)[1] linear double-stranded DNA genome and the components of the early transcription system are packaged within the core of the infectious virus particle so that mRNA synthesis begins immediately after entry into the cytoplasm (3, 4). The subsequent synthesis of DNA polymerase and other viral early proteins leads to DNA replication, which is followed by the intermediate and late phases of gene expression (5–8). The considerable progress that has been made in identifying the cis- and trans-acting components of the regulatory cascade is described in this review.

Early Transcription

Methods of Study—Vaccinia virus transcription can be studied conveniently *in vivo* or *in vitro*. Structurally and functionally, the RNA species made *in vitro* by permeabilized purified virus particles resemble those made early in infection (9–11). ATP hydrolysis is required for initiation (12) and elongation (13) of transcription. The transcripts accumulate transiently in the virus core, and high concentrations of ATP with a hydrolyzable β–γ bond are required for extrusion of the RNA (14). The average time for synthesis and release of mRNAs is about 2.2 min (15).

Highly active template-dependent transcription systems were obtained by disruption of the virions and removal of insoluble structural proteins and nucleic acids (16, 17). The ability to faithfully initiate and terminate transcription of added early gene templates is retained in these extracts (18) which also serve as a source of individual enzymes and factors involved in the synthesis and processing of mRNA.

The ability to ligate natural or mutated transcriptional regulatory signals to reporter genes and integrate them into the genome of infectious virus particles (19) has provided a reliable way of correlating structure with function *in vivo*.

mRNA—The mRNAs produced *in vitro* by virus cores or in infected cells are capped (10, 20), polyadenylated (9), and of discrete length with no indication of splicing. Some early mRNAs also contain short 5′ poly(A) leaders (21, 22) that are synthesized by an RNA editing process that will be discussed in conjunction with late mRNAs.

Promoters—The promoters for several early genes were found to extend only about 30 bp upstream of the RNA start sites. Within one early promoter, the effects of all single nucleotide substitutions were measured by expression of the β-galactosidase reporter gene (Fig. 2) and confirmed by *in vivo* and *in vitro* transcription (23). On the basis of these results, the promoter was divided into three regions relative to the RNA start site at +1: a 15-bp A-rich critical region (−13 to −28) in which many single nucleotide substitutions have a major effect, separated by 11 bp of a less critical T-rich sequence from a 7-bp region within which initiation at a purine occurs. Within the critical region, A residues are essential at certain locations and optimal at some others. A G residue is needed at −21 and Ts are important at −22 or −23. Similar to the TATA box of higher eukaryotic RNA polymerase II promoters, the critical region specifies the distance to the transcription initiation site downstream. Most natural promoters do not have optimal nucleotides in all positions, and variability in promoter strength may provide a way of regulating gene expression. When the promoter sequences of a large number of early genes were lined up, however, the predominant nucleotide at each position corresponded closely to the optimal one determined by mutagenesis.

Termination Signal—The DNA sequence corresponding to the eukaryotic processing signal AAUAAA (24), which appears about 20 nucleotides before the polyadenylation site of eukaryotic mRNAs, is not present near the 3′ ends of poxvirus genes. Instead, the ends of vaccinia virus early mRNAs occur 20–50 bp downstream of the sequence TTTTTNT, in which N can be any nucleotide including T (25). Termination, rather than RNA processing, is believed to occur for kinetic reasons and because of an absence of specific endonuclease activity (18). Incorporation of halogenated UTP derivatives into nascent transcripts prevented termination in a template-dependent system, suggesting that the signal is actually recognized in the RNA as UUUUUNU (15a). The same derivatives also blocked RNA extrusion from cores and led to the accumulation therein of long transcripts (15b).

As predicted, TTTTTNT occurs near the ends of most early genes, but in its absence mRNAs with ends that are coterminal with those of downstream mRNAs may be formed. *In vivo* studies suggest that termination is less than 100% efficient (26), and the occasional occurrence of TTTTTNT sequences within early genes may therefore down-regulate but not entirely prevent expression.

Virus Core-associated Enzymes

DNA-dependent RNA Polymerase—The α-aminitin-resistant RNA polymerase from vaccinia virions has an apparent molecular mass of nearly 500 kDa with two large and many small subunits (27), thus resembling its eukaryotic counterpart. Nevertheless, all of the subunits appear to be virus-encoded (28). The genes for the two large (*rpo147* and *rpo132*) and five of the small (*rpo35, rpo30, rpo22, rpo19,* and *rpo18*) polypeptides (22, 29, 30–32) have been identified (Table I). The large subunits are homologous to the corresponding subunits of eukaryotes and have significant but less sequence similarity to those of *Escherichia coli*. The *rpo147* and *rpo132*

[1] The abbreviations used are: bp, base pair(s); VETF, vaccinia virus early transcription factor; VLTF, vaccinia virus late transcription factor.

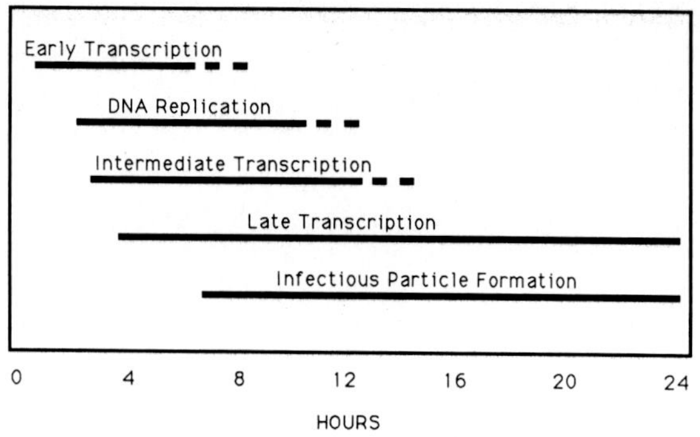

FIG. 1. **Temporal regulation of vaccinia virus gene expression.**

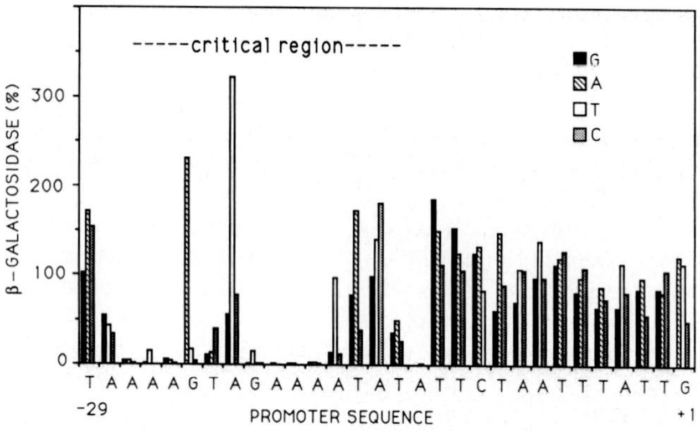

FIG. 2. **Effect of single nucleotide substitutions on early promoter activity.** Recombinant vaccinia viruses that contain the indicated natural P7.5 early promoter sequence, or derivatives with the single nucleotide substitutions, regulating expression of β-galactosidase were used to infect cells in the presence of cytosine arabinoside, an inhibitor of DNA replication. The activities obtained with the mutated promoters are given relative to the 100% value obtained with the natural promoter. Data are from Ref. 23.

TABLE I
Vaccinia virus RNA polymerase genes

Gene	Homolog	Sequence motif	Ref.
rpo147	RPB1 (yeast)	Zinc finger	29
rpo132	RPB2 (yeast)	Zinc finger, leucine zipper	30, 31
rpo35			[a]
rpo30	SII (mouse)	Zinc finger	32
rpo22		Leucine zipper	29
rpo19		Leucine zipper, acidic N terminus	[b]
rpo18			22

[a] B. Amegadzie, unpublished data.
[b] B-Y. Ahn, unpublished data.

subunits both have zinc finger motifs, but the *rpo147* subunit lacks the heptapeptide C-terminal repeats of the related RNA polymerase II subunit. Interestingly, the *rpo30* subunit appears to be a homolog of eukaryotic transcription elongation factor SII (32). All of the vaccinia virus RNA polymerase subunit genes have early promoters and some have late promoters in addition. A role for the host RNA polymerase, or some of its subunits, in expression of poxvirus genes has been suggested (33–35).

Early Transcription Factor—Purified vaccinia virus RNA polymerase lacks the ability to transcribe double-stranded DNA templates unless complemented with an early transcription factor called VETF that was isolated from virion extracts (36). DNA affinity-purified VETF is a heterodimeric protein that works in conjunction with all early promoters tested. The DNA binding of VETF is specific and dependent on the critical region of the early promoter, being abrogated by single nucleotide substitutions that decrease transcription (37). VETF has DNA-dependent ATPase activity that might account for the ATP hydrolysis requirement for initiation of RNA synthesis (38). The viral genes encoding the 70- and 82-kDa subunits of VETF have been identified (Table II), and both are expressed late in infection (39, 40). The protein sequence predicted for the smaller of the two subunits contains a motif associated with ATP binding or ATPase activity consistent with the properties of VETF; the larger polypeptide contains variations of the canonical zinc finger and leucine zipper motifs (39).

Capping and Methylating Enzymes—Capping and methylation of vaccinia virus mRNA is accomplished by a multifunctional enzyme complex composed of 97- and 33-kDa subunits that catalyze the removal of the terminal or γ-phosphate from an RNA chain, the transfer of the GMP moiety of GTP to the now diphosphate-ended RNA, and finally the transfer of a methyl group from *S*-adenosylmethionine to the 7-position of the added guanosine which stabilizes the final product against reversal by pyrophosphate ion (41, 42). The 97-kDa subunit reacts with GTP to form a covalent lysine-GMP intermediate (43, 44). Since short oligoribonucleotides can serve as cap acceptors, capping might occur shortly after transcription initiation, consistent with the finding of capping enzyme associated with the transcription complex (45). The genes encoding both subunits of capping enzyme have been identified (46, 47) and expressed as a functional heterodimer in *Escherichia coli* (48–50). The large subunit alone has RNA triphosphatase and guanylytransferase but not methyltransferase activities.

A 38-kDa RNA (nucleoside-2′-*O*-)methyltransferase catalyzes the transfer of a methyl group from *S*-adenosylmethionine to the ribose of the first encoded nucleotide, which is separated by a triphosphate bridge from the terminal 7-methylguanosine (51). Uncapped polyribonucleotides or even capped ones lacking the 7-methyl group are not methyl acceptors, indicating that 2′-*O*-methylation is the final step in the formation of the cap structure $m^7G(5')pppNm$- (52). The gene encoding this enzyme has not been identified nor has the role of ribose methylation been determined.

Termination Factor—Transcripts made by purified RNA polymerase and VETF fail to terminate without the addition of a protein that co-purifies with (and most probably is) capping enzyme (53). There is evidence that capping enzyme exists in a complex with RNA polymerase and VETF (45). Nevertheless, capping *per se* is not required for termination, and the precise role of this multifunctional protein in this process remains to be determined.

Poly(A) Polymerase—Purified poly(A) polymerase from vaccinia virus cores (54) or infected cells (55) has a M_r of approximately 80,000 and is associated with 55- and 33-kDa polypeptides. The enzyme is primer-dependent but with little evident sequence specificity. It is selective for ATP, although in its absence other ribonucleoside triphosphates can be polymerized inefficiently (56). Kinetic studies revealed that polyadenylation is a biphasic reaction. Viral genes for both

TABLE II
Vaccinia virus transcription/transactivation factor genes

Factor	Sequence motif	Ref.
VETF (82 kDa)	Leucine zipper, zinc finger	39
VETF (70 kDa)	Helicase	39, 40
VLTA[a] (30 kDa)		69
VLTA (26 kDa)	Zinc finger	69
VLTA (17 kDa)	Zinc finger	69

[a] VLTA, vaccinia virus late transactivator.

polypeptides associated with purified poly(A) polymerase have been identified.[2]

Additional Enzymes—Other enzymes that may have roles in transcription have been purified from vaccinia virus cores. These include nucleoside-triphosphate phosphohydrolase I, a virus-encoded DNA-dependent ATPase of M_r 61,000 (57–59). A conditionally lethal temperature-sensitive nucleoside-triphosphate phosphohydrolase I mutant is defective in intermediate and late gene expression suggesting that the enzyme may be involved in transcription (60, 61).

Nucleoside-triphosphate phosphohydrolase II, a M_r 68,000 protein that is immunologically distinct from nucleoside-triphosphate phosphohydrolase I, is stimulated by a wider range of nucleic acids than nucleoside-triphosphate phosphohydrolase I and hydrolyzes all four nucleoside triphosphates rather than just ATP (62). The gene encoding nucleoside-triphosphate phosphohydrolase II has not yet been identified nor is there information regarding its function.

The vaccinia virus-encoded DNA topoisomerase is able to relax both positively and negatively supercoiled DNA and has the properties of a cellular type I enzyme (63, 64). Although failed attempts to insertionally inactivate the topoisomerase gene suggest that it is essential (65), the enzyme is not required for transcription of linear DNA templates *in vitro* (36). The topoisomerase might be required with more topologically constrained natural templates *in vivo* or have an entirely unrelated function. Additional core-associated enzymes (reviewed in Ref. 2) include a protein kinase and deoxyribonuclease/ligase.

Intermediate Transcription

Although the existence of at least two temporal classes of postreplicative genes was predicted by analyzing the time course of polypeptide synthesis in vaccinia virus-infected cells (7, 8, 66), direct evidence for a class of intermediate genes, distinct from late, was only recently obtained (67). Whereas viral DNA replication is required for expression of true late genes regardless of whether they are genomic or present in transfected plasmids (68), transfected copies of intermediate genes are expressed in the presence of inhibitors of DNA replication (67). The latter data suggest that transacting factors required for intermediate expression are present prior to replication and hence are early viral and/or cellular proteins. There are now five known examples of intermediate genes (67, 69), but a promoter consensus sequence has not been determined.

Late Transcription

mRNA—Late mRNAs have several unusual features. Transcription termination signals for late transcription have not been recognized, and the 3′ ends of late mRNAs are heterogeneous in length (70, 71). Moreover, since both strands of DNA are transcribed, isolated late RNAs can be annealed with other late RNAs or with early RNAs to form ribonuclease-resistant hybrids (72, 73). Whether anti-sense RNA has a role *in vivo* is unknown, however. The discrete 3′ end of one major cowpox virus transcript provides an exception to the rule of length heterogeneity of late RNA (74).

A capped 5′ poly(A) tract of approximately 35 nucleotides is a characteristic feature of late mRNAs (74–77) whereas a shorter one is present in a minority of early mRNAs (21, 22). The poly(A) leader is not encoded as such within the genome and probably arises by RNA polymerase slippage when initiating within the highly conserved TAAAT sequence, as discussed below. Some speculations regarding the role of the capped 5′ poly(A) leader include a binding site for initiation factors and the 40 S ribosomal subunit, which would then scan, unimpeded by anti-sense RNA, to the first AUG (usually located immediately after the poly(A) leader) where ribosome assembly and translation occur.

Promoters—The late promoter may be considered in terms of three regions: an upstream sequence of about 20 bp with some consecutive T or A residues, separated by a region of about 6 bp from a highly conserved TAAAT element within which transcription initiates (78). Mutations within the A triplet of TAAAT drastically decreased transcription (78–80), and substitution of the flanking T residues also had a negative effect but to a degree that depended inversely on the strength of the promoter as determined by upstream sequences (78). Immediately downstream of the TAAAT, a G is optimal for promoter activity and an A is second best. Single nucleotide substitutions within the 6 bp upstream and 3 bp downstream of TAAAT had relatively modest effects on promoter strength (78). The region upstream of −7 is essential for late promoter function and usually contains runs of A or T residues with the latter having a much greater activating effect. A very strong synthetic promoter was constructed with runs of 18–20 T residues (78). Regardless of promoter strength, however, initiation always occurred within the A triplet.

The 5′ poly(A) leader was diminished in length when the T residues of TAAAT were mutated, and the leader was absent or limited to a few nucleotides when any of the three A residues themselves were mutated (78). Shortening of poly(A) also was noted when mutations further downstream were made (81). These data suggested that the poly(A) leader is formed by a mechanism involving backward slippage of the RNA polymerase.

In Vitro Transcription System—Extracts prepared from cells at late times after infection are capable of transcribing vaccinia virus late genes, and the resulting RNAs contain 5′ poly(A) leaders (82, 83). After passage of the cytoplasmic extract through a phosphocellulose column and stepwise elution, three fractions were obtained which together but not separately could transcribe late promoter templates (84). One of the factors called VLTF-1 was partially purified. It seems likely that the RNA polymerases used for early and late transcription are similar or at least share subunits since conditionally lethal mutations in the large subunit and in a small subunit include decreased expression of late genes (85, 86).

Transactivators of Late Gene Expression—A novel transfection approach was used to identify intermediate genes needed for late transcription (69). Earlier studies had indicated that intermediate and late promoter controlled reporter genes were expressed from transfected plasmids provided the cells were infected with vaccinia virus (67, 68). Significantly, however, only the intermediate promoters were active when viral DNA replication was inhibited. This replication block to expression of a late promoter-controlled reporter gene on a plasmid could be bypassed by also transfecting naked virion DNA (69). The role of the transfected virion DNA was investigated by substituting a library of cloned vaccinia virus DNA fragments. It was found, by repeated subcloning and transfection, that three intermediate class genes encoding polypeptides of 17, 26, and 30 kDa were both necessary and sufficient for expression of a transfected late gene in the presence of an inhibitor of DNA replication. Further studies revealed that the 30-kDa protein corresponds to VLTF-1 and is thus a transcription factor.[3]

There are several possible reasons why the DNA in the

[2] P. D. Gershon, B.-Y. Ahn, M. Garfield, and B. Moss, manuscript in preparation.

[3] C. Wright, J. Keck, and B. Moss, manuscript in preparation.

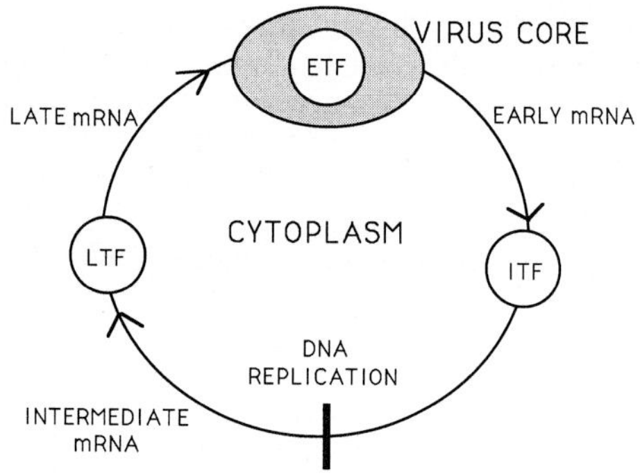

FIG. 3. **Transcriptional regulatory cycle.** *ETF,* early transcription factor; *ITF,* intermediate transcription factor(s); *LTF,* late transcription factor(s).

infecting virus particle only serves as a template for early gene expression whereas replicated or transfected DNA serves as a template for intermediate and late gene expression. Initially, the virion proteins may sequester the input DNA from newly synthesized enzymes and factors required for the expression of later classes of genes. Any input DNA that becomes more accessible to exogenous proteins may rapidly associate with the replication apparatus. Alternatively, expression of the intermediate and late classes of genes might be prevented by the presence of specific repressor proteins on the packaged DNA.

Regulatory Cycle

An oversimplified diagram of the transcriptional regulatory cycle is presented in Fig. 3. Poxvirus gene expression begins with the entry of the virus into the cytoplasm of the cell. The virus core, with its associated enzymes including RNA polymerase, VETF, guanylyltransferase, methyltransferases, and poly(A) polymerase, is programmed to express early genes and produce functional mRNAs. The translation products of these early mRNAs include RNA polymerase subunits, DNA polymerase, and putative factors for specific transcription of intermediate genes. The parental DNA does not serve as a template for intermediate gene expression, perhaps because its location in the core or commitment to DNA replication restricts access by a newly synthesized transcription complex or because of specific repressors of intermediate genes. With the occurrence of DNA replication, the cycle enters its second phase and templates are available for transcription of the three intermediate genes identified as transactivators of late gene expression. In the third phase of the cycle, the late genes encoding virion enzymes, early transcription factor, and structural proteins are expressed. Finally, the progeny viral particles are assembled and some are released ready to begin the cycle again.

REFERENCES

1. Moss, B. (1990) in *Virology* (Fields, B. N., Knipe, D. M., Chanock, R. M., Hirsch, M. S., Melnick, J., Monath, T. P., and Roizman, B. eds) pp. 2079–2112, Raven Press, New York
2. Moss, B. (1990) *Annu. Rev. Biochem.* **59,** 661–688
3. Munyon, W. H., and Kit, S. (1966) *Virology* **29,** 303–306
4. Kates, J. R., and McAuslan, B. (1967) *Proc. Natl. Acad. Sci. U. S. A.* **57,** 314–320
5. Oda, K., and Joklik, W. K. (1967) *J. Mol. Biol.* **27,** 395–419
6. Salzman, N. P., and Sebring, E. (1967) *J. Virol.* **1,** 16–23
7. Moss, B., and Salzman, N. P. (1968) *J. Virol.* **2,** 1016–1027
8. Pennington, T. H. (1974) *J. Gen. Virol.* **25,** 433–444
9. Kates, J., and Beeson, J. (1970) *J. Mol. Biol.* **50,** 19–23
10. Wei, C. M., and Moss, B. (1975) *Proc. Natl. Acad. Sci. U. S. A.* **72,** 318–322
11. Cooper, J. A., and Moss, B. (1978) *Virology* **88,** 149–165
12. Gershowitz, A., Boone, R. F., and Moss, B. (1978) *J. Virol.* **27,** 399–408
13. Shuman, E., Spencer, E., Furneaux, H., and Hurwitz, J. (1980) *J. Biol. Chem.* **255,** 5396–5403
14. Veomett, G. E., and Kates, J. R. (1973) *ICN-UCLA Symp. Mol. Cell. Biol.* 127–142

15a. Shuman, S., and Moss, B. (1988) *J. Biol. Chem.* **263,** 6220–6225

15b. Shuman, S., and Moss, B. (1989) *J. Biol. Chem.* **264,** 21356–21360

16. Golini, F., and Kates, J. R. (1985) *J. Virol.* **53,** 205–213
17. Rohrmann, G., and Moss, B. (1985) *J. Virol.* **56,** 349–355
18. Rohrmann, G., Yuen, L., and Moss, B. (1986) *Cell* **46,** 1029–1035
19. Mackett, M., Smith, G. L., and Moss, B. (1984) *J. Virol.* **49,** 857–864
20. Boone, R. F., and Moss, B. (1977) *Virology* **79,** 67–80
21. Ink, B. S., and Pickup, D. J. (1990) *Proc. Natl. Acad. Sci. U. S. A.* **87,** 1536–1540
22. Ahn, B.-Y., Jones, E. V., and Moss, B. (1990) *J. Virol.* **64,** 3019–3024
23. Davison, A. J., and Moss, B. (1989) *J. Mol. Biol.* **210,** 749–769
24. Proudfoot, N. J., and Brownlee, G. G. (1976) *Nature* **263,** 211–214
25. Yuen, L., and Moss, B. (1987) *Proc. Natl. Acad. Sci. U. S. A.* **84,** 6417–6421
26. Earl, P. L., Hügen, A. W., and Moss, B. (1990) *J. Virol.* **64,** 2448–2451
27. Baroudy, B. M., and Moss, B. (1980) *J. Biol. Chem.* **255,** 4372–4380
28. Jones, E. V., Puckett, C., and Moss, B. (1987) *J. Virol.* **61,** 1765–1771
29. Broyles, S. S., and Moss, B. (1986) *Proc. Natl. Acad. Sci. U. S. A.* **83,** 3141–3145
30. Patel, D. D., and Pickup, D. J. (1989) *J. Virol.* **63,** 1076–1086
31. Amegadzie, B., Holmes, M., Cole, N. B., Jones, E. V., Earl, P. L., and Moss, B. (1991) *Virology,* **180,** 88–98
32. Ahn, B.-Y., Gershon, P. D., Jones, E. V., and Moss, B. (1990) *Mol. Cell. Biol.* **10,** 5433–5441
33. Silver, M., McFadden, G., Wilton, S., and Dales, S. (1979) *Proc. Natl. Acad. Sci. U. S. A.* **76,** 4122–4125
34. Morrison, D. K., and Moyer, R. W. (1986) *Cell* **44,** 587–596
35. Wilton, S., and Dales, S. (1986) *Virus Res.* **5,** 323–341
36. Broyles, S. S., Yuen, L., Shuman, S., and Moss, B. (1988) *J. Biol. Chem.* **263,** 10754–10760
37. Yuen, L., Davison, A. J., and Moss, B. (1987) *Proc. Natl. Acad. Sci. U. S. A.* **84,** 6069–6073
38. Broyles, S. S., and Moss, B. (1988) *J. Biol. Chem.* **263,** 10761–10765
39. Gershon, P. D., and Moss, B. (1990) *Proc. Natl. Acad. Sci. U. S. A.* **87,** 4401–4405
40. Broyles, S. S., and Fesler, B. S. (1990) *J. Virol.* **64,** 1523–1529
41. Martin, S. A., and Moss, B. (1976) *J. Biol. Chem.* **251,** 7313–7321
42. Venkatesan, S., Gershowitz, A., and Moss, B. (1980) *J. Biol. Chem.* **255,** 903–908
43. Shuman, S., and Hurwitz, J. (1981) *Proc. Natl. Acad. Sci. U. S. A.* **78,** 187–191
44. Roth, M. J., and Hurwitz, J. (1984) *J.Biol. Chem.* **259,** 13488–13494
45. Broyles, S. S., and Moss, B. (1987) *Mol. Cell. Biol.* **7,** 7–14
46. Morgan, J. R., Cohen, L. K., and Roberts, B. E. (1984) *J. Virol.* **52,** 206–214
47. Niles, E. G., Lee Chen, G.-J., Shuman, S., Moss, B., and Broyles, S. S. (1989) *Virology* **172,** 513–522
48. Guo, P., and Moss, B. (1990) *Proc. Natl. Acad. Sci. U. S. A.* **87,** 4023–4027
49. Shuman, S. (1990) *J. Biol Chem* **265,** 11960–11966
50. Shuman, S., and Morham, S. G. (1990) *J. Biol. Chem.* **265,** 11967–11972
51. Barbosa, E., and Moss, B. (1978) *J. Biol. Chem.* **253,** 7692–7697
52. Barbosa, E., and Moss, B. (1978) *J. Biol. Chem.* **253,** 7698–7702
53. Shuman, S., Broyles, S. S., and Moss, B. (1987) *J. Biol. Chem.* **262,** 12372–12380
54. Moss, B., Rosenblum, E. N., and Gershowitz, A. (1975) *J. Biol. Chem.* **250,** 4722–4729
55. Nevins, J. R., and Joklik, W. K. (1977) *J. Biol. Chem.* **252,** 6939–6947
56. Shuman, S., and Moss, B. (1988) *J. Biol. Chem.* **263,** 8405–8412
57. Paoletti, E., Rosemond-Hornbeak, H., and Moss, B. (1974) *J. Biol. Chem.* **249,** 3273–3280
58. Rodriguez, J. F., Kahn, J. S., and Esteban, M. (1986) *Proc. Natl. Acad. Sci. U. S. A.* **83,** 9566–9570
59. Broyles, S. S., and Moss, B. (1987) *J. Virol.* **61,** 1738–1742
60. DeLange, A. M. (1989) *J. Virol.* **63,** 2437–2444
61. Kunzi, M. S., and Traktman, P. (1989) *J. Virol.* **63,** 3999–4010
62. Paoletti, E., Cooper, N., and Moss, B. (1974) *J. Biol. Chem.* **249,** 3281–3286
63. Shaffer, R., and Traktman, P. (1987) *J. Biol. Chem.* **262,** 9309–9315
64. Shuman, S., Golder, M., and Moss, B. (1988) *J. Biol. Chem.* **263,** 16401–16407
65. Shuman, S., Golder, M., and Moss, B. (1989) *Virology* **170,** 302–306
66. Opperman, H., and Koch, G. (1976) *J. Gen. Virol.* **32,** 261–273
67. Vos, J. C., and Stunnenberg, H. G. (1988) *EMBO J.* **7,** 3487–3492
68. Cochran, M. A., Mackett, M., and Moss, B. (1985) *Proc. Natl. Acad. Sci. U. S. A.* **82,** 19–23
69. Keck, J. G., Baldick, C. J., and Moss, B. (1990) *Cell* **61,** 801–809
70. Cooper, J. A., Wittek, R., and Moss, B. (1981) *J. Virol.* **39,** 733–745
71. Mahr, A., and Roberts, B. E. (1984) *J. Virol.* **49,** 510–520
72. Colby, C., Jurale, C., and Kates, J. R. (1971) *J. Virol.* **7,** 71–76
73. Boone, R. F., Parr, R. P., and Moss, B. (1979) *J. Virol.* **30,** 365–374
74. Patel, D. D., and Pickup, D. J. (1987) *EMBO J.* **6,** 3787–3794
75. Bertholet, C., Van Meir, E., ten Heggeler-Bordier, B., and Wittek, R. (1987) *Cell* **50,** 153–162
76. Schwer, B., Visca, P., Vos, J. C., and Stunnenberg, H. G. (1987) *Cell* **50,** 163–169
77. Ahn, B.-Y., and Moss, B. (1989) *J. Virol.* **63,** 226–232
78. Davison, A. J., and Moss, B. (1989) *J. Mol. Biol.* **210,** 771–784
79. Bertholet, C., Stocco, P., Van Meir, E., and Wittek, R. (1986) *EMBO J.* **5,** 1951–1957
80. Hänggi, M., Bannwarth, W., and Stunnenberg, H. G. (1986) *EMBO J.* **5,** 1071–1076
81. de Magistris, L., and Stunnenberg, H. G. (1988) *Nucleic Acids Res.* **16,** 3141–3156
82. Wright, C. F., and Moss, B. (1987) *Proc. Natl. Acad. Sci. U. S. A.* **84,** 8883–8887
83. Schwer, B., and Stunnenberg, H. G. (1988) *EMBO J.* **7,** 1183–1190
84. Wright, C. F., and Moss, B. (1989) *J. Virol.* **63,** 4224–4233
85. Ensinger, M. J. (1987) *J. Virol.* **61,** 1842–1850
86. Hooda-Dhingra, U., Thompson, C. L., and Condit, R. C. (1989) *J. Virol.* **63,** 714–729

THE JOURNAL OF BIOLOGICAL CHEMISTRY
Vol. 266, No. 4, Issue of February 5, pp. 2005-2008, 1991
Printed in U.S.A.

Minireview

Metal-catalyzed Oxidation of Proteins

PHYSIOLOGICAL CONSEQUENCES

Earl R. Stadtman‡ and Cynthia N. Oliver§

From the Laboratory of Biochemistry, National Heart, Lung, and Blood Institute, National Institutes of Health, Bethesda, Maryland 20892

Detailed studies of the oxidative modification of glutamine synthetase (GS), in extracts of *Escherichia coli* (1, 3) and *Klebsiella aerogenes* (2, 4) established that the oxidation is dependent upon NAD(P)H, O_2, and Fe(III) or Cu(II). The requirements for O_2 and an auxiliary electron donor (*i.e.* a donor other than GS itself) indicated that the oxidation is promoted by a mixed function oxidation (MFO)[1] type mechanism. In the meantime, it was found that many enzymes are readily inactivated by MFO systems (5–7) and that this renders them susceptible to proteolytic degradation by a variety of exogenous and endogenous proteinases (1, 8–12). It became evident also that the oxidation of proteins is likely implicated in the killing of bacteria by neutrophils (13), and in the generation of altered (inactive or less active) forms of enzymes that accumulate during animal aging (6, 14–16) and under several pathological conditions (17–23). It is the intent of this review to summarize the results of studies on the mechanism of protein oxidation by MFO systems and the biological implications of such oxidations.

The Mechanism of Metal Ion-catalyzed Protein Oxidation

The current view of how MCO[2] systems mediate protein oxidation is illustrated in Fig. 1. In this representation, the electron donor system is needed only to catalyze the reduction of O_2 to H_2O_2 and of Fe(III) to Fe(II). Depending upon the electron donor system used, the reduction of O_2 can proceed by a two-electron mechanism yielding H_2O_2 directly, or by way of two sequential one-electron transfer processes leading first to superoxide anion followed by its dismutation to H_2O_2 and O_2. Similarly, the reduction of Fe(III) to Fe(II) can occur directly or via the intermediate formation of $O_2^{\cdot-}$, which can react directly with Fe(III) to form Fe(II) and O_2. It is believed that the Fe(II) then binds to a metal binding site on the protein (P) after which the protein·Fe(II) complex reacts with the H_2O_2 to generate *in situ* an activated oxygen species ($\dot{O}H$, ferryl ion), which reacts with the side chains of amino acid residues at the metal binding site. Among other modifications, some amino acid residues are converted to carbonyl derivatives (25, 26). After oxidative modification, the protein becomes highly sensitive to proteolytic degradation, and in the case of enzymes they are converted to catalytically inactive or less active, more thermolabile forms (14).

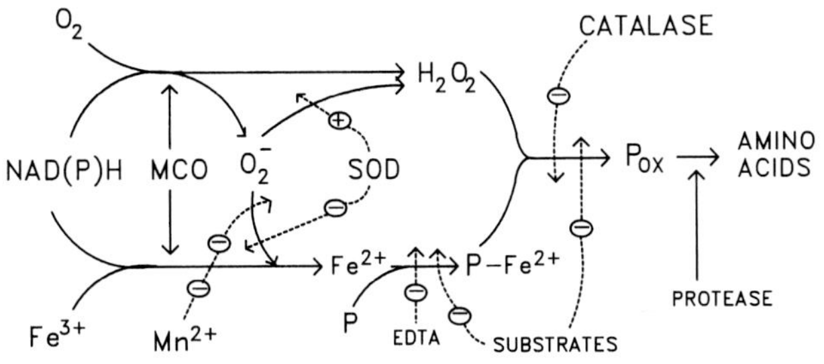

FIG. 1. **Mechanism of enzyme oxidation and degradation.** *SOD*, superoxide dismutase; *P*, protein; P_{ox}, oxidized protein. The symbols ⊕ and ⊖ indicate stimulation and inactivation, respectively.

Site-specific Nature of the MCO-catalyzed Reactions

According to Fig. 1, the MCO-catalyzed oxidation of proteins is a site-specific process involving the interaction of H_2O_2 and Fe(II) at metal binding sites on the protein. The site-specific nature of the reaction is indicated by the following facts. (*a*) The inactivation of enzymes by MCO systems is relatively insensitive to inhibition by free radical scavengers (formate, ethanol, and mannitol) (5, 6, 27). (*b*) Only one or at most only a few amino acid residues in a protein can be modified by MCO systems (25, 26) whereas almost all amino acid residues can be modified when proteins are subjected to free radicals obtained by radiolysis (28–30). (*c*) Most of the enzymes that are highly sensitive to modification by MCO systems require metal ions for catalytic activity. Therefore, they must contain a metal ion binding site. (*d*) In the case of *E. coli* glutamine synthetase, the loss of catalytic activity correlates with the loss of a single histidyl and a single arginyl residue per subunit, both of which are situated in close proximity to one of two divalent metal binding sites on the enzyme (31, 32).[3]

A plausible mechanism for the site-specific modification of a lysyl residue at the metal binding site of a protein is illustrated in Fig. 2. In this mechanism the reduction of Fe(III) (step *a*) is followed by binding of the Fe(II) to the enzyme (step *b*) to form a coordination complex in which the ε-amino group of a lysyl residue at the metal binding site serves as one of several ligands to which the Fe(II) is bound. The H_2O_2 produced by the reduction of O_2 (step *c*) may react with Fe(II) in the complex to form $\dot{O}H$, OH^-, and a Fe(III)·enzyme complex (step *d*). The $\dot{O}H$ thus formed abstracts a hydrogen atom from the carbon atom bearing the ε-amino group to form a carbon-centered radical (step *e*), which then donates its unpaired electron to Fe(III) in the complex to regenerate Fe(II) and coincidently converts the ε-amino group to an imino derivative (step *f*). Finally, the imino derivative undergoes spontaneous hydrolysis whereby NH_3, Fe(II) are released and an aldehyde derivative of the lysyl residue is generated (step *g*). The derivatized protein is thus rendered susceptible to proteolytic degradation. Upon acid hydrolysis, 2-aminoadipic semialdehyde would be one of its products. In this representation, the reaction of H_2O_2 with the Fe(II)·enzyme complex is shown to yield $\dot{O}H$ as a primary product. Similar

‡ To whom correspondence should be addressed: National Institutes of Health, 9000 Rockville Pike, Bldg. 3, Rm. 222, Bethesda, MD 20892. Tel.: 301-496-4096.

§ Present address: Merck Sharp and Dohme Research Laboratories, Bldg. 16-100, West Point, PA 19486.

[1] The abbreviations used are: MFO, mixed function oxidation; MCO, metal-catalyzed oxidation; PBN, *N-tert*-butyl-α-phenylnitrone; LDL, low density lipoprotein.

[2] To avoid confusion with mixed-function *oxidases* (24), MFO systems are now referred to as metal-catalyzed oxidation MCO systems.

[3] I. Climent and R. L. Levine, personal communication.

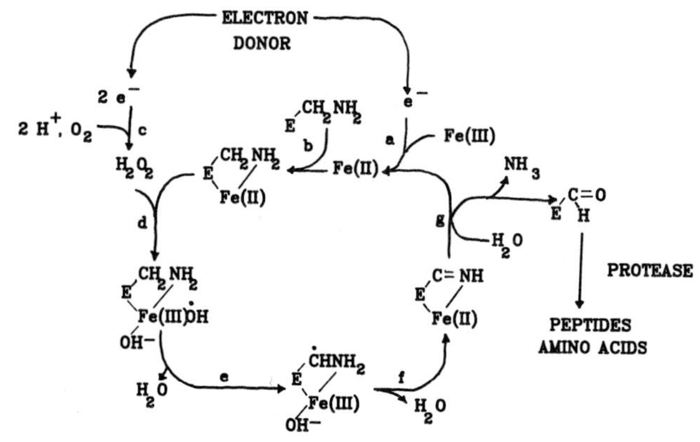

FIG. 2. **Site-specific metal ion-catalyzed protein oxidation.**

TABLE I
MCO systems that catalyze protein modification

1. NAD(P)H oxidase/NAD(P)H/O_2/Fe(III)
2. Xanthine oxidase/xanthine/Fe(III)/O_2
3. Cytochrome P-450/cytochrome P-450 reductase/Fe(III)/O_2
4. Cytochrome P-450_{cam}/*Putida redoxin*/*P. redoxin* reductase/ Fe(III)/NADH/O_2
5. Ascorbate/O_2/Fe(III)
6. Fe(II)/O_2
7. Fe(II)/H_2O_2
8. RSH/Fe(III)/O_2

mechanisms in which $(FeO)^{2+}$ or $(Fe(OH)_2)^{2+}$, or $(FeOH)^{3+}$ + OH^- is the active intermediate might also be considered. Nevertheless, the reaction is perceived as a "caged" reaction in which the activated oxygen species does not escape but reacts preferentially with functional groups at the metal binding site of the enzyme. This would account for the failure of radical scavengers to inhibit the reaction.

It follows from the mechanisms depicted in Figs. 1 and 2 that any system capable of producing H_2O_2 and of reducing Fe(III) or Cu(II) can provoke site-specific modifications of proteins possessing a metal binding site. Several enzymic and nonenzymic systems which have been shown to catalyze the oxidative modification of enzymes are listed in Table I. Perhaps the most physiologically relevant of these are those for which NADH or NADPH serves as the auxiliary electron donor, *e.g.* the NAD(P)H oxidases and the cytochrome P-450 systems.

Of the nonenzymic MCO systems, those in which ascorbate and mercaptans serve as electron donors are particularly important. The ascorbate system, originally developed by Udenfriend *et al.* (33), has been considered a model for mixed function oxidases that catalyze the O_2-dependent hydroxylation of aromatic compounds (drugs) (24). In the meantime, it was found to promote the oxidation of lipids (34), nucleic acids (35), and proteins (25–27, 36). Indeed, studies of Levine and co-workers on the oxidation of glutamine synthetase (25, 27) and amino acid homopolymers (26) by the ascorbate MCO system have contributed greatly to our understanding of the mechanism of site-specific metal ion-catalyzed oxidation of proteins. The biological significance of the RSH/Fe(III)/O_2 system seems assured by the discovery of a protein in yeast and in rat tissues that protects proteins from damage by this MCO system (37, 38).

Oxidation Marks Proteins for Degradation

The proposition that metal-catalyzed oxidation marks proteins for degradation stemmed from the observations that the oxidized form of glutamine synthetase is degraded by cell-free extracts of *E. coli* more rapidly than is the native enzyme (1, 3). Confirmation of this hypothesis was obtained by Rivett (9, 10) and by Roseman and Levine (11) showing that highly purified preparations of neutral alkaline proteinases from rat liver and *E. coli* catalyze rapid degradation of the oxidized forms of glutamine synthetase but have little or no ability to degrade the unoxidized enzymes. The concept gained additional support from the studies of Davies and co-workers showing that the degradation of endogenous proteins in *E. coli* (39), red blood cells (40, 41), and liver or heart mitochondria (42) is greatly enhanced following exposure of the cells to oxygen radicals or H_2O_2, and also by their studies showing that prior exposure of highly purified proteins to oxygen radical generation systems *in vitro* makes them highly susceptible to degradation by proteases in extracts of *E. coli* (39) or red blood cells (40).

The discovery that carbonyl derivatives of some amino acid residues are among the products of oxygen radical damage to proteins provides a means of assessing the extent of such damage under various physiological conditions. To this end, several methods have been developed for the quantitation of protein carbonyl groups (43).

Protein Oxidation and Aging

Catalytically inactive or less active, more thermolabile forms of enzymes accumulate in cells during aging (6, 44). To test the possibility that the age-related changes are due in part to metal-catalyzed oxidation reactions, the concentrations of protein carbonyl groups were measured in three different aging models, *viz.* in human erythrocytes, in cultured human dermal fibroblasts, and in rats of different ages. In the erythrocyte model advantage is taken of the fact that the density of erythrocytes increases with cell age. Therefore, cells of different ages can be separated from a single batch of blood by means of density gradient sedimentation. As shown in Fig. 3*A*, *upper panel*, the level of protein carbonyl groups in various fractions increased with cell density, *i.e.* with cell age (14). Moreover, the levels of glyceraldehyde-3-P dehydro-

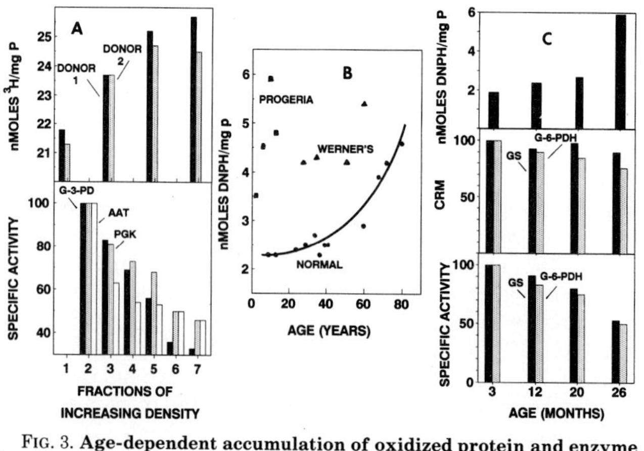

FIG. 3. **Age-dependent accumulation of oxidized protein and enzyme inactivation.** *A*, the human erythrocyte model: *upper panel*, the *black* and *shaded bars* refer to the protein carbonyl content of red cells from human donors 1 and 2, respectively; *lower panel*, the *black*, *shaded*, and *open bars* refer to the relative catalytic activities of glyceraldehyde-3-P dehydrogenase (*G-3-PD*), aspartate aminotransferase (*AAT*), and phosphoglycerate kinase (*PGK*), respectively, of human red cells. *B*, the cultured fibroblast model shows the carbonyl content of protein from cultured human fibroblasts. *C*, the rat model: *upper panel*, the *bars* refer to the amount of carbonyl groups in protein from rat liver extracts; *middle panel*, the *black* and *shaded bars* refer to the relative amount of material in rat liver extracts (*CRM*) that cross-reacts with antibodies to purified glutamine synthetase (*GS*) and glucose-6-P dehydrogenase (*G-6-PDH*), respectively; *lower panel*, the *black* and *shaded bars* refer to the relative catalytic activities of glutamine synthetase and glucose-6-P dehydrogenase, respectively. The data in *A* and *B* are from experiments as described in Oliver *et al.* (14). The data in *C* are from Starke-Reed and Oliver (16). The carbonyl content per mg of protein (*P*) is expressed either as ^3H per mg of protein obtained by reduction of carbonyl groups with $[^3H]BH_4$ or DNPH per mg of protein, where DNPH refers to the carbonyl content as a measured reaction with DNPH.

genase (G-3-PD), aspartate aminotransferase (ATT), and phosphoglycerate kinase (PGK) declined with cell density, in confirmation of earlier studies showing the activities of these enzymes decrease with cell age. In the fibroblast model, the carbonyl content of protein in cultured fibroblasts from individuals of different ages and from individuals with premature aging diseases (patients with progeria or Werner's syndrome) was measured using the 2,4-dinitrophenylhydrazone assay procedure (14). As shown in Fig. 3*B*, there was an almost exponential increase in the level of protein carbonyl groups with donor age over the range of 10–80 years. Moreover, the carbonyl content of protein in fibroblasts from individuals with either progeria or Werner's syndrome was very much higher than that in fibroblasts of normal individuals of the same age. In fact, the levels in fibroblasts from young donors with these aging diseases was about the same as that found in cultured fibroblasts from normal 80-year-old donors. Finally, as shown in Fig. 3*C*, *upper panel*, the carbonyl content of protein in hepatocytes from rats was found to increase gradually with animal age over the range of 3–20 months and then increased rapidly over the range of 20–26 months (16). Data in Fig. 3*C*, *lower panel*, show also that the activities of GS and glucose-6-P dehydrogenase (Glc-6-PD) decreased with animal age to values only about 50% of that found in young animals; moreover, the loss in enzyme activity was not associated with a comparable loss of antibody-specific immunoreactive protein (Fig. 3*C*, *middle panel*). Furthermore, the level of neutral alkaline protease(s) activity in hepatocytes also declines with animal age (Fig. 4) suggesting that the age-dependent accumulation of altered enzyme forms is due partly to a loss of those proteases that degrade oxidized proteins. This possibility is supported further by an analysis of the data in Fig. 4 which shows a replot of some of the data in Fig. 3*C*. The *shaded area* in Fig. 4 represents the difference between catalytically active Glc-6-P dehydrogenase and total Glc-6-P dehydrogenase protein as measured by specific antibody titration. This shaded fraction therefore is a measure of inactive Glc-6-PD. As shown in the *inset* there is a linear inverse relationship between the level of inactive Glc-6-PD and the level of neutral alkaline protease activity. Taken together the data in Figs. 3*C* and 4 show that the level of oxidized protein increases with animal age and that the concomitant accumulation of altered Glc-6-PD is related to a loss of the proteinases that are responsible for the degradation of altered (oxidized?) protein.

It should be emphasized that the 2–3-fold increase in levels

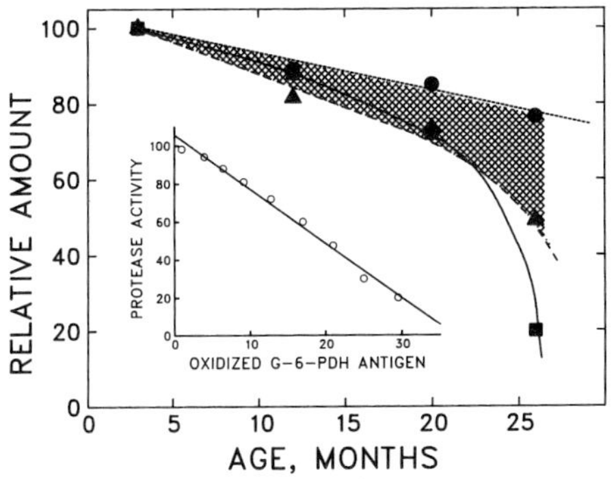

FIG. 4. **Relationship between protease activity and the accumulation of catalytically inactive glucose-6-P dehydrogenase protein.** The data are replots of data presented by Starke-Reed and Oliver (16). The *curves* are as follows: ■, neutral alkaline protease activity; ▲, Glc-6-PD (*G-6-PDH*) activity; ●, protein that reacts with antibodies against highly purified Glc-6-PD. The *shaded area* is a measure of the catalytically inactive Glc-6-PD antigen which accumulated. The *inset* represents the relationship between the relative amounts of observed neutral protease activity and the amount of inactive Glc-6-PD as determined by the difference between the total immunologically reactive Glc-6-PD protein and the catalytically active Glc-6-PD enzyme.

of oxidized proteins which occurs during aging (Fig. 3, *B* and *C*) is by no means trivial. It represents a substantial increase in the fraction of total protein that is oxidized. Based on the observation that on the average only one carbonyl group is introduced per enzyme subunit and the assumption that the average molecular weight of enzyme subunits is 50,000, it can be calculated that the amount of oxidized protein ranges from 10% of the total protein in young animals to between 23 and 30% in old animals (16). These are clearly minimal values since not all metal-catalyzed oxidation reactions yield carbonyl derivatives, *viz.* histidyl, prolyl, and methionyl residues yield asparaginyl, pyroglutamyl, and methionyl sulfoxide residues, respectively.

Protein Oxidation and Disease

Inflammatory Diseases—It is becoming apparent that MCO-mediated protein oxidation is an early indicator of tissue damage and that the formation of protein carbonyl derivatives is associated with pathological conditions both in humans and in animal model systems. Studies in this laboratory (45) have shown that activated neutrophils generate diffusible products (*viz.* H_2O_2, O_2^-, etc.) which in the presence of Fe(III) will cause inactivation of enzymes within intact bacterial cells by an endocytotic independent mechanism. Moreover, during periods of oxidative burst some endogenous neutrophil proteins are converted to carbonyl derivatives and endogenous enzymes are inactivated. This may account for the fact that neutrophils lose activity and undergo lysis after activation. These results raise the possibility that recruitment and activation of neutrophils at extravascular sites may provoke protein oxidation and thus contribute to the tissue damage that occurs during chronic inflammation. Indeed, Chapman *et al.* (21) have shown that the levels of protein carbonyl groups in the synovial fluid of patients with rheumatoid arthritis are higher than in patients with osteoarthritis.

Atherosclerosis—It is now generally believed that metal ion-catalyzed oxidation of low density lipoprotein (LDL) by endothelial cells and the subsequent uptake of the oxidized LDL by monocytes/macrophages to form foam cells are important events in atherogenesis (for review see Steinberg *et al.* (46)). Most attention has been focused on the oxidation of the lipid moieties of LDL (47, 48). However, the likelihood that protein oxidation is also involved is implied by the fact that modifications of apoprotein B include fragmentation and the loss of histidyl, lysyl, and prolyl residues; such changes are characteristic of MCO reactions (49).

Ischemia Reperfusion Tissue Damage—The proposal of McCord and co-workers (50) that oxygen free radicals are involved in ischemia reperfusion injury is supported by results of more recent studies with spin trapping techniques showing that free radicals are generated during the reperfusion phase (51, 52).

A role of protein oxidation in ischemia reperfusion injury was verified by the finding that during reperfusion for 60 min following ischemia in the gerbil brain the level of protein carbonyl groups increased from 6 to 13 nmol/mg protein; coincidentally, the specific activity of glutamine synthetase declined to about one-half of its original value (52). That free radicals are involved in these changes is indicated by the fact that prior treatment of the animals with the spin trap *N-tert*-butyl-α-phenylnitrone (PBN) partially protected them against protein oxidation and loss of glutamine synthetase provoked by ischemia reperfusion.

Neurologic Disorders—A role of protein oxidation in some neurological disorders is indicated by (*a*) the studies of Konat and Wiggins (53) showing that myelin proteins are lost when rat myelin preparations are incubated with Cu(II) and H_2O_2; (*b*) subsequent studies showing that when myelin preparations are incubated with MCO systems, protein carbonyl groups are generated and the myelin is rapidly degraded by myelin proteases[4]; and (*c*) the studies of Mickel *et al.* (23) showing that rapid correction of vasopressin-induced hyponatremia in rats is associated with increased protein carbonyl content and myelinolysis. In addition, Murphy and Kehrer (19) have reported that the level of protein carbonyl groups is elevated in the pectoralis muscle of dystrophic chickens.

Cataractogenesis—Finally, Garland and co-workers (54, 55) showed that there is a small but significant age-dependent increase in the protein carbonyl content of human lens, and that the level of oxidized proteins in cataractous lenses reaches values several times that of normal lenses. They showed also that exposure of bovine crystallins to MCO systems *in vitro* produces physical changes similar to those observed in human crystallins *in vivo*.

Conclusion

It would appear from the above that the accumulation of oxidized proteins may be an early indication of oxygen radical-mediated tissue damage, and in some pathological states may account for 50–70% of the total cellular protein (16). Since the intracellular level of oxidized proteins reflects the balance between the rates of oxidation and the rate of degradation of oxidized protein, the accumulation of oxidized protein is a complex function of the numerous factors that govern the synthesis and oxidation of proteins and the activities of various proteases that selectively degrade the oxidized forms. Except under special conditions (*viz.* smoking or excessive exposure to high energy radiation), MCO reactions are likely responsible for most of the free radical damage. The processes that govern the levels of H_2O_2 and the availability of Fe(III) and Cu(II) may be fundamental to the etiology of various metabolic disorders. The level of H_2O_2 is dictated by the availability of O_2 and the activities of numerous enzymic and nonenzymic MFO systems that generate H_2O_2 on the one hand, and of those enzymes that catalyze the decomposition of H_2O_2 (peroxidases, catalase) on the other. Factors that control the availability of Fe(III) and Cu(II) are still poorly understood, but the release of Fe(II) from ferritin and the release of Fe(III) from Fe(III)·enzyme complexes by the degradation of hemoglobin as occurs under certain physiological and pathological conditions are the most likely sources. Factors that govern the levels and activities of multicatalytic proteases and other proteases that degrade oxidized protein are even less well understood. It is likely, however, that the loss of these protease activities is largely responsible for the accumulation of oxidized enzymes during aging and oxidative stress. The possibility that the loss of protease activity is also due to oxidative free radical damage deserves attention.

REFERENCES

1. Oliver, C. N. (1981) *The Regulation of Glutamine Synthetase Degradation.* Ph.D. thesis, The Johns Hopkins University
2. Levine, R. L., Oliver, C. N., Fulks, R. M., and Stadtman, E. R. (1981) *Proc. Natl. Acad. Sci. U. S. A.* **78,** 2120–2124
3. Oliver, C. N., Levine, R. L., and Stadtman, E. R. (1981) in *Metabolic Interconversion of Enzymes* (Holzer, H., ed) pp. 259–268, Springer-Verlag, Berlin
4. Fulks, R. M., and Stadtman, E. R. (1985) *Biochim. Biophys. Acta* **843,** 214–229
5. Stadtman, E. R., and Wittenberger, M. E. (1985) *Arch. Biochem. Biophys.* **239,** 379–387
6. Fucci, L., Oliver, C. N., Coon, M. J., and Stadtman, E. R. (1983) *Proc. Natl. Acad. Sci. U. S. A.* **80,** 1521–1525
7. Chevion, M. (1988) *Free Radical Biol. & Med.* **5,** 27–37
8. Farber, J. M., and Levine, R. L. (1982) *Fed. Proc.* **41,** 865
9. Rivett, A. J. (1985) *J. Biol. Chem.* **260,** 12600–12606
10. Rivett, A. J. (1985) *J. Biol. Chem.* **260,** 300–305
11. Roseman, J. E., and Levine, R. L. (1987) *J. Biol. Chem.* **262,** 2101–2110
12. Davies, K. J. A., and Goldberg, A. L. (1987) *J. Biol. Chem.* **262,** 8227–8234
13. Oliver, C. N. (1987) *Arch. Biochem. Biophys.* **253,** 62–72
14. Oliver, C. N., Ahn, B.-W., Moerman, E. J., Goldstein, S., and Stadtman, E. R. (1987) *J. Biol. Chem.* **262,** 5488–5491
15. Garland, D. (1990) *Exp. Eye Res.* **50,** in press
16. Starke-Reed, P. E., and Oliver, C. N. (1989) *Arch. Biochem. Biophys.* **275,** 559–567
17. Parthasarathy, S., Wieland, E., and Steinberg, D. (1989) *Proc. Natl. Acad. Sci. U. S. A.* **86,** 1046–1050
18. Dicker, E., and Cedarbaum, A. I. (1988) *FASEB J.* **2,** 2901–2906
19. Murphy, M. E., and Kehrer, J. P. (1989) *Biochem. J.* **260,** 359–364
20. Griffiths, H. R., Lunec, J., Gee, C. A., and Wilson, R. L. (1988) *FEBS Lett.* **230,** 155–158
21. Chapman, M. L., Rubin, B. R., and Gracy, R. W. (1989) *J. Rheumatol.* **16,** 15–18
22. Merritt, T. A., Boynton, B. R., and Northway, W. H., Jr. (1988) *Bronchopulmonary Dysplasia,* Blackwell Scientific/Year Book Medical Publishers, Inc., Chicago
23. Mickel, H. S., Starke-Reed, P. E., and Oliver, C. N. (1990) *Biochem. Biophys. Res. Commun.* **172,** 92–97
24. Mason, H. S. (1957) *Adv. Enzymol.* **19,** 79–233
25. Levine, R. L. (1983) *J. Biol. Chem.* **258,** 11823–11827
26. Amici, A., Levine, R. L., Tsia, L., and Stadtman, E. R. (1989) *J. Biol. Chem.* **264,** 3341–3346
27. Levine, R. L. (1983) *J. Biol. Chem.* **258,** 11828–11833
28. Swallow, A. J. (1960) in *Radiation Chemistry of Organic Compound* (Swallow, A. J., ed) pp. 211–224, Pergamon Press, New York
29. Garrison, W. M., Jayko, M. E., and Bennett, W. (1962) *Radiation Res.* **16,** 483–502
30. Davies, K. J. A., Delsignore, M. E., and Lin, S. W. (1987) *J. Biol. Chem.* **262,** 9902–9907
31. Farber, J. M., and Levine, R. L. (1986) *J. Biol. Chem.* **261,** 4574–4578
32. Colombo, G., and Villafranca, J. J. (1986) *J. Biol. Chem.* **261,** 10587–10591
33. Udenfriend, S., Clark, C. T., Axelrod, J., and Brodie, B. B. (1954) *J. Biol. Chem.* **208,** 731
34. Samokyszyn, V. M., and Aust, S. D. (1989) in *Medical, Biochemical and Chemical Aspects of Free Radicals* (Hayaishi, O., Niki, E., Kondo, M., and Yoshikawa, T., eds) pp. 41–48, Elsevier, Tokyo
35. Chevion, M. (1988) *Free Radical Biol. & Med.* **5,** 27–37
36. Deshpande, V. V., and Joshi, J. G. (1985) *J. Biol. Chem.* **260,** 757–764
37. Kim, K., Rhee, S. G., and Stadtman, E. R. (1985) *J. Biol. Chem.* **260,** 15394–15397
38. Kim, I. H., Kim, K., and Rhee, S. G. (1989) *Proc. Natl. Acad. Sci. U. S. A.* **86,** 6018–6022
39. Davies, K. J. A., and Lin, S. W. (1988) *Free Radical Biol. & Med.* **5,** 215–223
40. Davies, K. J. A. (1987) *J. Biol. Chem.* **262,** 9895–9901
41. Pacifici, R. E., Salo, D. C., and Davies, K. J. A. (1989) *Free Radical Biol. & Med.* **7,** 521–536
42. Marcillat, O., Zhang, Y., Lin, S. W., and Davies, K. J. A. (1988) *Biochem. J.* **254,** 677–683
43. Levine, R. L., Garland, D., Oliver, C. N., Amici, A., Climent, I., Lenz, A.-G., Ahn, B.-W., Shaltiel, S., and Stadtman, E. R. (1990) *Methods Enzymol.* **186,** 464–478
44. Oliver, C. N., Fulks, R., Levine, R. L., Fucci, L., Rivett, A. J., Roseman, J. E., and Stadtman, E. R. (1984) in *Molecular Basis of Aging* (Roy, A. K., and Chatterjee, B., eds) pp. 235–262, Academic Press, New York
45. Oliver, C. N. (1987) *Arch. Biochem. Biophys.* **253,** 62–72
46. Steinberg, D., Parthasarathy, S., Carew, T. E., Khoo, J. C., and Witztum, J. W. (1989) *N. Engl. J. Med.* **320,** 915–924
47. Esterbauer, H., Joigens, G., Quehenberger, O., and Koller, E. (1987) *J. Lipid Res.* **28,** 495–509
48. Steinbrecher, U. P. (1987) *J. Biol. Chem.* **262,** 3603–3608
49. Steinbrecher, U. P., Witztum, J. L., Parthasarathy, S., and Steinberg, D. (1987) *Arteriosclerosis* **1,** 135–143
50. Parks, D. A., Bulkley, G. B., Gunger, N., Hamilton, S. R., and McCord, J. M. (1982) *Gastroenterology* **82,** 9–15
51. Floyd, R. A. (1990) *FASEB J.* **4,** 2587–2597
52. Oliver, C. N., Starke-Reed, P., Stadtman, E. R., Liu, G. T., Carney, J. M., and Floyd, R. A. (1990) *Proc. Natl. Acad. Sci. U. S. A.* **87,** 5144–5147
53. Konat, G. W., and Wiggins, R. C. (1985) *J. Neurochem.* **45,** 1113–1138
54. Garland, D., Russell, P., and Zigler, J. S., Jr. (1988) in *Oxygen Radicals in Biology and Medicine* (Simic, M. G., Taylor, K. S., Ward, J. F., and von Sonntag, C., eds) Vol. 49, pp. 347–353, Plenum Press, New York
55. Garland, D., Zigler, J. S., Jr., and Kinoshita, J. (1986) *Arch. Biochem. Biophys.* **251,** 771–776

[4] J. Muller and C. N. Oliver, unpublished results.

Minireview

THE JOURNAL OF BIOLOGICAL CHEMISTRY
Vol. 266, No. 5, Issue of February 15, pp. 2681-2684, 1991
Printed in U.S.A.

The Interleukin-2 Receptor

Thomas A. Waldmann

From the Metabolism Branch, National Cancer Institute, National Institutes of Health, Bethesda, Maryland 20892

The human body defends itself against foreign invaders such as bacteria and viruses by a defense system that involves antibodies and thymus-derived lymphocytes (T cells). A major advance over the past decade has been the demonstration that the diverse cells within the immunological network communicate with each other not only by cell/cell contact but also by an array of hormone-like proteins termed interleukins or lymphokines. Of all the lymphokines involved in this network of cellular communication, the most is known about the structure and function of interleukin-2 (IL-2)[1] and its specific receptor. IL-2 plays a pivotal role in the growth and function of T lymphocytes. In addition, it affects the activities of B lymphocytes, monocytes, natural killer, and lymphokine-activated killer cells.

T cells mediate important regulatory functions such as help and suppression as well as effector functions including the production of lymphokines and the cytotoxic destruction of antigen-bearing cells. Successful T cell-mediated immune responses require that these cells change from a resting to an activated state which requires two sets of signals from cell surface receptors to the nucleus. The first signal is initiated when an appropriately processed and presented antigen interacts with the heterodimeric T cell receptor for that specific antigen. Following this encounter, T cells enter a program of activation leading to the *de novo* synthesis and secretion of IL-2 (1–3). Resting T cells do not express high affinity IL-2R, but receptors are rapidly expressed on T cells after activation with antigen or mitogen (3, 4). The interaction of IL-2 with its induced cellular receptor then triggers proliferation of these cells culminating in the emergence of effector T lymphocytes mediating helper, suppressor, or cytotoxic functions.

The study of the interleukin-2 receptor (IL-2R) and its lymphokine has provided a framework for understanding the cellular and molecular events that are involved in the regulation of the immune response. Furthermore, an understanding of IL-2 and its receptor has opened the way for novel therapeutic approaches for a wide array of clinical conditions including leukemia/lymphoma, autoimmune disorders, and organ allograft rejection. The intent of this article is to summarize the information on the IL-2R with special reference to the most seminal discoveries and recent advances. Comprehensive recent reviews are available to provide more detailed information (3, 5–9).

Structure of the Multisubunit IL-2R

The IL-2Rα Subunit—There are three forms of cellular receptors for IL-2 based on their affinity for ligand: one with a very high affinity (dissociation constant K_d 10^{-11} M), one with an intermediate affinity (10^{-9} M), and one with a lower affinity (10^{-8} M) (5, 7–11). Monoclonal antibodies and affinity cross-linking studies with radiolabeled IL-2 were used to chemically characterize the multiple subunits of this receptor. Utilizing purified biosynthetically labeled IL-2, Robb and coworkers (4) demonstrated saturable high affinity binding sites on mitogen- and alloantigen-activated T cells. Further progress in the analysis of the structure, function, and expression of the human IL-2R was facilitated by the production of the anti-Tac (T cell activation) monoclonal antibody by Uchiyama *et al.* (12, 13). Anti-Tac was shown to recognize the human receptor for IL-2 since it blocks the binding of radiolabeled IL-2 to its receptor thus blocking IL-2-induced proliferation (14–17). The IL-2R protein identified by anti-Tac was characterized as a densely glycosylated and sulfated structure with an apparent molecular mass of 55 kDa (18, 19). This Tac protein is now termed the p55 IL-2Rα chain. Three research groups have cloned cDNAs for this IL-2R protein (20–22). The deduced amino acid sequence indicates that it is composed of 251 amino acids as well as an NH_2-terminal 21-amino acid signal peptide cleaved *in vivo*. The 219 terminal amino acids make up an extracellular domain that contains two potential *N*-linked glycosylation sites and multiple possible *O*-linked carbohydrate sites. A second 19-amino acid domain that contains a single hydrophobic region near the COOH terminus presumably represents a membrane-spanning region. The third and final domain is a very short (13-amino acid) cytoplasmic domain that contains several positively charged amino acids.

The IL-2Rβ Subunit—A series of issues was difficult to resolve when only the 55-kDa protein was considered. Specifically, the p55 protein (now termed IL-2Rα) identified by the anti-Tac monoclonal antibody was shown to participate in both the high affinity (10^{-11} M) and low affinity (10^{-8} M) forms of the IL-2R. In addition, the cloned cDNA for the IL-2Rα protein revealed a cytoplasmic domain that was too short to independently transduce receptor signals to the nucleus (20–22). Furthermore, it had been shown that certain cells not expressing IL-2Rα (including large granular lymphocytes which are precursors of activated natural killer and lymphokine-activated killer cells) can be activated by IL-2 to become efficient killers (23, 24). Finally, a cell line MLA-144 was defined that did not express the Tac protein yet manifested approximately 4,000 IL-2-binding sites with intermediate affinity. These observations led us to consider the possibility that the high affinity IL-2R was not a single protein but rather a receptor complex that included the Tac protein as well as novel non-Tac proteins (11, 25, 26). We used radiolabeled IL-2 in cross-linking studies to define the size of the IL-2R protein on MLA-144 and thereby identified a novel 70/75-kDa IL-2 binding protein (now termed IL-2Rβ) on this cell line. Unlike many multisubunit receptors, the two subunits of the IL-2R can be individually expressed in the absence of the other. We showed that fusion of cell membranes from a low affinity IL-2R line bearing the IL-2Rα alone with membranes from a line with intermediate affinity receptors bearing the IL-2Rβ protein alone generated hybrid membranes expressing high affinity receptors (26). In light of these observations, we proposed a multisubunit model for the high affinity IL-2R (11, 25–27). In this model an independently existing IL-2Rα or IL-2Rβ subunit would generate low or intermediate affinity receptors, respectively, whereas high

[1] The abbreviations used are: IL, interleukin; IL-2R, interleukin-2 receptor; MHC, major histocompatibility complex.

affinity receptors would be formed when both receptors are expressed and noncovalently associated in a receptor complex. In independent studies, Sharon and co-workers (28) proposed a similar model.

Kinetic binding studies with IL-2 provided an interesting perspective on how the two separate IL-2 binding chains cooperate to form the high affinity receptor (9, 29, 30). Each chain reacts very differently with IL-2, with distinct kinetic and equilibrium binding constants. The kinetics of association and dissociation of IL-2 to the IL-2Rα protein are rapid ($t_{1/2}$ = 4 and 6 s, respectively) (1 nM IL-2 was used in these studies), while the IL-2 association and dissociation to the IL-2Rβ protein are markedly slower ($t_{1/2}$ = 45 and 290 min, respectively) (9, 29, 30). The kinetic binding data obtained when the high affinity receptors are analyzed show that the association rate of this receptor depends on the fast associating IL-2Rα chain whereas the dissociation rate is derived from the slow dissociating IL-2Rβ chain. Because the affinity of binding at equilibrium is determined by the ratio of dissociation and association rate constants, this kinetic cooperation between the low and intermediate affinity ligand binding sites results in a receptor with a high affinity for IL-2.

Using an antibody Mikβ_1 developed by Tsudo and co-workers (31) that blocks IL-2 binding to IL-2Rβ, Hatakeyama and co-workers (32) have obtained a full-length cDNA encoding the IL-2Rβ chain that contains an open reading frame which defines a mature protein of 525 amino acids. The extracellular domain of 214 amino acids contains four potential sites for *N*-glycosylation and 8 half-cystine residues. A hydrophobic stretch of 25 amino acids (from residue 215 to 239 of the mature protein) appears to form the membrane-spanning region of the receptor. The cytoplasmic domain of 286 amino acids of the IL-2Rβ chain, which is much larger than that of the IL-2Rα chain, contains serine- (30 of 286), acidic-, and proline-rich (30 of 286) regions (32). This region does not contain a consensus sequence for a tyrosine kinase. Nevertheless, using cytoplasmic deletion mutants, Hatakeyama and co-workers (33) have shown that it retains a 46-residue region (between residues 267 and 312) just below the transmembrane region that is required for growth signal transduction but not for ligand binding and internalization.

Relationship between IL-2Rβ and Members of the Hematopoietic Receptor Superfamily—D'Andrea and co-workers (34), on cloning the erythropoietin receptor, defined several structural features shared by the erythropoietin receptor and the IL-2Rβ subunit including significant sequence homology. The highest degree of identity between the two molecules is present in the cytoplasmic region discussed above (33). Subsequently, the external domains of receptors for IL-2 (β subunit), IL-3, IL-4, IL-6, IL-7, granulocyte macrophage colony-stimulating factor, granulocyte colony-stimulating factor, prolactin, growth hormone, and erythropoietin were shown to contain two conserved motifs that are separated by a sequence of 90–100 amino acids (34–38). The first motif is composed of approximately 60 amino acids and contains four conserved half-cystines and one conserved tryptophan residue. The second motif located close to the transmembrane domain is composed of approximately 30 amino acids including a shared Trp-Ser-*X*-Trp-Ser motif. Furthermore, the cytoplasmic domains of IL-2Rβ, IL-3R, IL-4R, and the erythropoietin receptor also share homology with high percentages of proline and serine. It is conceivable that the cytoplasmic conserved regions interact with the same sets of proteins involved in a common signal transduction pathway among these receptors. Thus as reviewed by Bazan (39) and Cosman *et al.* (40), the IL-2Rβ appears to be a member of a new family of growth and differentiation factor receptors, the hematopoietic receptor superfamily, that is distinct from G protein-associated hormone receptors, receptors expressing a tyrosine kinase in their cytoplasmic domains, as well as those that belong to the immunoglobulin superfamily.

IL-2R-associated Proteins—Functional and structural evidence suggests a more complex subunit structure that involves proteins in addition to the IL-2Rα and IL-2Rβ IL-2-binding proteins. When the IL-2Rβ chain is expressed in large granular lymphocytes or is expressed by transfection in lymphoid cells, these cells are capable of binding IL-2 at intermediate affinity (10^{-9} M). However, when this chain is expressed alone in certain nonlymphoid cells, virtually no IL-2 binding capacity is induced although the β chain identified by specific monoclonal antibodies is demonstrable on the cell surface (32). Furthermore, the purified IL-2Rβ protein has only minimal IL-2 binding ability (41). These observations suggest that another protein cooperates with the IL-2Rβ protein to alter its conformation thus permitting intermediate affinity IL-2 binding. Further evidence supports a more complex structure that involves proteins in addition to the IL-2Rα and IL-2Rβ IL-2-binding proteins. With the use of coprecipitation analysis, radiolabeled IL-2 cross-linking procedures, flow cytometric resonance energy transfer measurement, and recovery following photobleaching, a series of additional proteins of 22, 35–40, 75 (non-IL-2 binding), and 95–105 kDa have been associated with the two IL-2-binding proteins (42–49) (Fig. 1). These associated proteins may play a role: 1) in changing the conformation of the p75 β chain thus altering its ligand binding affinity; 2) in receptor-mediated endocytosis of IL-2; or 3) in the transduction of the IL-2 signal. Furthermore, an association between IL-2R and Class I HLA molecules on T cells has been demonstrated (42, 43, 50, 51). Finally, using flow cytometric resonance energy transfer and fluorescent photobleaching recovery measurements, we have demonstrated that on cells expressing both molecules the IL-

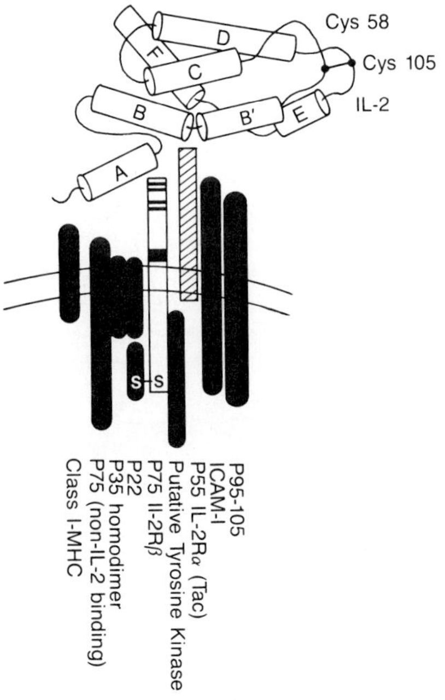

FIG. 1 **Proposed model of multisubunit structure of the IL-2R.** Two IL-2-binding proteins, p55 or IL-2Rα and p75 or IL-2Rβ, bind to different epitopes of the IL-2 molecule. The associated molecules, p22, p35, p75 (non-IL-2 binding), p95-105, Class I MHC, ICAM-1, and a putative tyrosine kinase, are represented in an arbitrary position relative to the two IL-2 binding chains. (IL-2 structure indicated by the entire clear area is reproduced by permission, with modifications, from B. J. Brandhuber, T. Boone, W. C. Kenney, and D. B. McKay (54).) The *capital letters* in the IL-2 figure define the different helical segments.

2Rα chain is in proximity and interacts physically with the intracellular adhesion molecule 1 (ICAM-1, CD54) (42, 43, 52). ICAM-1 binds to the lymphocyte function-associated antigen 1 (LFA-1, CD11a/CD18) promoting cell adhesion in immune inflammatory reactions (53).

Structure-Function Relationship of Interaction of IL-2 and IL-2R

The regions of IL-2 involved in binding to the multisubunit IL-2R proteins are being defined (Fig. 1). The three-dimensional structure of IL-2 has been determined to 3.0-Å resolution (54). It has 133 amino acids and six short α helical segments with no segments of β structure. Monoclonal antibodies directed toward defined regions of IL-2, deletion analysis, and site-specific mutagenesis of IL-2 in neutralization and binding assays have aided in the analysis of the different epitopes of human IL-2 involved in binding to the two IL-2R proteins (54–59). A short helical segment (helix A, amino acids 11–19) and the neighboring amino acid Asp^{20} are required for biological activity, and its normal conformation is required for binding to the IL-2Rβ subunit. In parallel studies, substitution of a glutamine or alanine for arginine at position 38 in the second α helix has only minimal effects on IL-2 proliferative activity but largely abrogates the binding of the resultant IL-2 analog to the α subunit of the IL-2R. Finally, the carboxyl-terminal residues 131–133 and two of the half-cystine residues at positions 58 and 105 that form a disulfide bridge are required for biological activity and binding.

IL-2-induced Signal Transduction

One of the major issues that is not resolved concerns the biochemical mechanisms involved in IL-2-induced signal transduction. The IL-2Rα chain is not sufficient to transduce the IL-2 signal. In contrast, when sufficient concentrations of IL-2 are present, the IL-2Rβ chain in large granular lymphocytes without the α protein permits signal transduction leading to activation of the Na^+/H^+ antiport, myc and myb oncogene expression, IL-2Rα induction, an increase in killer cell activity, and cellular proliferation. The IL-2/IL-2R signal transduction does not utilize protein kinase C, calcium mobilization, or phosphatidylinositol turnover which are pivotal for initial phases of T and B lymphocyte responses to antigen. Neither the IL-2Rα nor the IL-2Rβ express the consensus sequence (Gly-*X*-Gly-*X*-*X*-Gly) for the ATP-binding site of a tyrosine kinase (32). Nevertheless, in various cell lines, including those that express only the IL-2Rβ subunit, the interaction of IL-2 with this receptor induces tyrosine phosphorylation of the IL-2β chain as well as proteins of approximately 92, 73, 63, 57, and 42 kDa (61–68). A series of 56- (possibly $p56^{lck}$) to 63-kDa proteins has been proposed as potential IL-2Rβ chain-associated tyrosine kinases that could play a role in IL-2 signal transduction presumably through interactions with the cytoplasmic domain of IL-2Rβ.

Clinical Aspects of the Studies of IL-2R

Disorders of IL-2R Expression in Malignant and Autoimmune Diseases—The majority of resting T cells, B cells, or monocytes in the circulation do not display the IL-2Rα subunit (12). However, most T and B lymphocytes can be induced by antigenic stimulation to express this IL-2R protein (69–71). Furthermore, the IL-2Rα subunit has been detected on activated cells of the monocyte/macrophage series. Rubin and co-workers (72) demonstrated that activated normal peripheral blood mononuclear cells and certain lines of T or B cell origin release a soluble 45-kDa form of the IL-2Rα chain into the culture medium. Using an enzyme-linked immunosorbent assay, they showed that normal individuals have measurable amounts of this IL-2R chain in their plasma. The definition of elevated plasma levels of such IL-2Rα chains provides a very valuable noninvasive approach to detect both normal and disease-associated lymphocyte activation *in vivo*.

In contrast to the lack of IL-2Rα expression in normal resting mononuclear cells, this receptor subunit is expressed by a proportion of the abnormal cells in certain forms of lymphoid neoplasia, some autoimmune diseases, and in individuals rejecting allografts (73–80). Many of the abnormal cells in these diseases express the IL-2Rα or Tac antigen on their surface. Furthermore, the serum concentrations of the 45-kDa released soluble form of the IL-2Rα are elevated (73). In terms of the neoplasias, certain T cell, B cell, monocytic, and even granulocytic leukemias express IL-2Rα. For example, human T cell lymphotrophic virus I (HTLV-I)-associated adult T cell leukemia cells constitutively express many receptors (76, 80). Recent studies have linked this disordered expression of IL-2R in this retrovirus-induced T cell leukemia with the action of a transactivator gene product (tax-1) encoded by this virus (81–83). The tax protein that acts indirectly on cellular as well as viral proteins may play an important role in the early phases of HTLV-I-induced adult T cell leukemia by deregulating the expression of cellular genes encoding IL-2 and IL-2R that are involved in the control of T cell proliferation.

The IL-2R as a Target for Immunotherapy—The expression of IL-2Rα chains identified by anti-Tac in T cells from patients with certain lymphoid malignancies, select autoimmune diseases, and allograft rejection but not in normal resting cells provided the scientific basis for a therapeutic strategy that involves agents that eliminate IL-2R expressing cells. To exploit the difference in Tac antigen expression between normal resting cells and the abnormal T cells in leukemia, we have performed a clinical trial to evaluate the efficacy of intravenously administered unmodified anti-Tac monoclonal antibody in the treatment of patients with adult T cell leukemia (84). None of the patients suffered any untoward reactions, and 7 of 20 treated patients manifested transient, mixed (1), partial (3), or complete (3) remissions lasting from 1 to over 13 months following therapy (84).[2] To improve the effectiveness of IL-2R-directed therapy, different approaches have been initiated to modify the antibody for clinical purposes. For example, radiometal chelates of anti-Tac have been targeted to IL-2R expressing cells for use in radioimaging and radioimmunotherapy. Specifically, α- and β-emitting radionuclides (*e.g.* ^{212}Bi and ^{90}Y) were conjugated to anti-Tac by use of bifunctional chelates and were shown to be effective and specific immunocytotoxic agents for the elimination of IL-2Rα expressing T cells (85, 86).

Finally, "humanized" anti-Tac molecules (anti-Tac-H) have been prepared by genetic engineering in which the molecules are entirely human IgG_1-κ except for the small complementarity determining regions that are retained from the mouse antibody (87). The "humanized" anti-Tac may lack T helper cell recognition units and has been shown to be much less immunogenic in primates than the murine monoclonal antibody. Furthermore, the humanized IgG_1 antibody is able to perform antibody-dependent cytotoxicity with human mononuclear cells, a critical function that was absent in the original mouse monoclonal antibody (88).

In summary, our present understanding of the IL-2/IL-2R system opens the possibility for more specific immune intervention. The IL-2R may prove to be an extraordinarily versatile therapeutic target. The clinical application of IL-2R-

[2] T. A. Waldmann, unpublished observations.

directed therapy represents a new perspective for the treatment of certain neoplastic diseases, select autoimmune disorders, and for the prevention of allograft rejection.

Acknowledgment—I am grateful to Barbara Holmlund for her excellent editorial assistance.

REFERENCES

1. Morgan, D. A., Ruscetti, F. W., and Gallo, R. C. (1976) *Science* **193,** 1007–1008
2. Smith, K. A. (1980) *Immunol. Rev.* **51,** 337–357
3. Waldmann, T. A. (1986) *Science* **232,** 727–732
4. Robb, R. J., Munck, A., and Smith, K. A. (1981) *J. Exp. Med.* **154,** 1455–1474
5. Smith, K. A. (1988) *Science* **240,** 1169–1176
6. Greene, W. C., and Leonard, W. J. (1986) *Annu. Rev. Immunol.* **4,** 69–95
7. Waldmann, T. A. (1989) *Annu. Rev. Biochem.* **58,** 875–911
8. Waldmann, T. A. (1989) *J. Natl. Cancer Inst.* **81,** 914–923
9. Kuziel, W. A., and Greene, W. C. (1990) *J. Invest. Dermatol.* **94,** 27S–32S
10. Robb, R. J., Greene, W. C., and Rusk, C. M. (1984) *J. Exp. Med.* **160,** 1126–1146
11. Tsudo, M., Kozak, R. W., Goldman, C. K., and Waldmann, T. A. (1986) *Proc. Natl. Acad. Sci. U. S. A.* **83,** 9694–9698
12. Uchiyama, T., Broder, S., and Waldmann, T. A. (1981) *J. Immunol.* **126,** 1393–1397
13. Uchiyama, T., Nelson, D. L., Fleisher, T. A., and Waldmann, T. A. (1981) *J. Immunol.* **126,** 1398–1403
14. Leonard, W. J., Depper, J. M., Uchiyama, T., Smith, K. A., Waldmann, T. A., and Greene, W. C. (1982) *Nature* **300,** 267–269
15. Leonard, W. J., Depper, J. M., Robb, R. J., Waldmann, T. A., and Greene, W. C. (1983) *Proc. Natl. Acad. Sci. U. S. A.* **80,** 6957–6961
16. Miyawaki, T., Yachie, A., Uwadana, N., Ohzeki, S., Nagaoki, T., and Taniguchi, N. (1982) *J. Immunol.* **129,** 2474–2478
17. Robb, R. J., and Greene, W. C. (1983) *J. Exp. Med.* **158,** 1332–1337
18. Leonard, W. J., Depper, J. M., Waldmann, T. A., and Greene, W. C. (1984) in *Receptors and Recognition: Antibodies to Receptors* (Cuatrecasas, P., and Greaves, M., eds). pp. 45–66, Chapman and Hall, London
19. Shackelford, D. A., and Trowbridge, I. S. (1984) *J. Biol. Chem.* **259,** 11706–11712
20. Leonard, W. J., Depper, J. M., Crabtree, G. R., Rudikoff, S., Pumphrey, J., Robb, R. J., Krönke, M., Svetlik, P. B., Peffer, N. J., Waldmann, T. A., and Greene, W. C. (1984) *Nature* **311,** 626–631
21. Cosman, D., Cerretti, D. P., Larsen, A., Park, L., March, C., Dowler, S., Gillis, S., and Urdal, D. (1984) *Nature* **312,** 768–771
22. Nikaido, T., Shimizu, A., Ishida, N., Sabe, H., Teshigawara, K., Maeda, M., Uchiyama, T., Yodoi, J., and Honjo, J. (1984) *Nature* **311,** 631–635
23. Ortaldo, J. R., Mason, A. T., Gerard, J. P., Henderson, L. E., Farrar, W., Hopkins, R. F., Herberman, R. B., and Rabin, H. (1984) *J. Immunol.* **133,** 779–783
24. Grimm, E. A., and Rosenberg, S. A. (1984) in *Lymphokines* (Hopkins, R. F., Herberman, R. B., and Rabin, H., eds) Vol. 9, pp. 279–311, Academic Press, New York
25. Waldmann, T. A., Kozak, R. W., Tsudo, M., Oh-ishi, T., Bongiovanni, K. F., and Goldman, C. K. (1986) in *Progress in Immunology VI* (Cinadar, B., and Miller, R. G., eds) pp. 553–562, Academic Press, Orlando, FL
26. Tsudo, M., Kozak, R. W., Goldman, C. K., and Waldmann, T. A. (1987) *Proc Natl. Acad. Sci. U. S. A.* **84,** 4215–4218
27. Waldmann, T. A. (1988) *Harvey Lect.* **82,** 1–17
28. Sharon, M., Klausner, R. D., Cullen, B. R., Chizzonite, R., and Leonard, W. J. (1986) *Science* **234,** 859–863
29. Lowenthal, J. W., and Greene, W. C. (1987) *J. Exp. Med.* **166,** 1156–1161
30. Wang, H.-M., and Smith, K. A. (1987) *J. Exp. Med.* **166,** 1055–1069
31. Tsudo, M., Kitamura, F., and Miyasaka, M. (1989) *Proc. Natl. Acad. Sci. U. S. A.* **86,** 1982–1986
32. Hatakeyama, M., Tsudo, M., Minamoto, S., Kono, T., Doi, T., Miyata, T., Miyasaka, M., and Taniguchi, T. (1989) *Science* **244,** 551–556
33. Hatakeyama, M., Mori, H., Doi, T., and Taniguchi, T. (1989) *Cell* **39,** 837–845
34. D'Andrea, A. D., Fasman, G. D., and Lodish, H. F. (1989) *Cell* **58,** 1023–1024
35. Mosley, B., Beckmann, M. P., March, C. J., Idzerda, R. L., Gimpel, S. D., VandenBos, T., Friend, D., Alpert, A., Anderson, D., Jackson, J., Wignall, J. M., Smith, C., Gallis, B., Sims, J. E., Urdal, D., Widmer, M. B., Cosman, D., and Park, L. S. (1989) *Cell* **59,** 335–348
36. Yamasaki, K., Taga, T., Hirata, Y., Yawata, H., Kawanishi, Y., Seed, B., Taniguchi, T., Hirano, T., and Kishimoto, T. (1988) *Science* **241,** 825–828
37. Itoh, N., Yonehara, S., Schreurs, J., Gorman, D. M., Maruyama, K., Ishii, A., Yahara, I., Arai, K.-I., and Miyajima, A. (1990) *Science* **247,** 324–327
38. Gearing, D. P., King, J. A., Gough, N. M., and Nicola, N. A. (1989) *EMBO J.* **8,** 3667–3676
39. Bazan, J. F. (1990) *Proc. Natl. Acad. Sci. U. S. A.* **18,** 6934–6938
40. Cosman, D., Lyman, S. D., Idzerda, R. L., Beckmann, M. P., Park, L. S., Goodwin, R. G., and March, C. J. (1990) *Trends Biochem. Sci.* **15,** 265–269
41. Tsudo, M., Karasuryama, H., Kitamura, F., Tanaka, T., Kubo, S., Yamamura, Y., Tamatani, T., Hatakeyama, M., Taniguchi, T. M., and Miyasaka, M. (1990) *J. Immunol.* **145,** 599–606
42. Szöllösi, J., Damjanovich, S., Goldman, C. K., Fulwyler, M., Aszalos, A. A., Goldstein, G., Rao, P., Talle, M. A., and Waldmann, T. A. (1987) *Proc. Natl. Acad. Sci. U. S. A.* **84,** 7246–7251
43. Edidin, M., Aszalos, A., Damjanovich, S., and Waldmann, T. A. (1988) *J. Immunol.* **141,** 1206–1210
44. Herrmann, F., and Diamantstein, T. (1987) *Immunobiology* **175,** 145–158
45. Saragovi, H., and Malek, T. R. (1987) *J. Immunol.* **139,** 1918–1926
46. Nakamura, Y., Inamoto, T., Sugie, K., Masutani, H., Shindo, T., Tagaya, Y., Yamauchi, A., Ozawa, K., and Yodoi, J. (1989) *Proc. Natl. Acad. Sci. U. S. A.* **86,** 1318–1322
47. Sharon, M., Gnarra, J. R., and Leonard, W. J. (1990) *Proc. Natl. Acad. Sci. U. S. A.* **87,** 4869–4873
48. Colamonici, O. R., Neckers, L. M., and Rosolen, A. (1990) *J. Immunol.* **145,** 155–160
49. Saragovi, H., and Malek, T. R. (1990) *Proc. Natl. Acad. Sci. U. S. A.* **87,** 11–15
50. Sharon, M., Gnarra, J. R., Baniyash, M., and Leonard, W. J. (1988) *J. Immunol.* **141,** 3512–3515
51. Harel-Bellan, A., Krief, P., Rimsky, L., Farrar, W. L., and Mishal, Z. (1990) *Biochem. J.* **268,** 35–40
52. Burton, J., Goldman, C. K., Rao, P., Moos, M., and Waldmann, T. A. (1990) *Proc. Natl. Acad. Sci. U. S. A.* **87,** 7329–7333
53. Springer, T. A., Dustin, M. L., Kishimoto, T. K., and Marlin, S. D. (1987) *Annu. Rev. Immunol.* **5,** 223–252
54. Brandhuber, B. J., Boone, T., Kenney, W. C., and McKay, D. B. (1987) *Science* **238,** 1707–1709
55. Ju, G., Collins, L., Kaffka, K. L., Tsien, W.-H., Chizzonite, R., Crow, R., Bhatt, R., and Kilian, P. L. (1987) *J. Biol. Chem.* **262,** 5723–5731
56. Kuo, L.-M., and Robb, R. J. (1986) *J. Immunol.* **137,** 1538–1543
57. Collins, L., Tsien, W.-H., Seals, C., Hakimi, J., Weber, D., Bailon, P., Hoskings, J., Greene, W. C., Toome, V., and Ju, G. (1988) *Proc. Natl. Acad. Sci. U. S. A.* **85,** 7709–7713
58. Zurawski, S. M., and Zurawski, G. (1989) *EMBO J.* **8,** 2583–2590
59. Weigel, U., Meyer, M., and Sebald, W. (1989) *Eur. J. Biochem.* **180,** 295–300
60. Ju, G., Sqve, K., Bailon, P., Tsien, W.-H., Lin, P., Hakimi, J., Toome, V., and Greene, W. C. (1990) in *Proceedings of the International Symposium on the Biology and Clinical Application of Interleukin-2* (Rees, R. C., ed) Oxford University Press, New York
61. Gaulton, G. N., and Eardley, D. D. (1986) *J. Immunol.* **136,** 2470–2477
62. Benedict, S. H., Mills, G. B., and Gelfand, E. W. (1987) *J. Immunol.* **139,** 1694–1697
63. Ishii, T., Takeshita, T., Numata, N., and Sugamura, K. (1988) *J. Immunol.* **141,** 174–179
64. Evans, S. W., and Farrar, W. L. (1987) *J. Biol. Chem.* **262,** 4624–4630
65. Saltzman, E. M., Thom, R. R., and Casnellie, S. E. (1988) *J. Biol. Chem.* **263,** 6956–6959
66. Farrar, W. L., and Ferris, D. K. (1989) *J. Biol. Chem.* **264,** 12562–12567
67. Mills, G. B., May, C., McGill, M., Fung, M., Baker, M., Sutherland, R., and Greene, W. C. (1990) *J. Biol. Chem.* **265,** 3561–3567
68. Asao, H., Takeshita, T., Nakamura, M., Nagata, K., and Sugamura, K. (1990) *J. Exp. Med.* **171,** 637–644
69. Depper, J. M., Leonard, W. J., Krönke, M., Noguchi, P. D., Cunningham, R. E., Waldmann, T. A., and Greene, W. C. (1984) *J. Immunol.* **133,** 3054–3061
70. Waldmann, T. A., Goldman, C. K., Robb, R. J., Depper, J. M., Leonard, W. J., Sharrow, S. O., Bongiovanni, K. F., Korsmeyer, S. J., and Greene, W. C. (1984) *J. Exp. Med.* **160,** 1450–1466
71. Loughnan, M. S., and Nossal, G. J. V. (1989) *Nature* **340,** 75–79
72. Rubin, L. A., Kurman, C. C., Biddison, W. E., Goldman, N. D., and Nelson, D. L. (1985) *Hybridoma* **4,** 91–102
73. Nelson, D. L., Rubin, L. A., Kurman, C. C., Fritz, M. E., and Boutin, B. (1986) *J. Clin. Immunol.* **6,** 114–120
74. Williams, J. M., Kelley, V. E., Kirkman, R. L., Tilney, N. L., Shapiro, M. E., Murphy, J. R., and Strom, T. B. (1988) *Immunol. Invest.* **16,** 687–723
75. Diamantstein, T., and Osawa, H. (1986) *Immunol. Rev.* **92,** 5–27
76. Waldmann, T. A., Greene, W. C., Sarin, P. S., Saxinger, C., Blayney, D. W., Blattner, W. A., Goldman, C. K., Bongiovanni, K., Sharrow, S., Depper, J. M., Leonard, W., Uchiyama, T., and Gallo, R. C. (1984) *J. Clin. Invest.* **73,** 1711–1718
77. Waldmann, T. A. (1988) *J. Autoimmun.* **1,** 641–653
78. Sheibani, K., Winberg, C. D., Van de Velde, S., Blayney, D. W., and Rappaport, H. (1987) *Am. J. Pathol.* **127,** 27–37
79. Schwarting, R., Gerdes, J., and Stein, H. (1985) *J. Clin. Pathol.* **38,** 1196–1197
80. Uchiyama, T., Hori, T., Tsudo, M., Wano, Y., Umadome, H., Tamori, S., Yodoi, J., Maeda, M., Sawami, H., and Uchino, H. (1985) *J. Clin. Invest.* **76,** 446–453
81. Suzuki, N., Matsunami, N., Kanamori, H., Ishida, N., Shimizu, A., Yaoita, Y., Nikaido, T., and Honjo, T. (1987) *J. Biol. Chem.* **262,** 5079–5086
82. Ruben, S., Poteat, H., Tan, T.-H., Kawakami, K., Roeder, R., Hasteltine, W., and Rosen, C. A. (1988) *Science* **241,** 89–92
83. Siekevitz, M., Feinberg, M. B., Holbrook, N., Wong-Staal, F., and Greene, W. C. (1987) *Proc. Natl. Acad. Sci. U. S. A.* **84,** 5389–5393
84. Waldmann, T. A., Goldman, C. K., Bongiovanni, K. F., Sharrow, S. O., Davey, M. P., Cease, K. B., Greenberg, S. L., and Longo, D. (1988) *Blood* **72,** 1805–1816
85. Kozak, R. W., Atcher, R. W., Gansow, O. A., Friedman, A. M., Hines, J. J., and Waldmann, T. A. (1986) *Proc. Natl. Acad. Sci. U. S. A.* **83,** 474–478
86. Kozak, R. W., Raubitschek, A., Mirzadeh, S., Brechbiel, M. W., Junghans, R., Gansow, O. A., and Waldmann, T. A. (1989) *Cancer Res.* **49,** 2639–2644
87. Queen, C., Schneider, W. P., Selick, H. E., Payne, P. W., Landolfi, N. F., Duncan, J. F., Avdalovic, N. M., Levitt, M., Junghans, R. P., and Waldmann, T. A. (1989) *Proc. Natl. Acad. Sci. U. S. A.* **86,** 10029–10033
88. Junghans, R. P., Waldmann, T. A., Landolfi, N. F., Avdalovic, N. M., Schneider, W. P., and Queen, C. (1990) *Cancer Res.* **50,** 1495–1502

Minireview

THE JOURNAL OF BIOLOGICAL CHEMISTRY
Vol. 266, No. 6, Issue of February 25, pp. 3357–3360, 1991

Printed in U.S.A.

Antigenic Structures Recognized by Cytotoxic T Lymphocytes

Theodore J. Tsomides and Herman N. Eisen

From the Department of Biology, Center for Cancer Research, Massachusetts Institute of Technology, Cambridge, Massachusetts 02139

The antigenic structures recognized by T lymphocytes differ fundamentally from those recognized by B lymphocytes. B cells, through antibody molecules embedded in their cell surface membrane, collectively recognize an enormous diversity of antigens in solution or on cell surfaces, including native and unfolded proteins, peptides, polysaccharides, nucleic acids, steroids, and small organic molecules. The antigen-specific receptors on T cells are very similar to antibodies structurally and in the genetic basis for their diversity. Nevertheless, the antigenic structures recognized by T cells consist almost exclusively of peptides associated with integral membrane glycoproteins known as MHC[1] proteins (because they are encoded in a genetic region called the major histocompatibility complex). Thus T cells recognize protein antigens in a fragmented (or "processed") form on the surface of other cells expressing appropriate MHC molecules (a phenomenon termed MHC "restriction" of the T cell response (1)).

The MHC genetic region (also called HLA in the human and H-2 in the mouse) contains many highly polymorphic, tightly linked genes, and a particular set of alleles defines a haplotype. Any given individual in a population has an array of MHC molecules on the surface of its cells (one haplotype from each parent, codominantly expressed) which serves as a signature for that individual (or for the genetically uniform mice or rats of inbred strains). The essence of MHC restriction is that T cells normally recognize antigens on cells having the same haplotypes as the T cells themselves but not on cells of different haplotypes. (However, in the context of organ transplantation between individuals of different haplotypes, nonself MHC molecules are targets for a T cell-mediated rejection response. It was this "allogeneic" reaction which led to the discovery of the MHC and to the origin of the term histocompatibility.)

Mature T cells can be divided into two groups based on their cell surface expression of either CD8 or CD4 glycoproteins. Each group interacts with a different set of MHC molecules, called class I or class II, respectively. MHC molecules of both classes are noncovalently associated heterodimers on the cell surface, but class I is composed of a membrane-spanning 45-kDa α (or heavy) chain and a soluble, nonpolymorphic 12-kDa subunit (β_2-microglobulin or light chain), whereas class II consists of two transmembrane chains, α (35 kDa) and β (29 kDa). Only certain "antigen-presenting" cells (APC) express MHC-II (macrophages, dendritic cells, B cells, and a few others). When $CD4^+$ T cells recognize class II-expressing APC that have reacted with appropriate antigens, they usually respond by secreting lymphokines that influence lymphocytes and other cell types; hence these T cells are designated helper T cells. However, most nucleated cells express MHC-I, so $CD8^+$ T cells can potentially respond to virtually all cell types. In the presence of a suitable antigen, class I-expressing target cells are specifically destroyed by $CD8^+$ T cells; hence the latter are termed killer cells or cytotoxic T lymphocytes (CTL).

The division between class I and class II proteins is important not only for the different T cell responses provoked, but because they define different pathways through which peptides are presented to T cells (2). In particular, extracellular proteins (also termed exogenous) are taken up by APC, partially digested in a low pH endosomal compartment, and the resulting peptides bound by MHC-II for transport to the cell surface where they can stimulate $CD4^+$ T cells. In contrast, antigens synthesized within a cell (endogenous), such as viral proteins, tumor antigens, or so-called minor histocompatibility antigens, are degraded intracellularly to peptides and associate with MHC-I for transport to the cell surface where they flag the cell for recognition and lysis by $CD8^+$ CTL. The separation between the two antigen-processing pathways (Table I) is not absolute, but it offers a useful framework for probing the cell biology and biochemistry of each pathway and for evaluating exceptions. This minireview will concentrate on the class I pathway and in particular on the molecular interactions between antigen and MHC-I prerequisite to recognition by $CD8^+$ CTL.

$CD8^+$ CTL Recognize Peptides Derived from Endogenous Antigens

Viral infection elicits a vigorous CTL response leading to the specific lysis of infected cells. Experimental studies with influenza virus have revealed important features of the antiviral CTL response (3). Whereas anti-influenza antibodies recognize subtype-specific, virally encoded glycoproteins expressed on the surface of infected cells, CTL from both mice (4) and humans (5) cross-react with target cells infected by different influenza virus subtypes. This cross-reactivity was explained by the subsequent finding that most anti-influenza CTL are specific for conserved, intracellular viral proteins undetectable at the cell surface (*e.g.* nucleoprotein, matrix protein). Thus, cells transfected with the nucleoprotein gene served as targets for influenza-specific CTL despite the absence of cell surface nucleoprotein, giving rise to the suggestion that the antigen is converted into fragments (peptides) which are transported to the cell surface for CTL recognition (6). This model was confirmed by experiments showing that incubation of cells with synthetic peptides mimicking parts of the nucleoprotein rendered the cells susceptible to lysis by CTL (7). Incubation with whole nucleoprotein was ineffective, and only certain peptides from the nucleoprotein sequence were active. Presumably such peptides, substituting for peptides naturally produced from endogenously synthesized antigen, somehow associate directly with MHC-I molecules to form the complex recognized by CTL.

Subsequent studies with mouse and human CTL by many workers have identified at least 40–50 different synthetic peptides recognized in association with various MHC-I molecules. In addition to virally derived peptides, peptides corresponding to sequences from other endogenously synthesized proteins are recognized specifically by CTL (8), as well as peptides corresponding to mutated sites responsible for tumor

[1] The abbreviations used are: MHC, major histocompatibility complex; APC, antigen-presenting cells; CTL, cytotoxic T lymphocytes; ER, endoplasmic reticulum; β_2m, β_2-microglobulin.

TABLE I

Dichotomy in antigen recognition by T cells

T cell phenotype	Major function	MHC restriction	Target cell	Antigen
$CD4^+$ $CD8^-$	Helper cell	Class II	Specialized antigen-presenting cell	Extracellular (exogenous)
$CD8^+$ $CD4^-$	Killer cell	Class I	Any nucleated cell	Intracellular (endogenous)

immunogenicity in mouse cell lines (9). Thus many tumor-specific antigens (and minor histocompatibility antigens) are probably peptide fragments recognized at the cell surface by MHC-restricted CTL, explaining the historical failure of antibodies to detect these antigens.

Evidence for Peptide·MHC-I Complexes

A dramatic and compelling image of how a peptide·MHC-I complex might look was provided by the 3.5-Å x-ray crystallographic structure of a human MHC-I molecule, HLA-A2 (10, 11). The structure consists of four domains of approximately 90 amino acid residues each, three derived from the α-chain (α_1, α_2, and α_3) and one comprising β_2-microglobulin (transmembrane and cytosolic portions of the α-chain were removed by papain digestion early in the purification). The membrane proximal α_3 and β_2-microglobulin domains have tertiary structures resembling immunoglobulin folds, consistent with their known amino acid sequences. The α_1 and α_2 domains each have a nearly identical but novel structure containing four antiparallel β-strands and a long α-helix, such that when paired these domains form a single platform of eight β-strands topped by two α-helices. A long groove overlying the β-sheet and between the two α-helices was proposed to be the peptide binding site based on several considerations. 1) Most of the polymorphic residues of HLA cluster in or near the proposed binding site, accounting for the ability of different MHC molecules to interact with different peptides. 2) Residues known to be critical for T cell recognition through the use of natural or engineered MHC variants are similarly in or near the site. 3) The dimensions of the groove (25 Å long × 10 Å wide × 11 Å deep) could accommodate a peptide of between 8 amino acids (if fully extended) and 20 amino acids (if fully coiled into a helix), in accord with the lengths of synthetic peptides used to elicit T cell responses. 4) Most intriguingly, a continuous region of electron density not attributable to the HLA-A2 sequence is present in the groove; this "extra" electron density is believed to represent peptide or a mixture of peptides still bound to HLA. A second x-ray structure, that of HLA-Aw68, confirmed these features and also displayed unassignable electron density in the corresponding site; notably, this electron density was distinguishable from that seen with HLA-A2, suggesting a different composition of putative peptide occupant(s) (12). Even after structural refinement to 2.6-Å resolution the extra electron densities in the HLA grooves have resisted sequence assignment, supporting the idea that a mixture of peptides may be included.

Consistent with this inference, the biology of the system points to considerable degeneracy in peptide binding to MHC glycoproteins. The cells of an individual vertebrate organism express only a handful of different MHC-I proteins, probably up to six in humans (two each at the HLA-A, -B, and -C loci). Yet this small number of proteins mediates the MHC-restricted recognition of many thousands of peptide antigens by an equally large number of different T cells. Evidently, each MHC protein must have the capacity to interact with an extremely large diversity of peptide sequences. Nevertheless, binding is not totally degenerate, as MHC proteins vary in their abilities to present specific peptides (13). In this context, the extra electron densities seen in the two HLA structures may be interpreted as different mixtures of tightly bound peptides which co-purified and co-crystallized with their respective HLA proteins. Such degenerate binding contrasts profoundly with the highly specific "lock-and-key" mechanism considered a hallmark of immunological recognition and epitomized by the reaction between antigen and antibody (14) (and indeed probably also between peptide·MHC complex and T cell receptor (15)).

What Peptides Bind to MHC-I?

Efforts to understand the structural basis for binding between peptides and MHC-I proteins have taken several directions, including: 1) tabulation of synthetic peptides recognized by MHC-restricted T cells in order to discern common features or motifs and develop predictive binding algorithms; 2) structural characterization of naturally occurring peptides recognized by MHC-restricted T cells; 3) *in vitro* binding studies using synthetic peptides and purified MHC-I molecules; 4) reconstitution of peptide-free MHC-I with a known, homogeneous peptide for x-ray crystallography; and 5) elucidation of the physiological pathway for generation of peptide·MHC-I complexes by genetic and cell biological approaches.

Examination of several dozen peptides which elicit MHC-restricted T cell responses (both class I and class II) led to the formulation of predictive schemes based on common structural elements (16–19). The resulting algorithms often reflect some form of recurrent hydrophobicity (20). While several predicted synthetic peptides were biologically active in T cell assays, other peptides found by random screening to have similar activities did not fit the expected patterns. The relatively small (but growing) data base of available peptide sequences may have limited this approach, especially considering the distinctions in peptide binding by different MHC proteins. Furthermore, in most cases it is not known which residues of a given peptide are actually required for MHC binding, which interact primarily with a T cell receptor, which serve a role other than for specific binding, and which are superfluous or artifactual due to their synthetic origin.

Until recently the relationship between synthetic and naturally occurring peptides which bind MHC-I was largely uncertain, probably because characterization of natural peptides is challenging from a technical standpoint in view of their likely heterogeneity and consequent low abundance. Gel filtration fractionation of cell extracts followed by assay of the fractions for T cell recognition first gave clues as to natural peptide size (21, 22) and later led to the identification of two viral peptides (nonamers) whose sequences are contained within known biologically active synthetic peptides (23). A third viral peptide recovered directly from immunoprecipitated MHC-I molecules similarly represents a portion (octamer) of a known synthetic peptide (24). Thus in all likelihood natural peptides are relatively small peptides with no particular chemical modifications or structural themes yet evident.

Direct binding studies require a source of purified MHC molecules, which are available in low milligram amounts by detergent solubilization or papain digestion of membranes from a large number of cultured cells (*e.g.* 10^{11} cells, representing 50–100 liters of culture) (25–27). Conventional binding assays such as equilibrium dialysis and gel filtration have yielded abundant binding data for MHC-II systems, with demonstrable association constants for peptides on the order of 10^6 M^{-1} (28, 29). However, synthetic peptides known to interact with MHC-I on target cells fail to reveal appreciable

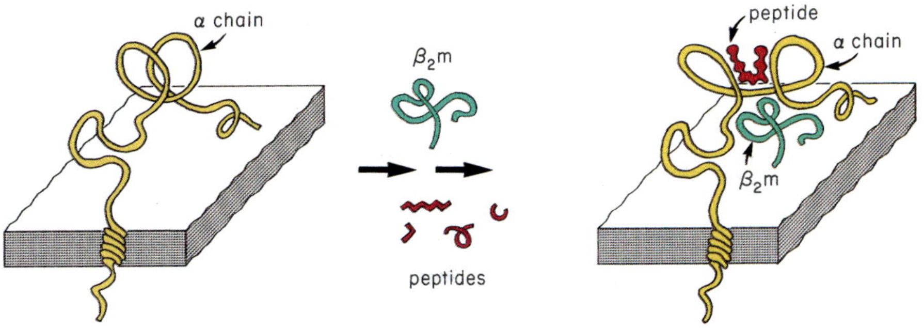

FIG. 1. **Peptides and β_2m associate with membrane-bound α-chain to form stable α·β_2m·peptide heterotrimers.** This multistep process occurs intracellularly, probably in the endoplasmic reticulum, and may also occur at the cell surface when peptides and β_2m are present in the extracellular medium.

binding to the appropriate purified MHC molecules by either gel filtration (30) or equilibrium dialysis (31). Substantial binding of detergent-solubilized MHC-I to peptides immobilized on plastic surfaces can be detected, but this binding appears largely indiscriminate with respect to MHC specificity, and it is unclear whether the solid phase environment favors binding by particular peptides due to physical properties other than intrinsic binding affinities (32–34). A similar report, showing more specificity, involved the addition of peptides to immobilized MHC-I to evoke a T cell response (35).

It may be that the low binding observed between peptides and MHC-I in solution accurately reflects an intrinsically low affinity. If the association constants are as low as 10^4 M^{-1} (consistent with available data), cell surface class I molecules will be 1% saturated at equilibrium given a peptide concentration of 10^{-6} M to induce half-maximal CTL lysis, and this might result in sufficient peptide·MHC-I complexes (100–500/target cell) for CTL recognition. However, this scenario is by no means the only plausible one. If the x-ray structures of HLA are taken to include tightly bound peptides, their dissociation rates must be extraordinarily slow, since purification procedures take at least 7–10 days. Therefore the purified HLA used in binding assays may similarly be occupied by unknown peptides which reduce the concentration of available binding sites. If such prior complexes exist in a large (perhaps even near-stoichiometric) proportion, failure to detect binding of labeled peptides by equilibrium dialysis is understandable and constitutes further support for exceptionally slow dissociation rates, since otherwise some exchange would be expected to occur. Assuming cell surface MHC-I is also tightly complexed with unknown peptides, sensitization of target cells for CTL lysis may involve binding to a small proportion of peptide-free MHC-I sites on the cell, again yielding relatively few antigenic complexes per cell. Whether all purified MHC-I molecules harbor unknown peptides is not yet known and in fact cannot be determined by x-ray crystallography, as only a small proportion of purified MHC-I (<5%) crystallizes.[2]

Where Do Peptide·MHC-I Complexes Form?

Recent insights into the assembly of peptide·MHC-I complexes derive from studies of a mutant mouse cell line, RMA-S, having what appears to be a defect in the generation or transport of endogenous peptides into the endoplasmic reticulum (ER) (36). Ordinarily, nascent class I α-chains and β_2-microglobulin (β_2m) associate into heterodimers in the ER and are transported to the cell surface within 30–60 min of their biosynthesis (37, 38). In RMA-S cells, α-chains and β_2m are synthesized normally but largely fail to assemble, so that cell surface MHC-I expression is only about 5% of wild-type levels under standard culture conditions (37 °C). When infected with influenza virus, RMA-S cells are not recognized by class I-restricted CTL, but when incubated with relatively high concentrations of suitable synthetic peptides from the virus, they become good targets for CTL and, surprisingly, express 2–5-fold more MHC-I on their surface (39).

These findings were explained by hypothesizing that synthetic peptides are taken up by RMA-S, somehow gain access to nascent α-chains and β_2m in the ER, and promote the proper folding, assembly, and transport to the cell surface of peptide·MHC-I complexes. The model implies that cell surface expression of MHC-I normally depends on the presence of peptides in the ER to help α and β_2m associate, and that RMA-S lacks this function. Subsequently, however, increased surface expression of MHC-I was fortuitously demonstrated by incubating RMA-S at reduced temperatures (23–31 °C), and these MHC-I molecules were unstable at 37 °C unless an appropriate synthetic peptide was added (40). Thus enhanced MHC-I expression in the presence of peptides might also be explained by a stabilization of molecules already at the cell surface (but not otherwise detected because they are conformationally altered and/or short-lived). Such labile and evidently peptide-free MHC-I molecules could either reach the cell surface without peptides or could release bound peptides via dissociation (41).

Further evidence that added peptides can interact with MHC-I at the cell surface comes from studies using inhibitors of protein export. Treatment of non-mutant cells with the compound brefeldin A (which blocks transport from the ER to the Golgi complex) or the viral product E19 (which specifically binds to and retains MHC intracellularly) prevents the presentation of endogenous (*e.g.* viral) antigens but does not interfere with the presentation of synthetic peptides to CTL (42–44). Similarly, if cells are "fixed" using glutaraldehyde (which stops protein synthesis and turnover), presentation of endogenous antigen but not of added peptide is abrogated (45). On the other hand, the increased MHC-I detected on RMA-S cells in the presence of added peptide could result from either peptide-induced stabilization of cell surface MHC-I or from association of peptide with MHC-I intracellularly, much like endogenously arising peptides.

Whether the effect of exogenous peptides on MHC-I takes place at the cell surface or in the ER (or perhaps both, depending upon peptide concentration), the properties of RMA-S and of a comparable human mutant cell line (46, 47) indicate that most stable MHC-I molecules at 37 °C include peptide as an integral partner with α-chain and β_2m. Moreover, β_2m can also stabilize MHC-I structure, as shown by an increased association between peptides and MHC-I when free β_2m is added to cells under limiting conditions of peptide (48). In addition, *in vitro* reconstitution of MHC-I from separated α-chains and β_2m is more efficient in the presence of either excess β_2m (49) or specific peptides (50); similar effects are obtained using detergent lysates from RMA-S cells (51). Finally, distinct MHC-I molecules are known to differ in their

[2] D. Wiley, personal communication.

requirements for β_2m to form stable cell surface structures (52, 53), and they may differ in their requirements for peptide as well (54).

How Are Peptides Generated in Vivo?

To elicit an effective class I-restricted CTL response, an endogenous antigen must be degraded into peptide fragments which can interact with MHC-I (α-chains and β_2m) to form functional complexes. Since MHC-I is co-translationally inserted into the ER (55), and intracellular antigens which do not enter the ER are recognized by CTL, a cytosolic proteolytic apparatus as well as a mechanism for peptide transport into the ER must be involved. It is not clear whether peptide fragments are generated by known or as yet unknown proteolytic systems, or how peptides are translocated into the ER, although recently a member of the protein family designated "ABC transporters" (for ATP-binding cassette) (56) has been implicated in the RMA-S mutation (57).

An alternative hypothesis provides for proteolysis within the ER, and according to the unexpected finding that mouse cells identical at all genes but MHC-I possess different peptide profiles, MHC molecules themselves may be involved in the generation or selection of presented peptides (58). For instance, MHC-I could shield bound peptides from degradation or serve as a template for proteolysis by other unknown proteins in the ER. Finally, an unusual pathway proposed for the generation of peptides avoids proteolysis altogether by theorizing the direct transcription and translation of short subgenic regions called "peptons" (59).

From Heterodimer to Heterotrimer

However they come about, it is increasingly clear that stable MHC-I molecules are most often heterotrimers consisting of α-chain, β_2m, and a short peptide of variable sequence. Association between peptide and MHC-I appears not to arise from the equilibrium binding of two stable entities, according to the classical lock-and-key paradigm, but rather as the end point of a complex assembly process involving three chains. The natural order of assembly of the heterotrimer and the factors which govern its stability remain open to question, but once formed, loss of either soluble component (peptide or β_2m) can soon lead to collapse of the overall structure, although cell surface β_2m is known to exchange with β_2m in the medium (60). Given the strong similarity between the x-ray structures of β_2m as a monomer (61) and as part of HLA complexes, and the lack of direct contact between β_2m and bound peptides, it appears that conformational changes which accompany heterotrimer formation primarily involve α-chain and peptide (Fig. 1). Direct physical studies may have to await the more ready availability of large quantities of MHC-I protein, but in the meantime the cell biology of this system remains an extremely active and fruitful area for investigation.

REFERENCES

1. Zinkernagel, R. M., and Doherty, P. C. (1974) *Nature* **251,** 547–548
2. Braciale, T. J., Morrison, L. A., Sweetser, M. T., Sambrook, J., Gething, M.-J., and Braciale, V. L. (1987) *Immunol. Rev.* **98,** 95–114
3. Townsend, A., and Bodmer, H. (1989) *Annu. Rev. Immunol.* **7,** 601–624
4. Zweerink, H. J., Courtneidge, S. A., Skehel, J. J., Crumpton, M. J., and Askonas, B. A. (1977) *Nature* **267,** 354–356
5. McMichael, A. J., and Askonas, B. A. (1978) *Eur. J. Immunol.* **8,** 705–711
6. Townsend, A. R. M., McMichael, A. J., Carter, N. P., Huddleston, J. A., and Brownlee, G. G. (1984) *Cell* **39,** 13–25
7. Townsend, A. R. M., Rothbard, J., Gotch, F. M., Bahadur, G., Wraith, D., and McMichael, A. J. (1986) *Cell* **44,** 959–968
8. Maryanski, J. L., Pala, P., Corradin, G., Jordan, B. R., and Cerottini, J.-C. (1986) *Nature* **324,** 578–579
9. Lurquin, C., Van Pel, A., Mariame, B., De Plaen, E., Szikora, J.-P., Janssens, C., Reddehase, M. J., Lejeune, J., and Boon, T. (1989) *Cell* **58,** 293–303
10. Bjorkman, P. J., Saper, M. A., Samraoui, B., Bennett, W. S., Strominger, J. L., and Wiley, D. C. (1987) *Nature* **329,** 506–512
11. Bjorkman, P. J., Saper, M. A., Samraoui, B., Bennett, W. S., Strominger, J. L., and Wiley, D. C. (1987) *Nature* **329,** 512–518
12. Garrett, T. P. J., Saper, M. A., Bjorkman, P. J., Strominger, J. L., and Wiley, D. C. (1989) *Nature* **342,** 692–696
13. Carreno, B. M., Anderson, R. W., Coligan, J. E., and Biddison, W. E. (1990) *Proc. Natl. Acad. Sci. U. S. A.* **87,** 3420–3424
14. Davies, D. R., Padlan, E. A., and Sheriff, S. (1990) *Annu. Rev. Biochem.* **59,** 439–473
15. Davis, M. M., and Bjorkman, P. J. (1988) *Nature* **334,** 395–402
16. Rothbard, J. B., and Taylor, W. R. (1988) *EMBO J.* **7,** 93–100
17. Margalit, H., Spouge, J. L., Cornette, J. L., Cease, K. B., DeLisi, C., and Berzofsky, J. A. (1987) *J. Immunol.* **138,** 2213–2229
18. Stille, C. J., Thomas, L. J., Reyes, V. E., and Humphreys, R. E. (1987) *Mol. Immunol.* **24,** 1021–1027
19. Claverie, J.-M., Kourilsky, P., Langlade-Demoyen, P., Chalufour-Prochnicka, A., Dadaglio, G., Tekaia, F., Plata, F., and Bougueleret, L. (1988) *Eur. J. Immunol.* **18,** 1547–1553
20. Reyes, V. E., Fowlie, E. J., Lu, S., Phillips, L., Chin, L. T., Humphreys, R. E., and Lew, R. A. (1990) *Mol. Immunol* **27,** 1021–1027
21. Heath, W. R., Hurd, M. E., Carbone, F. R., and Sherman, L. A. (1989) *Nature* **341,** 749–752
22. Rotzschke, O., Falk, K., Wallny, H.-J., Faath, S., and Rammensee, H.-G. (1990) *Science* **249,** 283–287
23. Rotzschke, O., Falk, K., Deres, K., Schild, H., Norda, M., Metzger, J., Jung, G., and Rammensee, H.-G. (1990) *Nature* **348,** 252–254
24. Van Bleek, G. M., and Nathenson, S. G. (1990) *Nature* **348,** 213–216
25. Parham, P. (1983) *Methods Enzymol.* **92,** 110–138
26. Lopez de Castro, J. A. (1984) *Methods Enzymol.* **108,** 582–600
27. Mescher, M. F., Stallcup, K. C., Sullivan, C. P., Turkewitz, A. P., and Herrmann, S. H. (1983) *Methods Enzymol.* **92,** 86–109
28. Babbitt, B. P., Allen, P. M., Matsueda, G., Haber, E., and Unanue, E. R. (1985) *Nature* **317,** 359–361
29. Buus, S., Sette, A., Colon, S. M., Jenis, D. M., and Grey, H. M. (1986) *Cell* **47,** 1071–1077
30. Chen, B. P., and Parham, P. (1989) *Nature* **337,** 743–745
31. Tsomides, T., and Eisen, H. N. (1990) *FASEB J.* **4,** A1696
32. Bouillot, M., Choppin, J., Cornille, F., Martinon, F., Papo, T., Gomard, E., Fournie-Zaluski, M.-C., and Levy, J.-P. (1989) *Nature* **339,** 473–475
33. Chen, B. P., Rothbard, J., and Parham, P. (1990) *J. Exp. Med.* **172,** 931–936
34. Frelinger, J. A., Gotch, F. M., Zweerink, H., Wain, E., and McMichael, A. J. (1990) *J. Exp. Med.* **172,** 827–834
35. Kane, K. P., Vitiello, A., Sherman, L. A., and Mescher, M. F. (1989) *Nature* **340,** 157–159
36. Ljunggren, H.-G., Paabo, S., Cochet, M., Kling, G., Kourilsky, P., and Karre, K. (1989) *J. Immunol.* **142,** 2911–2917
37. Krangel, M. S., Orr, H. T., and Strominger, J. L. (1979) *Cell* **18,** 979–991
38. Owen, M. J., Kissonerghis, A. M., and Lodish, H. F. (1980) *J. Biol. Chem.* **255,** 9678–9684
39. Townsend, A., Ohlen, C., Bastin, J., Ljunggren, H.-G., Foster, L., and Karre, K. (1989) *Nature* **340,** 443–448
40. Ljunggren, H.-G., Stam, N. J., Ohlen, C., Neefjes, J. J., Hoglund, P., Heemels, M.-T., Bastin, J., Schumacher, T. N. M., Townsend, A., Karre, K., and Ploegh, H. L. (1990) *Nature* **346,** 476–480
41. Schumacher, T. N. M., Heemels, M.-T., Neefjes J. J., Kast, W. M., Melief, C. J. M., and Ploegh, H. L. (1990) *Cell* **62,** 563–567
42. Nuchtern, J. G., Bonifacino, J. S., Biddison, W. E., and Klausner, R. D. (1989) *Nature* **339,** 223–226
43. Yewdell, J. W., and Bennink, J. R. (1989) *Science* **244,** 1072–1075
44. Cox, J. H., Yewdell, J. W., Eisenlohr, L. C., Johnson, P. R., and Bennink, J. R. (1990) *Science* **247,** 715–718
45. Hosken, N. A., Bevan, M. J., and Carbone, F. R. (1989) *J. Immunol.* **142,** 1079–1083
46. Salter, R. D., and Cresswell, P. (1986) *EMBO J.* **5,** 943–949
47. Cerundolo, V., Alexander, J., Anderson, K., Lamb, C., Cresswell, P., McMichael, A., Gotch, F., and Townsend, A. (1990) *Nature* **345,** 449–452
48. Rock, K. L., Rothstein, L. E., Gamble, S. R., and Benacerraf, B. (1990) *Proc. Natl. Acad. Sci. U. S. A.* **87,** 7517–7521
49. Elliott, T. J., and Eisen, H. N. (1990) *Proc. Natl. Acad. Sci. U. S. A.* **87,** 5213–5217
50. Kvist, S., and Hamann, U. (1990) *Nature* **348,** 446–448
51. Townsend, A., Elliott, T., Cerundolo, V., Foster, L., Barber, B., and Tse, A. (1990) *Cell* **62,** 285–295
52. Beck, J. C., Hansen, T. H., Cullen, S. E., and Lee, D. R. (1986) *J. Immunol.* **137,** 916–923
53. Allen, H., Fraser, J., Flyer, D., Calvin, S., and Flavell, R. (1986) *Proc. Natl. Acad. Sci. U. S. A.* **83,** 7447–7451
54. Lie, W.-R., Myers, N. B., Gorka, J., Rubocki, R. J., Connolly, J. M., and Hansen, T. H. (1990) *Nature* **344,** 439–441
55. Dobberstein, B., Garoff, H., Warren, G., and Robinson, P. J. (1979) *Cell* **17,** 759–769
56. Higgins, C. F., Hyde, S. C., Mimmack, M. M., Gileadi, U., Gill, D. R., and Gallagher, M. P. (1990) *J. Bioenerg. Biomembr.* **22,** 571–592
57. Monaco, J. J., Cho, S., and Attaya, M. (1990) *Science* **250,** 1723–1726
58. Falk, K., Rotzschke, O., and Rammensee, H.-G. (1990) *Nature* **348,** 248–251
59. Boon, T., and Van Pel, A. (1989) *Immunogenetics* **29,** 75–79
60. Bernabeu, C., van de Rijn, M., Lerch, P. G., and Terhorst, C. P. (1984) *Nature* **308,** 642–645
61. Becker, J. W., and Reeke, G. N., Jr. (1985) *Proc. Natl. Acad. Sci. U. S. A.* **82,** 4225–4229

Minireview

THE JOURNAL OF BIOLOGICAL CHEMISTRY
Vol. 266, No. 7, Issue of March 5, pp. 4025-4028, 1991

The Cholinesterases

Palmer Taylor

From the Department of Pharmacology, University of California, San Diego, La Jolla, California 92093

In 1914 Sir Henry Dale (1) suggested that an enzyme which degrades the esters of choline played a role in neurotransmission within the autonomic and somatic motor nervous systems and that this enzyme, acetylcholinesterase, was the target of action of the drug, physostigmine (eserine). In the intervening 75 years, inhibition of the enzyme has been used to augment both the nicotinic and muscarinic actions of acetylcholine. Cholinesterase inhibitors are frequently used as therapeutic agents and are employed worldwide as insecticides.

Over 20 years ago Jean Massoulié and François Reiger uncovered an unusual structural polymorphism in the cholinesterases (2). Subsequent studies have shown that the polymorphism is reflected in multiple modes of attachment of the enzyme to the outside surface of the cell. Accordingly, the precise localization of cholinesterases within synaptic junctions and the high catalytic potential of the enzyme are both critical to the fidelity of cholinergic synaptic function. Within the last 5 years primary structures of several cholinesterases have been determined, and this has enabled investigators to ascertain the structural basis of the polymorphism and the organization of the genes encoding the cholinesterases.

Cholinesterases may be classified as acetylcholinesterase (AChE,[1] EC 3.1.1.7) and butyrylcholinesterase (BuChE, EC 3.1.1.8) on the basis of differential specificity for acetylcholine and butyrylcholine hydrolysis. Several inhibitors have also been shown to be selective for one of the two enzymes. The AChEs are primarily associated with nerve and muscle, typically with localization at synaptic contacts, while BuChE is synthesized in liver with substantial amounts appearing in serum. The physiologic function of BuChE remains obscure; in fact, deficiencies of this enzyme in man only became apparent when succinylcholine was administered to produce neuromuscular blockade (3). The population distribution of BuChE mutations which yield an active enzyme resistant to certain inhibitors suggests a primary role for this enzyme to be in detoxification of plant esters ingested in the diet (4).

The characterization of AChE as solely an enzyme of the cholinergic nervous system is an oversimplification. First, BuChE is often found in the developing nervous system while AChE appears at later stages of development. Second, both AChE and BuChE are found in hematopoietic cells, and their presence can be correlated with cell differentiation. Finally, AChE carries novel cell surface antigens which could play a role in cell-cell communication (5).

Molecular Species of Cholinesterase

The cholinesterases exist in multiple molecular forms which may be distinguished by their subunit associations and hydrodynamic properties (Fig. 1) (6). The catalytic subunits associate with either a lipid-linked or a collagen-like subunit to form distinct heteromeric species. The species containing collagen-like structural units consist of tetramers of catalytic subunits, each of which is disulfide-linked to a single strand of a triple helical collagen-containing subunit (6–10). Similar to basement membrane collagens and procollagen, the collagen-like subunit has non-collagenous sequences at its amino and carboxyl termini (11) and associates with acidic basal laminar components within the synapse (12, 13). The dimensional asymmetry imparted by the filamentous collagen-like unit led to the designation of asymmetric or A forms with a numerical subscript specifying the number of attached catalytic subunits.

The lipid-linked subunit is approximately 20 kDa in mass (14), has covalently attached fatty acids, and tethers the catalytic subunits to the outside surface of the cell. It, too, links to a tetramer of catalytic subunits through disulfide bonds. Based on the precedence in *Torpedo*, where AChE abundance allows the gene product to be characterized, it is likely that the structural subunits are linked only with the hydrophilic catalytic subunits.

The homomeric forms of AChE typically exist as dimers and tetramers; occasionally monomeric species are also found. Their hydrodynamic properties led to the classification of globular or G forms, and they may be subdivided into hydrophilic G or amphiphilic (hydrophobic) G forms. The amphipathic character of the later species arises from the attachment of a glycophospholipid to the carboxyl terminus which localizes the enzyme at the outer surface of the membrane (15–17). The glycophospholipids found on cholinesterase are amide-bonded through ethanolamine to the carboxyl-terminal amino acid. A series of neutral and amino sugars links the ethanolamine to the distal inositol-containing phospholipid (15, 16). The inositol hydroxyl groups may also be acylated with a long chain fatty acid precluding dissociation of AChE from the cell surface after treatment with phospholipase C. Phospholipase C resistance arising from this acylation appears to be cell type-specific (17).

Less polymorphism is seen with BuChE, and to date, only hydrophilic and asymmetric forms of BuChE have been identified (6).

Structure of the Cholinesterases

Since the first primary structure of a cholinesterase was deduced, it has become clear that the enzyme defines a unique family of serine hydrolases (18). The sequence similarity of the cholinesterases with thyroglobulin, neurotactin, and glutactin also revealed that this family includes secreted and membrane-associated proteins of similar structure lacking hydrolase activity (Fig. 2). Moreover, the sequences suggest that the cholinesterases and homologous proteins play a role as cell surface proteins in cell adhesion and morphogenetic events in the nervous system (19, 20). Molecular cloning of cholinesterases and related esterases from mammalian, lower vertebrate, and invertebrate species has established the following characteristics of this protein family (18–26):

1) Sequence similarity between the cholinesterases and the serine hydrolases which largely serve as proteases (*i.e.* trypsin and subtilisin) is limited to only 4 residues immediately around the active site serine, Ser-200, rather than global similarity within the molecule. Residue identity around the catalytic histidine (His-440) is even more limited and may only be characterized by a glycine and perhaps an acidic amino acid on its carboxyl-terminal side (27).

2) The rank ordering of these serine and histidine positions within the linear sequence indicates that the cholinesterase family emerged as a consequence of convergent evolution rather than diverging from one of the other known hydrolase families. The trypsin family contains residues in the order:

[1] The abbreviations used are: AChE, acetylcholinesterase; BuChE, butyrylcholinesterase.

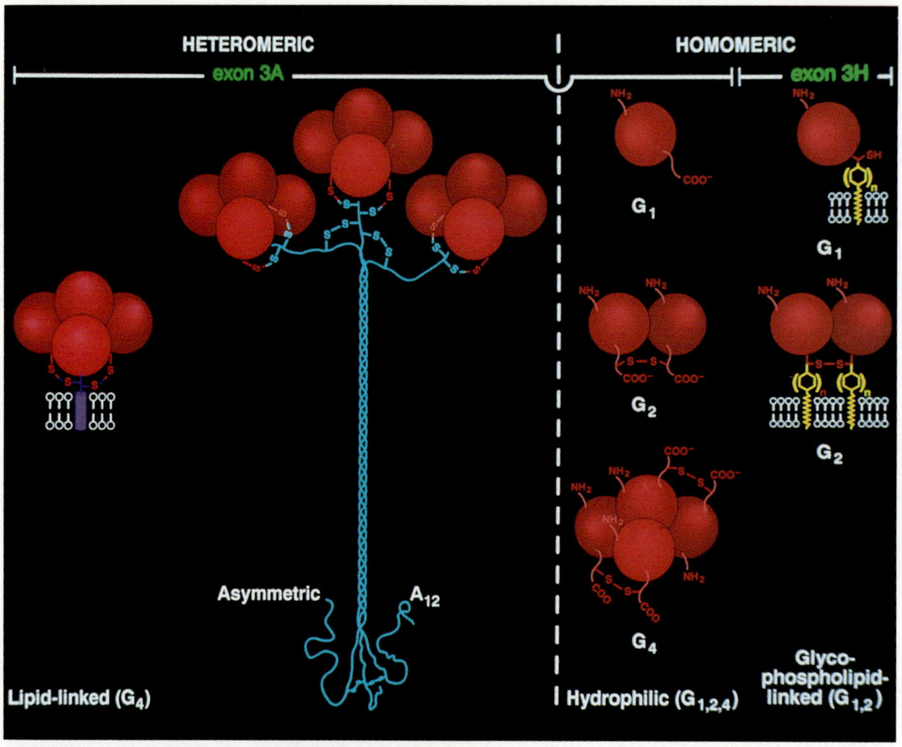

FIG. 1. **Molecular species of the cholinesterases.** In the heteromeric forms the hydrophilic catalytic subunit associates with two discrete types of structural subunits through disulfide bonds to form a lipid-linked or collagen-containing (asymmetric) species. The homomeric forms consist of oligomers of a hydrophilic or amphiphilic catalytic subunit. The amphiphilic AChE species employ an alternatively spliced exon (3H) giving a nascent peptide chain with a hydrophobic carboxyl-terminal sequence. The hydrophobic terminus of this peptide is cleaved concomitant with the addition of a glycophospholipid. All of the other forms characterized are encoded by exon 3A in their carboxyl-terminal region giving rise to a more hydrophilic sequence which is not processed. The precise geographic arrangement of tetramers, although shown here as tetrahedral, is unknown.

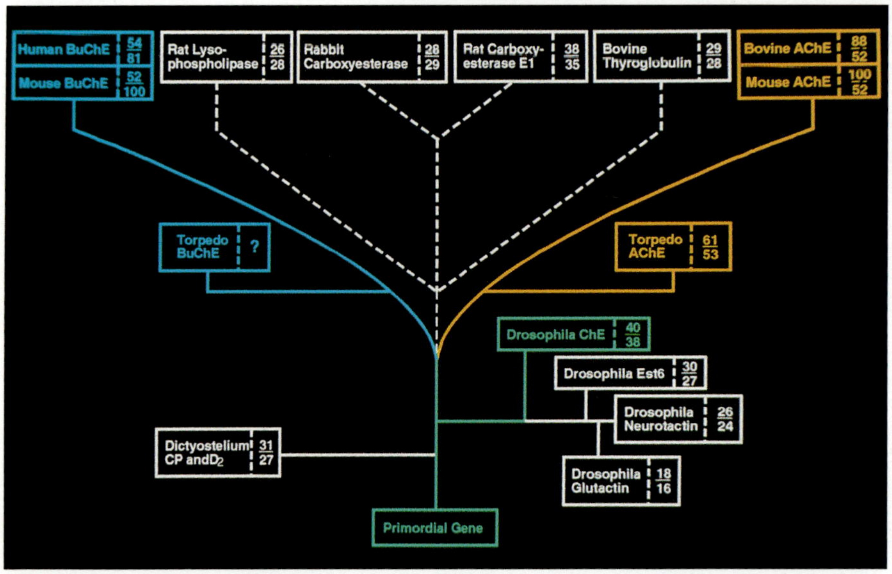

FIG. 2. **Sequence relationships of proteins showing sequence identity to acetylcholinesterase.** Numbers reflect the fractional amino acid identities with mouse AChE (*top*) and BuChE (*bottom*) determined with an ALLIGN program (49) using sequences recorded in GenBank. Comparisons between all cholinesterases except *Drosophila* were made over the entire open reading frame of the mature protein. For *Drosophila* cholinesterase and the other esterase and non-esterase proteins, the cholinesterase sequence encoded by exons 1 and 2 was used. Thyroglobulin, glutactin, and neurotactin are larger proteins than AChE and the regions between residues 2229–2769, 60–518, and 350–864, respectively, were used for sequence comparison. Several esterases from *Dictyostelium* to mammals fall in this family as do other high molecular weight-secreted and membrane-associated proteins such as thyroglobulin (precursor for thyroid hormone), neurotactin, and glutactin (respective integral and basement membrane, cell attachment proteins in *Drosophila*).

Asp-101, His-57, Ser-195; the subtilisin family has the order: Asp-32, His-64, and Ser-221.

3) Conservation of disulfide loops, A, B, and C, in many members of the family (28) (*i.e.* all of the cholinesterases, thyroglobulin, neurotactin, and certain other esterases) and loops A and B in a few of the smaller carboxylesterases and in the homologous cell attachment protein found in basement membranes, glutactin, suggests common secondary and probably tertiary structures for this family of proteins (Fig. 3).

4) Primary structures of the cholinesterases are well conserved with greater than 60% residue identity between *Torpedo* and mammalian AChEs. By comparison, the BuChEs and AChEs in the same or phylogenetically close species show about 50–52% identity in residues (22–24).

5) Torpedo AChE and mammalian BuChE contain eight cysteines while mammalian AChE contains only seven (24). Six of the cysteines found at conserved positions form intrasubunit disulfide bonds while the seventh located near the carboxyl-terminal end forms the linkage between catalytic subunits (28). Since some AChEs only contain seven cysteines, it is likely the carboxyl-terminal cysteine is employed in linkages between two catalytic subunits and between catalytic and the structural subunits. The eighth cysteine, when present, is found at variable locations in the cholinesterase sequences.

Gene Structure and Cholinesterase Biosynthesis

Single genes encode the catalytic subunits of AChE and BuChE (29, 30). Extensive polymorphism of the gene product results from alternative mRNA processing of the primary gene transcripts of AChE and subsequent posttranslational modifications. The first two exons within the open reading frame of the gene (labeled *1* and *2* in Fig. 4) encode the first 535 amino acids. Alternative third exons (*3A* and *3H*) encode the very carboxyl-terminal ends of the two species of AChE

(29). The glycophospholipid-containing and hydrophilic catalytic subunits of *Torpedo* AChE have identical amino acid sequences up through residue 535 after which the former species contains two unique amino acids in its sequence and the conjugated glycophospholipid (Fig. 3). The hydrophilic catalytic subunits continue for 40 residues of distinct sequence beyond the divergence point (31). S-1 nuclease (32) and RNase (33) protection studies showed the divergence point in the encoding mRNAs to correspond to this location in the gene product. A cDNA clone with its 3′ end encoding the sequence appropriate for attachment of a glycophospholipid and corresponding to the two unique amino acids in the gene product was also identified (32). A comparison of the encoding sequence with the gene product indicated that, as in other glycophospholipid-containing proteins, the terminal 28 amino acids in the nascent peptide are processed with the addition of the glycophospholipid. Subsequently, the positions of the two alternative exons were located in the gene (29).

Direct ligation of exon 3H from genomic DNA to exons 1 and 2 and transfection of the plasmid construct into COS-7 cells shows expression of AChE at the cell surface. Upon treatment with phosphatidylinositol-specific phospholipase C, active AChE is dissociated from the cell surface and released into the medium (34). Construction of a cDNA encoding a truncated AChE which lacks the sequence encoded by exon 3H upon transfection yields a protein which is secreted into the medium whereas a fusion protein encoded by a sequence consisting of exon 1 directly linked to exon 3H is retained on the cell surface (34). Immunoprecipitates of the latter fusion protein and the wild-type glycophospholipid-containing enzyme contained incorporated ethanolamine, whereas it was absent in the secreted recombinant DNA gene product. Thus, it appears that the sequence encoded by exon 3H is both necessary and sufficient for permitting the posttranslational modification in which the carboxyl-terminal sequence is cleaved and glycophospholipid is added.

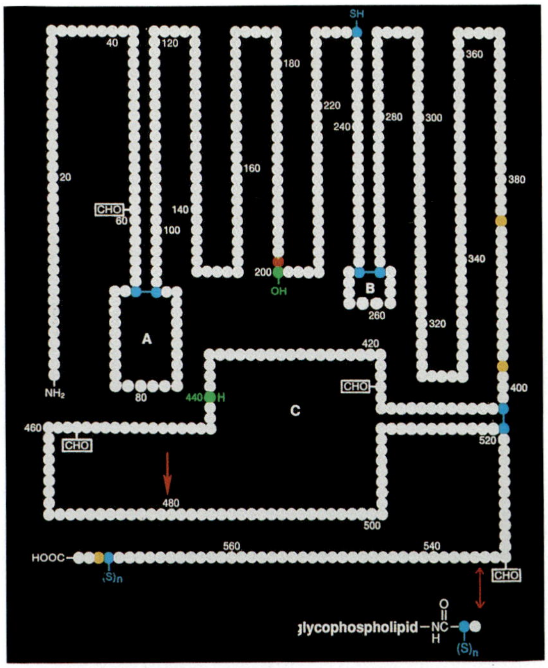

FIG. 3. **Sequence features and disulfide bond arrangement of the two alternatively spliced forms of acetylcholinesterase.** A similar disulfide pattern is maintained in homologous proteins. The active site serine (200) and histidine (440) are shown in *green*. Glu-199, a residue which can be modified to alter substrate specificity is shown in *red* (27). Shown in *orange* are the corresponding positions in thyroglobulin where the tyrosines which become iodinated are found. Cysteines (shown in *blue*) are denoted as: *SH*, free sulfhydryl; —, intrasubsunit disulfide; $-(S)_n$, intersubunit disulfides. The *single* and *double arrows* denote junctions between exons and alternatively spliced exons, respectively. The glycophospholipid attached species contains two unique amino acids in the processed gene product while the hydrophilic species contains 40 unique residues due to alternative mRNA splicing at the *double arrow*. The first coding exon extends from the 5′-non-translated region to the *single arrow* while the second extends from this point to the *double arrow*.

Since only two alternative exons have been identified to date, it appears that the catalytic subunits of the heteromeric, asymmetric, and lipid-linked forms and the homomeric soluble AChE form are encoded by exons 1, 2, and 3A while the glycophospholipid-containing form is encoded by 1, 2, and 3H. Such considerations are consistent with the observations that alternatively spliced exon 3A is the predominant one expressed in mouse muscle and brain (24). In contrast to the lower vertebrates (fish and amphibians), little evidence is found for the presence of the glycophospholipid-containing species in excitable tissues in mammals. Divergence from this distribution of encoding mRNA was only found in mammalian cells obtained from bone marrow and cultured hematopoietic cells (24), cell types where expression of the glycophospholipid-containing species prevails (15–17).

For the amphiphilic species encoded by exons 1, 2, and 3H, cleavage of a hydrophobic carboxyl-terminal peptide occurs immediately after translation (31, 32, 34). Through a transamidation reaction the glycophospholipid is conjugated through ethanolamine concomitant with cleavage of the peptide (35). Addition of *N*-linked oligosaccharides also occurs immediately following translation. Based on antigenic specificity, glycosylation or, more likely, its processing differs in the two forms of AChE in *Torpedo* (5, 36). Posttranslational processing steps for the hydrophilic subunit do not involve modification of its carboxyl-terminal end. Association of structural subunits is also critical to the molecular diversity of AChE. Disulfide bonding between catalytic and structural subunits occurs at a mid-Golgi stage (37).

In chick muscle, the rates of biosynthesis of the nascent AChE peptide greatly exceed new expression of active enzyme on the cell surface (38). A precursor pool of inactive enzyme with a rapid turnover rate exists intracellularly and may provide residual precursor capacity for expression of AChE at particular stages of tissue differentiation. Synapse formation is accompanied by deposition of a specialized basal lamina and its associated AChE. The slow turnover of the basal lamina and its associated AChE should place a minimal demand on synthesis of the asymmetric form of AChE in a mature synapse (39).

Functional Sites

The active site serine and the histidine involved in a putative catalytic triad function in a fashion similar to other serine

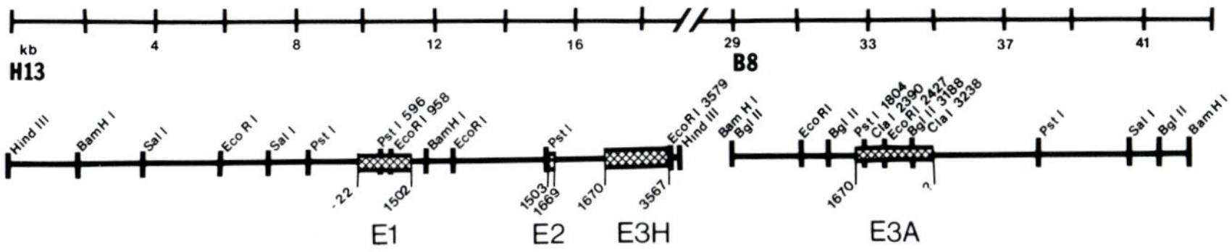

FIG. 4. **Organization of the *Torpedo* acetylcholinesterase gene.** Six exons have been defined 5′ of which are a TATA box and consensus regulatory regions (29). In this numbering system Exon (−1) and Exon (−2) (not shown) are in the 5′-non-translated region (T. J. Ekström and P. Taylor, submitted for publication). Exons 1, 2, 3A, and 3H (shown here) define the open reading frame where alternative splicing of 3A or 3H gives rise to mRNAs encoding two distinct carboxyl-terminal sequences. The *question mark* denotes the lack of identification of a polyadenylation signal in Exon 3A.

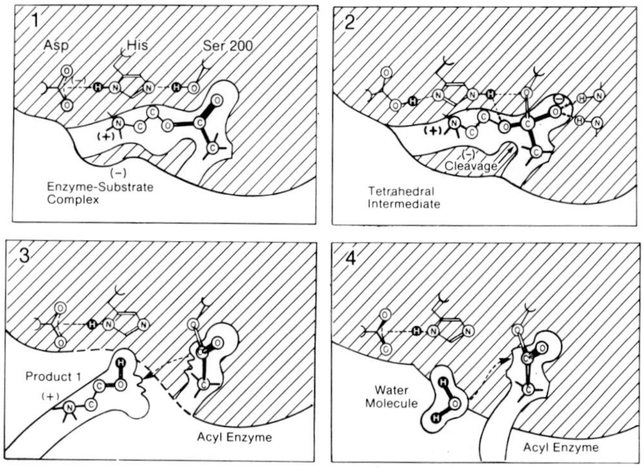

FIG. 5. **Steps in substrate hydrolysis catalyzed by acetylcholinesterase.** A putative catalytic triad of Ser-200, His-440, and an unknown Glu or Asp is shown. The substrate acetylcholine associates (*1*) leading to formation of tetrahedral transition state (*2*). The transition state collapses to form the trigonal acyl enzyme by linkage with the serine and loss of choline (*3*). The acyl enzyme rapidly deacylates (*4*) yielding active enzyme.

hydrolases. Although the roles and positions of serine and histidine have been defined in mutagenesis studies (27), the involvement of an aspartate in a catalytic triad and its role in catalysis remain open questions. Quinn and colleagues (40) have argued on the basis of proton inventory studies with varying molar ratios of H_2O/D_2O that proton transfer is not evident in AChE catalysis. Yet the absence of proton transfer does not preclude an inductive effect of a carboxylate side chain or a triad of amino acid side chains positioned so that the serine hydroxyl and a carboxylate from aspartate or glutamate are in apposition with the δ-1 and ϵ-2 imidazole nitrogens of the histidine (Fig. 5). The distinguishing feature of ester hydrolysis with cholinesterase is the unusually high turnover rate ($k_{cat} = 1.6 \times 10^4\ s^{-1}$ for AChE hydrolysis of acetylcholine) (41). Studies where the acyl enzyme was trapped during catalysis indicate comparable rates for the enzyme acylation and deacylation steps in the hydrolysis of acetylcholine (42). Acylation rates for acetylcholine appear to be diffusion-limited, while a conformational change accommodating the fit of substrate in the complex may be rate-limiting in the hydrolysis of neutral esters (41, 43).

Early studies revealed that carbamoyl esters, alkylphosphates and alkanesulfonates, are either substrates, or in the case of the slow deacylating alkanesulfonates and alkylphosphates, hemisubstrates for the cholinesterases (44, 45). The proper apposition of a positive charge in the substrate within the active center can greatly enhance association (41, 44). This principle of site direction was also elegantly employed in the development of oximes which enhanced deacylation of the phosphoryl enzymes (45). Such agents have proven to be useful antidotes in cholinesterase poisoning.

Inhibition mechanisms by reversible inhibitors can be quite complex since both the active center and a peripheral anionic site are involved. Moreover, inhibitors at the active center may affect the acylation or deacylation step, thereby achieving competitive or non-competitive inhibition behavior. Based on fluorescence energy transfer the peripheral site is some 20 Å removed from the active center (46). The role of the peripheral site in physiologic function of AChE remains uncertain as no endogenous inhibitors of AChE other than the substrate have been identified. However, substrate inhibition at high acetylcholine concentrations is well described in the AChEs. The substrate inhibition and peripheral anionic sites could be the same or overlapping entities. With primary structure of a large family of AChEs and BuChEs now described and with site-directed labeling (47) and crystallographic (48) and mutagenesis studies (27) ongoing, a detailed description of molecular structure of AChE and its correlation with function should emerge rapidly in the foreseeable future.

REFERENCES

1. Dale, H. H. (1914) *J. Pharmacol. Exp. Ther.* **6,** 147–190
2. Massoulié, J., and Reiger, F. (1969) *Eur. J. Biochem.* **11,** 441–451
3. LaDu, B. N. (1989) *Trends Pharmacol. Sci.* **10,** 309–313
4. Soreq, H., and Zakut, H. (1990) *Monographs Hum. Genet.* **13,** 1–102
5. Bon, S., Meflah, K., Musset, F., Grassi, J., and Massoulié, J. (1987) *J. Neurochem.* **49,** 1720–1731
6. Massoulié, J., and Toutant, J.-P. (1988) *Handbook of Experimental Pharmacology* (Whittaker, V. P., ed) Vol. 86, pp. 167–223, Springer-Verlag, Berlin
7. Lwebuga-Mukasa, J., Lappi, S., and Taylor, P. (1975) *Biochemistry* **15,** 1425–1435
8. Cartaud, J., Bon, S., and Massoulié, J. (1978) *J. Cell Biol.* **77,** 315–322
9. Anglister, L., and Silman, I. (1978) *J. Mol. Biol.* **125,** 293–311
10. Rosenberry, T. L., and Richardson, J. M. (1978) *Biochemistry* **16,** 3550–3558
11. Massoulié, J. (1991) *Proceedings of the 3rd International Conference on Cholinesterase* (Massoulié, J., Bacou, F., Barnard, E. A., Doctor, B. P., and Quinn, D. M., eds) American Chemical Society, Wash., D. C., in press
12. McMahan, U. J., Sanes, J. R., and Marshall, L. M. (1978) *Nature* **271,** 172–174
13. Brandan, E., Maldonado, M., Garrido, J., and Inestrosa, N. C. (1985) *J. Cell Biol.* **101,** 985–992
14. Inestrosa, N. C., Roberts, W. L., Marshall, T. L., and Rosenberry, T. L. (1987) *J. Biol. Chem.* **262,** 4441–4444
15. Silman, I., and Futerman, A. H. (1987) *Eur. J. Biochem.* **170,** 11–22
16. Roberts, W. L., Kim B. H., and Rosenberry, T. L. (1987) *Proc. Natl. Acad. Sci. U. S. A.* **84,** 7817–7821
17. Toutant, J. P., Richards, M. K., Krall, J. A., and Rosenberry, T. L. (1990) *Eur. J. Biochem.* **187,** 31–38
18. Schumacher, M., Camp, S., Maulet, Y., Newton, M., MacPhee-Quigley, K., Taylor, S. S., Friedmann, T., and Taylor, P. (1986) *Nature* **319,** 407–409
19. Olsen, P. F., Fessler, L. I., Nelson, R. E., Campbell, A. G., and Fessler, J. H. (1990) *EMBO J.* **9,** 3593–3601
20. De la Escalera, S., Backamp, E.-O., Moya, F., Piovant, M., and Jimenez, F. (1990) *EMBO J.* **9,** 3593–3601
21. Hall, L. M. C., and Spierer, P. (1986) *EMBO J.* **5,** 2947–2954
22. Prody, C. A., Zevin-Sonkin, D., Gnott, A., Goldberg, O., and Soreq, H. (1987) *Proc. Natl. Acad. Sci. U. S. A.* **54,** 3555–3559
23. Lockridge, O., Bartels, C. F., Vaughan, T. A., Wong, C. K., Norton, S. E., and Johnson, L. L. (1987) *J. Biol. Chem.* **262,** 549–557
24. Rachinsky, T. L., Camp, S., Li, Y., Ekström, T. J., Newton, M., and Taylor, P. (1990) *Neuron* **5,** 317–327
25. Doctor, B. P., Chapman, T. C., Christner, C. E., Deal, C. D., DeLa Hoz, D. M., Gentry, M. K., Ogert, R. A., Rush, R. S., Smyth, K. K., and Wolfe, A. D. (1990) *FEBS Lett.* **266,** 123–127
26. Chatonnet, A., and Lockridge, O. (1989) *Biochem. J.* **260,** 625–634
27. Gibney, G., Camp, S., Dionne, M., MacPhee-Quigley, K., and Taylor, P. (1990) *Proc. Natl. Acad. Sci. U. S. A.* **87,** 7546–7550
28. MacPhee-Quigley, K. M., Vedvick, T. S., Taylor, P., and Taylor, S. S. (1986) *J. Biol. Chem.* **26,** 13565–13570
29. Maulet, Y., Camp, S., Gibney, G., Rachinsky, T. L., Ekström, T. J., and Taylor, P. (1990) *Neuron* **4,** 289–301
30. Arpagaus, M., Kott, M., Vatsu, K. P., Bartels, C. F., LaDu, B. N., and Lockridge, O. (1990) *Biochemistry* **29,** 124–131
31. Gibney, G., MacPhee-Quigley, K., Thompson, B., Vedvick, T., Low, M., Taylor, S. S., and Taylor, P. (1988) *J. Biol. Chem.* **263,** 1140–1145
32. Sikorav, J. L., Duval, N., Ansetmet, A., Bon, S., Krejei, E., Legay, C., Osterlund, M., Reimund, B., and Massoulié, J. (1988) *EMBO J.* **7,** 2983–2993
33. Schumacher, M., Maulet, Y., Camp, S., and Taylor, P. (1988) *J. Biol. Chem.* **263,** 18979–18987
34. Gibney, G., and Taylor, P. (1990) *J. Biol. Chem.* **265,** 12576–12583
35. Ferguson, M. A. S., and Williams, A. F. (1988) *Annu. Rev. Biochem.* **67,** 285–320
36. Abramson, S. N., Ellisman, M., Deerinck, T. J., Maulet, Y., Gentry, M. K., Doctor, B. P., and Taylor, P. (1989) *J. Cell Biol.* **108,** 2301–2311
37. Rotundo, R. L. (1984) *Proc. Natl. Acad. Sci. U. S. A.* **81,** 479–483
38. Rotundo, R. L. (1988) *J. Biol. Chem.* **263,** 19398–19406
39. McMahan, U. J., and Wallace, B. G. (1989) *Dev. Neurosci.* **11,** 227–247
40. Quinn, D. M. (1987) *Chem. Rev.* **87,** 955–979
41. Rosenberry, T. L. (1975) *Adv. Enzymol Relat. Areas Mol. Biol.* **43,** 103–218
42. Froede, H. C., and Wilson, I. B. (1984) *J. Biol. Chem.* **259,** 11010–11013
43. Rosenberry, T. L. (1975) *Proc. Natl. Acad. Sci. U. S. A.* **72,** 3834–3838
44. Froede, H. C., and Wilson, I. B. (1971) in *The Enzymes* (Boyer, P. D., ed) Vol. 5, pp. 87–114, Academic Press, New York
45. Wilson, I. B. (1969) *Fed. Proc.* **18,** 752–758
46. Berman, H. A., Yguerabide, J., and Taylor, P. (1980) *Biochemistry* **19,** 2226–2235
47. Weise, C., Kreinenkamp, H. J., Raba, R., Pedak, A., Aaviksaar, A., and Hucho, F. (1990) *EMBO J.* **9,** 3885–3888
48. Sussman, J. L., Harel, M., Frolow, F., Varon, L., Toker, L., Futerman, A. H., and Silman, I. (1988) *J. Mol. Biol.* **203,** 821–823
49. Myers, E. W., and Miller, W. (1988) *Comput. Appl. Biosci.* **4,** 11–17

Minireview

THE JOURNAL OF BIOLOGICAL CHEMISTRY
Vol. 266, No. 8, Issue of March 15, pp. 4661-4664, 1991

Printed in U.S.A.

Lipid Activation of Protein Kinase C*

Robert M. Bell and David J. Burns

From the Department of Biochemistry, Duke University Medical Center, Durham, North Carolina 27710

Protein kinase C functions as a critical component of the signal transduction pathways that cells utilize to recognize and respond to a variety of extracellular agents (1, 2). These external stimuli cause the level of *sn*-1,2-diacylglycerols (DAG)[1] to increase; DAG then functions as a second messenger by binding to and activating protein kinase C (1–5). Phorbol esters and other tumor promoters are potent activators of protein kinase C (1, 2). Protein kinase C activation elicits a variety of cellular responses by phosphorylating target proteins on serine and threonine residues. There are numerous and excellent reviews on the stimuli, the receptors, the coupling mechanisms, the phospholipase systems, second messengers, the substrates of protein kinase C, and the cell types involved (1–11). This review will focus on the structure, function, and regulation of protein kinase C.

Protein kinase C has become the subject of intense investigation as a result of its critical functions within cells. As the transducer of lipid second messengers and the receptor for phorbol ester and other tumor promoters, protein kinase C has become a focal point of scientific inquiry by investigators once working in divergent fields of research. For the lipid biochemist and enzymologist, the role of lipids in protein kinase C regulation profoundly illustrates the functions of the cellular glycerolipids and sphingolipids and their breakdown products in cellular regulation. Both the glycero- and sphingolipids appear to exist as reservoirs for the production of active breakdown products or metabolites (3, 12–14). Recent studies on lipid second messengers and mediators explain, in part, why there is such great diversity in the cellular lipids. The emphasis of this minireview will be on the role of lipids as essential cofactors and as lipid activators of protein kinase C (Fig. 1). The review seeks to answer the question, how is protein kinase C regulated by lipids?

The Protein Kinase C Family—Nine members of the protein kinase C family have been identified to date by molecular cloning (15–23, 49, 77) (for review see Refs. 6–9); these serine/threonine protein kinases of similar structure are regulated by phospholipids and DAG (phorbol esters) (Fig. 2). The emerging data on structure and function, and tissue and cell distribution of the individual family members provide an emphatic statement that these are not mere isoenzymes of identical function but rather separate enzymes likely to have distinct and distinguishable functions (6–9). Individual members of the protein kinase C family likely function to elicit specific responses. Much work remains to be done to define the exact physiological role of the individual protein kinase C family members in cellular regulation.

Domain Structure of Protein Kinase C—Comparison of the primary structures of the members of the protein kinase C family inferred from cDNA sequence determinations revealed four conserved and five variable regions (6, 8, 9, 49, 77) (Fig. 2). The regulatory domain (V_1-C_2) contains the pseudosubstrate (an amino acid sequence that closely resembles protein kinase C substrates) (24) and one or two cysteine-rich regions that resemble the characteristics of the zinc finger motif (5, 8, 9, 25). This domain contains the high affinity phorbol ester binding site, the DAG binding site, and segments which interact with phospholipid by either Ca^{2+}-dependent or -independent mechanisms. The C2 region appears to confer the calcium dependence of the α, β-I, β-II, and γ subtypes (26); this region is conspicuously absent from the δ, ϵ, ζ, η, and L subtypes whose activities are Ca^{2+}-independent. The catalytic domain possesses the ATP binding and protein substrate binding sites. Inhibitors of protein kinase C can interact at either the regulatory or catalytic sites; calphostin (27) and sphingosine (12) would interact within the regulatory domain, while H-7 (28) and staurosporin (29) would interact at the catalytic site.

Activation Mechanism—The pseudosubstrate is thought to bind to the protein substrate binding site when the kinase is inactive (24) (see Fig. 3). This occurs when the enzyme exists in the cytoplasm or when the enzyme is bound to the membrane by attachment to its phospholipid cofactors by either calcium-dependent, or -independent mechanisms. Binding of DAG, or other lipid activators, is thought to produce conformational changes resulting in dislocation of the pseudosubstrate from the active site. This event then renders the kinase active. Considerable flexibility exists around the V_3 hinge region since three well separated regions of the enzyme undergo autophosphorylation by an intrapeptide mechanism (30). How do the lipids cause this critical change in conformation resulting in activation? What is known about the specificity of the phospholipid cofactor requirements? How specific are the interactions with DAG? Are there other lipid activators?

Activation of Protein Kinase C by Lipid—The interaction of lipids with protein kinase C proved critical to its discovery. Early studies defined factors present in the membrane which activated the kinase in a Ca^{2+}-dependent manner (31). These factors were lipid-soluble; purified phosphatidylserine (PS) substituted for the lipid-soluble factors more effectively than any other naturally occurring phospholipid (31, 32). At low levels of Ca^{2+}, crude lipid extracts from brain proved more effective than PS alone in activating protein kinase C (33). This led to the finding of a neutral lipid that activated protein kinase C in a PS- and Ca^{2+}-dependent manner; the neutral lipid was, of course, DAG, a new second messenger. This discovery linked protein kinase C to major signal transduction pathways involving the phosphatidylinositol cycle. Thus, protein kinase C was shown to be regulated physiologically by phospholipid cofactors and DAG, a lipid activator-second messenger (34).

Methods Used to Study Protein Kinase C-Lipid Interactions—Until 1985, studies on protein kinase C regulation by phospholipid cofactors and DAG activators utilized physically undefined preparations of phospholipid vesicles. Since the smallest unilamellar vesicle would contain 4,000 molecules of phospholipid, it was difficult to examine specificity and impossible to investigate the stoichiometry of these interactions. Since vesicles containing PS are known to fuse spontaneously in the presence of Ca^{2+}, the physical state of the phospholipid cofactor and lipid activation were impossible to control and define in physical terms. In that year, our laboratory reported on the development and application of mixed micellar methods to investigate the specificity and stoichiometry of protein kinase C-lipid interactions (35). These methods allowed systematic and independent variation of the number of phospholipid cofactors and lipid activators present in a physically defined entity, the mixed micelle. Molecular insights into mechanism and specificity emerged using the mixed micellar methods that were not attainable using phospholipid vesicles or dispersions (36). Also, in 1985 methods employing physically stable and defined phospholipid vesicles containing PS were employed to further characterize protein kinase C-lipid interactions (37). Both experimental approaches have proven valuable.

Inferences from Mixed Micellar Analysis—Systematic variation

* This work was supported by National Institutes of Health Grant GM38737.

[1] The abbreviations used are: DAG, *sn*-1,2-diacylglycerol; PS, phosphatidylserine; [^{3}H]PDBu, radiolabeled phorbol ester-phorbol dibutyrate; PIP_2, phosphatidylinositol 4,5-bisphosphate.

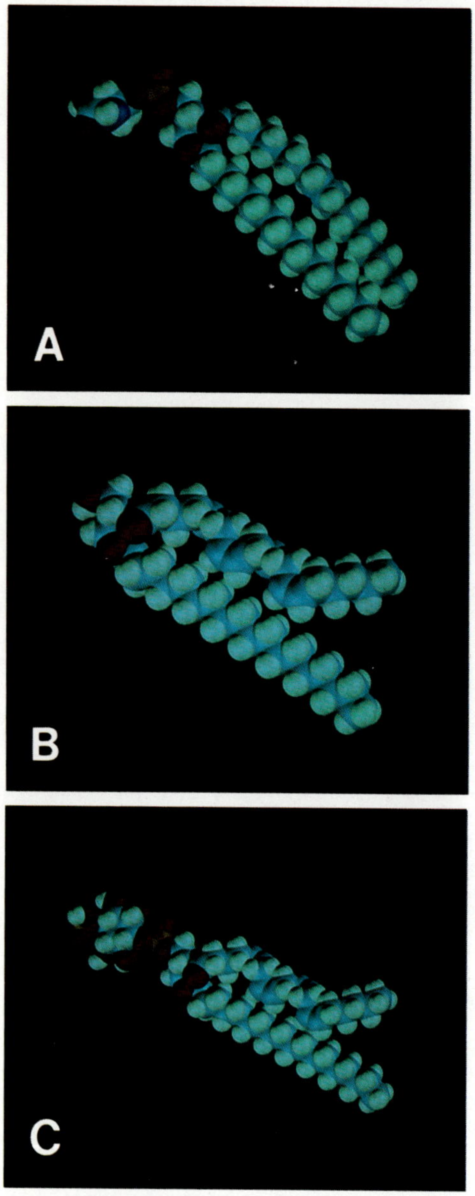

FIG. 1. **Lipids which function to activate protein kinase C.** The space-filled structural models of three lipid activators of protein kinase C are depicted to highlight their structural similarities and differences, and to identify the molecular features necessary for enzyme activation. *A*, PS; *B*, DAG; *C*, PIP_2. The interaction of protein kinase C with each of these molecules is detailed in the text. These molecules were constructed on an Evans & Sutherland work station using the Sybyl molecular modeling program. Atom legend: hydrogen, *gray*; carbon, *light blue*; oxygen, *red*; phosphorus, *orange*, nitrogen, *blue*.

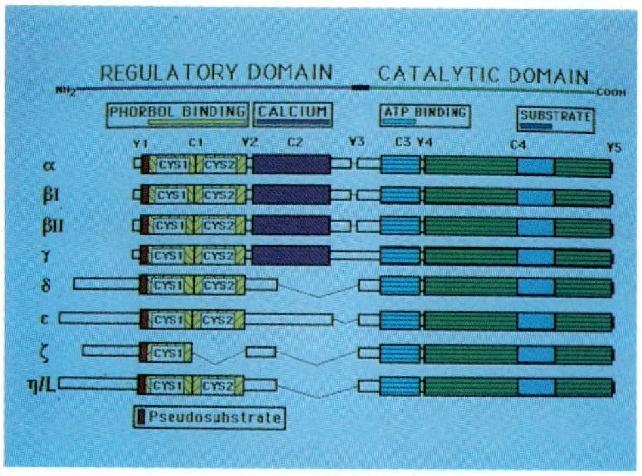

FIG. 2. **The protein kinase C family.** The domain structures of protein kinase C family members are depicted (5, 7, 9, 49, 77). Protein kinase C η and protein kinase C L are shown as one species since they are almost identical at the amino acid sequence level (49, 77). Protein kinase C L lacks several conserved amino acid residues in its second cysteine-rich region (most notably, two of the six conserved cysteine residues) compared with all other protein kinase C subtypes except ζ (49). Another potential subspecies ε′, may be derived from alternative splicing of a single mRNA transcript (22). *V*, variable regions of the PKC molecule; *C*, conserved regions of the PKC molecule.

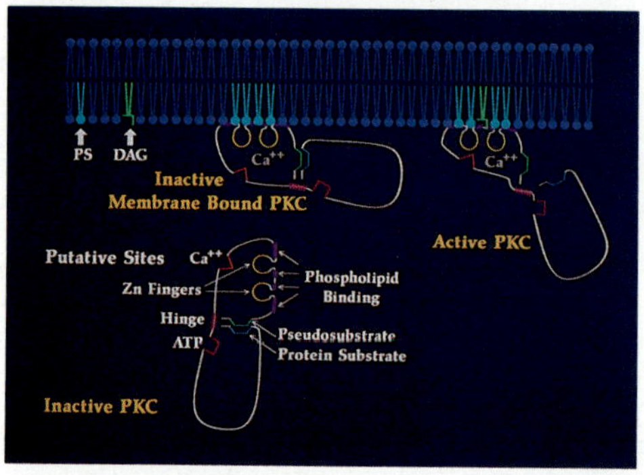

FIG. 3. **Schematic illustration of the mechanism of protein kinase C activation by lipids.** The significant sites on a Ca^{2+}-dependent member of the protein kinase C are depicted. These are labeled as putative sites since definitive evidence for their existence and location does not exist in some circumstances. The enzyme depicted apart from the membrane is inactive. The interaction of protein kinase C with the membrane is depicted to occur via binding to the phospholipid cofactor, PS, and is illustrated to be Ca^{2+}-dependent. The calcium binding sites do not exist in the absence of PS. Next, the critical DAG-dependent activation of PKC is illustrated. As discussed in the text, this leads to displacement of the pseudosubstrate. Since PKC contains two phorbol ester binding regions, the possibility exists that monomeric protein kinase C may bind 2 mol of $[^3H]$PDBu and, therefore, may be able to bind 2 mol of DAG. Only 1 mol of DAG is shown.

of the mole % DAG in Triton X-100 mixed micelles containing PS demonstrated that full activation of monomeric protein kinase C occurred when a single molecule of DAG was present (35, 36). Monomeric protein kinase C is also capable of binding $[^3H]$PDBu with high affinity (38). The PS dependence of protein kinase C activation and of $[^3H]$PDBu binding is highly cooperative demonstrating that several molecules of PS are required (35, 36, 38). Studies on the PS dependence of autophosphorylation (39, 40) showed a similar high degree of cooperativity to that observed with histone substrates. This would suggest that the observed PS dependence is not an artifact of its interaction with cationic substrates (41). Micelles containing 12 or more molecules of PS appear capable of fully activating protein kinase C (35, 36, 40, 42). The high degree of cooperativity in the PS dependence of activation first observed using mixed micellar methods is also observed using vesicles containing increasing mole % PS (42). A clear advantage of the mixed micellar methods is that the degree of activation by DAG or phorbol esters is far greater than in vesicles. Mixed micellar measurements also show a higher degree of phospholipid specificity than vesicular methods (35, 43). The superiority of the mixed micellar methods may reflect a lessening of artifactual Ca^{2+}-dependent activation by the anionic surfaces of phospholipid vesicles.

Phospholipid Cofactor Specificity—Since PS was the most effective of all the naturally occurring phospholipids, 16 analogues were prepared and tested for their ability to support protein kinase C activity, $[^3H]$PDBu binding, and binding to mixed micelles (43). Use of these analogues and the mixed micellar methods revealed a high degree of specificity. Several functional groups within the phospho-L-serine polar head group proved to be necessary for binding and activation of protein kinase C. Both the carboxyl and amino functional groups are essential. The distance between the phosphate moiety and these functional groups is also critical (phosphatidyl L-homoserine is not active). The recognition is stereospecific for L-serine, whereas there is no stereospecificity with respect to the glycerol backbone. From these data, three points of contact between protein kinase C and the phospho-L-serine head group of PS were inferred (43). Interactions with the amino, carboxyl, and phosphate moieties were envisioned (Fig. 1); these three bonds are sufficient for stereospecificity. Interestingly, the phospholipid head group was envisioned to help form the calcium binding site with protein kinase C (43, 44). This hypothesis was supported by direct binding measurements with calcium (45) and by the absence of potential calcium binding sites in all but one of the protein kinase C family members (7, 9, 16).

Only the anionic phospholipids display any phospholipid cofactor activity (31, 32). The activation of protein kinase C by phosphatidylglycerol and cardiolipin in vesicles does not carry over well into mixed micelles and is likely to be of limited physiological relevance given the location of these lipids in mitochondria (36). Phosphatidic acid is capable of partially supporting activity and $[^3H]$PDBu binding in mixed micelles; phosphatidic acid may be a special case of physiologic relevance since it is produced during signal transduction processes (43).

Overall, the weight of the evidence implies that PS functions in the physiological environment as the phospholipid cofactor. This high degree of specificity is further supported by the observations that several anionic lipids can replace only a portion of the PS molecules required for activation[2] (42). When protein kinase C binds to phospholipid monolayers, an increase in surface pressure is observed arguing that critical domains are inserted into the hydrocarbon region (46).

Specificity of Activation by DAG—Activation of protein kinase C by DAG analogues is amazingly specific. Over 50 analogues have been prepared and tested to date[3] (44, 47, 48). Activation is stereospecific and requires *sn*-1,2-DAG; 1,3-DAG and *sn*-2,3-DAGs were not effective (35, 44, 48). Both of the oxygen esters and the primary (3) hydroxyl are essential (44). The acyl chain composition does not appear critical as long as the aggregate chain length causes the DAG to associate with the membrane or micelle; *sn*-1,2-dibutyroylglycerol is inactive (44). Cyclic analogues restricting the functional groups have largely proven inactive (47, 48). This high degree of specificity is consistent with the second messenger role of DAG. Thus, at least three points of contact occur between protein kinase C and DAG. Hydrogen bonds with the two carbonyls and the hydroxyl moieties are inferred (Fig. 1).

Other Lipid Regulators—PIP_2 (see Fig. 1) activates protein kinase C in a PS- and Ca^{2+}-dependent manner (50, 51) that resembles, in part, activation by DAG (50, 51). The ability of PIP_2 to activate protein kinase C is unlike any other naturally occurring phospholipid. Since the level of PIP_2 decreases in stimulated cells, PIP_2 activation may have physiological significance. A major point requiring clarification is whether PIP_2 inhibits $[^3H]$PDBu binding (52) or interacts at a distinct site (51). The mechanistic analyses are complicated due to the ability of PIP_2 to function as a lipid activator and as a phospholipid cofactor to replace some PS molecules (51).

Arachidonic acid (53), cis-unsaturated fatty acids (54–56), and lipoxin A (57) activate protein kinase C *in vitro* by mechanisms independent of PS and DAG. Synergy between oleic acid and DAG in the presence of calcium was observed (56). Activation of protein kinase C occurs in cells exposed to cis-unsaturated fatty acids; whether these fatty acids activate protein kinase C directly or indirectly is not known (56). The physiological significance of these observations will require clarification.

The effects and physiological significance of other potential lipid effectors on protein kinase C are beyond the scope of this review. These potential modulators include sphingosine/lysosphingolipids (12), lysophosphatidylcholine (58), and ether lipids (59). The number of potentially physiologically significant lipid effectors of protein kinase C points to the possibility of multiple and distinct mechanisms of protein kinase C regulation by lipids.

Regions of Significant Lipid-Protein Interactions—Biochemical and genetic studies have established that the amino-terminal region (C_1-C_2 region) functions as the regulatory domain and contains the phorbol ester binding domain (60–62). Recent studies have shown that the C_1 region containing the cysteine-rich regions is sufficient for high affinity $[^3H]$PDBu binding[4] (61, 62). When expressed in *Escherichia coli* (61) or insect cells,[4] the Cys 1 and Cys 2 regions both possess $[^3H]$PDBu binding activity, albeit of reduced affinity[4] (10–20-fold). Most members of the protein kinase C family possess two cysteine-rich regions and, therefore, likely contain two phorbol ester binding sites (Fig. 2); formal proof will require the demonstration of the binding of 2 mol of $[^3H]$PDBu/mol of protein kinase C. The presence of two nonequivalent cyclic nucleotide binding sites within the regulatory subunits of the cAMP- and cGMP-dependent protein kinase families is established (63). These sites, in the cAMP-dependent protein kinase, differ in their off rates for cyclic nucleotides and cyclic nucleotide analogues can discriminate between the two sites (63). The physiological significance of two second messenger binding sites in these kinases and in protein kinase C is not clear. Agents such as the bryostatins (64) may interact more strongly at one site than another. The binding of large agents, such as high molecular weight tumor promoters, may sterically block access to the other site.

Single or tandem cysteine-rich motifs, similar to zinc finger structures, exist within the C_1 region of all protein kinase C cDNAs (25). As few as 83 amino acids from the second cysteine-rich region (Cys 2) of protein kinase C γ are sufficient for $[^3H]$PDBu binding.[4] This binding is strongly dependent on added phospholipids. For purposes of illustration and discussion, this 83-amino acid region of protein kinase C γ is shown in Fig. 4 as a C_4 zinc finger similar to the CI region of steroid receptors (66). This same region is also depicted as a Zn(II) C_6 binuclear cluster identified in the GAL4 transcriptional factor in Fig. 4 (67).

The existing mutagenesis studies are consistent with the zinc structures of protein kinase C being essential for phorbol binding[4] (61, 62). No direct measurements of zinc binding to protein kinase C have been reported, although added zinc can modulate activity (65). Also, *n*-chimaerin, a novel phospholipid-dependent phorbol ester receptor, is completely dependent upon zinc for $[^3H]$PDBu binding activity (71). However, the structure of these cysteine-rich regions and their dependence on zinc must be considered conjecture at this point.

The interactions between zinc fingers and DNA have been the focus of much recent research. The function of these putative zinc motifs in protein kinase C regulation is not clear although some protein kinase C is found in the nucleus (68) and protein kinase C has been reported to bind DNA (69). Structural similarities between DNA and anionic phospholipid surfaces are known; both possess phosphodiester linkages and hydrophobic regions. Lupus antibodies react with DNA and with anionic phospholipids (see Ref. 70 for details). The loops of the zinc fingers may function in regulation or in phorbol ester binding. Point mutations, additions, and deletions within the loops could help define these roles.

The regions flanking the putative zinc fingers contain a number of charged residues (Fig. 4). One of these regions shows homology with a portion of the active site of phospholipase A_2 (73). Presum-

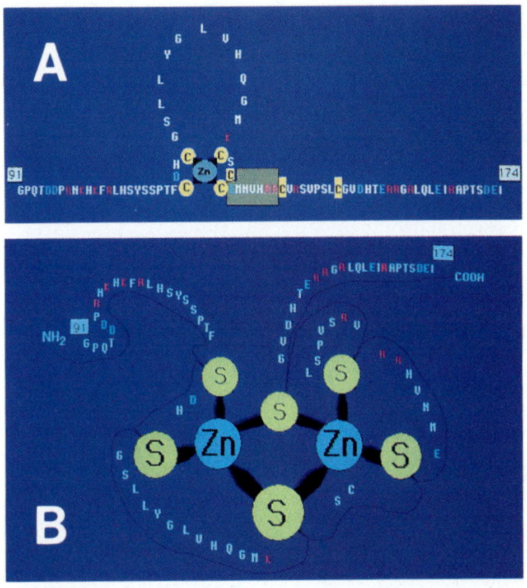

FIG. 4. **Possible zinc structural motifs within the phorbol ester binding domain.** *A*, the Cys 2 region of protein kinase C γ (amino acids 91–174) is depicted in a zinc finger motif similar to the CI region of steroid receptors (66). *B*, this same region is also depicted as a Zn(II) C_6 binuclear cluster identified in the GAL4 transcriptional factor (67). The illustration was adapted to this region of protein kinase C from that published for GAL4 (67). Positively charged residues are shown in *pink*. Negatively charged residues are shown in *light blue*. Cysteine residues are shown in *yellow*. The region of the protein kinase C molecule homologous to the PLA_2 is *bracketed* in *light brown*.

[2] M-H. Lee and R. M. Bell, unpublished data.
[3] J. Walsh and R. M. Bell, unpublished data.
[4] D. Burns and R. M. Bell, submitted for publication.

ably, these shared regions are involved in phospholipid-protein interactions. Interestingly, many of the amino acids in these flanking regions are present in *n*-chimaerin (71, 72) which also binds phorbol esters, but not in flanking regions of the zinc finger motifs present in transcriptional control factors or steroid receptors. Given the inferences that protein kinase C stereospecifically recognizes phosphatidyl-L-serine as a phospholipid cofactor, these charged amino acids could form ion pairs with the carboxyl, amino, and phosphate moieties. Each flanking region contains two or more pairs of 2 positively and 1 negatively charged residues which could interact with phospho-L-serine as described. These interactions are consistent with the data but will remain speculative until detailed structures are elucidated. Further work will be required to determine the exact amino acids interacting with phorbol esters, DAG and PS.

Prospectus—The critical involvement of lipids in the regulation of protein kinase C activity and cellular regulation is clear. Phospholipids, notably phosphatidylserine, function as essential cofactors. *sn*-1,2-Diacylglycerol functions as an activator of protein kinase C. Protein kinase C displays surprising specificity with respect to both phospholipid cofactor and lipid activators. The high cooperativity observed with PS argues convincingly for several molecules forming a highly ordered complex with protein kinase C. Activation occurs upon DAG binding to this complex. The complexity of lipid-protein interactions necessary to activate protein kinase C provides the necessary control; less specific interactions might allow unwanted and uncontrolled activation. The stereospecificity of the process is reassuring. While the number and sites of significant lipid-protein interactions are coming into focus, there is great need to obtain definitive structural data.

With the development of the baculovirus/insect cell expression system for individual members of the protein kinase C family (74–76), sufficient quantities of highly purified enzyme are likely to become available for detailed physical studies. Crystallization of protein kinase C, its catalytic fragment, or the lipid binding regulatory domain would represent a major step forward in structure elucidation. However, the critical details of lipid-protein interactions will be difficult to approach crystallographically. Structural analysis by two-dimensional NMR of the phorbol ester binding domains (83 amino acids) in mixed micelles containing PS and phorbol esters may reveal the precise sites of PS and phorbol ester interaction. Much remains to be done to elucidate the detailed mechanism by which lipids regulate members of the protein kinase C family.

Acknowledgments—The assistance of Dr. Roy Borchardt is gratefully acknowledged in the preparation of this review. He and Dr. Neal Tweedy prepared the computer-generated lipid molecules shown in Fig. 1.

REFERENCES

1. Nishizuka, Y. (1986) *Science* **233,** 305–312
2. Nishizuka, Y. (1988) *Nature* **334,** 661–665
3. Bishop, W., and Bell, R. M. (1988) *Oncogene Res.* **2,** 205–218
4. Woodgett, J. R., Hunter, T., and Gould, K. L. (1987) in *Cell Membranes: Methods and Reviews* (Elson, E. L., Frazier, W. A., and Glaser, L., eds) Vol. 3, pp. 215–340, Plenum Press, New York
5. Bell, R. M. (1986) *Cell* **45,** 631–632
6. Nishizuka, Y. (1989) *Annu. Rev. Biochem.* **58,** 31–44
7. Huang, K-P. (1989) *Trends Neurosci.* **12,** 425–432
8. Farago, A., and Nishizuka, Y. (1990) *FEBS Lett.* **268,** 350–354
9. Parker, P. J., Kour, G., Marais, R., Mitchell, F., Pears, C., Schaap, D., Stabel, S., and Webster, C. (1989) *Mol. Cell. Endocrinol.* **65,** 1–11
10. Rhee, S. G., Suh, P-G., Ryu, S-H., and Lee, S. Y. (1989) *Science* **244,** 546
11. Majerus, P., Connolly, T., Deckmyn, H., Ross, T., Bross, T., Ishii, H., Bansal, V., and Wilson, D. (1986) *Science* **234,** 1519–1526
12. Hannun, Y., and Bell, R. M. (1989) *Science* **243,** 500–507
13. Okazaki, T., Bielawaska, A., Bell, R. M., and Hannun, Y. (1990) *J. Biol. Chem.* **265,** 15823–15831
14. Exton, J. (1990) *J. Biol. Chem.* **265,** 1–4
15. Parker, P., Coussens, L., Totty, N., Rhee, L., Young, S., Chen, E., Stabel, S., Waterfield, M., and Ullrich, A. (1986) *Science* **233,** 853–859
16. Coussens, L., Parker, P., Rhee, L., Yang-Feng, T., Chen, E., Waterfield, M., Franke, U., and Ullrich, A. (1986) *Science* **233,** 859–866
17. Ono, Y., Kurokawa, T., Fujii, T., Kawahara, K., Igarashi, K., Kikkawa, U., Ogita, K., and Nishizuka, Y. (1986) *FEBS Lett.* **206,** 347–352
18. Knopf, J., Lee, M-H., Sultzman, L., Kriz, R., Loomis, C., Hedwick, R., and Bell, R. M. (1986) *Cell* **46,** 491–502
19. Ohno, S., Kawasaki, H., Imajoh, S., Suzuki, K., Inagaki, M., Yokokura, H., Sakoh, T., and Hidaka, H. (1987) *Nature* **325,** 161–166
20. Ono, Y., Kikkawa, U., Ogita, K., Fujii, T., Kurokawa, T., Asaoka, Y., Sekiguchi, K., Ase, K., Igarashi, K., and Nishizuka, Y. (1987) *Science* **236,** 1116–1120
21. Ono, Y., Fujii, T., Ogita, K., Kikkawa, U., Igarashi, K., and Nishizuka, Y. (1987) *FEBS Lett.* **226,** 125–128
22. Ono, Y., Fujii, T., Ogita, K., Kikkawa, U., Igarashi, K., and Nishizuka, Y. (1988) *J. Biol. Chem.* **263,** 6927–6943
23. Ohno, S., Akita, Y., Konno, Y., Imajoh, S., and Suzuki, K. (1988) *Cell* **53,** 731–741
24. House, C., and Kemp, B. (1990) *Cell Signalling* **2,** 187–190
25. Klug, A., and Rhodes, D. (1987) *Trends Biochem. Sci.* **12,** 464–469
26. Ono, Y., Fujii, T., Ogita, K., Kikkawa, U., Igarashi, K., and Nishizuka, Y. (1989) *Proc. Natl. Acad. Sci. U. S. A.* **86,** 3099–3103
27. Kobayashi, E., Nakano, H., Morimoto, M., and Tamaoki, T. (1989) *Biochem. Biophys. Res. Commun.* **159,** 548–553
28. Hidaka, H., Inagaki, M., Kawamoto, S., and Sasaki, Y. (1984) *Biochemistry* **23,** 5036–5041
29. Gross, J., Herblin, W., Do, U., Pounds, J., Buenaga, L., and Stephens, L. (1990) *Biochem. Pharmacol.* **40,** 343–350
30. Flint, A., Paladini, R., and Koshland, D. (1990) *Science* **249,** 408–411
31. Takai, Y., Kishimoto, A., Iwasa, Y., Kawahara, Y., Mori, T., and Nishizuka, Y. (1979) *J. Biol. Chem.* **254,** 3692–3695
32. Takai, Y., Kishimoto, A., Iwasa, Y., Kawahara, Y., Mori, T., Nishizuka, Y., Tamura, A., and Fujii, T. (1979) *J. Biochem. (Tokyo)* **86,** 575–578
33. Kaibuchi, K., Takai, Y., and Nishizuka, Y. (1981) *J. Biol. Chem.* **256,** 7146–7149
34. Takai, Y., Kishimoto, A., Kikkawa, U., Mori, T., and Nishizuka, Y. (1979) *Biochem. Biophys. Res. Commun.* **91,** 1218–1224
35. Hannun, Y., Loomis, C., and Bell, R. M. (1985) *J. Biol. Chem.* **260,** 10039–10043
36. Hannun, Y., Loomis, C., and Bell, R. M. (1986) *J. Biol. Chem.* **261,** 7184–7190
37. Boni, L., and Rando, R. (1985) *J. Biol. Chem.* **260,** 10819–10825
38. Hannun, Y., and Bell, R. M. (1986) *J. Biol. Chem.* **261,** 9341–9347
39. Newton, A., and Koshland, D. (1990) *Biochemistry* **29,** 6656–6661
40. Hannun, Y., and Bell, R. M. (1990) *J. Biol. Chem.* **265,** 2962–2972
41. Bazzi, M., and Nelsestuen, G. (1987) *Biochemistry* **26,** 1974–1982
42. Newton, A., and Koshland, D. (1989) *J. Biol. Chem.* **264,** 14909–14915
43. Lee, M-H., and Bell, R. M. (1989) *J. Biol. Chem.* **264,** 14797–14805
44. Ganong, B., Loomis, C., Hannun, Y., and Bell, R. M. (1986) *Proc. Natl. Acad. Sci. U. S. A.* **83,** 1184–1188
45. Bazzi, M., and Nelsestuen, G. (1990) *Biochemistry* **29,** 7624–7630
46. Bazzi, M., and Nelsestuen, G. (1988) *Biochemistry* **27,** 6776–6783
47. Rando, R. (1988) *FASEB J.* **2,** 2348–2355
48. Bonser, R., Thompson, N., Hodson, H., Beams, R., and Garland, L. (1988) *FEBS Lett.* **234,** 341–344
49. Bacher, N., Zisman, Y., Berent, T., and Livneh, E. (1991) *Mol. Cell. Biol.* **11,** 126–133
50. Chauhan, V., and Brockerhoff, H. (1988) *Biochem. Biophys. Res. Commun.* **155,** 18–23
51. Lee, M-H., and Bell, R. M. (1991) *Biochemistry* **30,** 1041–1049
52. Chauhan, A., Chauhan, V., Deshmukh, D., and Brockerhoff, H. (1989) *Biochemistry* **28,** 4952–4956
53. McPhail, L., Clayton, C., and Snyderman, R. (1984) *Science* **224,** 622–625
54. Murakami, K., Chan, S., and Routtenberg, A. (1986) *J. Biol. Chem.* **261,** 15424–15429
55. Sekiguchi, K., Tsukuda, M., Ogita, K., Kikkawa, U., and Nishizuka, Y. (1987) *Biochem. Biophys. Res. Commun.* **145,** 797–802
56. El Touny, S., Khan, W., and Hannun, Y. (1990) *J. Biol. Chem.* **265,** 16437–16443
57. Hasson, A., Serhan, C., Haeggstrom, J., Ingelman-Sundberg, M., and Samuelsson, B. (1986) *Biochem. Biophys. Res. Commun.* **134,** 1215–1222
58. Oishi, K., Raynor, R., Charp, P., and Kuo, J. (1988) *J. Biol. Chem.* **263,** 6865–6871
59. Daniel, L., Small, G., Schmift, J., Marasco, C., Ishaq, K., and Piantadosi, C. (1988) *Biochem. Biophys. Res. Commun.* **151,** 291–297
60. Lee, M-H., and Bell, R. M. (1986) *J. Biol. Chem.* **261,** 14867–14870
61. Ono, Y., Fujii, T., Igarashi, K., Kuno, T., Tanaka, C., Kikkawa, U., and Nishizuka, Y. (1989) *Proc. Natl. Acad. Sci. U. S. A.* **86,** 4868–4871
62. Kaibuchi, K., Fukumoto, Y., Oku, N., Takai, Y., Arai, K., and Muromatsu, M. (1989) *J. Biol. Chem.* **264,** 13489–13496
63. Taylor, S. (1989) *J. Biol. Chem.* **264,** 8443–8446
64. Kraff, A., Reeves, J., and Ashendel, C. (1988) *J. Biol. Chem.* **263,** 8437–8442
65. Murakami, K., Whiteley, M., and Routtenberg, A. (1987) *J. Biol. Chem.* **262,** 13902–13906
66. Hard, T., Kellenbach, E., Boelens, R., Maler, B., Dahlman, K., Freedman, L., Carlstedt-Duke, J., Yamamoto, K., Gustafsson, J-A., and Kaptein, R. (1990) *Science* **249,** 157–160
67. Pan, T., and Coleman, J. (1990) *Proc. Natl. Acad. Sci. U. S. A.* **87,** 2077–2081
68. Leach, K., Powers, E., Ruff, V., Jaken, S., and Kaufmann, S. (1989) *J. Cell Biol.* **109,** 685–695
69. Testori, A., Hii, C., Fournier, A., Burgoyne, L., and Murray, A. (1988) *Biochem. Biophys. Res. Commun.* **156,** 222–227
70. Hannun, Y., Foglesong, R., and Bell, R. M. (1989) *J. Biol. Chem.* **264,** 9960–9966
71. Ahmed, S., Kozma, R., Monfries, C., Hall, C., Lim, H., Smith, P., and Lim, L. (1990) *Biochem. J.* **272,** 767–773
72. Hall, C., Monfries, C., Smith, P., Lim, H., Kozma, R., Ahmed, S., Vanniasingham, V., Leung, T., and Lim, L. (1990) *J. Mol. Biol.* **211,** 11–16
73. Maraganore, J. (1987) *Trends Biochem. Sci.* **12,** 176–177
74. Patel, G., and Stabel, S. (1989) *Cell Signalling* **1,** 227–240
75. Burns, D., Bloomenthal, J., Lee, M-H., and Bell, R. (1990) *J. Biol. Chem.* **265,** 12044–12051
76. Stabel, S., Schaap, D., and Parker, P. (1991) *Methods Enzymol.*, in press
77. Osada, S., Mizuno, K., Saido, T. C., Akita, Y., Suzuki, K., Kurok, T., and Ohno, S. (1990) *J. Biol. Chem.* **265,** 22434–22440

Minireview

THE JOURNAL OF BIOLOGICAL CHEMISTRY
Vol. 266, No. 9, Issue of March 25, pp. 5355–5358, 1991

Helical Interactions in Homologous Pairing and Strand Exchange Driven by RecA Protein

Charles M. Radding

From the Departments of Human Genetics and Molecular Biophysics and Biochemistry, Yale University School of Medicine, New Haven, Connecticut 06510

In a search for mutants that affect genetic recombination, Clark and Margulies (1) discovered the recA gene in *Escherichia coli*, but subsequently recA mutants were found to be pleiotropic, affecting not only recombination but repair, mutagenesis, phage induction, and cell division (2). Not surprisingly, many years passed before the basis of this complex phenotype was understood. A milestone was the discovery by Roberts *et al.* (3) that the product of the *rec*A gene catalyzes the cleavage of the λ repressor and thereby inactivates it. There followed the discoveries that by proteolytic cleavage RecA protein also catalyzes the *inactivation* of the cellular LexA repressor, which derepresses more than 20 genes, and the *activation* of the UmuD protein, which is required for UV-induced mutagenesis (3–6). Especially after the discovery that RecA protein derepresses other genes, it appeared likely that the role of RecA protein in genetic recombination was indirect, via its effect on the expression of those other genes. Experiments on a temperature-sensitive mutant of *rec*A showed that the derepression of other genes was not a sufficient explanation for the role of *rec*A in recombination (7), but these experiments could not exclude the possible activation of recombination proteins by cleavage, as for example in the case of UmuD protein whose cleavage activates its role in UV mutagenesis. However, the cloning of the *rec*A gene (8) and the further finding that RecA protein has DNA-dependent ATPase activity (9, 10) led to the discovery that RecA protein also has remarkable recombination activities *in vitro* (11–13). Although mutants exist that are proficient in cleavage of LexA repressor and yet are defective in recombination (14), the observations on the recombination activities of RecA protein *in vitro* provide the only compelling evidence that it plays a direct role in recombination, for the reason just described (15–18).

Recombination Reactions of RecA Protein

RecA protein will pair two DNA molecules if one of them is single-stranded or partially single-stranded, the latter including duplex molecules with gaps or single-stranded tails (Fig. 1). The pairing of two completely duplex molecules has not been observed. The pairing of single-stranded or partially single-stranded DNA with fully duplex DNA that has suitable ends leads to a further reaction, the exchange of strands involving kilobase pairs of DNA (Fig. 1, *b* and *c*, Fig. 2). When a partially single-stranded molecule reacts with a fully duplex molecule that has suitable ends, strand exchange is reciprocal and produces the classical recombination intermediate first postulated by Holliday (71) (Fig. 1*c*).

Reactions of circular viral single strands with linear duplex DNA have provided a convenient model for elucidating the three distinct phases by which RecA protein carries out recombination *in vitro* (Fig. 2). In the presence of ATP, RecA protein first polymerizes on the single-stranded DNA to form a helical presynaptic nucleoprotein filament. During the second phase, called *synapsis*, the nucleoprotein filament binds naked duplex DNA in a second binding site and randomly searches for homology. This search is facilitated by the concentration of DNA within large networks that are formed by multiple short-lived contacts of the filament with naked duplex DNA. Homologous alignment produces a *joint molecule* in which the two partners are noncovalently joined by the hydrogen bonds of paired bases. The formation of such a *joint* can occur anywhere within the extent of the two homologous partners (Fig. 3, *b*–*d*); it does not require the presence of homologous ends in any of the chains involved (Fig. 1, *d* and *e*) (16, 19–22). Once homologous alignment has taken place, however, ends are rapidly found, and in the model system under consideration the complementary strand from duplex DNA is transferred directionally, starting from its 3′ end (Fig. 3*a*) into the nucleoprotein filament, progressively creating *heteroduplex* DNA within the filament and a displaced linear strand, which will also be coated with RecA protein when sufficient RecA protein is present (Fig. 2). At 2–10 base pairs/s, strand exchange is slow, but it can traverse kilobase pairs of DNA and, moreover, with surprising efficiency it can traverse heterologous insertions as long as 50–100 base pairs (23).

In the presence of sufficient RecA protein, the products of this model reaction are not naked DNA molecules but rather are nucleoprotein filaments, one containing the linear single strand that has been displaced from the original linear homoduplex molecule and the other containing nicked circular heteroduplex DNA (24, 25).

The reaction promoted by RecA protein is initiated by the interaction of a nucleoprotein filament and a naked DNA molecule. Within the time required for pairing and strand exchange at physiological pH, RecA protein binds preferentially and stoichiometrically to single-stranded DNA; it does not bind directly to duplex DNA. However, the impediment to the direct binding of RecA protein to duplex DNA is kinetic (18, 26). RecA protein can gain access to duplex DNA via the recombination reaction itself, via a contiguous single-stranded region, or via special conditions such as acid pH or the presence of ATPγS,[1] a nonhydrolyzable analog.

These properties enable us to explain key features of the recombination reactions. Thus, a single strand is required to initiate the reaction; when that single strand is a tail or a gap in the duplex DNA, RecA protein can invade the entire duplex molecule from that initiating site to form a nucleoprotein filament. Two DNA molecules that are completely coated with RecA protein cannot pair. Therefore, the pairing of complementary single strands (Fig. 1*a*) only occurs when there is not enough RecA protein to coat all strands completely; and reversal of the reaction of a single strand with duplex DNA, *i.e.* a second cycle of pairing and exchange, is inhibited when residual RecA protein is present on all products of the reaction (see Fig. 2) (16, 17, 25).

Single-stranded DNA binding proteins such as *E. coli* SSB play an auxiliary role in presynapsis and strand exchange, a role that does not require specific protein-protein interactions, since genetically unrelated binding proteins act indistinguishably. The combined interactions of RecA protein and *E. coli* SSB with DNA are complex and are not fully understood (16, 17). The role of SSB is clearest in the presynaptic phase where it helps RecA protein to coat single-stranded DNA completely by removing secondary structure (27, 28). In the model reaction of circular single-stranded DNA with linear duplex DNA, SSB also speeds strand exchange and improves the yield of products that can be resolved by gel electrophoresis: it does so, in part, by inhibiting the reinvasion of duplex DNA by the 5′ end of a partially displaced linear strand (25). The mechanism of the inhibition is not completely clear, but one contributing factor has been identified. Since RecA protein appears to nucleate randomly on single-stranded DNA but polymerizes in the 5′ to 3′ direction, 5′ ends appear to be less completely coated by RecA protein. As a result, 5′ ends are subject to preferential binding by SSB and consequent inactivation as starting points for strand exchange (29–31).

[1] The abbreviation used is: ATPγS, adenosine 5′-*O*-(thiotriphosphate).

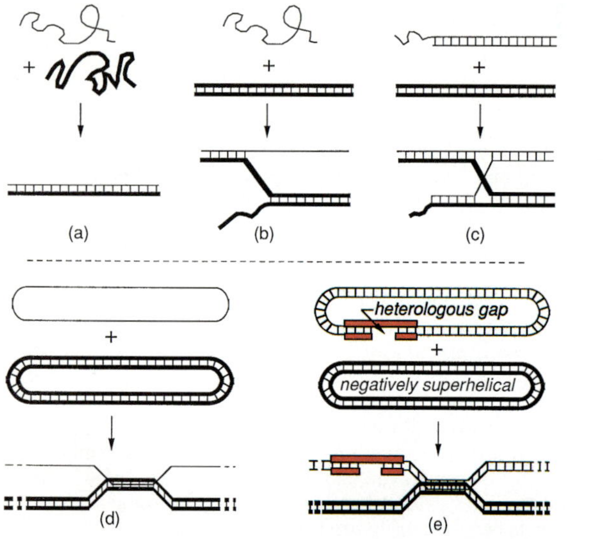

FIG. 1. **Recombination reactions promoted by RecA protein.** *a*, renaturation of complementary single strands; *b*, *asymmetric* (nonreciprocal) strand exchange following the pairing of single-stranded and double-stranded DNA; *c*, *symmetric* (reciprocal) strand exchange following the pairing of duplex DNA and partially single-stranded DNA; *d*, formation of a *three-stranded paranemic* joint; *e*, formation of a *four-stranded* paranemic joint. *Stippled areas* represent heterologous DNA.

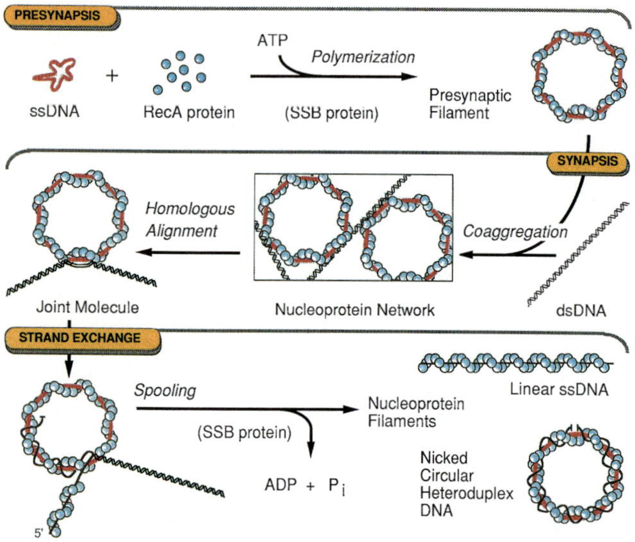

FIG. 2. **An overview of three phases of homologous pairing and strand exchange promoted by RecA protein acting on the model substrates, circular single-stranded (*ss*) DNA, and linear duplex DNA.** *ds*, double-stranded.

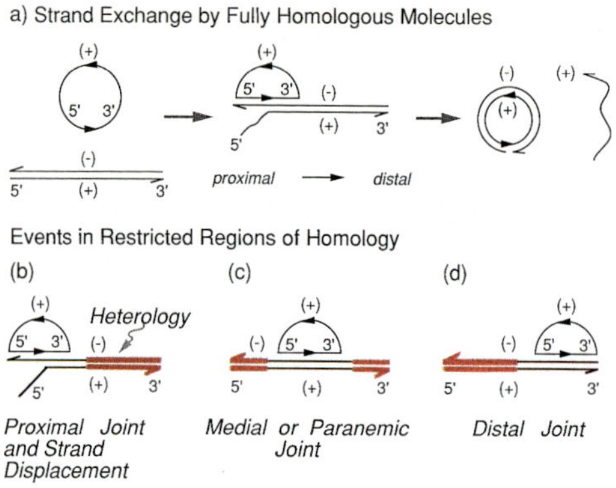

FIG. 3. **Homologous joints restricted to different sites in linear duplex DNA have different properties and fates.** As indicated, such joints may be designated according to their relationship to ends and the polarity of strand exchange.

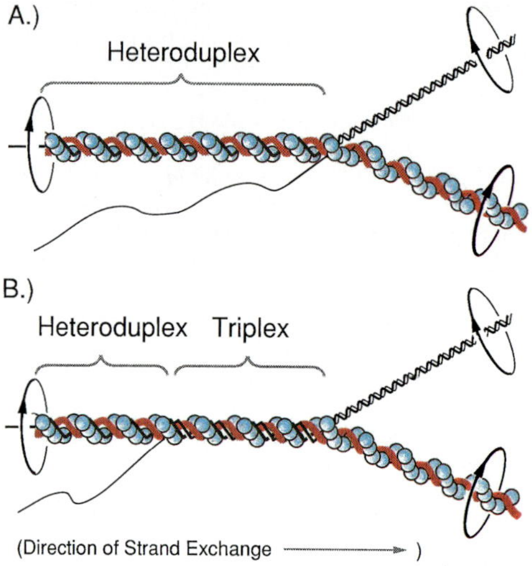

FIG. 4. **Alternative models of the RecA recombination fork.** *A*, by the simultaneous rotations indicated, a strand spools from the duplex DNA into the nucleoprotein filament while its complement simultaneously exits. *B*, the same rotary motions meld the helical duplex and the helical nucleoprotein filament, and one strand of the original duplex subsequently exits some distance away from the point of fusion.

A Right-handed Helical Nucleoprotein Filament

Unlike replication and transcription which require multiple different polypeptide chains with aggregate molecular masses between 10^5 and 10^6 daltons, the central events of homologous recombination, pairing, and strand exchange can be accomplished by a single polypeptide chain of M_r 37,800. This is so because RecA protein polymerizes on DNA to form a right-handed helical nucleoprotein filament that is specifically suited to the requirements of searching for homology and extensively exchanging helically intertwined strands. A single small polypeptide chain suffices because it can build a right-handed helix and because sequence specificity is governed not by a set of specialized polypeptides, but rather by the single-stranded (or double-stranded) DNA that comprises the core of the nucleoprotein filament.

Our understanding of the structure of RecA nucleoprotein filaments has come largely from electron microscopy (16, 32–36). Many of the observations and all of the analyses by image reconstruction have been made on filaments formed on double-stranded DNA in the presence of ATPγS, whose functional significance has not been demonstrated. Such analyses can be related to enzymatically active filaments formed on single-stranded DNA in the presence of ATP, because the isolation and characterization of active filaments have demonstrated a similar structure (16, 36, 37).

By contrast with the structure of chromatin, which may be characterized figuratively as DNA helically wrapped around beads of protein, the structure of the RecA presynaptic filament may be characterized as beads of protein helically wrapped around DNA. Single-stranded DNA, which lies well within the perimeter of the filament, is bound by its sugar-phosphate backbone with its bases exposed in a wide helical groove. The filament formed on double-stranded DNA in the presence of ATPγS, 115 Å in diameter with a pitch of 95 Å, has about six molecules of RecA protein per turn and 18.6 base pairs (33). A key feature of interest is the axial spacing of bases, which, in the active filament formed on single-stranded DNA in the presence of ATP, is 1.5 times greater than that between the bases of B-form DNA examined under the same conditions (36).

A second, more compressed form of the filament that is found in the absence of RecA protein appears to be inactive in forming joint molecules (38, 39).

Mutual Recognition by Two Helical Structures

Thus, in the pairing and exchange reactions driven by RecA protein, both participants in the homologous interaction are helical. The single-stranded partner is converted from a folded, disordered state to an extended, ordered form, which has a surprising feature; the spacing between bases in the presynaptic filament is greater than that in naked duplex DNA. How is recognition accomplished?

The first part of an answer is that the presynaptic filament has a second DNA binding site. The existence of this site was first inferred from observations on the striking colligative properties of the presynaptic filament, *i.e.* its ability to link DNA molecules together in large networks without regard to homology (40, 41). Binding studies involving either fixation or biophysical methods have further confirmed the existence of a second binding site (42, 43). Clearly, such a site should not bind the second molecule of DNA too tightly. In the limiting case of tight binding, the search for homology would be over after the first collision; and in fact, although the formation of nucleoprotein networks concentrates the DNA and speeds homologous pairing, it also limits the range of a search for homology. In this regard, the *in vitro* system seems a poor model for the efficient search that occurs during recombination *in vivo* (44).

The existence of a second binding site makes it easier to understand how the axial spacing of bases in the filament and duplex DNA are adjusted to test for complementarity. The protein may bind duplex DNA in a conformation that reduces twist locally and increases the spacing between bases. That view is supported by observations which reveal that a limited reduction in twist is associated with contacts between RecA presynaptic filaments and heterologous duplex DNA (45).[2] Indirectly, the stoichiometry of binding in the second site also supports this conclusion (42, 43).

Three and Four DNA Strands in Joints

The ability of RecA protein to pair a circular single strand and homologous circular duplex DNA (Fig. 1*d*) revealed that no DNA ends are required at all to form joints and thus signaled the existence of an interesting structure in which RecA protein holds three strands in apposition (16). Because these substrates lack any ends, they are topologically constrained; if the joint were a right-handed triplex, there would necessarily exist compensatory left-handed interwrapping of the duplex DNA and the nucleoprotein filament; these compensatory wraps could be loose solenoidal turns. Joints are also formed by circular single strands and linear duplex DNA with heterologous ends (Fig. 3*c*). The linearity of the duplex partner allows the potential formation of a right-handed triplex intermediate that has no topological need for compensatory left-handed wraps; and yet when paired with a circular single strand, either the circular duplex or the linear duplex with heterologous ends yields unstable joint molecules that dissociate promptly when RecA protein is removed (46, 47). Such joints, called *paranemic* joints, are thought to represent the first paired intermediate. Electron microscopic studies of paranemic joints, those formed by circular single strands plus linear duplex DNA with heterologous ends (as in Fig. 3*c*), revealed nucleoprotein structures encompassing three strands of DNA that were up to several thousand residues in length, although quite variable (48, 49).

Recent experiments from three laboratories have shown that homologous recognition and the formation of a joint can even occur by the direct interaction of naked superhelical DNA with a region of RecA nucleoprotein filament that contains duplex DNA. This was accomplished by making the RecA filament on circular duplex DNA with a single-stranded gap in a heterologous region (Fig. 1*e*). The requirement that the naked DNA be superhelical, as well as other observations, suggests that such joints are even less stable than those formed between circular single strands and circular duplex DNA (19–22). Instability notwithstanding, these represent paranemic four-stranded joints; the topology forbids any net swapping of strands (such as illustrated in Fig. 1, *b* and *c*), and homologous recognition occurs at sites where there are two intact duplex DNA molecules.

Strand Exchange by Two Helical Structures

The raison d'être of the helical nucleoprotein filament is seen most readily in strand exchange, which is the transfer of a strand from one helix to another. By analogy, a thread can be transferred from one spool to another by rotating both spools around their long axes or by moving one spool about the other (17, 50). Experiments designed to test these alternatives in strand exchange showed that duplex DNA and the RecA nucleoprotein filament each rotates about its long axis during strand exchange, a conclusion based on the effects of positive torsional stress and its release (50). One can imagine a neat exchange in which the two spools that we are considering have little contact at the *recombination fork* (Fig. 4*A*). Several electron microscopic studies, however, indicated a more complex situation in which naked duplex DNA was seen to enter the nucleoprotein filament at a certain point, and a single strand, variably coated with protein, exited at some other point, as indicated schematically in Fig. 4*B* (51, 52). The latter picture of the recombination fork was supported by observations which showed that the final physical separation of one strand from the nucleoprotein complex (*cf.* Fig. 4*B*) was delayed many minutes after the complete formation of nascent heteroduplex DNA and, moreover, that these penultimate intermediates, when deproteinized, were exceptionally thermostable and contained 1–2 kilobases of the nascent displaced strand in a form resistant to endonucleolytic and exonucleolytic digestion. Similar evidence of an extensive region of three-stranded DNA in deproteinized intermediates was found when the completion of strand exchange was blocked by a long heterologous region, as in Fig. 3*b* (25, 53).

The foregoing observations support the view first proposed in a model by Howard-Flanders *et al.* (54) that strand exchange involves the melding of duplex DNA with the RecA nucleoprotein filament to form helical regions that contain three or four strands of DNA (Fig. 4*B*).

Triplex DNA

In the reaction of a circular viral or plus single strand with linear duplex DNA, the ends of the duplex molecule may be identified as *proximal* or *distal* in relation to the known polarity of strand exchange; joints can be localized to either end or the middle of the duplex DNA by limiting the location of homology (Fig. 3, *b–d*). Medial or paranemic joints rapidly dissociate when they are deproteinized, whereas joints formed between completely homologous substrates leading to strand exchange and heteroduplex formation (Fig. 3*a*) become progressively more stable as a function of time (55). Since strand exchange is directional, extensive exchange should not occur in distal joints (Fig. 3*d*), a conclusion supported by both old and new data (56, 57). Two recent studies have shown, however, that deproteinized distal joints are surprisingly stable, about as stable as duplex DNA. Hsieh *et al.* (58) observed such stability in distal joints even when the region of homology was only 26 base pairs long. They contrasted this with the stability of branched intermediates made by thermal annealing, which dissociated at a temperature that was 30 °C lower. Rao *et al.* (59) directly compared distal joints with proximal joints, both of which were several kilobases in length, and found that the bulk of distal joints survived heating to 80 °C whereas half of proximal joints dissociated at 55 °C. A further comparison of the sensitivity of proximal and distal joints to nucleases showed that all three strands in distal joints lacked the sensitivity of single strands, whereas the proximal joints had the expected sensitivity of the branched structure shown in Fig. 3*a*. In both studies, the completeness of deproteinization was confirmed by sensitive immunological tests. The extra stability of distal joints has been interpreted as evidence of extra hydrogen bonds in a region of three-stranded DNA, as contrasted with the lower stability of branched proximal joints that are subject to dissociation by branch migration.

These observations are consistent with those which show that at all stages of pairing and strand exchange, three or four strands of DNA are apposed in a helical nucleoprotein structure. The observations on various deproteinized joints indicate further that pairing creates a stable triplex structure in natural DNA.

[2] E. Rould, K. Muniyappa, and C. Radding, unpublished observations.

The Role of ATP

Under the conditions of the recombination reactions, RecA protein hydrolyzes about 30 mol of ATP/min. Single-stranded DNA is an essential cofactor. The formation of an active presynaptic filament requires ATP, dATP, or ATPγS (17, 18). Since RecA protein hydrolyzes ATPγS extremely slowly, hydrolysis is clearly not required for homologous pairing (60). Under the usual and optimal conditions of strand exchange, the addition of ATPγS promptly halts further exchange (61), which led to the view that hydrolysis is required for strand exchange.

However, Menetski *et al.* (62) found that, over a narrow range of concentration of Mg^{2+}, around 4–5 mM, ATPγS supported apparent strand exchange involving the usual circular single-stranded and linear duplex substrates, but in this case exchange fell far short of completion. Rosselli and Stasiak (63) observed that a duplex oligonucleotide could undergo a complete exchange reaction with a RecA nucleoprotein filament in the presence of ATPγS, *i.e.* formation of a short heteroduplex segment and displacement (after deproteinization) of the other short chain. These authors (62, 63) suggested that the hydrolysis of ATP may be required to recycle RecA protein.

Nucleoprotein Filaments in Recombination

Although RecA protein appears to be universal in prokaryotes (18), an analog in eukaryotes that is clearly similar has not been identified: several eukaryotic pairing activities lack a requirement for ATP; none has been shown yet to form a nucleoprotein filament (64, 65, 67–70).

In the recombination reactions promoted by RecA protein, the prominence of nucleoprotein intermediates at every stage puts the molecular basis of homologous recombination in a new light. To the extent that the reactions of RecA protein are general, the creation of heteroduplex DNA is related to the properties of nucleoprotein filaments and may have little relation at all to spontaneous branch migration and the behavior in solution of branched DNA intermediates that are devoid of protein. Still further, branched intermediates in recombination must be resolved by other enzymes, which presumably encounter nucleoprotein filaments, not naked DNA (66). The recent solution of the crystal structure of RecA protein promises further progress in understanding how the nucleoprotein filament works.[3]

REFERENCES

1. Clark, A. J., and Margulies, A. D. (1965) *Proc. Natl. Acad. Sci. U. S. A.* **53,** 451–459
2. Walker, G. C. (1984) *Microbiol. Rev.* **48,** 60–93
3. Roberts, J. W., Roberts, C. W., and Craig, N. L. (1978) *Proc. Natl. Acad. Sci. U. S. A.* **75,** 4714–4718
4. Burckhardt, S. E., Woodgate, R., Scheuermann, R. H., and Echols, H. (1988) *Proc. Natl. Acad. Sci. U. S. A.* **85,** 1811–1815
5. Nohmi, T., Battista, J. R., Dodson, L. A., and Walker, G. C. (1988) *Proc. Natl. Acad. Sci. U. S. A.* **85,** 1816–1820
6. Shinagawa, H., Iwasaki, H., Kato, T., and Nakata, A. (1988) *Proc. Natl. Acad. Sci. U. S. A.* **85,** 1806–1810
7. Kobayashi, I., and Ikeda, H. (1978) *Mol. Gen. Genet.* **166,** 25–29
8. McEntee, K., Hesse, J. E., and Epstein, W. (1976) *Proc. Natl. Acad. Sci. U. S. A.* **73,** 3979–3983
9. Roberts, J. W., Roberts, C. W., Craig, N. L., and Phizicky, E. M. (1978) *Cold Spring Harbor Symp. Quant. Biol.* **43,** 917–920
10. Ogawa, T., Wabiko, H., Tsurimoto, T., Horii, T., Masukata, H., and Ogawa, H (1978) *Cold Spring Harbor Symp. Quant. Biol.* **43,** 909–915
11. Weinstock, G. M., McEntee, K., and Lehman, I. R. (1979) *Proc. Natl. Acad. Sci. U. S. A.* **76,** 126–130
12. Shibata, T., DasGupta, C., Cunningham, R. P., and Radding, C. M. (1979) *Proc. Natl. Acad. Sci. U. S. A.* **76,** 1638–1642
13. Cassuto, E., West, S. C., Mursalim, J., Conlon, S., and Howard-Flanders, P. (1980) *Proc. Natl. Acad. Sci. U. S. A.* **77,** 3962–3966
14. Tessman, E. S., and Peterson, P. K. (1985) *J. Bacteriol.* **163,** 688–695
15. Radding, C. M. (1982) *Annu. Rev. Genet.* **16,** 405–437
16. Radding, C. M. (1988) in *Genetic Recombination* (Kucherlapati, R., and Smith, G. R., eds) pp. 193–229, American Society for Microbiology, Washington, D. C.
17. Cox, M. M., and Lehman, I. R. (1987) *Annu. Rev. Biochem.* **56,** 229–262
18. Roca, A. I., and Cox, M. M. (1990) *CRC Crit. Rev.,* **25,** 415–456
19. Conley, E. C., and West, S. C. (1989) *Cell* **56,** 987–995
20. Conley, E. C., and West, S. C. (1990) *J. Biol. Chem.* **265,** 10156–10163
21. Lindsley, J. E., and Cox, M. M. (1990) *J. Biol. Chem.* **265,** 10164–10171
22. Chiu, S-K., Wong, B. C., and Chow, S. A. (1990) *J. Biol. Chem.* **265,** 21262–21268
23. Bianchi, M. E., and Radding, C. M. (1983) *Cell* **35,** 511–520
24. Pugh, B. F., and Cox, M. M. (1987) *J. Biol. Chem.* **262,** 1337–1343
25. Chow, S. A., Rao, B. J., and Radding, C. M. (1988) *J. Biol. Chem.* **263,** 200–209
26. Pugh, B. F., and Cox, M. M. (1987) *J. Biol. Chem.* **262,** 1326–1336
27. Muniyappa, K., Shaner, S. L., Tsang, S. S., and Radding, C. M. (1984) *Proc. Natl. Acad. Sci. U. S. A.* **81,** 2757–2761
28. Kowalczykowski, S. C., and Krupp, R. A. (1987) *J. Mol. Biol.* **193,** 97–113
29. Register, J. C., III, and Griffith, J. (1985) *J. Biol. Chem.* **260,** 12308–12312
30. Konforti, B. B., and Davis, R. W. (1987) *Proc. Natl. Acad. Sci. U. S. A.* **84,** 690–694
31. Konforti, B. B., and Davis, R. W. (1990) *J. Biol. Chem.* **265,** 6916–6920
32. Stasiak, A., and DiCapua, E. (1982) *Nature* **299,** 185–186
33. Stasiak, A., and Egelman, E. H. (1988) in *Genetic Recombination* (Kucherlapati, R., and Smith, G. R., eds) pp. 265–307, American Society for Microbiology, Washington, D. C.
34. Griffith, J. D., and Harris, L. D. (1988) *CRC Crit. Rev. Biochem.* **23,** Suppl. 1, S43–S86
35. Heuser, J., and Griffith, J. (1989) *J. Mol. Biol.* **210,** 473–484
36. Flory, J., Tsang, S. S., and Muniyappa, K. (1984) *Proc. Natl. Acad. Sci. U. S. A.* **81,** 7026–7030
37. Tsang, S. S., Muniyappa, K., Azhderian, E., Gonda, D. K., Radding, C. M., Flory, J., and Chase, J. W. (1985) *J. Mol. Biol.* **185,** 295–309
38. Williams, R. C., and Spengler, S. J. (1986) *J. Mol. Biol.* **187,** 109–118
39. Kahn, R., and Radding, C. M. (1984) *J. Biol. Chem.* **259,** 7495–7503
40. Rusche, J. R., Konigsberg, W., and Howard-Flanders, P. (1985) *J. Biol. Chem.* **260,** 949–955
41. Tsang, S. S., Chow, S. A., and Radding, C. M. (1985) *Biochemistry* **24,** 3226–3232
42. Takashi, M., Kubista, M., and Norden, B. (1989) *J. Mol. Biol.* **205,** 137–147
43. Müller, B., Koller, T., and Stasiak, A. (1990) *J. Mol. Biol.* **212,** 97–112
44. Honigberg, S. M., Rao, B. J., and Radding, C. M. (1986) *Proc. Natl. Acad. Sci. U. S. A.* **83,** 9586–9590
45. Iwabuchi, M., Shibata, T., Ohtani, T., Natori, M., and Ando, T. (1983) *J. Biol. Chem.* **258,** 12394–12404
46. Bianchi, M., DasGupta, C., and Radding, C. M. (1983) *Cell* **34,** 931–939
47. Riddles, P. W., and Lehman, I. R. (1985) *J. Biol. Chem.* **260,** 165–169
48. Umlauf, S. W., Cox, M. M., and Inman, R. B. (1990) *J. Biol. Chem.* **265,** 16898–16912
49. Bortner, C., and Griffith, J. (1990) *J. Mol. Biol.* **215,** 623–634
50. Honigberg, S. M., and Radding, C. M. (1988) *Cell* **54,** 525–532
51. Stasiak, A., Stasiak, A. Z., and Koller, T. (1984) *Cold Spring Harbor Symp. Quant. Biol.* **49,** 561–570
52. Register, J. C., Christiansen, G., and Griffith, J. (1987) *J. Biol. Chem.* **262,** 12812–12820
53. Rao, B. J., Jwang, B., and Radding, C. M. (1990) *J. Mol. Biol.* **213,** 789–809
54. Howard-Flanders, P., West, S. C., and Stasiak, A. (1984) *Nature* **309,** 215–220
55. Kahn, R., Cunningham, R. P., DasGupta, C., and Radding, C. M. (1981) *Proc. Natl. Acad. Sci. U. S. A.* **78,** 4786–4790
56. Wu, A. M., Kahn, R., DasGupta, C., and Radding, C. M. (1982) *Cell* **30,** 37–44
57. Dutreix, M., Rao, B. J., and Radding, C. M. (1991) *J. Mol. Biol.,* in press
58. Hsieh, P., Camerini-Otero, C. S., and Camerini-Otero, R. D. (1990) *Genes & Dev.,* **4,** 1951–1963
59. Rao, B. J., Dutreix, M., and Radding, C. M. (1991) *Proc. Natl. Acad. Sci. U. S. A.,* in press
60. Honigberg, S. M., Gonda, D. K., Flory, J., and Radding, C. M. (1985) *J. Biol. Chem.* **260,** 11845–11851
61. Cox, M. M., and Lehman, I. R. (1981) *Proc. Natl. Acad. Sci. U. S. A.* **78,** 3433–3437
62. Menetski, J. P., Bear, D. G., and Kowalczykowski, S. C. (1990) *Proc. Natl. Acad. Sci. U. S. A.* **87,** 21–25
63. Rosselli, W., and Stasiak, A. (1990) *J. Mol. Biol.* **216,** 335–352
64. Heyer, W-D., and Kolodner, R. D. (1989) *Biochemistry* **28,** 2856–2862
65. Hamatake, R. K., Dykstra, C. C., and Sugino, A. (1989) *J. Biol. Chem.* **264,** 13336–13342
66. Connolly, B., and West, S. C. (1990) *Proc. Natl. Acad. Sci. U. S. A.* **87,** 8476–8480
67. Hsieh, P., Meyn, M. S., and Camerini-Otero, R. D. (1986) *Cell* **44,** 885–894
68. Moore, S. P., and Fishel, R. (1990) *J. Biol. Chem.* **256,** 11108–11117
69. Kucherlapati, R., and Moore, P. D. (1988) in *Genetic Recombination* (Kucherlapati, R., and Smith, G. R., eds) pp. 575–595, American Society for Microbiology, Washington, D. C.
70. Kmiec, E. B., and Holloman, W. K. (1983) *Cell* **33,** 857–864
71. Holliday, R. (1964) *Genet. Res.* **5,** 282–304

[3] R. Story and T. Steitz, personal communication.

Minireview

THE JOURNAL OF BIOLOGICAL CHEMISTRY
Vol. 266, No. 10, Issue of April 5, pp. 6007–6010, 1991

Printed in U.S.A.

Structure, Expression, and Regulation of Protein Kinases Involved in the Phosphorylation of Ribosomal Protein S6*

Raymond L. Erikson

From the Department of Cellular and Developmental Biology, Harvard University, Cambridge, Massachusetts 02138

With current methods, our image of the events that occur between mitogen interaction at the plasma membrane and cell division is akin to visualizing a guitar based solely on Braque's or Picasso's cubist studies. The fragments are there, but some of us cannot recreate the original orientations. Nevertheless, great progress has been made in understanding the initial steps, with the elucidation of the structure of peptide growth factors and their receptor protein tyrosine kinases. The molecular events that link the initial activation of growth factor receptors to changes in mRNA levels, the initiation of DNA synthesis, and subsequent mitosis are still largely obscure. Filling this gap in animal cells has required the identification of substrates of the receptor protein tyrosine kinase activity that are physiologically significant in the regulation of cell division. Recently, many phosphotyrosine-containing proteins have been identified in cells with anti-phosphotyrosine serum; however, it will take a major effort to determine the nature of these proteins and their significance as substrates. As the result of other experimental approaches, several known proteins have been demonstrated to be phosphorylated on tyrosine residues following mitogen stimulation of cells, including, but not limited to phospholipase Cγ, the GTPase-activating protein, $p34^{cdc2}$, and MAP[1] kinase (1–7). It is not within the scope of this communication to critically review these data.

Because assessment of the significance of tyrosine phosphorylation on a protein of unknown function is a daunting task, we and others have elected to study a biochemical marker in mitogen-stimulated cells, namely phosphorylation on serine residues of the 40 S ribosomal subunit protein S6. S6 phosphorylation was first demonstrated during liver regeneration (8), and numerous studies have shown that S6 is multiply phosphorylated on serine residues as the result of stimulation of a variety of cells by several mitogenic agents and oncogene products (9–13). S6 is also phosphorylated when oocytes enter meiotic cell division in response to progesterone or insulin (14). Although the phosphorylation of S6 is a highly conserved response to signals that stimulate cell proliferation, the actual function of phosphorylated S6 during protein synthesis is unclear. There is a possibility, but no strong evidence, that phosphorylation of 40 S subunits may result in more efficient initiation/translation of mRNAs in general or permit the selective translation of certain mRNAs (15, 16). Although no clarification of this issue is readily at hand, study of the regulation of the relevant protein kinases is of interest, since it may provide new insights about signal transduction and initiation of cell division. The regulation of these enzyme activities must take into account the fact that the initial events involve separate, unrelated stimuli: tyrosine phosphorylation by oncogene products or growth factor receptors, activation of protein kinase C by tumor-promoting phorbol ester, or, in the case of progesterone, a steroid hormone receptor linked to the adenylate cyclase system (17–23). Recent progress on the relevant protein kinases in this research area will be summarized in this article. Although there is evidence for the regulation of phosphoprotein phosphatases responsible for S6 dephosphorylation in mitogen-stimulated cells (24, 25), limited space does not permit review of those data.

Structure and Expression of S6 Kinases

S6KI/S6KII and pp65–70—Two major classes of S6 kinases, the RSK (for ribosomal S6 kinase) family containing S6KI and S6KII, and pp65–70, have been purified and partially characterized over the past several years. The first member of the RSK family, a M_r = 92,000 protein kinase, was purified from unfertilized *Xenopus* eggs and designated S6KII for the second peak of activity from the initial column used in its purification (26, 27). The first peak, S6KI, is related to S6KII and has a M_r = 90,000 (28). The amino acid sequences of tryptic peptides from S6KII were used to design oligonucleotide probes to obtain molecular clones from *Xenopus* cDNA libraries. The sequences of the cDNAs obtained predict a M_r = 83,000 protein kinase with the unusual feature of two apparent catalytic domains; the N-terminal half of the molecule is related to the catalytic subunit of cAMP-dependent protein kinase, whereas the C-terminal half is related to the catalytic subunit of phosphorylase *b* kinase (29). The role of each potential catalytic domain in phosphotransferase reactions is unclear. Although the catalytic subunits of protein kinase A and phosphorylase *b* kinase have associated regulatory proteins, there is no evidence for a regulatory subunit associated with S6KI or -II.

The RSK structure is highly conserved in other species; cDNA clones predict similar enzymes expressed in chicken, mice, and humans (30). Polyclonal antiserum to purified *Xenopus* S6KII immunoprecipitates S6 kinase activity from stimulated cultured cells, and antibody raised against recombinant *Xenopus* RSK protein detects in several species a M_r = 90,000 protein kinase that is structurally related to S6KII (21, 31–33). Therefore, the activities originally identified as S6KI and -II represent a family of protein kinases that have been highly conserved through evolution. Such conservation, it is commonly argued, indicates that a protein is crucial for orderly cellular behavior.

Although these data indicate that S6 kinases of M_r = 90,000 are present in mammalian and avian cells, purification of S6 kinase from these sources has yielded enzymes in the M_r = 65,000–70,000 range, denoted pp70 (34–39). An enzyme from this class of S6 kinases has recently been sequenced and cloned independently by two groups (40, 41). The cDNAs predict a protein kinase of M_r = 59,000 with a single catalytic domain 57% identical to the N terminus of the RSK gene product. Both S6KII and pp70 exhibit lower mobility on sodium dodecyl sulfate gels than predicted by the cDNA sequences. The larger apparent M_r values of the purified enzymes result in part from changes in migration because of phosphorylation during activation of the enzymes, as discussed below. The cDNAs for the pp70 enzyme predict that the C-terminal domain contains a 25-amino acid basic sequence with 5 serine/threonine residues. This sequence may be sufficiently related to the C terminus of the S6 substrate itself to represent a pseudosubstrate site (40) and may prevent substrate-enzyme interaction and/or ATP binding. Phosphorylation of this region of pp70 by an as yet unidentified mitogen-activated protein kinase would presumably relieve the constraints of this inhibitory domain, leading to activation of the enzyme. The relationship of S6 kinases and other enzymes is illustrated in Fig. 1.

To date, no pp70 S6 kinase has been reported to be in *Xenopus* oocytes or unfertilized eggs. In view of the highly conserved nature of S6KI and -II and of S6 phosphorylation, this is surprising.

* Research in my laboratory is supported by National Institutes of Health Grant CA42580 and by an American Cancer Society research professorship. The costs of publication of this article were defrayed in part by the payment of page charges. This article must therefore be hereby marked "*advertisement*" in accordance with 18 U.S.C. Section 1734 solely to indicate this fact.

[1] The abbreviation used is: MAP, microtubule-associated protein.

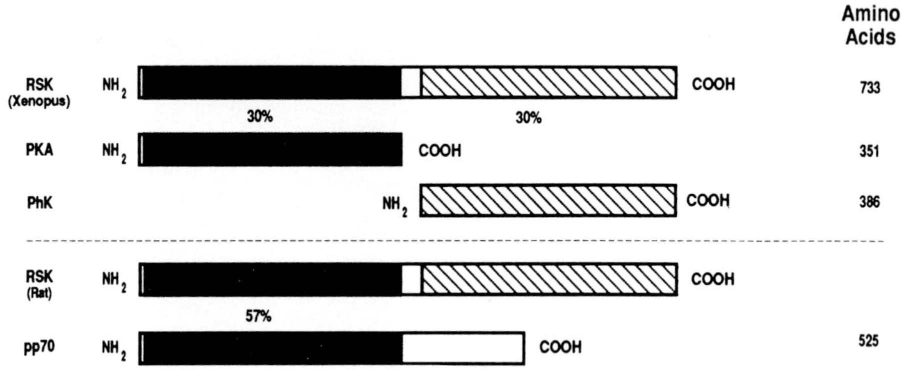

FIG. 1. **S6 kinases compared with other protein kinases.** The sequence of RSK (*Xenopus*) is compared with that of the catalytic subunits of bovine cAMP-dependent protein kinase (*PKA*) and bovine phosphorylase *b* kinase (*PhK*). The predicted sequence of $pp90^{RSK}$ is also highly similar to the catalytic domain of protein kinase C (29). The relationship of RSK (rat) to the catalytic domain of pp70 (rat) is shown (40). The percent identity is given in each case.

Examination of other amphibian tissues may reveal such an enzyme. The availability of cDNA and antibody for pp70 should facilitate detection of an amphibian pp70 gene product.

Relationship to Other Enzymes/Other Substrates—The presence of two catalytic domains in the RSK gene products suggests they have the potential to phosphorylate substrates other than S6. Protein kinase A and phosphorylase *b* kinase recognize the general site Arg-Arg-*X*-Ser and Arg-*X*-*X*-Ser, respectively. *In vitro*, S6KII phosphorylates the sequence Arg-*X*-*X*-Ser in several proteins (42) and phosphorylates lamin C at one of the sites phosphorylated *in vivo* (43). S6KII also phosphorylates the M_r = 160,000 (G) subunit of protein phosphatase 1, predominantly at site 1, the site phosphorylated *in vivo* in response to insulin (44). The physiological significance of the phosphorylation of substrates other than S6 is uncertain, because many protein kinases phosphorylate substrates *in vitro* that are not substrates *in vivo*. These observations suggest additional studies. For example, the activation of the RSK gene product, peaking within 10–15 min (see below), resembles that of the expression of genes such as *myc* and *fos*, which are regarded as important in the transition of the cell cycle (45, 46). It is known that cAMP and protein kinase A influence gene expression (47), and the RSK gene product may also act on the same substrates.

In this regard it seems relevant that two distantly related RSK genes with decidedly different modes of expression have been identified in mice. One is expressed abundantly in the intestine, a tissue undergoing rapid cell proliferation, but is expressed at a much lower level in the brain or heart, whereas a second gene shows a converse mode of expression (30). Since brain and heart are not regarded as undergoing active proliferation, the expression of some members of the RSK family may play a functional role not normally associated with cell division. These data suggest either that the RSK gene product has other significant substrates in the brain and heart or that in these tissues S6 phosphorylation has a role unrelated to cell proliferation. The substrate specificity of the RSK gene product appears to be somewhat broader than that thus far reported for the pp70 class of enzymes.

Regulation of Enzyme Activity

Activation in Vivo by Peptide Growth Factors/Phorbol Esters/ $pp60^{v\text{-}src}$—Initial characterization of soluble growth factor-stimulated S6 kinase activity in cultured cells demonstrated that full recovery of kinase activity required the presence of phosphatase inhibitors in the extraction buffer, suggesting that the enzyme(s) may be activated by phosphorylation (18). S6 kinase activation closely correlated with increased S6 phosphorylation *in vivo*, supporting the notion that the enzyme(s) being measured were indeed responsible for the phosphorylation of S6 in these cells. In these experiments quiescent cells were stimulated with agents known to act through receptors with protein tyrosine kinase activity. S6 kinase activity is also stimulated by the oncogene product $pp\text{-}60^{v\text{-}src}$, a protein tyrosine kinase, and by phorbol ester (17), an agent acting through the serine-threonine-specific protein kinase C. These observations raised questions about the number of pathways that lead to S6 phosphorylation. Since growth factor receptors stimulate production of the second messengers diacylglycerol and inositol trisphosphate (48), a variety of stimuli has the potential to mediate activities through protein kinase C. No conclusive results on the physiologically significant activation of protein kinase C under these conditions have been reported, however. S6 kinase activation by serum or $pp60^{v\text{-}src}$ can be dissociated from phorbol ester-stimulated activity in cells in which protein kinase C has been down-regulated (49, 50). Initial experiments did not distinguish the contribution of the two classes of S6 kinases, but more recent studies[2] show that both protein kinase C-dependent and -independent pathways may lead to the activation of both pp70 and $pp90^{RSK}$. Therefore at least two mechanisms can regulate S6 kinase(s); one utilizes protein kinase C, whereas the other(s) does not.

Two Distinct Enzymes Display S6 Kinase Activity in Cultured Cells—The stimulation of S6 kinase activity in Swiss 3T3 cells treated with epidermal growth factor follows a biphasic time course (51). There is an early phase of activity at 10–15 min, followed by a short period of decline and then a late, prolonged phase beginning after 30–60 min. These authors present data showing that a single peak of enzyme activity is detected upon Mono Q chromatography at both early and late times and conclude that the pp70 class of enzyme is responsible for all the S6 kinase activity in both phases. A time-dependent, biphasic activation of S6 kinase activity is also observed in chicken embryo fibroblasts in response to serum stimulation (32, 33). Studies with anti-RSK antiserum showed that $pp90^{RSK}$ is a major contributor to the early phase (5–15 min) of the response and, to a lesser extent, to the later phase. Partial purification of the S6-phosphorylating activity from cells at different times following stimulation resolved two distinct enzymes of M_r = 90,000 and 65,000. The activity of these enzyme fractions supported the conclusions regarding the relative contributions of each protein kinase at the various times. The M_r = 65,000 enzyme is apparently the chicken homolog of the pp70 class of S6 kinases. Thus, in contrast to 3T3 cells, in chicken embryo fibroblasts both the $pp90^{RSK}$ and the pp70 S6 kinases are activated but with distinct kinetics. A diagram to illustrate these events is presented in Fig. 2. It is unclear why no $pp90^{RSK}$ was detected in Swiss 3T3 cells (51); however, the RSK gene product is detected in another Swiss 3T3 line and in HeLa cells (33).

Both Classes of S6 Kinases Are Phosphorylated on Serine and Threonine and Are Inactivated by Phosphatases—Treatment of activated pp70 (52) or S6KII (53) with phosphatase 1 or 2A, which act primarily on phosphoserine and phosphothreonine, inactivates their capacity to phosphorylate S6. Also, the two enzyme classes become highly phosphorylated during activation in metabolically labeled cells. Phosphoamino acid analysis of S6KII from *Xenopus* oocytes and of its homologue from cultured cells, and of pp70 from serum-stimulated cells, revealed radiolabel predominantly in phosphoserine, with significantly less in phosphothreonine (28, 31, 54, 55). Thus, both classes of S6 kinase are activated by phosphorylation of serine/threonine residues, but since neither class yields phosphotyrosine they are unlikely to be directly activated by protein tyrosine kinases.

The RSK gene product expressed by recombinant baculovirus in insect cells is phosphorylated and activated by co-infection

[2] J. Blenis, personal communication.

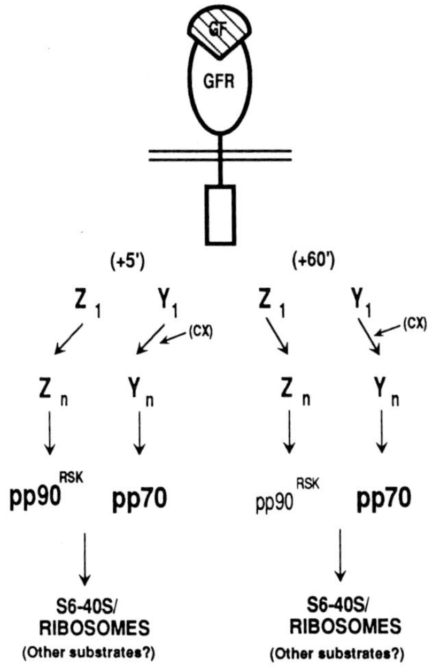

FIG. 2. **Activation of ribosomal S6 kinases by two independent pathways.** The activation of $pp90^{RSK}$ and pp70 by two independent pathways in cultured cells is illustrated. *Y* and *Z* represent putative protein kinases, and the *multiple arrows* indicate that the total number of steps involved is uncertain. It is also assumed all steps leading to pp90 and pp70 activation involve phosphorylation. The reduced size of $pp90^{RSK}$ at 60′ indicates that it is not a major contributor to S6 phosphorylation at that point. Although completely independent pathways are shown, the possibility remains that when all the components are identified and characterized, there will be steps in common. *GF*, growth factor; *GFR*, growth factor receptor; *CX*, cycloheximide.

with $pp60^{v\text{-}src}$-expressing virus (56). In this system only a fraction of the $pp90^{RSK}$ is activated, with only the most highly phosphorylated form showing increased capacity to phosphorylate S6 in 40 S subunits. Partial purification of the unactivated and activated enzyme and determination of specific activity indicate activation of at least 4000-fold. This degree of activation is far greater than that observed in the total population of enzyme under commonly employed cell culture conditions and changes our view of the total degree of activation possible for an individual molecule. The significance of such a vast potential for altering enzyme activity *in vivo* is an important question.

Both classes of S6 kinases undergo autophosphorylation *in vitro* without concomitant increased enzyme activity. The capacity to autophosphorylate closely parallels the capacity to phosphorylate an exogenous substrate, suggesting that autophosphorylation has no functional significance. Peptide maps of the enzymes metabolically labeled during activation *in vivo* are more complex than those of the autophosphorylated enzymes (28, 53, 55). It is likely some of the additional sites are of importance for activation of enzyme activity.

Cycloheximide Activates pp70 but Not pp90—During their early studies on S6 phosphorylation in regenerating liver, Gressner and Wool (57) noted that puromycin- or cycloheximide-treated animals also showed increased S6 phosphorylation. Similar results have been obtained in cultured cells. The finding that inhibitors of protein synthesis stimulate S6 phosphorylation, a process associated with increased protein synthesis and cell division, is an apparent paradox. Activated S6 kinase purified from rat liver after cycloheximide treatment appears in all respects to be identical to the M_r = 70,000 enzyme activated in cultured cells (55, 58).

As mentioned above, anti-S6KII antiserum is able to immunoprecipitate enzyme activity, due to the presence of $pp90^{RSK}$, from serum-stimulated cells. Treatment with phorbol ester or activation of $pp60^{v\text{-}src}$ also elevated activity measured in the immunocomplex. In contrast, in parallel experiments cycloheximide failed to elevate immunocomplex-associated S6 kinase activity, whereas direct assays revealed substantial activation in cell extracts.[3] Cycloheximide added together with mitogen did not inhibit the stimulation of $pp90^{RSK}$ activity as measured in the immunocomplex. These data indicate that in somatic cells pp70, but not $pp90^{RSK}$, is activated by cycloheximide and suggest that distinct pathways are operative. Furthermore, these data suggest that since S6 becomes phosphorylated *in vivo* after cycloheximide treatment, activation of pp70 is sufficient and $pp90^{RSK}$ is unnecessary for a full physiological response. Conversely, in *Xenopus* oocytes, activation of $pp90^{RSK}$ may be sufficient for phosphorylation of the S6 in the entire complement of 10^{12} ribosomes within each cell.

In contrast to the results with cycloheximide in cultured fibroblasts, evidence has been published showing antiserum raised against purified S6KII immunoprecipitates S6 kinase activity from *Xenopus* oocytes treated with protein synthesis inhibitors (59). In these experiments no evidence was presented regarding the molecular nature of the enzymes responsible, making interpretation of the observations uncertain.

S6 Kinase Kinases—The determination that both classes of S6 kinases are phosphorylated on serine and threonine but yield no detectable phosphotyrosine after metabolic labeling in mitogen-stimulated cells indicates they are not direct substrates for protein tyrosine kinases. Assuming they are activated solely by a phosphorylation mechanism, there must be at least one intervening kinase to transduce the tyrosine kinase signal. The protein kinase originally denoted MAP-2 (for microtubule-associated protein-2) kinase has emerged as a strong candidate for this role in one case. It is phosphorylated on both threonine and tyrosine after metabolic labeling in insulin-stimulated cells and thus is likely to be a protein tyrosine kinase substrate (6). Moreover, MAP kinase has been reported to be inactivated by dephosphorylation of either phosphothreonine or phosphotyrosine (60). As yet the enzyme(s) responsible for either the tyrosine or threonine phosphorylation of MAP kinase is unidentified. Although the threonine phosphorylation could be the result of autophosphorylation, the data argue against this possibility, since autophosphorylation of purified MAP kinase has not been reported. Therefore, a protein threonine kinase must also function in this pathway. A partial cDNA clone that may encode a MAP kinase has been obtained, although no evidence on the biochemical nature of the gene product was presented (61).

MAP kinase has been successfully used to partially reactivate phosphatase-inactivated S6KII *in vitro*, further supporting its claim as an intervening kinase between mitogen stimulus and activation of S6 kinase (53). An important unresolved question in these studies concerns whether the sites phosphorylated in S6KII by MAP kinase are the same as those phosphorylated *in vivo* during activation. The phosphopeptide map of S6KII radiolabeled *in vivo* is more complex than that of S6KII phosphorylated by MAP kinase *in vitro*, possibly indicative of the activity of additional S6 kinase kinases in oocytes (28, 53). Analysis of phosphorylation sites should resolve this issue.

Thus far attempts to reactivate the phosphatase-treated pp70 class of enzyme with MAP kinase have failed (55). The results on cycloheximide activation, together with the inability to activate the pp70 class with MAP kinase *in vitro*, support the notion of an independent pathway for the stimulation of this class of S6 kinase. Presumably cycloheximide-treated cells will yield a pp70 kinase kinase.

Reprise and Offing

The immediate goals appear to be identifying protein kinases other than MAP kinase that activate the two classes of S6 kinases. Strategies such as those used by Ahn and Krebs (62), whereby fractions from stimulated cells are assessed for their ability to activate kinase(s) in preparations from unstimulated cells, are under way in several laboratories. To date, *in vitro* activation has provided humble scraps of encouragement in the best of circumstances, but appropriate combinations of enzymes can be anticipated to yield better results. Identification of the sites of phosphorylation in each class of enzyme may give some

[3] E. Erikson and J. Blenis, unpublished data.

insight about the mechanism that leads to increased protein kinase activity. Perhaps, in the distant future, the three-dimensional structure of the active and inactive form of each enzyme will reveal more details of the mechanism of activation by reversible protein phosphorylation. The nature of the threonine and the tyrosine kinases responsible for activating MAP kinase is of major interest, as is the nature of putative analogues in the pathway to pp70 activation. The pp70 pathway also promises to reveal a novel mode of protein kinase activation involving inhibition of protein synthesis. Since the components of these activation pathways are major contributors to serine phosphorylation in animal cells, it causes us to consider that each enzyme may have substrates other than those identified thus far.

Although both classes of S6 kinases were originally purified based on their capacity to specifically phosphorylate S6 in 40 S ribosomal subunits *in vitro* and are active at times consistent with an *in vivo* role in S6 phosphorylation, one cannot conclude with certainty that S6 phosphorylation is their actual function in signal transduction or cell proliferation. The studies on the pp90 class of enzymes suggest a capacity to influence reversible phosphorylation of other substrates. Nevertheless, since there are redundant pathways for activation of these enzymes, it is not unlikely that cells have also evolved redundance in the enzymes that mediate the final step.

Acknowledgments—I thank Eleanor Erikson, John Blenis, and Manfred Lohka for their comments on the numerous versions of this communication, and Laurie Scott for her patience and effort in transforming the manuscript into its final form.

REFERENCES

1. Ellis, C., Moran, M., McCormick, F. & Pawson, T. (1990) *Natue* **343,** 377
2. Gould, K. L. & Nurse, P. (1989) *Nature* **342,** 39–45
3. Kaplan, D. R., Morrison, D. K., Wong, G., McCormick, F. & Williams, L. T. (1990) *Cell* **61,** 125–133
4. Margolis B, Rhee, S. G., Felder, S., Mervic, M., Lyall, R., Levitzki, A., Ullrich, A., Zilberstein, A. & Schlessinger, J. (1989) *Cell* **57,** 1101–1107
5. Meisenhelder J., Suh, P.-G., Rhee, S. G. & Hunter, T. (1989) *Cell* **57,** 1109
6. Ray, L. B. & Sturgill, T. W. (1988) *Proc. Natl. Acad. Sci. U. S. A.* **85,** 3753–3757
7. Wahl, M. I., Olashaw, N. E., Shunzo, N., Rhee, S. G., Pledger, W. J. & Carpenter, G. (1989) *Mol. Cell. Biol.* **9,** 2934–2943
8. Gressner, A. M. & Wool, I. G. (1974) *J. Biol. Chem.* **249,** 6917–6925
9. Decker, S. (1981) *Proc. Natl. Acad. Sci. U. S. A.* **78,** 4112–4115
10. Martin-Perez, J., Siegmann, M. & Thomas, G. (1984) *Cell* **36,** 287–294
11. Blenis, J. & Erikson, R. L. (1984) *J. Virol.* **50,** 966–969
12. Trevillyan, J. M., Kulkarni, R. K. & Byus, C. V. (1984) *J. Biol. Chem.* **259,** 897–902
13. Blenis, J., Spivack, J. G. & Erikson, R. L. (1984) *Proc. Natl. Acad. Sci. U. S. A.* **81,** 6408–6412
14. Nielsen, P. J., Thomas, G. & Maller, J. L. (1982) *Proc. Natl. Acad. Sci. U. S. A.* **79,** 2937–2941
15. Duncan, R. & McConkey, E. H. (1982) *Eur. J. Biochem.* **123,** 535–538
16. Thomas, G., Martin-Perez, J., Siegmann, M. & Otto, A. M. (1982) *Cell* **30,** 235–242
17. Blenis, J. & Erikson, R. L. (1985) *Proc. Natl. Acad. Sci. U. S. A.* **82,** 7621
18. Novak-Hofer, I. & Thomas, G. (1985) *J. Biol. Chem.* **260,** 10314–10319
19. Pelech, S. L., Olwin, B. B. & Krebs, E. G. (1986) *Proc. Natl. Acad. Sci. U. S. A.* **83,** 5968–5972
20. Smith, C. J., Rubin, C. S. & Rosen, O. M. (1980) *Proc. Natl. Acad. Sci. U. S. A.* **77,** 2641–2645
21. Erikson, E., Stefanovic, D., Blenis, J., Erikson, R. L. & Maller, J. L. (1987) *Mol. Cell. Biol.* **7,** 3147–3155
22. Martin-Perez, J., Rudkin, B. B., Siegmann, M. & Thomas, G. (1986) *EMBO J.* **5,** 725–731
23. Cicirelli, M. F., Pelech, S. L. & Krebs, E. G. (1988) *J. Biol. Chem.* **263,** 2009–2019
24. Olivier, A. R., Ballou, L. M. & Thomas, G. (1988) *Proc. Natl. Acad. Sci. U. S. A.* **85,** 4720–4724
25. Chan, C. P., McNall, S. J., Krebs, E. G. & Fischer, E. H. (1988) *Proc. Natl. Acad, Sci. U. S. A.* **85,** 6257–6261
26. Erikson, E. & Maller, J. L. (1985) *Proc. Natl. Acad. Sci. U. S. A.* **82,** 742
27. Erikson, E. & Maller, J. L. (1986) *J. Biol. Chem.* **261,** 350–355
28. Erikson, E. & Maller, J. L. (1989) *J. Biol. Chem.* **264,** 13711–13717
29. Jones, S. W., Erikson, E., Blenis, J., Maller, J. L. & Erikson, R. L. (1988) *Proc. Natl. Acad. Sci. U. S. A.* **85,** 3377–3381
30. Alcorta, D. A., Crews, C. M., Sweet, L. J., Bankston, L., Jones, S. W. & Erikson, R. L. (1989) *Mol. Cell. Biol.* **9,** 3850–3859
31. Sweet, L. J., Alcorta, D. A., Jones, S. W., Erikson, E. & Erikson, R. L. (1990) *Mol. Cell. Biol.* **10,** 2413–2417
32. Sweet, L. J., Alcorta, D. A. & Erikson, R. L. (1990) *Mol. Cell. Biol.* **10,** 2787–2792
33. Chen, R.-H. & Blenis, J. (1990) *Mol. Cell. Biol.* **10,** 3204–3215
34. Tabarini, D., Garcia de Herreros, A., Heinrich, J. & Rosen, O. M. (1987) *Biochem. Biophys. Res. Commun.* **144,** 891–899
35. Blenis, J., Kuo, C. & Erikson, R. L. (1978) *J. Biol. Chem.* **262,** 14373
36. Jeno, P., Ballou, L. M. & Thomas, G. (1988) *Proc. Natl. Acad. Sci. U. S. A.* **85,** 406–410
37. Jeno, P., Jaggi, N., Luther, H., Siegmann, M. & Thomas, G. (1989) *J. Biol. Chem.* **264,** 1293–1297
38. Price, D. J., Nemenoff, R. A. & Avruch, J. (1989) *J. Biol. Chem.* **264,** 13825–13833
39. Gregory, J. S., Boulton, T. G., Sang, B.-C. & Cobb, M. H. (1989) *J. Biol. Chem.* **264,** 18397–18401
40. Banerjee, P., Ahmad, M. F., Grove, J. R., Kozlosky, C., Price, D. J. & Avruch, J. (1990) *Proc. Natl. Acad. Sci. U. S. A.* **87,** 8550–8554
41. Kozma, S. C., Ferrari, S., Bassand, P., Siegmann, M., Totty, N. & Thomas, G. (1990) *Proc. Natl. Acad. Sci. U. S. A.* **87,** 7365–7369
42. Erikson, E. & Maller, J. (1988) *Sec. Mess. Phosphoprot.* **12,** 135–143
43. Ward, G. E. & Kirschner, M. W. (1990) *Cell* **61,** 561–577
44. Dent, P., Lavionne, A., Nakielny, S., Caudwell, F. B., Watt, P. & Cohen, P. (1990) *Nature* **348,** 302–308
45. Greenberg, M. E. & Ziff, E. B. (1984) *Nature* **311,** 433–438
46. Kelly, K., Cochran, B. H., Stiles, C. D. & Leder, P. (1983) *Cell* **35,** 603–610
47. Roesler, W. J., Vandenbark, G. R. & Hanson, R. W. (1988) *J. Biol. Chem.* **263,** 9063–9066
48. Berridge, M. J. (1987) *Annu. Rev. Biochem.* **56,** 159–193
49. Blenis, J. & Erikson, R. L. (1986) *Proc. Natl. Acad. Sci. U. S. A.* **83,** 1733
50. Pelech, S. & Krebs, E. (1987) *J. Biol. Chem.* **262,** 11598–11606
51. Susa, M., Olivier, A. R., Fabbro, D. & Thomas, G. (1989) *Cell* **57,** 817–824
52. Ballou, L. M., Jeno, P. & Thomas, G. (1988) *J. Biol. Chem.* **263,** 1188
53. Sturgill, T. W., Ray, L. B., Erikson, E. & Maller, J. L. (1988) *Nature* **334,** 715–718
54. Ballou, L. M., Siegmann, M. & Thomas, G. (1988) *Proc. Natl. Acad. Sci. U. S. A.* **85,** 7154–7158
55. Price, D. J., Gunsalus, J. R. & Avruch, J. (1990) *Proc. Natl. Acad. Sci. U. S. A.* **87,** 7944–7948
56. Vik, T. A., Sweet, L. J. & Erikson, R. L. (1990) *Proc. Natl. Acad. Sci. U. S. A.* **87,** 2685–2689
57. Gressner, A. M. & Wool, I. G. (1974) *Biochem. Biophys. Res. Commun.* **60,** 1482–1490
58. Kozma, S. C., Lane, H. A., Ferrari, S., Luther, H., Siegmann, M. & Thomas, G. (1989) *EMBO J.* **8,** 4125–4132
59. Stefanovic, D. & Maller, J. L. (1988) *Exp. Cell Res.* **179,** 104–114
60. Anderson, N. G., Maller, J. L., Tonks, N. K. & Sturgill, T. W. (1990) *Nature* **343,** 651–653
61. Boulton, T. G., Yancopoulos, G. D., Gregory, J. S., Slaughter, C., Moomaw, C., Hsu, J. & Cobb, M. H. (1990) *Science* **249,** 64–67
62. Ahn, N. G. & Krebs, E. G. (1990) *J. Biol. Chem.* **265,** 11495–11501

Minireview

THE JOURNAL OF BIOLOGICAL CHEMISTRY
Vol. 266, No. 11, Issue of April 15, pp. 6659-6662, 1991

DNA Topoisomerases: Why So Many?*

James C. Wang

From the Department of Biochemistry and Molecular Biology, Harvard University, Cambridge, Massachusetts 02138

Several new DNA topoisomerases have been discovered recently. In the yeast *Saccharomyces cerevisiae*, a gene has been identified to encode a protein homologous to eubacterial DNA topoisomerase I; this gene has been termed *TOP3* and its product DNA topoisomerase III (1). Sequence data suggest that a yeast gene *HPR1* may encode yet another DNA topoisomerase (2, 3). In *Escherichia coli*, two genes *parC* and *parE* have been found to code for a new topoisomerase termed DNA topoisomerase IV; the amino acid sequences of the *parC* and *parE* polypeptides are homologous to, respectively, those of the A and B subunit of DNA gyrase, also termed bacterial DNA topoisomerase II (4).

These new members of the DNA topoisomerase family are not merely stand-ins or substitutes of their more extensively studied relatives, because mutations in their structural genes effect distinct phenotypes. Yeast *top3* mutants are characterized by poor growth and higher frequency of recombination between a class of repetitive sequences termed δ sequences (1); *E. coli parC* temperature-sensitive mutants are known (5) and *parC* or *parE* mutants exhibit defects in the segregation of newly replicated chromosomes (4, 5). It is the functional distinctiveness of these new enzymes that leads to the question in the title of this review, why are there so many DNA topoisomerases?

Supercoiling and Relaxation of Intracellular DNA

Two decades ago, when the first DNA topoisomerase was discovered in extracts of *E. coli* cells (6), DNA replication topped the list of potential biological processes that might require such an enzyme. The separation of DNA strands in one region should lead to overwinding of the DNA in other regions; the bacterial enzyme was shown, however, to relax preferentially underwound or negatively supercoiled DNA (6). In 1976, a second DNA topoisomerase was discovered in *E. coli* (7). This enzyme, termed DNA gyrase or bacterial DNA topoisomerase II, is both a DNA topoisomerase and a DNA-dependent ATPase, and it couples ATP hydrolysis and DNA negative supercoiling. The enzymatic properties of DNA gyrase immediately led to a number of suggestions regarding its physiological roles. First, unlike *E. coli* DNA topoisomerase I, gyrase is capable of removing positive supercoils in the presence of ATP and thus appears to be more suitable in solving the problem of DNA overwinding during semiconservative replication. Second, the negative supercoiling of DNA by DNA gyrase hints strongly that DNA supercoiling might occur inside a living cell. Furthermore, because the actions of DNA gyrase and DNA topoisomerase I are diametric, they might form an opposing pair in the regulation of the degree of supercoiling of intracellular DNA (8–10).

The idea that DNA gyrase and DNA topoisomerase I regulate DNA supercoiling *in vivo* is supported by genetic evidence that the lethal phenotype of *E. coli topA* null mutants is compensated by mutations in *gyrA* or *gyrB* that reduce gyrase activity (11–13). Several experimental findings could not be accounted for, however, by the model. In 1983, Lockshon and Morris (14) found that inhibition of DNA gyrase in *E. coli* harboring pBR322 resulted in the accumulation of positive supercoils in the plasmid. It made no sense why inhibiting a negative supercoiling activity should lead to positive supercoiling. Pruss (15) also made an interesting finding in 1985 that pBR322 isolated from *E. coli topA* mutants is about twice as negatively supercoiled as the same plasmid from $topA^+$ strains, but samples of a derivative of pBR322 isolated from $topA^-$ and $topA^+$ strains show little difference in supercoiling. Subsequently, Pruss and Drlica (16) showed that the difference in supercoiling of pBR322 or its derivatives in $topA^-$ and $topA^+$ strains is strongly dependent on the transcription of the region of the plasmid encoding resistance to the drug tetracycline. Why should the degree of supercoiling of a plasmid depend on the expression of a gene?

The possibility that transcription might lead to the supercoiling of the DNA template under specific conditions was first raised in 1985 (17) prior to the experimental findings of Pruss and Drlica (16). In 1987, Liu and Wang (18) proposed a more general model to account for the results of Lockshon and Morris (14) as well as those of Pruss and Drlica (16). The essence of the model is illustrated in Fig. 1. According to this model, two oppositely supercoiled domains may accompany transcription or other processes involving the tracking of a macromolecular assembly along a DNA. The degree of supercoiling of a particular region of intracellular DNA is determined by how fast the supercoils are being generated by the tracking process, how fast they can be removed by the DNA topoisomerases, and how fast the oppositely supercoiled domains can neutralize each other through diffusional pathways (18). The last factor is probably important in the compensation of *topA* mutations by *gyrA* or *gyrB* mutations in *E. coli.*

The twin supercoiled domain model of transcription provides a rationale for the existence of two seemingly diametric DNA topoisomerases in eubacteria, DNA gyrase and topoisomerase I. Rather than countering the actions of each other, these two enzymes actually act jointly to relax the two oppositely supercoiled domains that may form during transcription, or more generally in processes involving the tracking of a macromolecular assembly along a DNA. Each of the two topoisomerases has evolved beautifully to accomplish its mission. The specificity of eubacterial DNA topoisomerase I for the negatively supercoiled domain is achieved by its unpairing of a short stretch of DNA in forming the active enzyme-DNA complex (6, 19, 20); in contrast, the right-handed wrapping of a DNA segment around DNA gyrase (21) causes the enzyme to preferentially bind to the positively supercoiled domain. In cells undergoing active transcription, gyrase might act as an enzyme that specifically removes positive supercoils rather than as one that introduces negative supercoils (18).

This division of labor between the two bacterial topoisomerases is, of course, not the only way of relaxing oppo-

* The bulk of the work on DNA topoisomerases in the author's laboratory has been supported by grants from the National Institutes of Health (GM24544 and its predecessor GM14621; CA47958).

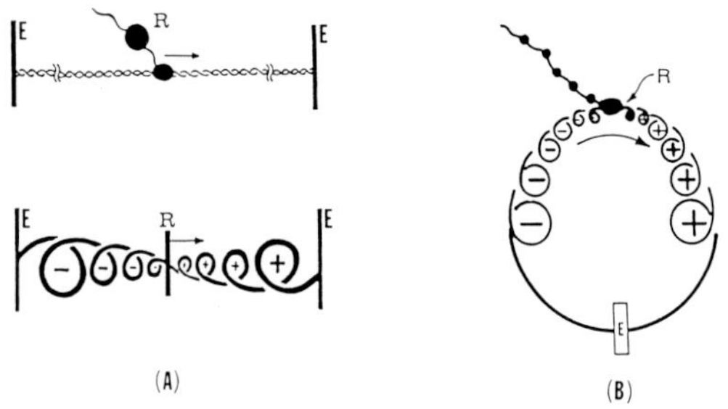

FIG. 1. *A*: *top*, a transcriptional ensemble R is shown on a DNA segment, the ends of which are anchored to a certain cellular entity E. As R tracks along the DNA, it would be expected to encircle around the template because of the helical geometry of DNA. If this encircling motion is prevented, through anchoring of any component of R to a cellular entity or the DNA itself, or because of a high viscous drag accompanying the movement of R in a cellular milieu, then the DNA must turn around its helical axis instead. Rotation of the DNA would in turn generate positive supercoils ahead of R and negative supercoils behind it. *Bottom*, the net effect of preventing the turning of R around the DNA is to squeeze all helical turns ahead of R into a shorter and shorter region as R advances, and the opposite occurs behind R. *B*, the anchoring of the ends of the DNA segment to E, depicted in *A*, is not an essential feature of the model. Here the terminal elements are combined to illustrate the situation for a plasmid; E could be non-existent, in which case the viscous drag against the rotation of the DNA itself between the oppositely supercoiled domains would provide the only retarding force against the merging of these domains; in the other extreme, E could be an immobile site on the cell membrane. Drawings are adopted from Refs. 18 and 26.

sitely supercoiled domains. In eukaryotes, the entrance of a new enzyme and the evolution of an old one have demonstrated clearly that it is possible for one enzyme to relax both supercoiled domains. Eukaryotic DNA topoisomerase I, which shows little sequence homology to either eubacterial DNA topoisomerase I or DNA gyrase and thus represents a new family of the topoisomerase clan, is well known for its relaxation of negatively and positively supercoiled domains with nearly equal efficiency (22); eukaryotic DNA topoisomerase II, which is homologous to bacterial DNA gyrase, similarly relaxes both oppositely supercoiled domains (23). While the bacterial type II enzyme binds preferentially to a positively rather than negatively supercoiled region due to the right-handed wrapping of a DNA segment around the enzyme, this DNA wrap is absent in the DNA-eukaryotic type II enzyme complex (23); probably as a consequence, the eukaryotic enzyme binds to oppositely supercoiled domains without an intrinsic bias.

Are All DNA Topoisomerases Involved in the Relaxation of Supercoiled Domains in Vivo?

Giaever and Wang (24) argue that in eukaryotes, DNA topoisomerases I and II are the only activities that can relax supercoiled intracellular DNA efficiently. In their experiments, an endogenous yeast plasmid, termed the 2μ plasmid, was shown to become positively supercoiled only in cells expressing *E. coli* DNA topoisomerase I and only when both yeast DNA topoisomerases I and II were inactivated. If there is a third yeast enzyme capable of relaxing positive supercoils efficiently, there should be no accumulation of positive supercoils in those experiments; if there is a third yeast enzyme capable of relaxing negative supercoils efficiently, the accumulation of positive supercoils in the plasmid should not depend on the expression of the *E. coli* enzyme.

In *E. coli*, Bliska and Cozzarelli (25) showed that intracellular plasmids become less negatively supercoiled upon inhibition of DNA gyrase by Norfloxacin in a *topA*$^+$ but not in a *topA*$^-$ strain. Thus it appears that DNA topoisomerase I is the major activity that relaxes negatively supercoiled DNA *in vivo*. Furthermore, because of the accumulation of positive supercoils in plasmids upon inhibition of DNA gyrase by Novobiocin (14, 26), gyrase appears to be the only enzyme capable of relaxing positive supercoils unless the other relaxation activities in *E. coli* are also inhibited by this drug. If DNA topoisomerases I and II in yeast, and DNA topoisomerase I and DNA gyrase in *E. coli* are the dominant activities in the relaxation of supercoiled domains *in vivo*, what are the functions of the other DNA topoisomerases?

DNA Topoisomerases and the Unraveling of Intertwined DNA Strands or Duplex DNA Pairs

As mentioned earlier in this review, in terms of the physiological roles of DNA topoisomerases the well known "swivel" problem for the semiconservative replication of two intertwined DNA chains dominated the thinking in the early years. From the molecular point of view, there are actually two distinct topological problems in the replication of DNA. The first accompanies the elongation stage of DNA replication and can be viewed as a special case of the twin supercoiled domain model involving the tracking of a macromolecular assembly along a DNA. As the replication fork advances, positive supercoils are generated ahead of it. Behind the fork, the separated parental strands can be viewed as a limiting case of negative supercoiling; separated strands represent the highest degree of negative supercoiling achievable in a duplex DNA. In *E. coli*, as described earlier, DNA gyrase would be the best candidate to fulfill the role of removing the positive supercoils, in agreement with experimental findings (14, 26). In yeast, either DNA topoisomerase I or II should do, which is again supported by experimental data (27).

The second problem occurs near the end of the elongation step and is illustrated in Fig. 2 for a pair of converging replication forks. Depending on the relative rate of unraveling

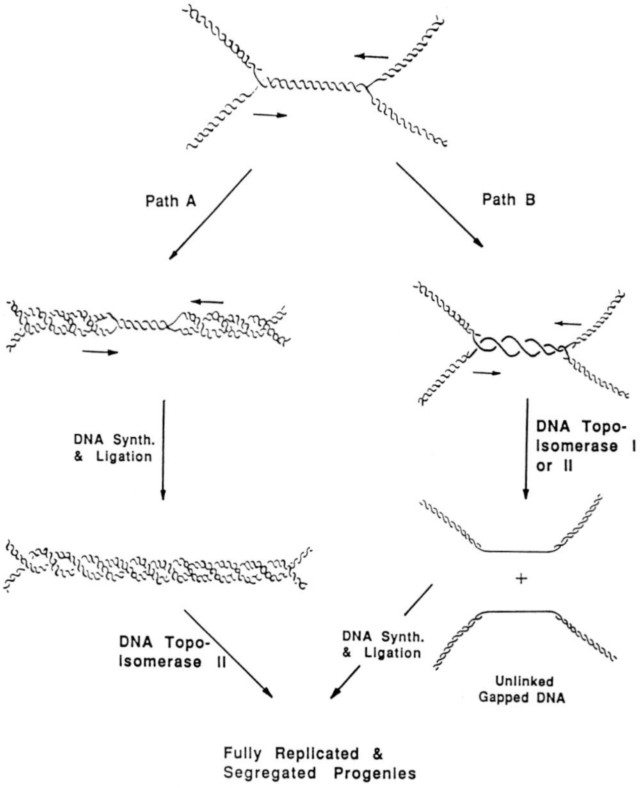

FIG. 2. **Two paths for the merging of a pair of converging replication forks.** This drawing was modified from Ref. 43.

the intertwined parental strands, which requires a DNA topoisomerase, and that for the completion of progeny strand synthesis, which involves the polymerizing machinery and DNA ligase, two extreme events may occur; in Path A, unraveling is slow and the replicating DNA ends up as a pair of multiply intertwined duplex molecules (Ref. 43; catenanes if the original DNA is in the form of a ring); in Path B, completion of progeny strand synthesis is slow, and a pair of gapped but unlinked progeny molecules is formed.

From the known enzymatic properties of type I and type II DNA topoisomerases, a type II enzyme is needed to unlink the intertwined duplex molecules that are formed in Path A. In the yeasts *S. cerevisiae* and *Schizosaccharomyces pombe*, Path A is apparently a major route near the end of replication, and the essentiality of DNA topoisomerase II rests most likely on its role in the unlinking of the intertwined duplex molecules (28). In *E. coli*, the Par or partition-defective phenotype, namely the formation of large nucleoids in the midcell, has been observed for both gyrase and DNA topoisomerase IV mutants, suggesting that both type II DNA topoisomerases are necessary for this step (4, 5, 29). Mutations in the *E. coli minB* locus have also been shown to affect nucleoid segregation and plasmid supercoiling, and the possibility that the *minB* products might be a topoisomerase or interacting with a topoisomerase involved in chromosomal partitioning has been raised (30). Why more than one type II DNA topoisomerase is needed for chromosomal segregation is unclear. In addition to binding to DNA, *E. coli* DNA topoisomerase IV has also been implicated to interact with the cell membrane (4, 5). The ways the various topoisomerases interact with the cell membrane or other cellular entities might underlie the requirement of multiple topoisomerases in the segregation of chromosomes.

Path A is unlikely to be the only path, however, and Path B provides an alternate route for segregating the pairs of progeny molecules (for examples, see Refs. 31–33). In Path B, either a type I or a type II DNA topoisomerase may unlink the intertwined structure. *E. coli* DNA topoisomerase I, for example, can link or unlink DNA containing single-stranded regions efficiently (9, 22). *E. coli* DNA topoisomerase III, as well as *E. coli* DNA topoisomerase I, has been shown to efficiently unlink gapped plasmid DNA near the end of a round of replication *in vitro* (31, 32).

It should be emphasized that although Paths A and B are parallel pathways, the use of type I DNA topoisomerases in Path B cannot salvage the lethal effect of blocking Path A through the inactivation of the type II DNA topoisomerases. This follows from chromosomal loss or breakage, which would occur if the intertwined DNA intermediate depicted in Path A is not unlinked before the partition of the chromosomes.

In *E. coli*, it has been suggested that DNA topoisomerases I and III may participate in Path B (31, 32). Similarly, the poor growth phenotype of yeast DNA topoisomerase III might be related to its role in such a path (1, 3). Inactivation of the type I enzymes does not necessarily lead to cell killing, however, because the type II enzymes can substitute for them in this path or provide an alternative path, namely Path A.

DNA Topoisomerases and Genome Stability

An exciting recent finding in the study of DNA topoisomerases is their involvement in the maintenance of genome stability. Christman *et al.* (34) found that yeast *top1* null mutants or *top2 ts* mutants at semipermissive temperatures exhibit a much higher frequency of recombination in the ribosomal DNA gene cluster than their wild-type controls and suggested that both DNA topoisomerases I and II are required to suppress recombination within the cluster. Kim and Wang (35) found that in yeast *Δtop1 top2 ts* cells grown at a permissive temperature, the rDNA gene cluster is unstable and excision of the genes occurs to form extracellular rDNA rings; expression of either DNA topoisomerase I or increasing the level of DNA topoisomerase II in this strain restores the stability of the chromosomal rDNA cluster, and the extrachromosomal rings are shown to reintegrate into their usual locus on the chromosome. As mentioned earlier, null mutations in the yeast *TOP3* gene increase the frequency of recombination between the δ sequences (1).

Although it has been suggested that the increase in recombination frequency in *top* mutants might be a result of these mutations on DNA supercoiling, this interpretation is only straightforward for the *Δtop1 top2 ts* double mutant results of Kim and Wang (35); the strain they used is completely devoid of DNA topoisomerase I, and even at a permissive temperature there is only a low level of DNA topoisomerase II. Thus, neither of the two major supercoil relaxation activities is present at adequate levels under their experimental conditions, and template supercoiling in the actively transcribed rDNA gene cluster may very well occur. The results of Christman *et al.* (34) with the *top1* cells can be fitted into the same scheme, as DNA topoisomerase I is the most potent relaxation activity in eukaryotic cells. Their results with *TOP1⁺ top2 ts* cells cannot be readily explained in the same way, however, as the presence of DNA topoisomerase I should be sufficient to solve the supercoiling problem. Because the experiments with the *TOP1⁺ top2 ts* strain were carried out at semipermissive temperatures, these results might be more complicated than those obtained with *top1 TOP2⁺* cells, and the molecular mechanisms underlying the increase in recombination in these two cases might not be the same.

The effects of *TOP3* mutations on recombination in particular are difficult to account for by the supercoiling explanation. *In vivo* and *in vitro*, the *TOP3* gene product is at best a rather weak relaxation activity. One could invoke DNA sequence specificity to argue that DNA topoisomerase III might be a better relaxation activity for certain DNA sequences, but this argument does not circumvent the difficulty of explaining the *top3* mutant phenotypes in the presence of other strong relaxation activities, namely DNA topoisomerases I and II. It is this dilemma that led Wang *et al.* (3) to postulate that there might be cellular processes that would normally minimize mitotic recombination by resolving inadventitiously paired DNA strands. Conceivably, DNA topoisomerases that are inefficient in relaxing supercoiled regions but efficient in unlinking intertwined strands, such as yeast or *E. coli* DNA topoisomerase III, might be participating in these processes.

DNA Topoisomerases and Chromosomal Folding

Those who have used a long garden hose would appreciate the problem of coiling such an object into an ordered form. Even in the early years of DNA topoisomerase studies, the plausible involvement of such an enzyme in chromosomal condensation and decondensation was recognized (36). Recently, genetic and cytological studies with *S. pombe* (37), and *in vitro* studies with cell extracts (38) have implicated a role of eukaryotic DNA topoisomerase II in chromosomal condensation and decondensation.

From the point of view of DNA topology, either eukaryotic DNA topoisomerase I or II should be sufficient to overcome the problems of twisting and coiling of a long DNA into an organized and compact structure (DNA topoisomerase I would not be able to substitute for DNA topoisomerase II only if interlocking of the DNA loops in the compact structure is necessary). One explanation of the DNA topoisomerase II requirement in the presence of DNA topoisomerase I is a structural role of the type II enzyme itself in the organization of the condensed chromosome. Immunomicroscopy has implicated eukaryotic DNA topoisomerase II as a major component of the chromosome scaffold (39–42).

Conclusion

One could view the ascendency of the DNA topoisomerases in nature as a consequence of the selection of double-stranded DNA as the genetic material. A plethora of problems accompanying the various transactions involving DNA is all deeply rooted in the bihelical geometry of this master molecule. Tracking of a macromolecular assembly along a DNA, in transcription or replication for example, may generate supercoiled domains that must be attended to. The unraveling of the complementary DNA strands near the end of a round of replication, and the formation of multiply intertwined double-stranded progeny molecules that may occur during this process, pose unique topological problems. Folding of a chromatin fiber into a compact form again poses problems that must be solved. The list can go on.

All the problems cited above would have disappeared if DNA existed as short linear pieces. Nature, however, has apparently decided that a bound volume is more easily manageable than many loose sheets. The DNA topoisomerases provide some of the necessary tools in the various manipulations of the long threadlike double-stranded DNA. It also seems likely that during the long history of evolution, the DNA topoisomerases, through their manipulation of DNA topology, have been assigned additional roles in the optimization of the intracellular state of DNA. DNA supercoiling, for example, has profound effects on DNA structure and interactions between DNA and other molecules; there is also ample evidence that DNA supercoiling affects many cellular processes.

As enzymes, the DNA topoisomerases are magicians among magicians; they open and close gates in DNA without leaving a trace, and they enable two DNA strands or duplexes to pass each other as if the physical laws of spatial exclusion do not exist. Because the biological functions of the DNA topoisomerases are deeply rooted in the double helix structure of DNA, it should not be surprising that these enzymes participate in nearly all biological processes involving DNA; the recent discovery of several new DNA topoisomerases has brought a deeper understanding of their many vital roles in living cells. Why are there so many DNA topoisomerases? The simple answer is that the selection of long double-stranded DNA as the genetic material has set the stage for their entrance.

Acknowledgment—I thank all my present and past co-workers for making the study of DNA topoisomerases a joyful undertaking.

REFERENCES

1. Wallis, J. W., Chrebet, G., Brodsky, G., Rolfe, M., and Rothstein, R. (1989) *Cell* **58,** 409–419
2. Aguilera, A., and Klein, H. L. (1990) *Mol. Cell. Biol.* **4,** 1439–1451
3. Wang, J. C., Caron, P. R., and Kim, R. A. (1990) *Cell* **62,** 403–406
4. Kato, J.-I., Nishimura, Y., Imamura, R., Niki, H., Hiraga, S., and Suzuki, H. (1990) *Cell* **63,** 393–404
5. Kato, J.-I., Nishimura, Y., Yamada, M., Suzuki, H., and Hirota, Y. (1988) *J. Bacteriol.* **170,** 3967–3977
6. Wang, J. C. (1971) *J. Mol. Biol.* **55,** 523–533
7. Gellert, M., Mizuuchi, K., O'Dea, M., and Nash, H. (1976) *Proc. Natl. Acad. Sci. U. S. A.* **73,** 3872–3876
8. Gellert, M. (1981) *Annu. Rev. Biochem.* **50,** 879–910
9. Wang, J. C. (1985) *Annu. Rev. Biochem.* **54,** 665–697
10. Maxwell, A., and Gellert, M. (1986) *Adv. Protein Chem.* **38,** 69–107
11. DiNardo, S., Voelkel, K., Sternglanz, R., Reynolds, A. E., and Wriight, A. (1982) *Cell* **31,** 43–51
12. Pruss, G. J., Manes, S. H., and Drlica, K. (1982) *Cell* **31,** 35–42
13. Raji, A., Zabel, D. J., Laufer, C. S., and Depew, R. E. (1985) *J. Bacteriol.* **162,** 1173–1179
14. Lockshon, D., and Morris, D. R. (1983) *Nucleic Acids Res.* **11,** 2999–3017
15. Pruss, G. J. (1985) *J. Mol. Biol.* **185,** 51–63
16. Pruss, G. J., and Drlica, K. (1986) *Proc. Natl. Acad. Sci. U. S. A.* **83,** 8952–8956
17. Wang, J. C. (1985) in *Interrelationship Among Aging, Cancer and Differentiation, Jerusalem Symposium on Quantum Chemistry and Biochemistry* (Pullman, B., Ts'o, P. O. P., and Schneider, E. L., eds) Vol. 18, pp. 173–181, D. Reidel, Holland
18. Liu, L. F., and Wang, J. C. (1987) *Proc. Natl. Acad. Sci. U. S. A.* **84,** 7024–7027
19. Kirkegaard, K., Pflugfelder, G., and Wang, J. C. (1984) *Cold Spring Harbor Symp. Quant. Biol.* **49,** 411–419
20. Kirkegaard, K., and Wang, J. C. (1985) *J. Mol. Biol.* **185,** 625–637
21. Liu, L. F., and Wang, J. C. (1978) *Cell* **15,** 979–984
22. Champoux, J. J. (1990) in *DNA Topology and Its Biological Effects* (Cozzarelli, N. R., and Wang, J. C., eds) pp. 217–242, Cold Spring Harbor Laboratory, Cold Spring Harbor, NY
23. Hseih, T. (1990) in *DNA Topology and Its Biological Effects* (Cozzarelli, N. R., and Wang, J. C., eds) pp. 243–263, Cold Spring Harbor Laboratory, Cold Spring Harbor, NY
24. Giaever, G. N., and Wang, J. C. (1988) *Cell* **55,** 849–856
25. Bliska, J. B., and Cozzarelli, N. R. (1987) *J. Mol. Biol.* **194,** 205–218
26. Wu, H. Y., Shyy, S.-H., Wang, J. C., and Liu, L. F. (1988) *Cell* **53,** 433–440
27. Kim, R. A., and Wang, J. C. (1989) *J. Mol. Biol.* **208,** 257–267
28. Yanagida, M., and Wang, J. C. (1987) in *Nucleic Acids and Molecular Biology* (Eckstin, F., and Lilley, D. M. J., eds) Vol. 1, pp. 196–209, Springer-Verlag, Berlin
29. Steck, T. R., and Drlica, K. (1984) *Cell* **36,** 1081–1088
30. Mulder, E., El'Bouhali, M., Pas, E., and Woldringh, C. L. (1990) *Mol. & Gen. Genet.* **221,** 87–93
31. Minden, J. S., and Marians, K. J. (1986) *J. Biol. Chem.* **261,** 11906–11917
32. DiGate, R. J., and Marians, K. J. (1988) *J. Biol. Chem.* **263,** 13366–13373
33. Weaver, D. T., Fields-berry, S. C., and Depamphilis, M. L. (1985) *Cell* **41,** 565–575
34. Christman, M. F., Dietrich, F. S., and Fink, G. R. (1988) *Cell* **55,** 413–425
35. Kim, R. A., and Wang, J. C. (1989) *Cell* **57,** 975–985
36. Baase, W. A., and Wang, J. C. (1974) *Biochemistry* **13,** 4299–4303
37. Uemura, T., Ohkura, H., Adachi, Y., Morino, K., Shiozaki, K., and Yanagida, M. (1987) *Cell* **50,** 917–925
38. Newport, J. (1987) *Cell* **48,** 205–217
39. Earnshaw, W. C., and Heck, M. M. S. (1985) *J. Cell Biol.* **100,** 1716–1725
40. Gasser, S. M., and Laemmli, U. K. (1986) *Cell* **46,** 521–530
41. Cockerill, P. N., and Garrard, W. T. (1986) *Cell* **44,** 273–282
42. Boy de la Tour, E., and Laemmli, U. K. (1988) *Cell* **55,** 937–944
43. Varshavsky, A., Levinger, L., Sundin, O., Barsoum, J., Ozkaynak, E., Swerdlow, P., and Finley, D. (1983) *Cold Spring Harbor Symp. Quant. Biol.* **47,** 511–528

Minireview

THE JOURNAL OF BIOLOGICAL CHEMISTRY
Vol. 266, No. 12, Issue of April 25, pp. 7313–7316, 1991

Printed in U.S.A.

Tumor Necrosis Factor

NEW INSIGHTS INTO THE MOLECULAR MECHANISMS OF ITS MULTIPLE ACTIONS*

Jan Vilček and Tae H. Lee

From the Department of Microbiology and Kaplan Cancer Center, New York University Medical Center, New York, New York 10016

The tumor necrosis factor (TNF)[1] "family" includes two structurally and functionally related proteins, TNF-α or cachectin (1, 2) and TNF-β or lymphotoxin (3, 4). TNF-α is produced mainly by monocytes and/or macrophages, whereas TNF-β is a product of lymphoid cells. The close relationship between these two proteins was not known until 1984 when cloning of the cDNAs for human TNF and lymphotoxin unexpectedly revealed that they are about 30% homologous at the amino acid level (5, 6). TNF-α and TNF-β bind to the same cell surface receptors (7), and they are very similar (though not identical) in the spectra of their activities. Nevertheless, the regulation of expression of the TNF-α and TNF-β genes and the processing of the two corresponding proteins by the producing cells are completely different.

Originally thought of as selective anti-tumor agents, TNF-α and TNF-β (especially the former) are now grouped among the major "inflammatory cytokines," *i.e.* they are characteristically produced at the sites of inflammation by infiltrating mononuclear cells. They play a beneficial role as immunostimulants and important mediators of host resistance to many infectious agents and, probably, malignant tumors (8, 9). On the other hand, TNF-α turned out to be identical to cachectin, postulated to mediate wasting during chronic infections (2). There is increasing evidence that overproduction of TNF-α during infections leads to severe systemic toxicity and even death, *e.g.* TNF-α is a major factor in the development of septic shock following infection with Gram-negative bacteria (10). TNF-α also has been implicated in the pathogenesis of some autoimmune disorders (11) and of graft-*versus*-host disease (12).

Among the hallmarks of TNF-α and TNF-β is the extremely pleiotropic nature of their actions. This multiplicity of actions can be ascribed to the following: (*a*) TNF receptors are present on virtually all cells examined, (*b*) TNF action leads to the activation of multiple signal transduction pathways, kinases, and transcription factors, and (*c*) TNF action leads to the activation of an unusually large array of cellular genes.

TNF-α and TNF-β Genes: Regulation of Expression

The TNF-α and TNF-β genes are single copy genes, closely linked within the cluster of major histocompatibility complex (MHC) genes, located on the short arm of human chromosome 6 (13, 14) and murine chromosome 17 (15) at the boundary of the class III and class I MHC regions (Table I). In all species examined (human, murine, and rabbit), the TNF-β gene is always 5′ to the TNF-α genes, the TNF-α and TNF-β genes are each approximately 3 kilobase pairs long, and each gene consists of four exons and three introns (Fig. 1; reviewed in Refs. 16 and 17). These similarities strongly suggest that the two genes were derived from a common ancestral gene by gene duplication. Whereas the organization of the TNF-α and TNF-β genes and their coding regions (especially the fourth exons, which code for 80–89% of the mature proteins) show a high degree of homology, little similarity is seen between the 5′-flanking regions of the two genes that are thought to contain most of the elements responsible for the regulation of transcription (13, 18).

A comparison of cell types shown to be capable of producing TNF-α or TNF-β mRNA or protein upon appropriate stimulation shows that TNF-α is inducible in a wider variety of cells than is TNF-β (Table II). Some cells can be stimulated to make both TNF forms, but it appears that different signals and different pathways account for TNF-α and TNF-β induction in a single cell (17, 26). The 5′-flanking region of the murine TNF-α gene, linked to the chloramphenicol acetyltransferase gene, was responsive to lipopolysaccharide (LPS) in cultured murine bone marrow macrophages (18). Gene deletion analysis and gel retardation assays suggested that several sequence motifs with close homology to the κB enhancer and a MHC class II-like "Y box" (Fig. 1) are involved in the LPS-induced transcriptional activation of the TNF-α gene. Nonetheless, a recent study showed that deletion of three of these κB-like sites from the TNF-α promoter had little influence on the induction of the gene by LPS or by Sendai virus (19). Several κB-like sequences have also been identified in the TNF-β genes of man, mouse, and rabbit (17, 20). However, little is known about how these and other enhancer sequences may cooperate in the tissue-specific inducible activation of the TNF-β gene.

In addition to transcriptional activation, there is evidence for the regulation of the TNF-α gene at the level of translation. Translational activation by LPS in a line of murine macrophages was shown to be regulated by the UA-rich sequence in the 3′-untranslated region of the human TNF-α mRNA (31). This region in the TNF mRNA strongly suppressed the translation of a reporter coding sequence to which it was attached. The suppressive effect of UA-rich sequences on translation was also demonstrated with IFN-β mRNA (32). Earlier, the UA-rich sequence, conserved in the 3′-untranslated region of many cytokine genes, was implicated in the regulation of mRNA stability (33).

Secretion of the TNF-α and TNF-β Proteins

Mature TNF-α consists of 157 amino acids (156 in the rat, rabbit, or mouse), whereas TNF-β contains 171 residues (34). Under denaturing conditions, the molecular weights of human TNF-α and TNF-β were approximately 17,000 and 25,000, respectively (35, 36). In addition to the 14 additional amino acids present at the amino terminus of mature TNF-β, the difference in size can be accounted for by the absence of *N*-glycosylation in human TNF-α, as opposed to the presence of one *N*-glycosylation site in TNF-β. Some other features of TNF-α and TNF-β are compared in Table I. In native form, both TNF-α and TNF-β are tightly packed trimers (37, 38), which is in agreement with the earlier molecular weight estimates of approximately 45,000 and 65,000 for TNF-α and TNF-β, respectively (21, 37, 38).

TNF-α and TNF-β differ strikingly in the mode of their secretion. The 34-amino acid long presequence of TNF-β shows characteristics typical for signal peptide sequences of secretory proteins, *i.e.* it is very hydrophobic and has charged residues near the amino terminus (13). These features suggest that the TNF-β polypeptide is processed and released from cells in a manner characteristic for most secreted proteins. In contrast, TNF-α has a 76-residue long precursor sequence containing both hydrophobic and hydrophilic regions. There is evidence that this long presequence serves to anchor the TNF-α precursor polypeptide in the plasma membrane (39–41). The 17-kDa released form of TNF-α is derived from the 26-kDa integral transmembrane TNF-

* Work done in this laboratory was supported by National Institutes of Health Grant CA-47304 and American Cancer Society Grant CD-477.

[1] The abbreviations used are: TNF, tumor necrosis factor; MHC, major histocompatibility complex; LPS, lipopolysaccharide; PKC, protein kinase C; PKA, protein kinase A; NF, nuclear factor; IFN, interferon; IL, interleukin; EGF, epidermal growth factor; IRF, interferon regulatory factor; NGF, nerve growth factor.

TABLE I

Comparison of some properties of human and murine TNF-α and TNF-β

Feature	TNF-α (cachectin)	TNF-β (lymphotoxin)
Chromosomal location	Chr. 6 (human) Chr. 17 (murine)	Chr. 6 (human) Chr. 17 (murine)
Gene organization	3 introns	3 introns
Amino acids in mature protein	157 (human) 156 (murine)	171 (human) 169 (murine)
Amino acids in presequence	76 (human) 79 (murine)	34 (human) 33 (murine)
Cysteine residues[a]	2 (human) 2 (murine)	None (human) 1 (murine)
Methionine residues[a]	None	3
N-Glycosylation	No (human) Yes (murine)	Yes (human) Yes (murine)
Released via secretory pathway	No	Yes
Quaternary structure	Trimer	Trimer

[a] In mature protein.

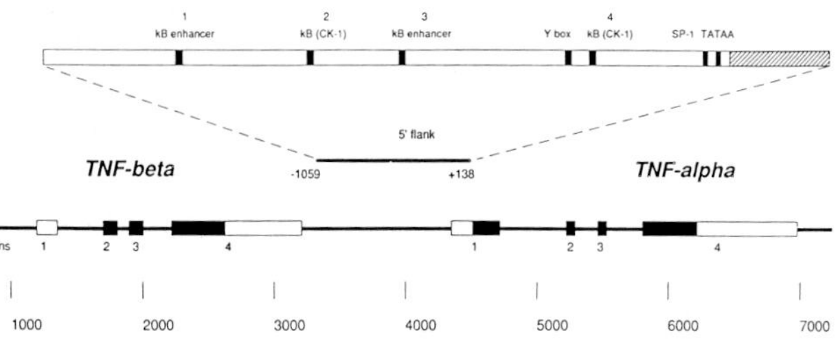

FIG. 1. **Structure of the murine TNF locus.** The coding portions of the exons are *shaded.* The *inset* shows location of elements of possible significance in transcriptional regulation identified in the 5′-flanking region of the TNF-α gene. Reproduced from Shakhov *et al.* (18) by copyright permission of the Rockefeller University Press.

TABLE II

Different types of cells demonstrated to be capable of producing TNF-α and/or TNF-β mRNA or protein upon appropriate stimulation

Cell type	Inducibility of TNF-α	Inducibility of TNF-β
Monocytes or macrophages	21[a]	
T cells (CD4+)	22	23
T cells (CD8+)	22	24
B cells	25	25
LAK cells	17	
NK cells	22	
Neutrophils	26	
Astrocytes	27	27
Endothelial cells	28	
Smooth muscle cells	29	
Miscellaneous nonhematopoietic tumor cell lines	16, 30	

[a] Numbers correspond to references listed at the end of this article. Blank spaces in the TNF-β column indicate that no evidence for inducibility is available.

α molecule by proteolytic cleavage (39), possibly involving a serine protease (42).

The hydrophobic region between residues −44 and −26 in the presequence represents the transmembrane domain, and the amino-terminal residues −76 to −50 may constitute the intracytoplasmic portion of the TNF-α precursor polypeptide. The region between residues −14 and −1 may constitute a "stem" from which the mature form of TNF-α is cleaved (33). Although in human TNF-α the cleavage usually occurs between alanine (−1) and valine (+1), additional cleavage sites appear to exist between residues +1 and +12 (41). A naturally occurring and apparently inactive released form of murine TNF-α was found to incorporate an additional 10 residues from the presequence (43).

Kriegler *et al.* (39) postulated that the portion of the transmembrane form of TNF-α exposed on the outside of the cell not only serves as a precursor of released TNF-α but also that this membrane-anchored molecule itself can bind to a receptor on an adjacent cell and thereby engage in intercellular communication. Recent evidence supports the idea that membrane-anchored TNF-α on the surface of monocytes and/or macrophages can mediate the lysis of TNF-susceptible tumor cell lines (39, 40) and that even a nonsecretable cell surface mutant of TNF-α can kill tumor cells or virus-infected cells (41). In addition, it is possible that membrane-anchored TNF-α expressed on the surface of monocytes and/or macrophages mediates many other immunomodulatory and inflammation-promoting actions of TNF through a process termed "juxtacrine stimulation." The latter term was coined by Massagué (44) to describe a similar mode of action characteristic of a group of membrane-anchored growth factor precursors, which includes transforming growth factor-α, epidermal growth factor, and several related molecules.

TNF Receptors

Protein purification and cDNA cloning studies have led to the recent identification of two distinct receptors, the 55-kDa (45, 46) and the 75-kDa (47, 48) TNF receptors. The predicted extracellular portions of these two receptors are similar in their sequence not only to each other but also to the extracellular domains of the nerve growth factor (NGF) receptor and several structurally related cell surface molecules, including CDw40, OX40 antigen, and the T2 antigen of Shope fibroma virus. The extracellular portions of these proteins form a receptor family containing four characteristic domains with regularly spaced cysteine residues (Fig. 2). In contrast, no significant homologies were found between the intracytoplasmic portions of the two TNF receptors, nor are these portions related to any other known protein sequences (48). The lack of relatedness of the intracellular portions suggests that the two receptors activate different intracellular signaling pathways. Alternatively, some yet unidentified associated protein, rather than the intracytoplasmic regions of the receptors themselves, could act as the signal transducer; the latter type of mechanism was found to operate in the signal transduction of the interleukin-6 (IL-6) receptor-induced responses (49).

It appears that the two TNF receptors function very differently from the earlier identified two-chain receptor for IL-2 in which cross-linking of the two receptors is required for high affinity binding (50). In contrast, the presence of either the 55- or 75-kDa TNF receptor alone seems to be sufficient for high affinity binding and full biological activity. This conclusion is based on the demonstration that cells expressing only one of the two receptors are fully responsive to TNF. However, based on studies with receptor-specific blocking antibodies, Thoma *et al.* (51) concluded that binding to the 75-kDa TNF receptor is not sufficient to initiate a variety of responses to TNF-α and that the 55-kDa receptor is essential for TNF receptor function in several cell lines. Each of the two receptors binds TNF-α or TNF-β with a similar high affinity (45, 47), although some experiments suggest the existence of epitopes specific for TNF-α and TNF-β (52). An elegant recent study has shown that cross-linking of receptors with a bivalent antibody specific for either the 55- or 75-kDa TNF receptor was sufficient to generate a broad spectrum of TNF's actions (53), providing the strongest evidence so far that each of the two receptors alone can transduce the signal needed for biological activity.

Amounts of the 55- and 75-kDa TNF receptors expressed on different cell lines vary. Some cell lines were found to express only the 55-kDa receptor (Hep2), whereas most other cell lines examined have both receptors, expressed in different proportions (47, 48). Synthesis of the two receptors appears to be independ-

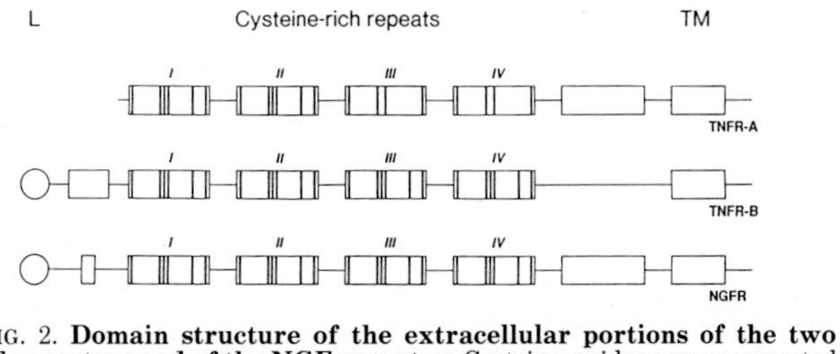

FIG. 2. **Domain structure of the extracellular portions of the two TNF receptors and of the NGF receptor.** Cysteine residues are represented by *vertical lines* within the four domains. *TNFR-A*, 75-kDa TNF receptor; *TNFR-B*, 55-kDa TNF receptor; *NGFR*, NGF receptor; *L*, predicted leader sequence; *TM*, transmembrane region. Reproduced from Dembic *et al.* (48).

ently regulated (47, 51). Although most studies so far have not revealed significant functional differences between the two types of TNF receptors with respect to the biological actions they mediate, more studies are needed to address this point.

Naturally occurring soluble fragments of the two TNF receptors have been found in human urine and serum (54–57). These fragments, originally identified as TNF inhibitory peptides, are now known to represent truncated portions of the extracellular domains of the two TNF receptors. The function of these TNF-binding peptides could be to regulate the bioavailability of TNF in the body. An interesting possibility is that the soluble TNF receptor fragments might function as ligands for the extracellular portion of the 26-kDa TNF precursor polypeptide spanning the plasma membrane in activated monocytes and/or macrophages (see above). Featuring extracellular, transmembrane, and intracytoplasmic regions (39), the 26-kDa TNF precursor resembles a receptor structurally, but whether it can actually function as a signaling receptor is not known.

Signal Transduction

Unlike some major growth factor receptors, there is no evidence that the TNF receptors possess intrinsic protein kinase activity. Phosphorylation of several distinct proteins was shown to occur within minutes after the exposure of cells to TNF (57–60); this is likely to be due to the activation of several major cellular kinases. Activation of protein kinase C (PKC) has been demonstrated in several cell lines (60, 61). A release of Ca^{2+} from internal stores and an increase in inositol 1,4,5-trisphosphate levels have not been seen after TNF treatment, and Krönke *et al.* (60) propose that diacylglycerol generated from phosphatidylcholine (rather than full phosphatidylinositol-4,5-bisphosphate) may be responsible for PKC activation. Alternatively, unsaturated fatty acids, *e.g.* arachidonic acid generated as a result of phospholipase A2 activation by TNF, might cause a Ca^{2+}-independent PKC activation (60). There is also evidence for a rapid increase in cAMP levels and resulting protein kinase A (PKA) activation (62). Shiroo and Matsushima (63) obtained evidence for the activation by TNF and IL-1 of a serine protein kinase that appears distinct from PKC, PKA, or casein kinase. Although several investigators found no evidence for increased tyrosine kinase activity, enhanced autophosphorylation on tyrosine of the EGF receptor (64) and rapid tyrosine phosphorylation of two cytosolic proteins (65) were seen in some cell lines.

In view of the extremely pleiotropic nature of TNF actions, it is perhaps not surprising that several alternate pathways can be activated by TNF (Table III). Several investigators showed that elimination of PKC activity did not block major TNF actions, suggesting that either PKC is not involved in these activities or that alternate pathways can fully compensate (60, 62). Similarly, activation of PKA is probably not essential for most TNF actions, although it might contribute to the activation of some genes by TNF (72).

Several transcription factors were shown to be activated or newly induced by TNF (Table III). NF-κB activation was shown to be involved in the induction of a number of TNF-responsive genes (69–72). Activation of preformed NF-κB occurs in the absence of protein synthesis (72), providing a mechanism for a rapid and relatively direct transcriptional activation. Most of the other transcription factors listed in Table III are newly synthesized upon stimulation with TNF, *e.g.* AP-1 (its components are products of the c-*fos* and c-*jun* genes, both of which are induced by TNF) and interferon regulatory factors 1 and 2 (IRF-1 and -2), originally identified as transcription factors involved in the regulation of IFN-α/β gene expression.

More than one signaling pathway or transcription factor can contribute to the activation of a single gene by TNF. For example, three different *cis*-acting elements were shown to be important in IL-6 induction by TNF or IL-1 (72, 75, 77). Similarly, optimal transcriptional activation of the IL-8 gene by TNF or IL-1 appears to involve the cooperative action of transcription factors on a NF-κB-like and a NF-IL-6-like binding site (78).

TABLE III

Signal transduction pathways likely involved in TNF actions

Membrane-associated events	Stimulation of GTPase and GTP binding (66)
	Activation of phospholipase A2 (67)
	Activation of phosphatidylcholine-specific phospholipase C (60)
Soluble intracellular mediators generated	cAMP (62)
	Diacylglycerol (60)
	Prostaglandins (68)
Protein kinases activated	Protein kinase C (61)
	Protein kinase A (62)
	EGF receptor tyrosine kinase (64)
	Novel serine kinase (63)
Transcription factors activated or induced	NF-κB (69–72)
	AP-1 (61)
	IRF-1 (73)
	IRF-2 (74)
	NF-IL-6 (75)
	NF-GMa (76)

TNF-regulated Genes and Proteins

Most of the pleiotropic biological actions of TNF can be attributed to its ability to activate a startling variety of genes in a multitude of target cells (Table IV). TNF and the functionally related cytokine IL-1 (81) are the only natural substances known to have such a large spectrum of target genes. The list in Table IV is far from complete. Among the genes not listed are several TNF-activated genes whose functions await identification (91, 96). A more extensive, but also incomplete, list of TNF-responsive genes has been compiled by Krönke *et al.* (60). Although most studies have concentrated on genes and proteins that are induced by TNF, it is apparent that the expression of other genes is inhibited, *e.g.* c-*myc* in some cells whose growth is suppressed by TNF (60), collagen in human fibroblasts (97), or thrombomodulin in human vascular epithelial cells (98).

Many genes regulated by TNF are similarly affected by other cytokines, especially IL-1 and, somewhat less commonly, the interferons or growth factors (60, 81, 91, 99). These similarities can be explained by the recent demonstrations that diverse cytokines share the ability to activate identical intracellular signaling pathways and transcription factors. For example, TNF and IL-1 share the ability to elevate intracellular cAMP levels in some cells (62), and they are both potent activators of NF-κB (71, 72). Interferon-α/β shares with TNF and IL-1 the ability to induce the transcription factors IRF-1 and IRF-2 (73, 74). This redundancy in the activation of intracellular signaling pathways and target genes by structurally dissimilar cytokines that bind to unrelated receptors helps to explain the recently recognized be-

TABLE IV

Partial list of genes or proteins induced by TNF[a]

Transcription factors	c-*fos* (79)
	c-*jun* (61)
	IRF-1 (73)
Cytokines	IL-1α and -β (80, 81)
	IL-6 (82)
	IL-8 (78)
	IFN-β (83)
	TNF-α (60)
Growth factors	PDGF (84)
	GM-CSF (85)
	M-CSF (86)
Receptors	IL-2 receptor, α chain (69)
	EGF receptor (87)
Cell adhesion molecules	ELAM-1 (88)
	ICAM-1 (89)
Inflammatory mediators	Tissue factor (90)
	Collagenase (61, 68)
	Stromelysin (91)
Acute phase proteins	α1-Acid glycoprotein (92)
	Haptoglobin (92)
	C3 complement (92)
Major histocompatibility complex proteins	Class I MHC (93)
	Class II MHC (11, 60)
Other proteins	2′-5′-Oligoadenylate synthetase (60)
	Manganese superoxide dismutase (94)
	Plasminogen activator inhibitor 1 and 2 (95)
Viruses	HIV-1 (70, 71)

[a] In a great majority of cases, only TNF-α was examined.

wildering redundancy in the biological functions of different cytokines.

Acknowledgments—We thank Ilene Toder for help with the preparation of the manuscript and Bharat B. Aggarwal and Jedd Wolchok for critical reading of the manuscript.

REFERENCES

1. Carswell, E. A., Old, L. J., Kassel, R. L., Green, S., Fiore, N., and Williamson, B. (1975) *Proc. Natl. Acad. Sci. U. S. A.* **72,** 3666–3670
2. Beutler, B., Greenwald, D., Hulmes, J. D., Chang, M., Pan Y.-C. E., Mathison, J., Ulevitch, R., and Cerami, A. (1985) *Nature* **316,** 552–554
3. Williams, T. W., and Granger, G. A. (1968) *Nature* **219,** 1076–1077
4. Ruddle, N. H., and Waksman, B. H. (1968) *J. Exp. Med.* **128,** 1267–1279
5. Gray, A. W., Aggarwal, B. B., Benton, C. V., Bringman, T. S., Henzel, W. J., Jarrett, J. A., Leung, D. W., Moffat, B., Ng, P., Svedersky, L. P., Palladino, M. A., and Nedwin, G. (1984) *Nature* **312,** 721–724
6. Pennica, D., Nedwin, G. E., Hayflick, J. S., Seeburg, P. H., Derynck, R., Paladino, M. A., Kohr, W. J., Aggarwal, B. B., and Goeddel, D. V. (1984) *Nature* **312,** 724–729
7. Aggarwal, B. B., Eessalu, T. E., and Hass, P. E. (1985) *Nature* **318,** 665
8. Old, L. J. (1985) *Science* **230,** 630–632
9. Havell, E. A. (1987) *J. Immunol.* **139,** 4225–4231
10. Tracey, K. J., Fong, Y., Hesse, D. G., Manogue, K. R., Lee, A. T., Kuo, G. C., Lowry, S. F., and Cerami, A. C. (1987) *Nature* **330,** 662–664
11. Pujol-Borrell, R., Todd, I., Doshi, M., Bottazzo, G. F., Sutton, R., Gray, D., Adolf, G. R., and Feldmann, M. (1987) *Nature* **326,** 304–306
12. Piguet, P.-F., Grau, G. E., Allet, B., and Vassalli, P. (1987) *J. Exp. Med.* **166,** 1280–1289
13. Nedwin, G. E., Naylor, S. L., Sakaguchi, A. Y., Smith, D., Jarrett-Nedwin, J., Pennica, D., Goeddel, D. V., and Gray, P. W. (1985) *Nucleic Acids Res.* **13,** 6361–6373
14. Spies, T., Blanck, G., Bresnahan, M., Sands, J., and Strominger, J. L. (1988) *Science* **243,** 214–217
15. Muller, U., Jongeneel, C. V., Nedospasov, S. A., Lindahl, K. F., and Steinmetz, M. (1987) *Nature* **325,** 265–267
16. Spriggs, D. R., Deutsch, S., and Kufe, D. W. (1991) in *Tumor Necrosis Factor: Structure, Function and Mechanism of Action* (Aggarwal, B. B., and Vilček, J., eds) pp. 3–34, Marcel Dekker, Inc.
17. Turetskaya, R. L., Fashena, S. J., Paul, N. L., and Ruddle, N. H. (1991) in *Tumor Necrosis Factor: Structure, Function and Mechanism of Action* (Aggarwal, B. B., and Vilček, J., eds) pp. 35–60, Marcel Dekker, Inc.
18. Shakhov, A. N., Collart, M. A., Vassalli, P., Nedospasov, S. A., and Jongeneel, C. V. (1990). *J. Exp. Med.* **171,** 35–47
19. Goldfeld, A. E., Doyle, C., and Maniatis, T. (1990) *Proc. Natl. Acad. Sci. U. S. A.* **87,** 9769–9773
20. Paul, N. L., Lenardo, M. J., Novak, K. D., Sarr, T., Tang, W.-L., and Ruddle, N. H. (1990) *J. Virol.* **64,** 5412–5419
21. Kelker, H. C., Oppenheim, J. D., Stone-Wolff, D. S., Henriksen-DeStefano, D., Aggarwal, B. B., Stevenson, H. C., and Vilček, J. (1985) *Int. J. Cancer* **36,** 69–73
22. Cuturi, M. C., Murphy, M., Costa-Giomi, M. P., Weinmann, R., Perussia, B., and Trinchieri, G. (1987) *J. Exp. Med.* **165,** 1581–1594
23. Eardley, D. D., Shen, F. W., Gershon, R. K., and Ruddle, N. H. (1980) *J. Immunol.* **124,** 1199–1202
24. Jongeneel, C. V., Nedospasov, S. A., Plaetinck, G., Naquet, P., and Cerottini, J. C. (1988) *J. Immunol.* **140,** 1916–1922
25. Sung, S.-S. J., Jung, L. K. L., Walters, J. A., Jeffes, E. W. B., Granger, G. A., and Fu, S. M. (1989) *J. Clin. Invest.* **84,** 236–243
26. Lindemann, A., Riedel, D., Oster, W., Ziegler-Heitbrock, H. W. L., Mertelsmann, R., and Herrmann, F. (1989) *J. Clin. Invest.* **83,** 1308–1312
27. Lieberman, A. P., Pitha, P. M., Shin, H. S., and Shin, M. L. (1989) *Proc. Natl. Acad. Sci. U. S. A.* **86,** 6348–6352
28. Libby, P., Ordovas, J. M., Auger, K. R., Robbins, A. H., Birinyi, L. K., and Dinarello, C. A. (1986) *Am. J. Pathol.* **124,** 179–185
29. Warner, S. J. C., and Libby, P. (1989) *J. Immunol.* **142,** 100–109
30. Rubin, B. Y., Anderson, S. L., Sullivan, S. A., Williamson, B. D., Carswell, E. A., and Old, L. J. (1986) *J. Exp. Med.* **164,** 1350–1355
31. Han, J., Brown, T., and Beutler, B. (1990) *J. Exp. Med.* **171,** 465–475
32. Kruys, V., Marinx, O., Shaw, G., Deschamps, J., and Huez, G. (1989) *Science* **245,** 852–855
33. Shaw, G., and Kamen, R. (1986) *Cell* **46,** 659–667
34. Fiers, W. (1991) in *Tumor Necrosis Factor: Structure, Function and Mechanism of Action* (Aggarwal, B. B., and Vilček, J., eds) pp. 79–92, Marcel Dekker, Inc.
35. Aggarwal, B. B., Moffat, B., and Harkins, R. N. (1984) *J. Biol. Chem.* **259,** 686–691
36. Aggarwal, B. B., Kohr, W. J., Hass, P. E., Moffat, B., Spencer, S. A., Henzel, W. J., Bringman, T. S., Nedwin, G. E., Goeddel, D. V., and Harkins, R. N. (1985) *J. Biol. Chem.* **260,** 2345–2354
37. Jones, E. Y., Stuart, D. I., and Walker, N. P. C. (1989) *Nature* **338,** 225
38. Eck, M. J., and Sprang, S. R. (1989) *J. Biol. Chem.* **264,** 17595–17605
39. Kriegler, M., Perez, C., DeFay, K., Albert, I., and Lu, S. D. (1988) *Cell* **53,** 45–53
40. Luettig, B., Decker, T., and Lohmann-Matthes, M. L. (1989) *J. Immunol.* **143,** 4034–4038
41. Perez, C., Albert, I., DeFay, K., Zachariades, N., Gooding, L., and Kriegler, M. (1990) *Cell* **63,** 251–258
42. Scuderi, P. (1989) *J. Immunol.* **143,** 168–173
43. Cseh, K., and Beutler, B. (1989) *J. Biol. Chem.* **264,** 16256–16260
44. Massagué, J. (1990) *J. Biol. Chem.* **265,** 21393–21396
45. Loetscher, H., Pan, Y.-C. E., Lahm, H.-W., Gentz, R., Brockhaus, M., Tabuchi, H., and Lesslauer, W. (1990) *Cell* **61,** 351–359
46. Schall, T. J., Lewis, M., Koller, K. J., Lee, A., Rice, G. C., Wong, G. H. W., Gatanaga, T., Granger, G. A., Lentz, R., Raab, H., Kohr, W. J., and Goeddel, D. V. (1990) *Cell* **61,** 361–370
47. Smith, C. A., Davis, T., Anderson, D., Solam, L., Beckmann, M. P., Jerzy, R., Dower, S. K., Cosman, D., and Goodwin, R. G. (1990) *Science* **248,** 1019–1023
48. Dembic, Z., Loetscher, H., Gubler, U., Pan, Y.-C. E., Lahm, H.-W., Gentz, R., Brockhaus, M., and Lesslauer, W. (1990) *Cytokine* **2,** 231–237
49. Taga, T., Hibi, M., Hirata, Y., Yamasaki, K., Yasukawa, K., Matsuda, T., Hirano, T., and Kishimoto, T. (1989) *Cell* **58,** 573–581
50. Hatakeyama, M., Tsudo, M., Minamoto, S., Kono, T., Doi, T., Miyata, T., Miyasaka, M., and Taniguchi, T. (1989) *Science* **244,** 551–556
51. Thoma, B., Grell, M., Pfizenmaier, K., and Scheurich, P. (1990) *J. Exp. Med.* **172,** 1019–1023
52. Espevik, T., Brockhaus, M., Loetscher, H., Nonstad, U., and Shalaby, R. (1990) *J. Exp. Med.* **171,** 415–426
53. Engelmann, H., Holtmann, H., Brakebusch, C., Avni, Y. S., Sarov, I., Nophar, Y., Hadas, E., Leitner, O., and Wallach, D. (1990) *J. Biol. Chem.* **265,** 14497–14504
54. Seckinger, P., Isaaz, S., and Dayer, J.-M. (1989) *J. Biol. Chem.* **264,** 11966–11973
55. Engelmann, H., Novick, D., and Wallach, D. (1990) *J. Biol. Chem.* **265,** 1531–1536
56. Olsson, I., Lantz, M., Nilsson, E., Peetre, C., Thysell, H., Grubb, A., and Adolf, G. (1989) *Eur. J. Haematol.* **42,** 270–275
57. Schütze, S., Scheurich, P., Pfizenmaier, K., and Krönke, M. (1989) *J. Biol. Chem.* **264,** 3562–3567
58. Robaye, B., Hepburn, A., Lecocq, R., Fiers, W., Boeynaems, J.-M., and Dumont, J. E. (1989) *Biochem. Biophys. Res. Commun.* **163,** 301–308
59. Marino, M. W., Pfeffer, L. M., Guidon, P. T., Jr., and Donner, D. B. (1989) *Proc. Natl. Acad. Sci. U. S. A.* **86,** 8417–8421
60. Krönke, M., Schütze, S., Scheurich, P., and Pfizenmaier, K. (1991) in *Tumor Necrosis Factor: Structure, Function and Mechanism of Action* (Aggarwal, B. B., and Vilček, J., eds) pp. 189–216, Marcel Dekker, Inc.
61. Brenner, D. A., O'Hara, M., Angel, P., Chojkier, M., and Karin, M. (1989) *Nature* **337,** 661–663
62. Zhang, Y., Lin, J.-X., Yip, Y. K., and Vilček, J. (1988) *Proc. Natl. Acad. Sci. U. S. A.* **85,** 6802–6805
63. Shiroo, M., and Matsushima, K. (1990) *Cytokine* **2,** 13–20
64. Donato, N. J., Gallick, G. E., Steck, P. A., and Rosenblum, M. G. (1989) *J. Biol. Chem.* **264,** 20474–20481
65. Kohno, M., Nishizawa, N., Tsujimoto, M., and Nomoto, H. (1990) *Biochem. J.* **267,** 91–98
66. Imamura, K., Sherman, M. L., Spriggs, D., and Kufe, D. (1988) *J. Biol. Chem.* **263,** 10247–10253
67. Clark, M. A., Chen, M.-J., Crooke, S. T., and Bomalaski, J. S. (1988) *Biochem. J.* **250,** 125–132
68. Dayer, J.-M., Beutler, B., and Cerami, A. (1985) *J. Exp. Med.* **162,** 2163
69. Lowenthal, J. W., Ballard, D. W., Bohnlein, E., and Greene, W. C. (1989) *Proc. Natl. Acad. Sci. U. S. A.* **86,** 2331–2335
70. Duh, E. J., Maury, W. J., Folks, T. M., Fauci, A. S., and Rabson, A. B. (1989) *Proc. Natl. Acad. Sci. U. S. A.* **86,** 5974–5978
71. Osborn, L., Kunkel, S., and Nabel, G. J. (1989) *Proc. Natl. Acad. Sci. U. S. A.* **86,** 2336–2340
72. Zhang, Y., Lin, J.-X., and Vilček, J. (1990) *Mol. Cell. Biol.* **10,** 3818–3823
73. Fujita, T., Reis, L. F. L., Watanabe, N., Kimura, Y., Taniguchi, T., and Vilček, J. (1989) *Proc. Natl. Acad. Sci. U. S. A.* **86,** 9936–9940
74. Reis, L. F. L., Fujita, T., Lee, T. H., Taniguchi, T., and Vilček, J. (1990) in *Molecular and Cellular Biology of Cytokines* (Oppenheim, J. J., Powanda, M. C., Kluger, M. J., and Dinarello, C. A., eds) pp. 1–6, Wiley-Liss, New York
75. Akira, S., Isshiki, H., Sugita, T., Tanabe, O., Kinoshita, S., Nishio, Y., Nakajima, T., Hirano, T., and Kishimoto, T. (1990) *EMBO J.* **9,** 1897
76. Shannon, M. F., Occhiodoro, F. S., Kuczek, E. S., Pell, L. M., and Vadas, M. A. (1990) *J. Cell. Biochem. Suppl.* **14B,** 217
77. Ray, A., Sassone-Corsi, P., and Sehgal, P. B. (1989) *Mol. Cell. Biol.* **9,** 5537–5547
78. Mukaida, N., Mahe, Y., and Matsushima, K. (1990) *J. Biol. Chem.* **265,** 21128–21133
79. Lin, J.-X., and Vilček, J. (1987) *J. Biol. Chem.* **262,** 11908–11911
80. Turner, M., Chantry, D., Buchan, G., Barrett, K., and Feldmann, M. (1989) *J. Immunol.* **143,** 3556–3561
81. Le, J., and Vilček, J. (1987) *Lab. Invest.* **56,** 234–248
82. Kohase, M., Henriksen-DeStefano, D., May, L. T., Vilček, J., and Sehgal, P. B. (1986) *Cell* **45,** 659–666
83. Jacobsen, H., Mestan, J., Mittnacht, S., and Dieffenbach, C. W. (1989) *Mol. Cell. Biol.* **9,** 3037–3042
84. Hajjar, K. A., Hajjar, D. P., Silverstein, R. L., and Nachman, R. L. (1987) *J. Exp. Med.* **166,** 235–245
85. Broudy, V. C., Kaushansky, K., Segal, G. M., Harlan, J. M., and Adamson, J. W. (1986) *Proc. Natl. Acad. Sci. U. S. A.* **83,** 7467–7471
86. Oster, W., Lindemann, A., Horn, S., Mertelsmann, R., and Herrmann, F. (1987) *Blood* **70,** 1700–1703
87. Palombella, V. J., Yamashiro, D. J., Maxfield, F. R., Decker, S. J., and Vilček, J. (1987) *J. Biol. Chem.* **262,** 1950–1954
88. Bevilacqua, M. P., Stengelin, S., Gimbrone, M. A., Jr., and Seed, B. (1989) *Science* **243,** 1160–1165
89. Pober, J. S., LaPierre, L. A., Stolpen, A. H., Brock, T. A., Springer, T. A., Fiers, W., Bevilacqua, M. P., Mendrick, D. L., and Gimbrone, M. A., Jr. (1987) *J. Immunol.* **138,** 3319–3324
90. Scarpati, E. M., and Sadler, J. E. (1989) *J. Biol. Chem.* **264,** 20705–20713
91. Lee, T. H., Lee, G. W., Ziff, E. B., and Vilček, J. (1990) *Mol. Cell. Biol.* **10,** 1982–1988
92. Gauldie, J., Richards, C., Northeman, W., Fey, G., and Baumann, H. (1989) *Ann. N. Y. Acad. Sci.* **557,** 46–59
93. Collins, T., Lapierre, L. A., Fiers, W., Strominger, J. L., and Pober, J. S. (1986) *Proc. Natl. Acad. Sci. U. S. A.* **83,** 446–450
94. Wong, G. H., and Goeddel, D. V. (1988) *Science* **242,** 941–944
95. Medcalfe, R. L., Kruithof, E. K. O., and Schleuning, W.-D. (1988) *J. Exp. Med.* **168,** 751–759
96. Dixit, V. M., Green, S., Sarma, V., Holzman, L. B., Wolf, F. W., O'Rourke, K., Ward, P. A., Prochownik, E. V., and Marks, R. M. (1990) *J. Biol. Chem.* **265,** 2973–2978
97. Solis-Herruzo, J., Brenner, D. A., and Chojkier, M. (1988) *J. Biol. Chem.* **263,** 5841–5845
98. Conway, E. M., and Rosenberg, R. D. (1988) *Mol. Cell. Biol.* **8,** 5588–5592
99. Vilček, J. (1990) in *Peptide Growth Factors and their Receptors, Handbook of Experimental Pharmacology* (Sporn, M. A., and Roberts, A. B., eds) pp. 3–38, Springer-Verlag, Berlin

THE JOURNAL OF BIOLOGICAL CHEMISTRY
Vol. 266, No. 13, Issue of May 5, pp. 7963–7966, 1991

Printed in U.S.A.

Minireview

Three Proteolytic Systems in the Yeast *Saccharomyces cerevisiae**

Elizabeth W. Jones

From the Department of Biological Sciences, Carnegie Mellon University, Pittsburgh, Pennsylvania 15213

The proteases of yeast that were first identified and characterized were detected biochemically (1–4) and proved to be of vacuolar (lysosomal) origin (5, 6). An additional nonvacuolar set was defined genetically by mutations that caused incomplete proteolytic processing of precursors to killer toxin and the pheromone α-factor (7–13). Once the dominant vacuolar proteases could be eliminated by mutation, a substantial number of new enzyme activities, including additional endoproteinases, carboxypeptidases, aminopeptidases, and dipeptidyl aminopeptidases, were detected biochemically (14–16).

I will concentrate here on three groups of proteases: the cytosolic proteasome, vacuolar proteases, and proteases located within the secretory pathway. Not only is a good deal known about the functions of these enzymes, but also, where zymogens are involved, several of the maturation pathways have been elucidated. The compartments and their contained enzymes are listed in Table I; maturation pathways for several of the enzymes are depicted in Fig. 1. For more extensive coverage of all of the proteases and their properties see Refs. 17 and 18.

Cytosolic Proteasome

In 1984, Achstetter *et al.* (19) reported purification of a very large peptidase, which proved to contain 10–12 nonidentical subunits and to show homology to the 20 S cylindrical particles common to many eukaryotic cells (20, 21). Variously called prosome, proteasome, multiprotease complex, or multicatalytic protease, among other names, the enzyme complex catalyzes hydrolysis of bonds on the carboxyl side of neutral/hydrophobic residues, basic residues, and acidic amino acids (19–21).

Five, possibly six, of the genes encoding proteasome subunits have been cloned, sequenced, and disrupted, mostly by reverse genetics (22–25). Four (possibly five) of the subunits are essential for cell viability (22–25). For one of the essential subunits, encoded by *PRE1*, nonlethal alleles are available. Study of the *pre1* mutants indicates that they have lost only a subset of the proteolytic activities of the proteasome, that mutant cells are temperature-sensitive for growth and for degradation of canavanine-containing proteins and accumulate ubiquitinated proteins upon temperature or canavanine stress, and that the homozygous mutant diploid cells are unable to sporulate (24).

Vacuolar Proteases

There are six lumenal proteases: two endoproteinases, proteinases A and B (PrA, PrB),[1] two carboxypeptidases, carboxypeptidases Y and S (CpY, CpS), and two aminopeptidases, aminopeptidases I and Co (ApI, ApCo). Dipeptidyl aminopeptidase B (DPAP-B) is found in the vacuolar membrane (26).

* This work was supported by United States Public Health Service Research Grant DK18090 from the National Institutes of Health.

[1] The abbreviations used are: Pr, proteinase; Cp, carboxypeptidase; Ap, aminopeptidase; DPAP, dipeptidyl aminopeptidase.

TABLE I
Proteases of the yeast vacuole and secretory pathway

Enzyme	Type	Gene[a]	Zymogen form
Vacuole			
PrA	Aspartic	*PEP4* (*PRA1*, *PHO9*)	Yes
PrB	Serine	*PRB1*	Yes
CpY	Serine	*PRC1*	Yes
CpS	Metallo-(Zn^{2+})	*CPS1* (*DUT1*)	Unknown
ApI	Metallo-(Zn^{2+})	*LAP4* (*APE1*)	Yes
ApCo	Metallo-(Co^{2+})		
DPAP-B	Serine	*DAP2* (*DPP2*)	No
Secretory pathway			
Signal peptidase		*SEC11* and others	Unknown
Kex2 protease	Serine	*KEX2*	Yes
Kex1 carboxypeptidase	Serine	*KEX1*	No
DPAP-A	Serine?[b]	*STE13*	Unknown; predict no
Yeast aspartyl protease III	Aspartic	*YAP3*	Unknown; predict yes

[a] Gene names are given according to genetic convention. The primary name is given first; assignments after the initial discovery are given in parentheses.

[b] DPAP-A shows marked homology to DPAP-B (56). Since DPAP-B is inhibited by phenylmethylsulfonyl fluoride, I infer DPAP-A will be as well, although evidence is lacking.

Proteinase A

PrA, encoded by the *PEP4* gene, is an aspartyl proteinase with similarity to the two-domain class of aspartyl proteinases that includes pepsin, renin, cathepsin D, and penicillopepsin (27, 28). This glycoprotein of 42 kDa carries two asparagine-linked glycosyl side chains; its four cysteines are thought to form disulfide bonds (29). Three proteolytic cleavages occur during the post-translational maturation of Pep4p. After removal of the signal sequence by signal peptidase (30), another 47 amino acids are removed from the NH_2 terminus late in the Golgi or in the vacuole (31). This intramolecular reaction is autocatalytic, since mutational change of either aspartate residue of the active site (Asp $\rightarrow$ Asn) results in nearly complete failure of processing.[2] The final cleavage removes 7 amino acids; it is catalyzed by PrB (32).[3] The final "intermediate" has not been detected kinetically; it accumulates in a *prb1*Δ mutant (32).

Proteinase B

PrB, encoded by the *PRB1* gene, is a member of the subtilisin family of serine proteases, which includes proteinase K (33). Despite the presence of three potential tripeptide acceptors for Asn-linked glycosyl chains within the catalytic region (five in the precursor), the 31-kDa enzyme carries no Asn-linked glycosyl side chains (34); since it is a glycoprotein we presume it carries *O*-linked mannose (35).

Four proteolytic cleavages occur during maturation of Prb1p (Fig. 1). After removal of the signal sequence by signal peptidase,[4] about 260 additional amino acids are removed from the NH_2 terminus in the endoplasmic reticulum (36, 37). This reaction is intramolecular and autocatalytic, for it does not occur if the active site serine (Ser $\rightarrow$ Ala) or aspartate (Asp $\rightarrow$ Asn) is changed by mutation (38).[4] The third cleavage, which converts the 40-kDa product of autocatalysis to a 37-kDa intermediate, is catalyzed by PrA (36). The final cleavages occur late, in either the late Golgi or, more likely, the vacuole (36). We infer that the fourth and final cleavage, which converts the 37-kDa species to the mature 31-kDa mature enzyme, is autocatalytic in nature, because a mutation, *prb1-628*, profoundly reduces the rate of production

[2] C. A. Woolford, J. D. Garman, and E. W. Jones, unpublished data.

[3] J. D. Garman, C. A. Woolford, M. F. Tam, and E. W. Jones, unpublished data.

[4] V. Nebes and E. Jones, unpublished data.

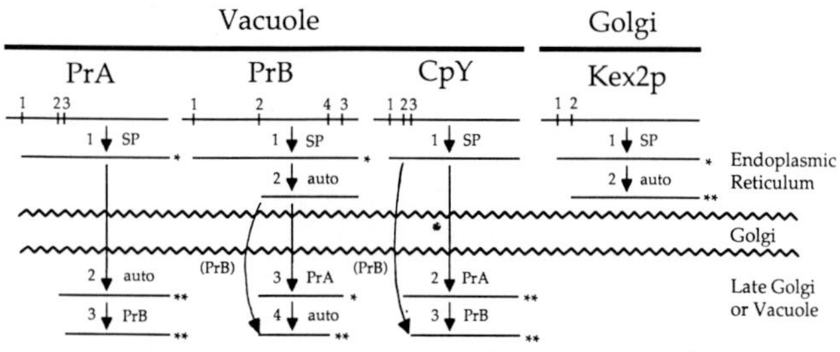

FIG. 1. **Maturation pathways for selected yeast proteases.** For each enzyme the *top line* designates the initial translation product; sites of proteolytic cleavage are *numbered* above the *line* and at the reaction step, in the order in which the cleavages normally occur. Abbreviations are: *SP*, signal peptidase; *auto*, autocatalytic; *Kex2p*, Kex2 endoproteinase; *, active in intramolecular autocatalysis; **, proteolytically active on other molecules. PrB is thought to catalyze two cleavages that bypass the requirement for PrA, giving rise to the phenomenon of phenotypic lag (42, 43, 50). The two suspected cleavages are signified by (PrB) adjacent to *curved* reaction *arrows*. Not to scale.

of 31-kDa PrB from the 37-kDa species (36). The *prb1-628* mutation results in a change of Ala^{152} to Thr^{152},[4] which lies close to the essential Asn^{155} of the oxyanion hole (39, 40) and within 10 Å of the active site serine (41). The defect is in active site function, not substrate properties, since the *prb1-628* mutation is completely recessive and all the antigen produced in the heterozygote is of the mature form.[4] We presume that protease B itself can also catalyze this cleavage in *trans*, albeit less efficiently, and that this is the molecular explanation for the phenomenon called phenotypic lag (42, 43).

Carboxypeptidase Y

CpY, encoded by the *PRC1* gene, is a serine protease. This 61-kDa glycoprotein shows broad substrate specificity and carries four Asn-linked glycosyl side chains (44, 45). Three proteolytic cleavages occur during maturation of Prc1p. After signal peptidase cleavage (46), an amino-terminal peptide is removed in the late Golgi or the vacuole (47–49) in a reaction catalyzed by PrA (48). *In vitro* data suggest that a third cleavage is catalyzed by PrB (50), a reaction that can bypass the PrA-catalyzed cleavage and is presumed to account for phenotypic lag (42, 43).

Carboxypeptidase S

CpS, encoded by the *CPS1* gene, is a metal ion-dependent carboxypeptidase. We know little of its synthesis or possible precursor(s). However, its synthesis is not dependent on PrA activity (51).

Aminopeptidase I

ApI, encoded by the *LAP4* gene, is a metalloexopeptidase. It is a glycoprotein of 640 kDa and contains 12 subunits (52). Because *pep4* mutations greatly reduce the activity of ApI (53), we expect a zymogen precursor to ApI, with production of active enzyme catalyzed by PrA. Whether the large antigenic species reported for cells in steady state (54) corresponds to such a precursor is not yet known. The *LAP4* gene does not appear to encode a signal sequence of the type normally responsible for entry into the endoplasmic reticulum (54).

Aminopeptidase Co

ApCo is a 100-kDa metalloexopeptidase that requires Co^{2+} (55). Nothing has been reported about its gene or its synthesis.

Dipeptidyl Aminopeptidase B

DPAP-B, encoded by the *DAP2* gene, is a 120-kDa integral membrane glycoprotein of the yeast vacuole (56). An apparently typical hydrophobic helical domain near the NH_2 terminus is postulated to function as an internal signal sequence and transmembrane anchor for this presumed type II integral membrane glycoprotein. Production of active DPAP-B does not require PrA activity (56).

Maturation Information

Ignoring the contributions of signal peptidase, one can summarize what is known about production of these vacuolar proteases as follows. Some (PrA, PrB, CpY, ApI) but not all (CpS, DPAP-B) of the vacuolar proteases are synthesized as inactive precursors. All of the proteolytic cleavages of the maturation pathways are self-catalyzed or catalyzed by PrA or PrB. Thus, all of the information required for the maturation cleavages and the ultimate activities of all of these proteases is self-contained within this set of protease precursors. One can extend this concept of self-sufficiency of maturation information to include other vacuolar hydrolases, since production of activity of the vacuolar species of alkaline phosphatase (51), trehalase (57), and RNase(s) (51) is also dependent on PrA. The roles of PrA and PrB are not equally important in these maturation pathways. PrA plays an essential role in the production of itself, PrB, CpY, and ApI, among proteases, and alkaline phosphatase, trehalase, and RNase(s), among other hydrolases. The requirement of PrB for production of activity is limited to production of PrB activity itself. In the other cases examined, PrB seems to function to "trim" the ends of enzymatically active species to give enzyme of "mature" size. The functional significance of such trimming is not known.

Functions of Vacuolar Proteases

Much of our understanding of the functions of these enzymes stems from analyses of mutant strains. Structural gene mutations for CpY, CpS, or PrB (*prc1*, *cps1* or *prb1*, respectively) eliminate only the activity of the enzyme encoded by the affected gene, making analysis straightforward. Mutations in *PEP4*, the structural gene for PrA, can give different phenotypes, depending upon the severity of the mutation. Insertion or nonsense mutations (*e.g. pep4-3*) completely prevent production of active PrA. Its absence ensures failure of the cell to produce all enzyme activities in whose maturation pathways PrA participates, including PrB, CpY, ApI, alkaline phosphatase, RNase(s), and vacuolar trehalase (32, 48, 51, 53, 57, 58). Missense mutations in *PEP4* can cause a greater range of phenotypes, ranging from completely pleiotropic (*pep4-11*) to an effect only on PrA activity (*pep4-625*) (32). Mature sized (presumably autoprocessed) PrA antigen that failed to hydrolyze hemoglobin was found in the *pep4-625* mutant. All other vacuolar hydrolase activities in this mutant were at wild-type levels, implying that PrA functioned properly in the maturation pathways (32, 59). On the surface, the *pep4-625* mutation allows one to assess the consequence of loss of activity of PrA alone. This seems unwarranted, however, since all *in vivo* assays of PrA function are positive and only the *in vitro* test of PrA activity is negative. Thus caution must be exercised in evaluating analyses employing *pep4-625* and, presumably, *pra1-1*, which also appears to lack only PrA activity (60) (*pra1-1* is a *pep4* allele and should be so called by genetic convention; the sequence change has not been reported).

One can with some confidence assign certain functions to the vacuolar proteases. None of them is essential for cell viability. The two endoproteinases, PrA and PrB, but principally PrA, catalyze cleavages that result in maturation of precursors and activation of zymogens of other vacuolar proteases and hydrolases (see above). Since the enzymes show little specificity in cleavage of denatured peptides (61, 62), the specificity manifested in the maturation pathways is presumed to reflect conformational constraints.

All of the proteases probably participate in degradation of some proteins and peptides that turn over as a normal part of the life cycle (Kex2 protease (63); α-factor (64)) but are not involved in the initial steps in turnover of nonsense fragments (65), missense proteins (65, 66), or unassembled ribosomal proteins (67) or enzyme subunits (66). It is very clear that PrA and PrB activities are required for degradation of analog-containing proteins, for starvation-induced vegetative protein degradation, and for the ability to survive nitrogen starvation (PrA is particularly important for the last effect) (66).

The vacuolar proteases account for the bulk of the protein degradation that takes place when cells are starved for nitrogen in sporulation medium (66, 68). In the absence of PrB activity, the frequency of ascospore formation can range from nil to high, depending on conditions of preculture and background genotype (66, 68). Pleiotropic *pep4* mutations eliminate ascospore formation (66, 68); whether this is caused by the low rates of protein degradation or the loss of other enzyme activities (RNase(s), trehalases, etc.) cannot be assessed. *pra1-1* homozygotes sporulate poorly (66), while *pep4-625* homozygotes sporulate rather well[5]; these latter results are difficult to evaluate, as discussed above.

It is unclear whether vacuolar proteases contribute to degradation of fructose 1,6-bisphosphatase during catabolite inactivation, since there are reports on both sides of the issue (66, 69, 70). Vacuolar proteases appear not to be involved in other cases of catabolite inactivation that have been tested (71, 72).

In sum, in growing cells the vacuolar proteases catalyze maturation of precursors and participate in a limited amount of protein turnover. In response to stress, they catalyze massive amounts of protein degradation that facilitates cellular restructuring, particularly during sporulation.

Proteases of the Secretory Pathway

Signal Peptidase—Signal peptidase is an integral membrane protein that contains at least four subunits, one of which is glycosylated (73). The 18-kDa subunit is the product of the *SEC11* gene, a not unexpected finding, since signal peptides of secreted proteins are not removed in the *sec11* mutant at high temperature (74) and the predicted sequence of Sec11p shows marked similarity to that of one subunit of canine signal peptidase (75).

Kex2 Endoprotease—Kex2 protease, encoded by the *KEX2* gene, is an endoproteinase that cleaves on the COOH-terminal side of Lys-Arg or Arg-Arg paired basic residues (9, 76). It is a serine protease of the subtilisin class (77) and requires Ca^{2+} for activity. This glycoprotein carries both *N*-linked and *O*-linked sugars. Kex2 protease is an integral membrane protein, anchored by a sub-COOH-terminal transmembrane domain in a compartment thought to be the late Golgi (78, 79). The *KEX2* gene was first identified by mutations that prevented production of killer toxin (killer expression) and caused sterility in cells of α mating type because of failure to produce the pheromone α-factor (7, 8). The α-factor precursor contains four copies of the pheromone peptide sequence separated by spacers (Fig. 2). The cleavage catalyzed by Kex2 protease is shown in Fig. 2.

Two proteolytic cleavages occur during production of Kex2 protease. After removal of the signal sequence by signal peptidase (63), additional amino acids are removed from the NH_2 terminus (63) before the protein leaves the endoplasmic reticulum. The processing is thought to be intramolecular and autocatalytic; Lys-Arg sequences are found at residues 79–80 and 108–109, in a region that precedes the subtilisin homology, consistent with an autocatalytic cleavage mechanism (63).

Kex1 Carboxypeptidase—Kex1 carboxypeptidase, encoded by the *KEX1* gene, is a serine protease (12, 13). It appears to be specific for basic residues (13). This glycoprotein contains *N*-linked sugars (13). Kex1 protease is an integral membrane protein. There are no indications of a zymogen form for this enzyme. Preliminary results suggest a late Golgi location (80).

The *KEX1* gene was first identified by mutations that prevented production of killer toxin (7, 8). Unlike *kex2* mutations, however, *kex1* mutations have no effect on fertility of cells of α mating type. The cleavages thought to be catalyzed by Kex1 protease are shown in Fig. 2. The fertility of *kex1* mutants of α mating type depends on α-factor production solely from the COOH-terminal α-factor repeat (12), proving that Kex1 protease, in keeping with its substrate specificity (13), catalyzes removal of the COOH-terminal Lys and Arg residues.

Dipeptidyl Aminopeptidase A—DPAP-A, encoded by the *STE13* gene (11), is a membrane protein (11). The gene has been sequenced.[6] The predicted polypeptide shows marked similarity to that of DPAP-B both in primary structure and in its gross topology (56). It is expected to be a type II glycoprotein, anchored in the Golgi by its sub-NH_2-terminal internal signal sequence,

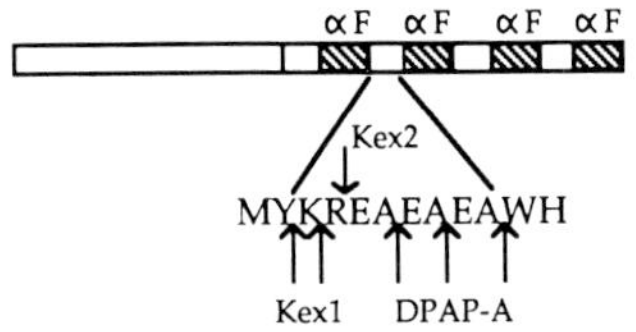

FIG. 2. **Proteolytic cleavages in maturation of the α-factor (αF) precursor.** Cleavages occur in the Golgi in the order Kex2 endoproteinase, Kex1 carboxypeptidase, DPAP-A (86). *Arrows* identify the bonds cleaved within one spacer unit that lies between the α-factor COOH terminus, Met-Tyr, and the next NH_2 terminus, Trp-His.

with a large lumenal domain and a small cytosolic domain. By analogy with what is known about production of DPAP-B, no zymogen form is expected. DPAP-A catalyzes cleavage of Glu-Ala or Asp-Ala dipeptides from the α-factor precursor (Fig. 2).

Yeast Aspartyl Protease III (Yap3 Protease)—A gene, *YAP3*, was described whose product, when overproduced, catalyzed cleavage of the α-factor precursor on the COOH-terminal side of paired basic residues when Kex2 protease activity was absent (81). Conceptual translation of *YAP3* yields a protein with sequence similarity to the two-domain group of aspartyl proteases. Its topological features resemble those of Kex2 protease. It is predicted to have an NH_2-terminal signal sequence and a sub-COOH-terminal transmembrane anchor; most of the protein should be within the compartmental lumen. In alignment with other aspartyl proteases, Yap3 protease has an extra ~45 amino acids at its NH_2 terminus (excluding the predicted signal sequence). Interestingly, there are two Lys-Arg pairs in this 45-amino acid stretch. The second pair is at residues 66–67, within one amino acid of the NH_2 terminus (at 68 or 69) predicted by alignment. I predict that Yap3 protease will have a zymogen form that will undergo autocatalytic activation, probably in the endoplasmic reticulum. Neither the synthesis nor localization of Yap3 protease has been studied.

Functions of Proteases of the Secretory Pathway

Since the *sec11* mutation causes temperature-sensitive lethality, signal peptidase is presumed to be an essential enzyme (74). None of the other proteases of the secretory pathway are essential for viability. These proteases act to process precursors to one or more secreted peptides, with α-factor pheromone and killer toxin being known end products. The normal substrate for Yap3 protease (when expressed at normal levels) is unknown.

Comparison of Synthesis of Vacuolar and Secretory Pathway Proteases

Comparison of the routes of production of vacuolar proteases and secretory pathway proteases, both of which traverse the endoplasmic reticulum and most of the Golgi during production, surfaces the interesting fact that all of the endoproteinases studied (Kex2 protease, PrA, PrB) are synthesized as zymogens that undergo an intramolecular, autocatalytic cleavage step. For Kex2 protease, autocatalysis removes an NH_2-terminal *pro*peptide, and the product is active before the protein leaves the endoplasmic reticulum. For PrB, autocatalysis likewise removes a long NH_2-terminal *pro*peptide within the endoplasmic reticulum, but a COOH-terminal 60-amino acid "silencer" peptide prevents the product from being active. It is only later in the pathway, in the late Golgi or vacuole, that removal of the "silencer" peptide is initiated and the enzyme is activated.

As has been reported for some prokaryotic proteases (82–85), the *pro*peptide is required if active PrB is to be produced. Its presence ensures that the three Asn receptors within the catalytic sequence do not become glycosylated and that PrA-catalyzed cleavage of the 40-kDa intermediate, rather than degradation, occurs.[4] In addition to any functions the *pro* sequences may provide in inducing conformational changes to allow further processing or activity, they in effect buy time in a spatial continuum. A substantial fraction of the enzymes that pass through the endoplasmic reticulum section of the secretory pathway has paired basic residues that could render them susceptible to Kex2

[5] E. Jones, unpublished data.
[6] C. A. Flanagan and J. Thorner, personal communication.

protease (PrA, PrB, ApI, DPAP-A and -B, alkaline phosphatase, invertase, Kex1 protease). Yet none are cleaved by Kex2 protease. Possibly these enzymes, immediately after translocation into the lumen of the (rough) endoplasmic reticulum, assume conformations that bury the basic pairs before Kex2 protease activates itself and becomes active on other molecules (possibly elsewhere in the endoplasmic reticulum). The fact that the PrA-catalyzed cleavages that finally lead to active PrA and PrB occur so late in the pathway, and possibly only within the vacuole itself, ensures that the secretory pathway itself is protected from the action of fairly nonspecific proteases that can be present at high levels.

Several of the vacuolar exopeptidases (and other hydrolases) are synthesized as zymogens, whereas none of the secretory exoproteases appears to be. It is noteworthy that the exopeptidases of the secretory pathway have very restricted substrate specificities whereas vacuolar exopeptidases show broad specificities. Possibly this difference, combined with the fact that vacuolar enzymes may be expressed to high levels, necessitates the existence of zymogens for the vacuolar exopeptidase species, since premature or inopportune expression of the vacuolar activities could be prejudicial to the secretory system.

Connections

Both the cytosolic proteasome and the vacuolar proteases participate in the massive protein degradations triggered by analog-containing proteins or by nitrogen starvation, especially in connection with meiosis and sporulation. It is unknown whether this co-participation simply reflects coincidence in timing of the responses of two separate systems or is an indication of a deeper interaction or interrelationship between the systems. And it is possible that the tantalizing hints of a proteolytic system within the endoplasmic reticulum that degrades unstable *pro* segments and unprocessed precursors[4] signify another connection to the vacuolar system.

We are beginning to get a clear picture of the roles and functions within cells of these proteolytic systems as single entities. We can expect to ferret out any connections between them and gain a better understanding of the totality in the future.

Acknowledgments—I thank the following members of my laboratory for their very thoughtful and helpful suggestions: J. David Garman, Vicki L. Nebes, Robert A. Preston, and Carol A. Woolford.

REFERENCES

1. Dernby, K. G. (1917) *Biochem. Z.* **81,** 107–208
2. Willstätter, R., and Grassman, W. (1926) *Hoppe-Seyler's Z. Physiol. Chem.* **153,** 250–282
3. Hata, T., Hayashi, R., and Doi, E. (1967) *Agric. Biol. Chem.* **31,** 150–159
4. Lenney, J. F., and Dalbec, J. M. (1967) *Arch. Biochem. Biophys.* **120,** 42
5. Lenney, J. F., Matile, P., Wiemken, A., Schellenberg, M., and Meyer, J. (1974) *Biochem. Biophys. Res. Commun.* **60,** 1378–1383
6. Wiemken, A., Schellenberg, M., and Urech, K. (1979) *Arch. Microbiol.* **123,** 23–35
7. Wickner, R. B. (1974) *Genetics* **76,** 423–432
8. Leibowitz, M. J., and Wickner, R. B. (1974) *Proc. Natl. Acad. Sci. U. S. A.* **73,** 2061–2065
9. Julius, D., Brake, A., Blair, L., Kunisawa, R., and Thorner, J. (1984) *Cell* **37,** 1075–1089
10. Sprague, G., Rine, J., and Herskowitz, I. (1981) *J. Mol. Biol.* **153,** 323
11. Julius, D., Blair, L., Brake, A., Sprague, G. F., Jr., and Thorner, J. (1983) *Cell* **32,** 839–852
12. Dmochowska, A., Dignard, D., Henning, D., Thomas, D. Y., and Bussey, H. (1987) *Cell* **50,** 573–584
13. Cooper, A., and Bussey, H. (1989) *Mol. Cell. Biol.* **9,** 2706–2714
14. Achstetter, T., Emter, O., Ehmann, C., and Wolf, D. H. (1984) *J. Biol. Chem.* **259,** 13334–13343
15. Wolf, D. H., and Ehmann, C. (1981) *J. Bacteriol.* **147,** 418–426
16. Achstetter, T., Ehmann, C., and Wolf, D. H. (1983) *Arch. Biochem. Biophys.* **226,** 292–305
17. Rendueles, P., and Wolf, D. (1988) *FEMS Microbiol. Rev.* **54,** 17–46
18. Jones, E. W. (1990) *Methods Enzymol.* **194,** 428–453
19. Achstetter, T., Ehmann, C., Osaki, A., and Wolf, D. H. (1984) *J. Biol. Chem.* **259,** 13344–13348
20. Tanaka, K., Yoshimura, T., Kumatori, A., Ichihara, A., Ikai, A., Nishigai, M., Kameyama, K., and Takagi, T. (1988) *J. Biol. Chem.* **263,** 16209–16217
21. Kleinschmidt, J., Escher, C., and Wolf, D. (1988) *FEBS Lett.* **239,** 35
22. Fujiwara, T., Tanaka, K., Orino, E., Yoshimura, T., Kumatori, A., Tamura, T., Chung, C. H., Nakai, T., Yamaguchi, K., Shin, S., Kakizuka, A., Nakanishi, S., and Ichihara, A. (1990) *J. Biol. Chem.* **265,** 16604–16613
23. Emori, Y., Tsukahara, T., Kawasaki, H., Ishiura, S., Sugita, H., and Suzuki, K. (1991) *Mol. Cell. Biol.* **11,** 344–353
24. Heinemeyer, W., Kleinschmidt, J. A., Saidowsky, J., Escher, C., and Wolf, D. H. (1991) *EMBO J.* **10,** 555–562
25. Balzi, E., Chen, W., Capieaux, E., McCusker, J., Haber, J., and Goffeau, A. (1989) *Gene (Amst.)* **83,** 271; correction (1990) *Gene (Amst.)* **89,** 151
26. Suárez Rendueles, M. P., Schwenke, J., Garcia Alvarez, N., and Gascon, S. (1981) *FEBS Lett.* **131,** 296–300
27. Woolford, C. A., Daniels, L. B., Park, F. J., Jones, E. W., Van Arsdell, J. N., and Innis, M. A. (1986) *Mol. Cell. Biol.* **6,** 2500–2510
28. Ammerer, G., Hunter, C. P., Rothman, J. H., Saari, G. C., Valls, L. A., and Stevens, T. H. (1986) *Mol. Cell. Biol.* **6,** 2490–2499
29. Dreyer, T., Halkier, B., Svendsen, I., and Ottesen, M. (1986) *Carlsberg Res. Commun.* **51,** 27–41
30. Klionsky, D., Banta, L., and Emr, S. (1988) *Mol. Cell. Biol.* **8,** 2105
31. Zubenko, G. S., Park, F. J., and Jones, E. W. (1983) *Proc. Natl. Acad. Sci. U. S. A.* **80,** 510–514
32. Jones, E. W., Woolford, C. A., Moehle, C. M., Noble, J. A., and Innis, M. A. (1989) *UCLA Symposium on Cellular Proteases and Control Mechanisms*, pp. 141–147, Alan R. Liss, Inc., New York
33. Moehle, C. M., Tizard, R., Lemmon, S. K., Smart, J., and Jones, E. W. (1987) *Mol. Cell. Biol.* **7,** 4390–4399
34. Mechler, B., Müller, M., Müller, H., Muessdoerffer, F., and Wolf, D. H. (1982) *J. Biol. Chem.* **257,** 11203–11206
35. Ulane, R., and Cabib, E. (1976) *J. Biol. Chem.* **251,** 3367–3374
36. Moehle, C., Dixon, C., and Jones, E. (1989) *J. Cell. Biol.* **108,** 309
37. Mechler, B., Hirsch, H. H., Müller, H., and Wolf, D. H. (1988) *EMBO J.* **7,** 1705–1710
38. Schiffer, H. H., Hirsch, H. H., Müller, H., and Wolf, D. H. (1990) *Yeast* **6,** S479 (abstr.)
39. Wells, J. A., Cunningham, B. C., Graycar, T. P., and Estell, D. A. (1986) *Philos. Trans. R. Soc. Lond. A* **317,** 415–423
40. Bryan, P., Pantoliano, M. W., Quill, S. G., Hsiao, H.-Y., and Poulos, T. (1986) *Proc. Natl. Acad. Sci. U. S. A.* **83,** 3743–3745
41. Wright, C. S., Alden, R. A., and Kraut, J. (1969) *Nature* **221,** 235–242
42. Zubenko, G. S., Park, F. J., and Jones, E. W. (1982) *Genetics* **102,** 679–690
43. Jones, E., Moehle, C., Kolodny, M., Aynardi, M., Daniels, L., and Garlow, S. (1986) in *UCLA Symposium on Yeast Cell Biology* (Hicks, J., ed) p. 505, Alan R. Liss, Inc., New York
44. Trimble, R. B., and Maley, F. (1977) *Biochem. Biophys. Res. Commun.* **78,** 935–944
45. Hayashi, R., Moore, S., and Stein, W. H. (1973) *J. Biol. Chem.* **248,** 2296
46. Blachly-Dyson, E., and Stevens, T. H. (1987) *J. Cell. Biol.* **104,** 1183–1191
47. Hasilik, A., and Tanner, W. (1978) *Eur. J. Biochem.* **85,** 599–608
48. Hemmings, B. A., Zubenko, G. S., Hasilik, A., and Jones, E. W. (1981) *Proc. Natl. Acad. Sci. U. S. A.* **78,** 435–439
49. Stevens, T., Esmon, B., and Schekman, R. (1982) *Cell* **30,** 439–448
50. Mechler, B., Müller, H., and Wolf, D. H. (1987) *EMBO J.* **6,** 2157–2163
51. Jones, E. W., Zubenko, G. S., and Parker, R. R. (1982) *Genetics* **102,** 665
52. Metz, G., and Röhm, K.-H. (1976) *Biochim. Biophys. Acta* **429,** 933–949
53. Trumbly, R. J., and Bradley, G. (1983) *J. Bacteriol.* **156,** 36–48
54. Chang, Y.-H., and Smith, J. A. (1989) *J. Biol. Chem.* **264,** 6979–6983
55. Achstetter, T., Ehmann, C., and Wolf, D. H. (1982) *Biochem. Biophys. Res. Commun.* **109,** 341–347
56. Roberts, C. J., Pohlig, G., Rothman, J. H., and Stevens, T. (1989) *J. Cell. Biol.* **108,** 1363–1373
57. Harris, S. D., and Cotter, D. A. (1987) *Curr. Microbiol.* **15,** 247–249
58. Jones, E. W. (1977) *Genetics* **85,** 23–33
59. Jones, E. W., Zubenko, G. S., Parker, R. R., Hemmings, B. A., and Hasilik, A. (1981) *Alfred Benzon Symp.* **16,** 182–198
60. Mechler, B., and Wolf, D. H. (1981) *Eur. J. Biochem.* **121,** 47–52
61. Dreyer, T. (1989) *Carlsberg Res. Commun.* **54,** 83–97
62. Kominami, E., Hoffschulte, H., Leuschel, L., Maier, K., and Holzer, H. (1981) *Biochim. Biophys. Acta* **661,** 136–141
63. Fuller, R. S., Brenner, C., Gluschankof, P., and Wilcox, C. A. (1991) in *Advances in Life Sciences* (Jörnvall, H., and Höög, J.-O., eds) Birkhäuser Verlag, Berlin, in press
64. Singer, B., and Riezman, H. (1990) *J. Cell. Biol.* **110,** 1911–1922
65. Zubenko, G. (1981) Ph.D. thesis, Carnegie Mellon University, Pittsburgh, PA
66. Teichert, U., Mechler, B., Müller, H., and Wolf, D. H. (1989) *J. Biol. Chem.* **264,** 16037–16045
67. Tsay, Y.-F., Thompson, J. R., Rotenberg, M. O., Larkin, J. C., and Woolford, J. L., Jr. (1988) *Genes & Dev.* **2,** 664–676
68. Zubenko, G. S., and Jones, E. W. (1981) *Genetics* **97,** 45–64
69. Funaguma, T., Toyoda, Y., and Sy, J. (1985) *Biochem. Biophys. Res. Commun.* **130,** 467–471
70. Schäfer, W., Kalisz, H., and Holzer, H. (1987) *Biochim. Biophys. Acta* **925,** 150–155
71. Hemmings, B. A., Zubenko, G. S., and Jones, E. W. (1980) *Arch. Biochem. Biophys.* **202,** 657–660
72. Zubenko, G., and Jones, E. (1979) *Proc. Natl. Acad. Sci. U. S. A.* **76,** 4581
73. YaDeau, J. T., Klein, C., and Blobel, G. (1991) *Proc. Natl. Acad. U. S. A.* **88,** 517–521
74. Böhni, P. C., Deshaies, R. J., and Schekman, R. W. (1988) *J. Cell. Biol.* **106,** 1035–1042
75. Greenburg, G., Shelness, G. S., and Blobel, G. (1989) *J. Biol. Chem.* **264,** 15762–15765
76. Fuller, R. S., Brake, A., and Thorner, J. (1989) *Proc. Natl. Acad. Sci. U. S. A.* **86,** 1434–1438
77. Mizuno, K., Nakamura, T., Ohshima, T., Tanaka, S., and Matsuo, H. (1988) *Biochem. Biophys. Res. Commun.* **156,** 246–254
78. Fuller, R. S., Brake, A. J., and Thorner, J. (1989) *Science* **246,** 482–486
79. Franzusoff, A., Redding, K., Crosby, J., Fuller, R. S., and Schekman, R. (1991) *J. Cell. Biol.* **112,** 27–37
80. Cooper, A., and Bussey, H. *Yeast* **6,** S488 (abstr.)
81. Egel-Mitani, M., Flygenring, H., and Hansen, M. (1990) *Yeast* **6,** 127
82. Power, S. D., Adams, R. M., and Wells, J. A. (1986) *Proc. Natl. Acad. Sci. U. S. A.* **83,** 3096–3100
83. Zhu, X., Ohta, Y., Jordan, F., and Inouye, M. (1989) *Nature* **339,** 483–484
84. Silen, J. L., Frank, D., Fujishige, A., Bone, R., and Agard, D. A. (1989) *J. Bacteriol.* **171,** 1320–1325
85. Silen, J. L., and Agard, D. A. (1989) *Nature* **341,** 462–464
86. Fuller, R., Sterne, R., and Thorner, J. (1988) *Annu. Rev. Physiol.* **50,** 345

THE JOURNAL OF BIOLOGICAL CHEMISTRY
Vol. 266, No. 14, Issue of May 15, pp. 8647–8650, 1991

Protein *N*-Myristoylation

Jeffrey I. Gordon‡, Robert J. Duronio‡, David A. Rudnick‡, Steven P. Adams§, and George W. Gokel¶

From the ‡Department of Molecular Biology and Pharmacology, Washington University School of Medicine, St. Louis, Missouri 63110, §Monsanto Company, St. Louis, Missouri 63198, and the ¶Department of Chemistry, University of Miami, Coral Gables, Florida 33124

Protein *N*-myristoylation refers to the co-translational (1) linkage of myristic acid (C14:0) via an amide bond to the NH_2-terminal Gly residues of a variety of eukaryotic cellular and viral proteins (2, 3). The reaction is catalyzed by myristoyl-CoA:protein *N*-myristoyltransferase (NMT).[1] *N*-Myristoylation appears to be irreversible (4) but exceptions may exist (5).

N-Myristoylproteins have diverse biological functions and diverse intracellular destinations (2, 3). Examples of NMT substrates include protein kinases such as the catalytic (C) subunit of cAMP-dependent protein kinase (PK-A) and $p60^{src}$, phosphatases such as calcineurin B, proteins involved in transmembrane signaling such as several guanine nucleotide-binding α subunits of heterotrimeric G proteins, the gag polyprotein precursors of a number of retroviruses (*e.g.* HIV-I and Moloney murine leukemia virus) as well as the capsid proteins of some papovaviruses and picornaviruses.

Deletion or substitution of the Gly^2 residue of *N*-myristoylproteins by site-directed mutagenesis prohibits their acylation, allowing the properties of the mutant, nonmyristoylated and wild-type, *N*-myristoylated species to be compared and contrasted. This experimental strategy has been used to show that myristate, a rare fatty acid comprising <2% of total cellular fatty acids, is needed for full expression of the biological function of several *N*-myristoylproteins. For example, $Gly^2 \rightarrow Ala^2$ mutagenesis of $p60^{v\text{-}src}$ prevents its targeting to the plasma membrane and blocks its ability to transform cells without affecting its tyrosine kinase activity. The interaction between $p60^{src}$ and a 32-kDa plasma membrane receptor appears to require *N*-myristoylation of the tyrosine kinase (6). Mutagenesis of the Gly^2 residue of HIV-1 $Pr55^{gag}$ blocks viral assembly (7, 8). Failure to *N*-myristoylate $G_{0\alpha}$ reduces its affinity for G protein $\beta\gamma$ (9). Site-directed mutagenesis of poliovirus VP4 (10, 11) as well as x-ray crystallographic studies (12) indicate that myristate plays a critical role in assembly by affecting protein-protein interactions among capsid components.

Different *N*-myristoylproteins are associated with distinct intracellular compartments and membranes (2), *e.g.* the endoplasmic reticulum in the case of cytochrome-b_5 reductase, the plasma membrane in the case of $p60^{v\text{-}src}$, the cytoplasm when the C subunit of PK-A is bound to its regulatory subunit, and the nucleoskeleton for polyoma and SV40 VP2. Based on these observations, it appears that myristate is not a signal that is sufficient, in and of itself, for achieving the final subcellular location of *N*-myristoylproteins.

Recent genetic, biochemical, organic chemical, molecular, and cell biological studies have yielded (i) a number of insights about how protein *N*-myristoylation is regulated and (ii) provided strategies for modulating this process *in vivo* that may have therapeutic implications.

[1] The abbreviations used are: NMT, myristoyl-CoA:protein *N*-myristoyltransferase; HIV-I, human immunodeficiency virus I; PK-A, catalytic subunit of cAMP-dependent protein kinase; O6, 6-oxatetradecanoic acid.

Analysis of the Peptide and Acyl-CoA Substrate Specificity of NMT in Vitro

The most intensively studied NMT is that produced in *Saccharomyces cerevisiae.* This 53-kDa monomeric enzyme has been purified 11,000-fold to apparent homogeneity (13).

The specificities of several NH_2-terminal modifying enzymes such as *N*-acetyltransferase and methionylaminopeptidase are apparently determined by the physicochemical properties of the first 2 residues of their protein substrates (14–17). In contrast, the peptide substrate specificity of *S. cerevisiae* NMT is profoundly affected by amino acids distributed over a broader region of the primary translation product. Analyses of >100 synthetic peptides (2, 13, 18) have indicated that *S. cerevisiae* NMT has no intrinsic methionylaminopeptidase activity and displays an absolute requirement for an NH_2-terminal Gly. Introduction of bulky hydrophobic residues at position 2 (the 3rd residue of the primary translation product) results in competitive inhibitors. Serine at position 5 promotes high affinity interactions with the acyltransferase. Basic residues in positions 7 and 8 result in better substrates than the corresponding neutral residue-substituted peptides which, in turn, are better substrates than those with acidic residues in these positions. Finally, for certain *N*-myristoylproteins (*e.g.* $p60^{src}$ and some G_α subunits), sequences beyond the NH_2-terminal eight amino acids appear to play a role in recognition by the enzyme (19, 20).

S. cerevisiae NMT displays a high degree of specificity for myristoyl-CoA *in vitro* (21). It is able to accommodate acyl chains that are one methylene group shorter or longer than C14:0, perhaps because there is no selective pressure to avoid these fatty acids which are so rare in yeast (22). However, the enzyme is highly selective against transfer of palmitate, which is abundantly represented in eukaryotic cells. Over 80 fatty acid analogs of myristate have been used to further probe the molecular determinants that govern the unique acyl-CoA substrate specificity of this enzyme (21, 23). Analogs with single oxygen or sulfur substitutions at C3–C13 are well accommodated by NMT (21, 23). These analogs are approximately equivalent in chain length to myristate but have markedly reduced hydrophobicity (equivalent to C10:0–C12:0 fatty acids). *In vitro* analyses of these single heteroatom-substituted analogs as well as naturally occurring C10:0–C16:0 fatty acids have demonstrated that the acyl-CoA specificity of *S. cerevisiae* NMT derives from the chain length and not the hydrophobicity of fatty acyl ligands *and* an apparent cooperativity between myristoyl-CoA and peptide binding (21). Surveys (23) of myristate analogs containing a single triple bond (alkynes, Y), cis (Z) or trans (E) double bonds (alkenes), or an aromatic substitution have suggested that the acyl chain of myristoyl-CoA is present in a bent conformation near C6 with the first four or five carbons of the fatty acid in an extended conformation: *e.g.* Y5 is not a substrate while Y4 and Y6 have activities comparable with C14:0; Z5 is a superior substrate to all the other 10 cis double bond isomers tested; E5 is inferior to E6 and E7 which, in turn, are comparable to myristate. These surveys (23) also show the acyl-CoA binding site possesses a complex sensor that measures the distance from the carboxyl to the omega end of the fatty acid as well as the steric volume at the omega terminus. The geometry of the terminal sensing device may be conical, *e.g.* 10-phenyldecanoic acid is a markedly better substrate *in vitro* than 11-phenylundecanoic acid but is equivalent to 10-(4-tolyl)decanoic acid. This latter compound has the same relative positions of phenyl and carboxyl as 10-phenyldecanoic acid but has an additional methyl group on the aromatic ring. The role of the CoA moiety in acyl-CoA recognition by NMT has only begun to be addressed, *e.g.* the 3′-phosphate group of CoA is not absolutely required but does enhance binding (24).

The Catalytic Mechanism of NMT

The mechanism of catalysis of *S. cerevisiae* NMT has been characterized (22, 24). Isoelectric focusing and tryptophan fluorescence quenching experiments (22) have indicated that a high affinity complex can form *in vitro* between the acyltransferase and myristoyl-CoA in the absence of a peptide substrate. The isoelectric focusing and fluorescence properties of this complex are restored to those of the apoenzyme upon addition of a peptide substrate. Addition of [1-^{14}C]myristoyl-CoA to apo-NMT followed by proteolysis, denaturing sodium dodecyl sulfate-polyacrylamide gel electrophoresis, hydroxylamine treatment, and NH_2-terminal sequence analysis have yielded results that are suggestive of a covalent oxy- or thioester-linked acyl-enzyme complex mediated by a serine, threonine, or cysteine residue located between Arg^{42} and Thr^{220} (22).

These data, together with the kinetic characterization of cooperativity between myristoyl-CoA and peptide binding (see above), are indicative of an ordered catalytic process in which myristoyl-CoA binds to NMT prior to peptide and thereby influences the interaction of NMT with peptide. The data do not, however, distinguish between ordered reaction mechanisms in which CoA product release precedes (ping-pong) or follows (sequential Bi Bi) peptide substrate binding. Additional binding studies using *S*-(2-oxo)pentadecyl-CoA, an analog of myristoyl-CoA that cannot be hydrolyzed due to a methylene bridge between the acyl carbonyl moiety and the sulfur atom of CoA (25), as well as myristoylpeptide and CoA, have indicated that CoA is necessarily retained in the acyl-NMT complex prior to peptide addition (24). This is inconsistent with a ping-pong reaction mechanism in which the first product of a two-substrate/two-product reaction releases prior to binding of the second substrate.

Product inhibition studies have demonstrated that myristoyl peptide is a competitive inhibitor for myristoyl-CoA and a noncompetitive inhibitor for peptide while CoA is a noncompetitive inhibitor of both myristoyl-CoA and peptide (24). The pattern of product inhibition observed in a two-substrate reaction, under nonsaturating conditions, can be diagnostic for the mechanism of catalysis by that enzyme (26). For Theorell-Chance and ping-pong mechanisms, each product is a competitive inhibitor for one substrate and a noncompetitive inhibitor for the other substrate. In random mechanisms, each product can be a competitive inhibitor for each substrate. Only an ordered Bi Bi reaction mechanism exhibits a pattern of inhibition in which one product is a noncompetitive inhibitor for both substrates (the other product is competitive and noncompetitive for the two substrates, respectively). Thus, the pattern of product inhibition of NMT is consistent with the structural data and diagnostic for an ordered Bi Bi reaction in which myristoyl-CoA binds to NMT prior to peptide and CoA release precedes myristoylpeptide release. The reaction mechanism characterized *in vitro* suggests that NMT binds myristoyl-CoA *in vivo* prior to identifying and acylating a nascent polypeptide chain containing the appropriate recognition sequence.

Reconstitution of Protein N-Myristoylation in Escherichia coli Provides an in Vivo Assay of NMT Activity

The structural gene encoding *S. cerevisiae* NMT (*NMT1*) has been cloned and sequenced (27). The 1365-nucleotide open reading frame of *NMT1* specifies a polypeptide of 455 residues with a calculated M_r of 52,387 that has no obvious primary sequence similarity to any other protein in current data bases (27).

E. coli has no endogenous NMT activity (28). Protein *N*-myristoylation can be reconstituted in this bacterium using a dual plasmid expression system (28). The *NMT1* gene and a cDNA encoding a known or putative *N*-myristoylprotein can be co-expressed in *E. coli* using separate plasmids containing (i) individually inducible promoters, (ii) different but compatible origins of replication, and (iii) different antibiotic resistance genes (29). Following sequential induction of NMT, then substrate protein synthesis, acylation can be detected by the addition of ^{3}H-fatty acid to the media. Several known eukaryotic *N*-myristoylproteins have been examined in this system. These include the Cα subunit of mouse PK-A (27), rat $G_{i\alpha 1}$ and rat $G_{o\alpha}$ (20), and the gag and nef proteins of HIV-1 (30). Exogenous [^{3}H]palmitate must be metabolically converted to myristate before incorporation, indicating that acylation in the co-expression system is specific for C14:0 fatty acids (27). *N*-Myristoylation in this system is also remarkably efficient (80–100% substitution is achieved) and requires a Gly^2 residue (27, 30). Finally, some oxygen-containing analogs of myristate that are substrates *in vitro* can also be incorporated into these proteins *in vivo* (20, 28–30).

The ability to reconstitute this eukaryotic protein modification in *E. coli* suggests that other eukaryotic proteins produced in *S. cerevisiae* are not required to achieve the stringent acyl-CoA and protein substrate specificities of NMT. However, the system can be used to demonstrate and explore the dependence of *N*-myristoylation on cellular activities common to both prokaryotes and eukaryotes. All *N*-myristoylproteins must be substrates for methionylaminopeptidase. A penultimate Gly allows for efficient recognition of nascent polypeptides by both yeast and *E. coli* methionylaminopeptidase (15, 16). *N*-Myristoylproteins must not be substrates for other NH_2-terminal processing enzymes, such as *N*-acetyltransferase, or must avoid interactions with them. The availability and metabolic processing of C14:0 *in vivo* may also affect *N*-myristoylation. *E. coli* depends upon the action of the *fadL*$^+$ long chain fatty acid transporter (31) and the acyl-CoA synthetase encoded by *fadD*$^+$ (32) to import exogenous fatty acids. NMT produced in *E. coli* must somehow avoid inhibition by palmitoyl-CoA (see above). Together these considerations suggest that *N*-myristoylation may be regulated not only by the intrinsic properties of NMT but also by temporal and spatial coordination of other enzyme activities and substrate pools.

The *E. coli* co-expression system provides a way for exploring structure/activity relationships in NMT and its protein substrates. The ability of *S. cerevisiae* NMT to acylate a variety of mammalian *N*-myristoylproteins underscores the degree of conservation of its substrate specificity over the course of eukaryotic evolution. The co-expression system will be useful in directly determining whether a protein is subject to this form of covalent lipid modification. The role of the myristoyl moiety in the function of a given protein can also be examined by comparing the biological properties of purified, *E. coli*-derived, nonmyristoylated, myristoylated, and analog-substituted species (*e.g.* Ref. 9). Finally, the activities of alternative substrates and inhibitors of NMT can be surveyed.

Modulation of NMT Activity in Vivo

Studies in S. cerevisiae—The *NMT1* locus has been mapped by physical and genetic methods to the right arm of chromosome XII, 22 centimorgans centromere proximal to *cdc42* (27, 33). *NMT1* is not allelic with any previously mapped genetic loci. Disruption or deletion of *NMT1* causes recessive lethality (27, 33), indicating that one or more cellular proteins requires its myristoyl moiety to fulfill an essential function and that no other cellular activity can substitute for NMT. Subsequent whole cell mutagenesis experiments designed to search for suppressors of the lethal *nmt1* null phenotype indicated that the frequency of such revertants must be less than 10^{-6}, suggesting that a single *S. cerevisiae* locus could not be altered (with ethyl methanesulfonate) such that it would compensate for an *nmt* null allele. These data imply that C14:0 fulfills an irreplaceable function for a single critical *N*-myristoylprotein or that there is more than one essential *N*-myristoylprotein in *S. cerevisiae.* Metabolic labeling studies with [^{3}H]myristate and subsequent one- or two-dimensional gel electrophoretic fractionation of cell lysates reveal that this budding yeast synthesizes 10–12 *N*-myristoylproteins (33, 34). Based on inspection of NH_2-terminal sequences, these probably include the GTP-binding GPA1 (35), GPA2 (36), ARF1, and ARF2 (37, 38) proteins, and the VPS15 serine/threonine kinase (39). GPA2 and VPS15 are not essential (36, 39). GPA1 is a component of a haploid-specific, essential G protein involved in mating pheromone signal transduction (35). However, *NMT1* is required for growth of diploid cells, suggesting that GPA1 is not the only essential *N*-myristoylprotein in *S. cerevisiae*, e.g.

cells with *arf1 arf2* double mutations are inviable (38).

Analysis of conditional lethal mutations of *NMT1* has yielded additional insights about enzyme structure/activity relationships as well as the contribution of other cellular factors to efficient *N*-myristoylation *in vivo*. Mutations in *S. cerevisiae* that produce auxotrophy for saturated fatty acids have been mapped to several unlinked loci (40, 41). They result in temperature-sensitive lethality that can be complemented with exogenous C14:0, C16:0, or C18:0 (40). The majority of these strains contain mutations of the fatty acid synthetase genes (*FAS1, FAS2*). Meyer and Schweizer (42) also characterized a single mutation that affects a locus other than *FAS1* or *FAS2*. This mutant (LK181) is specifically auxotrophic for myristate at 36 °C; growth occurs only if the media is supplemented with at least 500 μM C14:0. Other shorter or longer chain saturated fatty acids will not suffice (33, 42). This contrasts with FAS mutations that can be rescued with a variety of different chain length fatty acids (C12:0, C14:0, and C16:0 (40)). At intermediate temperatures (30 °C), the ability of different chain length saturated fatty acids to rescue growth of LK181 parallels the chain length specificity of *S. cerevisiae* NMT (22, 42). Subsequent meiotic mapping studies indicated that the mutation in LK181 is allelic with the *NMT1* locus (33). This mutation, now designated *nmt1–181*, causes supersensitivity to cerulenin, a compound that effectively eliminates *de novo* sources of long chain acyl-CoAs by inhibiting *S. cerevisiae* fatty acid synthetase. A functional acyl-CoA synthetase gene (*FAA1*) is required for myristate to complement growth of strains containing *nmt1–181* at the restrictive temperature (33). Together these data suggest that the *nmt1–181* mutation produces a sensitivity to the size of acyl-CoA pools available within this yeast. Little is currently known about how these pools are maintained. Even less is known about their intracellular locations. The $\alpha_6\beta_6$ fatty acid synthetase complex of *S. cerevisiae* synthesizes primarily palmitoyl-CoA and steroyl-CoA, although small amounts of myristoyl-CoA are also generated (43). *S. cerevisiae* also contains acyl chain elongating enzymes (44) capable of supplying acyl-CoA pools by extending exogenous shorter chain length, saturated fatty acids by 2 carbon units (Fig. 1).

Sequence analysis of the *nmt1–181* allele revealed a single missense mutation encoding an Asp for Gly substitution at position 451 of the 455-residue primary translation product (33). Introduction of *nmt1–181* into a *nmt1* null strain recapitulates the mutant phenotype (33). Metabolic labeling experiments with [^{3}H]myristate reveal a decrease in the incorporation of label into a subset of *N*-myristoylproteins after shifting to the restrictive temperature. Western blot studies indicate that steady state levels of the mutant and wild type enzyme are comparable at both permissive and restrictive conditions. Together these observations suggest that at 36 °C, some *S. cerevisiae* proteins are poorer substrates for nmt-181 than others.

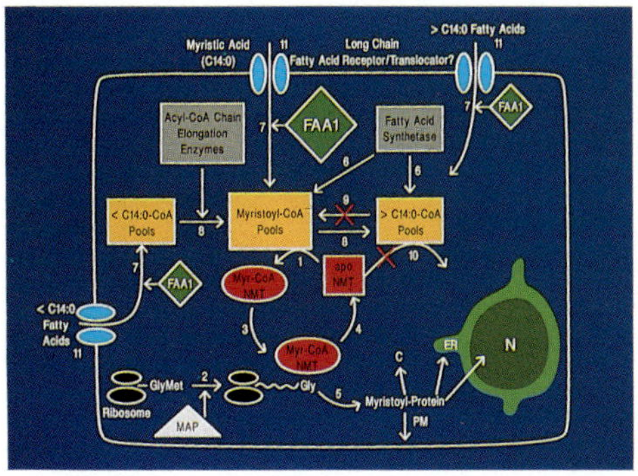

FIG. 1. **Regulation of protein *N*-myristoylation *in vivo*.** Analyses of the catalytic mechanism of NMT and characterization of *nmt1–181* have suggested that protein *N*-myristoylation in *S. cerevisiae* involves the following steps. 1) apo-NMT acquires its acyl-CoA substrate only from pools of myristoyl (*Myr*)-CoA; 2) the initiator Met is removed from nascent polypeptides containing a Gly^2 residue by methionylaminopeptidase (*MAP*); 3) myristoyl-CoA NMT identifies and binds nascent polypeptide chains that encode the appropriate NH_2-terminal consensus sequence; 4) apo-NMT is regenerated following co-translational transfer of myristate to the exposed Gly^2 residue of the acceptor polypeptide; 5) by analogy to higher eukaryotes, the resulting *N*-myristoylproteins can be targeted to a variety of intracellular destinations (cytoplasm (*C*); nucleus (*N*); plasma membrane (*PM*); endoplasmic reticulum (*ER*), etc.). Myristoyl-CoA pools within cells are supplied by *de novo* fatty acid biosynthesis (6), uptake of exogenous myristic acid and conversion to the CoA thioester by the *FAA1* acyl-CoA synthetase (7), and elongation of shorter chain length acyl-CoA (8). Acyl-CoAs with chain lengths longer than C14 do not efficiently supply myristoyl-CoA pools due to the low levels of β-oxidation in *S. cerevisiae* (9). Palmitoyl-CoA may be inaccessible by NMT *in vivo* since it can effectively inhibit NMT *in vitro* (10). The mechanism of entry of exogenous fatty acids into *S. cerevisiae* is not known and may involve specific acceptor/translocator proteins as in *E. coli* (11).

nmt-181 has been expressed in *E. coli* (33). *In vitro* kinetic analyses of the purified mutant enzyme demonstrated a 10-fold increase in myristoyl-CoA K_m at 36 °C relative to wild type that contributes to a 200-fold decrease in catalytic efficiency (acyl-CoA V_m/K_m). Whole cell mutagenesis with ethyl methanesulfonate allowed isolation of a number of intragenic, but no extragenic, suppressors of the temperature-sensitive auxotrophy caused by *nmt1–181*. Sequence analysis of these alleles after their recovery from the *S. cerevisiae* genome by double strand gap repair (45) revealed the following substitutions: Asp^{451} $\rightarrow$ Asn^{451}; Glu^{293} $\rightarrow$ Lys^{293}/Asp^{451} Glu^{167} $\rightarrow$ Lys^{167}/Asp^{451}. Engineering a Lys codon at position 451 of *NMT1* created an allele incapable of complementing either *nmt1–181* or *nmt1* null mutants (33). The catalytic activity at 36 °C of nmt-181 (Asp^{451}) and nmt-81 (Asn^{451}) in the *E. coli* co-expression system reflected the extent to which each protein could support growth in *S. cerevisiae* at the nonpermissive temperature (33).

Together these observations invite a number of conclusions. *First*, nmt-181 is a sensitive reporter of intracellular myristoyl-CoA pools. Because strains containing *nmt1–181* are affected by cerulenin, it appears that NMT can access myristoyl-CoA synthesized *de novo*. NMT also has access to exogenous C14:0 since this fatty acid can rescue the mutant phenotype. Longer chain length fatty acids (*e.g.* C16:0) and shorter chain length fatty acids (*e.g.* C12:0) failed to complement growth at 36 °C. This suggests that these fatty acids cannot expand cellular myristoyl-CoA pools by β-oxidation or chain elongation to an extent that is sufficient to meet the requirements of the mutant enzyme for *N*-myristoylation of critical cellular proteins at a level required to support vegetative growth. *Second*, the *in vivo* kinetics provide insights about the limits of reduction in affinity of the enzyme for its acyl-CoA ligand which are compatible with viability. *Third*, it appears that the physicochemical properties of this extreme COOH-terminal residue (Gly^{451}) can critically affect myristoyl-CoA binding. Given the ordered Bi Bi reaction mechanism, perturbations in myristoyl-CoA binding would be expected to produce catalytic alterations. Increases in intracellular myristoyl-CoA levels are likely to compensate for this reduced affinity and may also perturb the location of these pools (making them more accessible).

Some heteroatom analogs of myristate with single oxygen or sulfur for methylene substitutions can replace C14:0 in rescuing growth at 36 °C of strains containing *nmt1–181*, *e.g.* 4-oxa-, 5-thia-, 7-thia-, 8-thia-, 9-thia-, 10-thia-, or 13-thiatetradecanoic acids at 500 μM. Rescue of growth by these analogs requires a functional *FAA1* gene product and thus their conversion to CoA thioesters. However, the extent of their complementation of growth is always less than that achieved with myristate and varies between compounds (9-thiatetradecanoic acid is the best). Some analogs cannot support any growth of the mutant at 36 °C (*e.g.* 3-thia-, 6-thia-, 12-thiatetradecanoic acids plus analogs with oxygen substitutions at C5–C13). Surprisingly, several analogs prevent growth of *nmt1–181* strains at the permissive temperature while not affecting wild-type strains. This effect is most pronounced with 6-oxatetradecanoic acid (O6). The level of accumulation of tritiated O6 in wild-type and mutant yeast cells after a 30-min labeling is 1–5% that of myristate (33). Proteins that can be easily detected by metabolic labeling with [^{3}H]myristate cannot be detected under identical conditions using [^{3}H]O6 (33). Furthermore, O6 does not inhibit NMT *in vitro* (21, 30). These data suggest that O6 may affect a function before any interaction with NMT. Since depletion of endogenous myristoyl-CoA pools (*i.e.* by cerulenin treatment) prevents growth of the mutant at 24 °C, the observed phenotype is consistent with O6 interfering with the generation of sufficient cellular myristoyl-CoA pools for the mutant enzyme to function at a level necessary for adequate acylation of critical substrates. Any number of processes that contribute to steady state intracellular myristoyl-CoA levels, such as exogenous fatty acid uptake (by a receptor/translocator?) or activation by FAA1 or biosynthesis, may be inhibited by O6 (see Fig. 1).

Modulating Protein N-Myristoylation in Mammalian Cells—The ability of certain heteroatom-containing analogs of myristate to rescue growth of strains containing *nmt1-181* could reflect their ability to substitute for myristate in certain critical cellular *N*-myristoylproteins. Metabolic labeling studies in *S. cerevisiae* with a limited number of oxatetradecanoic acids indicate that they can enter cells and serve as substrates for acyl-CoA synthetase and NMT (34).

More extensive metabolic labeling experiments (46) using (i) tritiated oxatetradecanoic acids containing oxygen for methylene substitutions at C6, C11, and C13 and (ii) a variety of cultured mammalian cell lines have shown that these analogs are selectively incorporated into distinct subsets of cellular *N*-myristoylproteins. This selectivity may arise, in part, from the cooperativity that affects acyl-CoA and peptide binding. *In vitro* studies have shown that a given analog-CoA can produce modest perturbations in the enzyme's peptide binding site resulting in sequence-specific alterations in peptide V_m/K_m (34). Incorporation of an analog into a given protein produces analog-dependent and -specific effects on function, *e.g.* the degree of redistribution of $p60^{v\text{-}src}$ from membrane to cytosolic fractions depends upon the location of the oxygen substitution (34, 46). These selective effects have also been noted between homologous proteins, *e.g.* some G_α subunits undergo redistribution with a given analog while others do not (47). Many cellular *N*-myristoylproteins appear to be insensitive to the effects of heteroatom substitution, as judged by their cytoplasmic-membrane distribution, even though these compounds exhibit reductions in hydrophobicity equivalent to the removal of 2–4 methylene groups from C14:0. This suggests that different proteins have different dependences upon myristate for their function. Selective incorporation and selective effects on function probably account for the relative lack of toxicity of oxatetradecanoic acids (at concentrations up to 200 μM (30, 34, 46, 48)).

The ability of these small molecular weight (≈230–250) analogs to efficiently enter cells and serve as substrates for mammalian acyl-CoA synthetase and NMT provides an opportunity to explore the biological effects of altering the physicochemical properties of the acyl moiety of *N*-myristoylproteins. Recent studies (30, 48) have demonstrated that some oxatetradecanoic acid analogs (*e.g.* 13-oxatetradecanoic acid) can inhibit the replication of HIV-I without accompanying cellular toxicity in both acutely and chronically infected $CD4^+$H9 cells. This appears due, at least in part, to incorporation of the analog into $Pr55^{gag}$, an associated redistribution of the polyprotein from membrane to cytosolic fractions, and a reduction in its proteolytic processing. This latter effect may be due to the analog's effect on release of viral protease from gag-pol which requires dimerization (of gag-pol (49)). Exploiting cellular NMT activity to selectively deliver these alternate substrates may represent a generally useful strategy for defining the properties of myristate that contribute to the function of specific *N*-myristoylproteins and for perturbing this function.

Future Directions

The nature of, and relationships between, NMT's acyl-CoA and peptide binding sites will require definition of its atomic structure as well as analyses of NMTs from other eukaryotes. Understanding regulation of protein *N*-myristoylation *in vivo* will necessitate a better appreciation of the spatial and functional relationships between NMT and (i) its nascent polypeptide and acyl-CoA ligands, (ii) acyl-CoA-generating enzymes, and (iii) other NH_2-terminal modifying enzymes. The factors that allow for coordination of the activities of these enzymes need to be identified. Modulation of NMT activity *in vivo*, either through the use of specific inhibitors or alternative substrates, may have therapeutic applications not only for oncogenesis but for any pathogen that is dependent upon *N*-myristoylation of critical proteins (viruses, fungi, etc.). The integration of genetics, biochemistry, cell and molecular biology, together with organic chemistry should ensure that future explorations of this form of protein acylation will yield a global view of its significance and modulation.

REFERENCES

1. Wilcox, W., Hu, J.-S., and Olson, E. N. (1987) *Science* **238,** 1275–1278
2. Towler, D. A., Gordon, J. I., Adams, S. P., and Glaser, L. (1988) *Annu. Rev. Biochem.* **57,** 5382–5391
3. James, G., and Olson, E. N. (1990) *Biochemistry* **29,** 2623–2634
4. James, G., and Olson, E. N. (1989) *J. Biol. Chem.* **264,** 2623–2634
5. da Silva, A. M., and Klein, C. (1990) *J. Cell Biol.* **111,** 401–407
6. Resh, M. D. (1990) *Nature* **346,** 84–86
7. Gottlinger, H. G., Sodroski, J. G., and Haseltine, W. A. (1989) *Proc. Natl. Acad. Sci. U. S. A.* **86,** 5781–5785
8. Bryant, M., and Ratner, L. (1990) *Proc. Natl. Acad. Sci. U. S. A.* **87,** 523
9. Linder, M. E., Pang, I-H., Duronio, R. J., Gordon, J. I., Sternweis, P. C., and Gilman, A. G. (1991) *J. Biol. Chem.* **266,** 4654–4659
10. Marc, D., Drugeon, G., Haenni, A. L., Girard, M., and van der Werf, S. (1989) *EMBO J.* **8,** 2661–2668
11. Moscufo, N., Simons, J., and Chow, M. (1991) *J. Virol.*, in press
12. Chow, M. J., Newman, J. F. E., Filman, D., Hogle, J. M., Rowlands, D. J., and Brown, F. (1987) *Nature* **327,** 482–486
13. Towler, D. A., Adams, S. P., Eubanks, S. R., Towery, D. S., Jackson-Machelski, E., Glaser, L. and Gordon, J. I. (1987) *Proc. Natl. Acad. Sci. U. S. A.* **84,** 2140–2144
14. Lee, F.-J. S., Lin, L.-W., and Smith, J. A. (1988) *J. Biol. Chem.* **263,** 14948–14955
15. Hirel, P.-H., Schmitter, J.-M., Dessen, P., Fayat, G., and Glanquet, S. (1989) *Proc. Natl. Acad. Sci. U. S. A.* **86,** 8247–8251
16. Chang, Y.-H., Teichert, U., and Smith, J. (1990) *J. Biol. Chem.* **265,** 19892
17. Moerschell, R. P., Hosokawa, Y., Tsunasawa, S., and Sherman, F. (1990) *J. Biol. Chem.* **265,** 19638–19643
18. Towler, D. A., Adams, S. P., Eubanks, S. R., Towery, D. S., Jackson-Machelski, E., Glaser, L., and Gordon, J. I. (1988) *J. Biol. Chem.* **263,** 1784–1790
19. Glover, C. J., Goddard, C., and Felsted, R. L. (1988) *Biochem. J.* **250,** 485
20. Duronio, R. J., Rudnick, D. A., Adams, S. P., Towler, D. A., and Gordon, J. I. (1991) *J. Biol. Chem.*, in press
21. Heuckeroth, R. O., Glaser, L. and Gordon, J. I. (1988b) *Proc. Natl. Acad. Sci. U. S. A.* **85,** 8795–8799
22. Rudnick, D. A., McWherter, C. A., Adams, S. P., Ropson, I. J., Duronio, R. J., and Gordon, J. I. (1990) *J. Biol. Chem.* **265,** 13370–13378
23. Kishore, N. S., Lu, T., Knoll, L. J., Katoh, A., Rudnick, D. A., Mehta, P. P., Devadas, B., Huhn, M., Atwood, J. L., Adams, S. P., Gokel, G. W., and Gordon, J. I. (1991) *J. Biol. Chem.* **266,** 8835–8855
24. Rudnick, D. A., McWherter, C. A., Rocque, W. J., Lennon, P. J., Getman, D. P., and Gordon, J. I. (1991) *J. Biol. Chem.* **266,** 9732–9739
25. Paige, L. A., Zheng, G.-Q., DeFrees, S. A., Cassady, J. M., and Geahlen, R. L. (1989) *J. Med. Chem.* **32,** 1665–1667
26. Rudolph, R. B. (1979) *Methods Enzymol.* **63,** 411–436
27. Duronio, R. J., Towler, D. A., Heuckeroth, R. O., and Gordon, J. I. (1989) *Science* **243,** 796–800
28. Duronio, R. J., Jackson-Machelski, E., Heuckeroth, R. O., Olins, P. O., Devine, C. S., Yonemoto, W., Slice, L. W., Taylor, S. S., and Gordon, J. I. (1990) *Proc. Natl. Acad. Sci. U. S. A.* **87,** 1506–1510
29. Duronio, R. J., Rudnick, D. A., Johnson, R. L., Linder, M. E., and Gordon, J. I. (1991) *Methods: A Companion to Methods in Enzymology,* **1,** 253
30. Bryant, M., Ratner, L., Duronio, R., Kishore, N. S., Devadas, B., Adams, S., and Gordon, J. I. (1991) *Proc. Natl. Acad. Sci. U. S. A.* **88,** 2055
31. Kumar, G. B., and Black, P. N. (1991) *J. Biol. Chem.* **266,** 1348–1353
32. Kameda, K., and Nunn, W. D. (1981) *J. Biol. Chem.* **256,** 5702–5707
33. Duronio, R. J., Rudnick, D. A., Johnson, R. L., Johnson, D. R., and Gordon, J. I. (1991) *J. Cell Biol.*, in press
34. Heuckeroth, R. O., and Gordon, J. I. (1989) *Proc. Natl. Acad. Sci. U. S. A.* **86,** 5262–5266
35. Miyajima, I., Nakafuku, M., Nakayama, N., Brenner, C., Miyajima, A., Kaibuchi, K., Arai, K., Kaziro, Y., and Matsumoto, K. (1987) *Cell* **50,** 1011–1019
36. Nakafuku, M., Obara, T., Kaibuchi, K., Miyajima, I., Miyajima, A., Itoh, H., Nakamura, S., Arai, K.-I., Matsumoto, K., and Kaziro, Y. (1988) *Proc. Natl. Acad. Sci. U. S. A.* **85,** 1374–1378
37. Sewell, J. L., and Kahn, R. A. (1988) *Proc. Natl. Acad. Sci.* **85,** 4620
38. Stearns, T., Kahn, R. A., Botstein, D., and Hoyt, M. A. (1990) *Mol. Cell. Biol.* **10,** 6690–6699
39. Herman, P. K., Stack, J. H., DeModena, J. A., and Emr, S. D. (1991) *Cell* **64,** 425–437
40. Schweizer, E., and Bolling, H. (1970) *Proc. Natl. Acad. Sci.* **67,** 660
41. Schweizer, E., Werkmeister, K., and Jain, M. K. (1978) *Mol. Cell. Biochem.* **21,** 95–107
42. Meyer, K. H., and Schweizer, E. (1974) *J. Bacteriol.* **117,** 345–350
43. Lynen, F. (1969) *Methods Enzymol.* **14,** 17–33
44. Orme, T. W., McIntyre, J., Lynen, F., Kuhn, L., and Schweizer, E. (1972) *Eur. J. Biochem.* **24,** 407–415
45. Orr-Weaver, T. L., and Szostak, J. W. (1983) *Proc. Natl. Acad. Sci. U. S. A.* **80,** 4417–4421
46. Johnson, D. R., Cox, A. D., Solski, P. A., Devadas, B., Adams, S. P., Leimgruber, R. M., Heuckeroth, R. O., Buss, J. E., and Gordon, J. I. (1990) *Proc. Natl. Acad. Sci. U. S. A.* **87,** 8511–8515
47. Mumby, S. M., Heuckeroth, R. O., Gordon, J. I., and Gilman, A. G. (1990) *Proc. Natl. Acad. Sci. U. S. A.* **87,** 728–732
48. Bryant, M., Heuckeroth, R. O., Kimata, J. T., Ratner, L., and Gordon, J. I. (1989) *Proc. Natl. Acad. Sci. U. S. A.* **86,** 8655–8659
49. Naria, M. A., Fitzgerald, P. M. D., McKeever, B. M., Leu, C., Heimbach, J. L., Herber, W. K., Sigal, I. S., Darke, P. L., and Springer, J. P. (1989) *Nature* **337,** 615–620

Minireview

THE JOURNAL OF BIOLOGICAL CHEMISTRY
Vol. 266, No. 15, Issue of May 25, pp. 9339-9342, 1991

Printed in U.S.A.

Nitrogenases*

Robert H. Burris

From the Department of Biochemistry, University of Wisconsin, Madison, Wisconsin 53706

The activity of nitrogenase is central to maintaining the biogeochemical nitrogen cycle on earth. In a planet bathed with N_2, fixed nitrogen often is the limiting element for growth of microorganisms, plants, and animals. A limited number of prokaryotes has been blessed with the ability to fix N_2 (1), and the other creatures on earth exploit these microorganisms. The nitrogen fixers in turn must depend upon their own photosynthesis or the photosynthesis of green plants for the energy that is required in large amounts to reduce N_2. As usual, the animals are content to be consumers and to let the green plants and microorganisms do the work that keeps the animals alive.

Nomenclature

Nitrogenase is a fascinating enzyme system, as it consists of two distinct proteins. Dinitrogenase (also referred to as MoFe protein or protein I) is a protein, usually of molecular mass 220–240 kDa, that binds and reduces N_2 or other substrates. Dinitrogenase reductase (also referred to as Fe protein or protein II) has the specific role of passing electrons one at a time to dinitrogenase. Strong reductants will not reduce dinitrogenase directly but must function via dinitrogenase reductase.

Agents of Nitrogen Fixation

There are various prokaryotic microorganisms that are capable of fixing N_2, but despite claims in the literature for N_2 fixation by eukaryotes there is no established case for such fixation. However, eukaryotes can function in symbiosis with prokaryotes fixing N_2. Fixation of N_2 has been demonstrated for the photosynthetic cyanobacteria, photosynthetic bacteria, archaebacteria, aerobes, anaerobes, facultative anaerobes, microaerobic bacteria, actinomycetes in association with nonleguminous plants, root nodule bacteria in symbiosis with leguminous plants, and by microaerobic bacteria in a looser relationship referred to as associative fixation. In quantitative terms, the symbiosis between leguminous plants and their root nodule bacteria (*Rhizobium* sp.) is the most productive system. Despite the variety of N_2-fixing microorganisms, there is an amazing homology in their N_2-fixing systems (2).

The Enzyme System

The nature of the enzyme system could not be resolved until N_2 fixation was achieved with cell-free preparations from the N_2-fixing bacteria. Although N_2 fixation had been achieved earlier with cell-free preparations (3), the preparations were inconsistent. Consistent cell-free preparations first were recovered by Carnahan *et al.* in 1960 (4, 5), who used strictly anaerobic conditions, avoided low temperatures (to minimize cold lability) (6), and supported fixation with high concentrations of pyruvate. With cell-free preparations from *Clostridium pasteurianum*, Mortenson (7) was able to demonstrate that the system was composed of two proteins that we refer to as dinitrogenase and dinitrogenase reductase.

Early work centered on *C. pasteurianum*, and the electron donor proved to be ferredoxin (8). In certain other organisms flavodoxin can serve as reductant (9). In reconstructed systems one can use $Na_2S_2O_4$ as the reductant (10), and it can pass electrons to dinitrogenase reductase in the absence of ferredoxin or flavodoxin.

The transfer of electrons from dinitrogenase reductase to dinitrogenase requires the mediation of MgATP. The early papers (5) reported that ATP inhibited N_2 fixation, so when McNary and Burris (11) indicated a requirement for ATP the finding was greeted with great skepticism. Winter and Burris (12) established that 4ATPs were hydrolyzed for each pair of electrons transferred between dinitrogenase reductase and dinitrogenase.

The sequence of electron transfer in the nitrogenase system was established primarily by observations of the characteristic EPR signals of dinitrogenase and dinitrogenase reductase at low temperature. Dinitrogenase reductase is reduced by ferredoxin or flavodoxin (or by $Na_2S_2O_4$ in reconstructed systems) (13). It binds MgATP, and this lowers its potential by about 100 mV to around −400 mV (14). At that potential it can transfer electrons to dinitrogenase. The transfer is accompanied by the hydrolysis of MgATP to MgADP and P_i. As the ADP (not ATP) is inhibitory to the reaction, it must be converted back to ATP. One electron is transferred for each 2MgATP hydrolyzed. As a single electron is inadequate to reduce N_2, the cycle must be repeated until the dinitrogenase has accumulated adequate electrons to reduce N_2.

As the reduction of N_2 is accompanied by an obligatory reduction of $2H^+ \rightarrow H_2$, the overall reaction becomes as follows.

$$N_2 + 8H^+ + 8e^- \rightarrow 2NH_3 + H_2$$

As 2MgATP are required for each electron transferred, the reaction requires a minimum of 16MgATP under ideal conditions. Under normal physiological conditions the requirement is closer to 20–30 MgATP.

ATP hydrolysis accompanies electron transfer between dinitrogenase reductase and dinitrogenase, and it now is established that cleavage of MgATP and conversion to bound MgADP + P_i occurs before electron transfer from dinitrogenase reductase to dinitrogenase in preparations from *Klebsiella pneumoniae* (15). Thorneley and Lowe (16) have indicated that dissociation of the complex between dinitrogenase and dinitrogenase reductase after MgATP hydrolysis and electron transfer is rate-limiting for the overall nitrogenase reaction.

It generally is accepted that the FeMo center constitutes the active site in dinitrogenase. Shah and Brill (17) in 1977 isolated an Fe·Mo complex from dinitrogenase; it is referred to as FeMo-co. The synthesis of FeMo-co is mediated by a number of genes (18). The product of *nifQ* seems to be required for early stages of processing of Mo (19). Strains of *K. pneumoniae* with mutations in nifB, nifN, and nifE generate an apodinitrogenase that can be activated *in vitro* with FeMo-co (20). $NifV^-$ mutants are of special interest as they synthesize an altered form of FeMo-co. Dinitrogenase from $NifV^-$ mutants is effective in reducing protons and acetylene, but it does not reduce N_2 well. CO inhibits reduction of protons in these mutants, whereas proton reduction escapes such CO inhibition in organisms with wild type dinitrogenase (21).

Ugalde *et al.* (22) have constructed an *in vitro* system for

* Our research is supported by the College of Agricultural and Life Sciences, University of Wisconsin-Madison, and by Department of Energy Grant DE-FG02-87ER13707.

synthesis of FeMo-co, and this has permitted the demonstration that homocitrate is incorporated into normal FeMo-co. When modified forms of FeMo-co, *e.g.* those carrying citrate, isocitrate, or homoisocitrate are produced and incorporated into dinitrogenase, they alter substrate specificity and response to inhibitors (23, 24). Homocitrate accumulated in *K. pneumoniae* with an effective *nifV* gene (25). Madden *et al.* (26) have demonstrated that the addition of homocitrate to the medium in which $NifV^-$ mutants of *Azotobacter vinelandii* were grown cured the phenotype; fluorohomocitrate and homocitrate lactone did not effect a cure.

Substrates for Nitrogenase

The physiological substrate for nitrogenase is N_2. However, nitrogenase is versatile and can reduce a variety of substrates (1). Proton reduction to H_2 is an obligatory reaction of nitrogenase when it reduces N_2. N_2O, cyanide, methyl isocyanide, azide, acetylene, cyclopropene, cyanamid, and diazirine also serve as substrates. N_2O was the first compound other than N_2 to be demonstrated to serve as a substrate for nitrogenase (27). The most useful alternative substrate has been C_2H_2, as it furnishes a simple and highly sensitive test for nitrogenase. It is reduced to C_2H_4 which can be detected in very low concentrations with a gas chromatography unit equipped with a flame ionization detector.

Reactions with the nitrogenase system include the following.

$$2H^+ + 2e^- \rightarrow H_2$$

$$N_2 + 8H^+ + 8e^- \rightarrow 2NH_3 + H_2$$

$$N_2O + 2H^+ + 2e^- \rightarrow N_2 + H_2O$$

$$CN^- + 7H^+ + 6e^- \rightarrow CH_4 + NH_3$$

$$CH_3NC + 6H^+ + 6e^- \rightarrow CH_3NH_2 + CH_4$$

$$N_3^- + 3H^+ + 2e^- \rightarrow N_2 + NH_3$$

$$C_2H_2 + 2H^+ + 2e^- \rightarrow C_2H_4$$

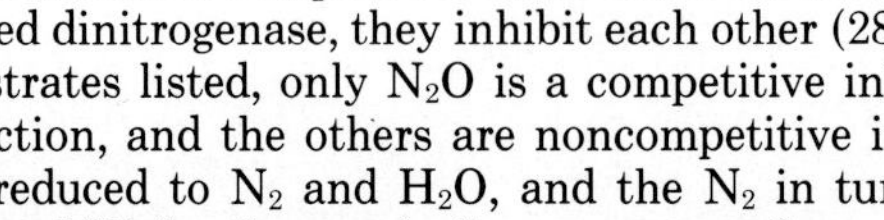

$$H_2N\text{—}CN + 6H^+ + 6e^- \rightarrow NH_3 + CH_3NH_2$$

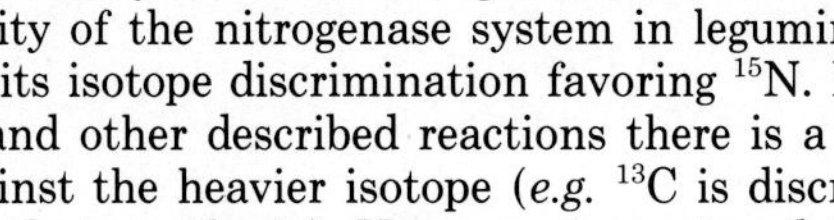

As all substrates compete for electrons from the same pool of reduced dinitrogenase, they inhibit each other (28). Among the substrates listed, only N_2O is a competitive inhibitor of N_2 reduction, and the others are noncompetitive inhibitors. N_2O is reduced to N_2 and H_2O, and the N_2 in turn can be reduced to NH_3 by the usual nitrogenase reaction.

A curiosity of the nitrogenase system in leguminous root nodules is its isotope discrimination favoring ^{15}N. In photosynthesis and other described reactions there is a discrimination against the heavier isotope (*e.g.* ^{13}C is discriminated against in photosynthesis). However, in root nodules ^{15}N is more concentrated. The more vigorously fixing plants usually have the more ^{15}N-enriched nodules (29, 30). There has been no satisfactory explanation for this unusual isotope discrimination (31).

Inhibitors of Nitrogenase

H_2 is a specific, competitive inhibitor of nitrogenase and H_2 is a product of proton reduction by nitrogenase. The inhibition by H_2 was an unexpected finding in the 1930s when N_2-fixing bacteria in association with red clover plants were used as the test system. Insight into the reaction has been provided by studies of the HD-forming reaction of nitrogenase. Kinetic studies suggested that an infinite concentration of N_2 would not block H_2 production completely (28). However, this prediction was not tested experimentally until Simpson and Burris (32) placed a purified nitrogenase system under 51 atmospheres of N_2 to determine whether the high pN_2 would block production of H_2. Under this extreme pN_2, 73% of the electrons were used to reduce N_2 and 27% were used to reduce H^+ to H_2. Apparently at infinite pN_2 75% of the electrons would be allocated for N_2 reduction and 25% for H^+ reduction.

H_2 production is the only known reaction of nitrogenase that escapes CO inhibition. Nitrogenase also can form H_2 by an alternative CO-sensitive pathway, termed the HD system. If one supplies D_2 to nitrogenase in the presence of N_2, the D_2 is dissociated and reassembled as 2HD (33). There have been reports of an N_2-independent pathway for HD formation (34), but Li (35) in a series of careful experiments could observe production of HD only in the presence of N_2.

Guth and Burris (36) performed kinetic studies of HD formation and found that the reaction was an ordered, sequential reaction in which D_2 had to bind before N_2. When D_2 binds to the nitrogenase enzyme before N_2, then the D_2 is dissociated, reassembles with $2H^+$ to form 2HD, and at this time the N_2 is released unchanged from the enzyme. When H_2 binds (instead of D_2) it dissociates, combines with $2H^+$, and each H_2 yields $2H_2$. If N_2 binds to nitrogenase before H_2, it is reduced to NH_3 in the normal fashion. As the pN_2 is increased, more N_2 binds first; as the pD_2 is increased more D_2 binds first as predicted for a competitive reaction. Any N_2 that binds after D_2 (or H_2) forms a complex that aborts and releases N_2 without reduction of the N_2.

We should mention a third (and little studied) reaction for formation of H_2 by nitrogenase. Liang and Burris (37) found that there is a reaction that releases a burst of H_2 when nitrogenase activity is initiated. This is a once-only reaction that lasts for a few seconds after components of the nitrogenase system are mixed, and it is followed by the slower, steady-state production of H_2. Interestingly, this reaction is stoichiometric with the molybdenum content of the nitrogenase system, and it terminates after the production of $1H_2$ per dinitrogenase molybdenum. The H_2 burst apparently is not a catalytic event but results from a once-only activation process.

The obligatory production of H_2 requires that nitrogenase must dissipate over 25% of its electrons in reducing protons to H_2. As the organisms have no way to eliminate production of H_2, their only recourse is to recycle the H_2 to recover some of its energy. With hydrogenase they can oxidize the H_2 back to H_2O and couple this reaction to ATP formation, or to the reduction of ferredoxin or flavodoxin which can donate electrons to dinitrogenase reductase. Evans and co-workers (38) have reported 5% or greater improvement in yields of soybeans from plants inoculated with hup^+ rhizobia (those having a hydrogenase) as compared with soybeans inoculated with hup^- strains.

Oxygen

As indicated, functional nitrogenase systems occur in strict anaerobes, facultative aerobes (growing anaerobically), microaerobic organisms, and strict aerobes. In every case, when nitrogenase components have been isolated from these organisms, the dinitrogenase and dinitrogenase reductase have proved to be sensitive to O_2. So organisms exposed to O_2 must have a mechanism to protect their nitrogenases. Strict anaerobes fix N_2 only in the absence of O_2. Microaerobic organisms function best on N_2 when the pO_2 is well below 1 kilopascals. Aerobes, like the *Azotobacter* strains, have the most rapid respiration of any known organisms, and this aids them in scavenging O_2; they also may undergo some conformational change that protects them when they are fixing N_2 (1). Root nodules harboring N_2-fixing symbiotic bacteria contain hemoglobin with a high affinity for O_2. The hemoglobin facilitates the diffusion of O_2 into the nodule tissue but releases the O_2 there at a concentration that is adequate to support ATP formation via the cytochrome system but is not concentrated enough to inactivate the nitrogenase components. Anyone contemplating introduction of the nitrogenase system into plants or other organisms must engineer a mechanism to protect the components against inactivation by O_2.

Genetics Related to Nitrogenase

The most intense area of study of nitrogenase in the past decade has centered on the genetics of the system. The most studied N_2 fixer has been *K. pneumoniae*, because this organism has much in common with *Escherichia coli*, whose genetic system has been studied exhaustively. In a relatively short time, 17 genes associated with N_2 fixation were described. Subsequently 3 or more have been added. There is a striking homology among the nitrogenases from various sources, so it is not surprising that their genetic arrays are similar. The alignment in *K. pneumoniae* is as shown by Gallon and Chaplin (39), and they discuss the function of the various gene products.

nifH codes for the structural unit of dinitrogenase reductase, and *nifD* and *nifK* for the structural units of dinitrogenase. However, these are not active until modified by the action of other gene products. Apparently six gene products are required for the production of FeMo-co.

In the investigation of electron transport for the reduction of N_2, only the system in *K. pneumoniae* has been defined both by genetic and biochemical analysis (39). Flavodoxin, a *nifF* gene product, and pyruvate:flavodoxin oxidoreductase, a *nifJ* gene product, are involved in reactions that couple the oxidation of pyruvate to support reduction of dinitrogenase reductase. *A. vinelandii* has a flavodoxin system that appears to be comparable with that from *K. pneumoniae.*

Homology among Nitrogenases

After nitrogenase had been separated into dinitrogenase and dinitrogenase reductase, the question arose whether one could combine the components from different nitrogenases and generate an active enzyme complex. Detroy *et al.* (40) first reported success in obtaining N_2 fixation by nonhomologous crosses of the nitrogenase components from four bacteria, *A. vinelandii*, *K. pneumoniae*, *Bacillus polymyxa*, and *C. pasteurianum*. Of the 12 possible nonhomologous crosses, 6 were active. This was verified by other investigators, but the most comprehensive tests were made by Emerich and Burris (2), who purified dinitrogenase and dinitrogenase reductase from eight organisms and found that about 85% of the nonhomologous crosses were catalytically active. It is apparent that remarkable homology has been preserved among the nitrogenases of aerobes, anaerobes, facultative organisms, and between free-living and symbiotic nitrogen fixers. In addition, one could point out that all the dinitrogenases are MoFe proteins (exceptions of alternative nitrogenases will be discussed later), the dinitrogenase reductases all are Fe proteins, they all require ATP and a strong reductant, they all are labile to oxygen, they are comparable in molecular mass, they all reduce a variety of substrates, and they exhibit comparable responses to inhibitors. The nonhomologous cross between *C. pasteurianum* dinitrogenase reductase and *A. vinelandii* dinitrogenase is inactive (41) because it forms a tight-binding complex that cannot dissociate.

Model for Nitrogenase Activity

The most thoroughly developed model for the action of nitrogenase (from *K. pneumoniae*) has been presented by the research group at Sussex (42, 43). Reduced Kp2 (dinitrogenase reductase) binds 2MgATP and then combines with Kp1 (dinitrogenase). In this complex, electrons are passed from dinitrogenase reductase to dinitrogenase. MgATP is hydrolyzed to MgADP + P_i, and then the complex dissociates to $Kp2_{ox}$, $(MgADP)_2$, and $Kp1_{red}$. The $Kp2_{ox}(MgADP)_2$ then is reduced with dithionite or another electron donor and the MgADP converted back to MgATP to regenerate $Kp2(MgATP)_2$ in the reduced state; reduction occurs before replacement of 2MgADP by 2MgATP (44). The model is developed in detail in the papers of Lowe and Thorneley (42) and Thorneley and Lowe (43).

In 1978 Thorneley and co-workers (45) reported that they had demonstrated spectrophotometrically an enzyme-bound dinitrogen hydride intermediate in a nitrogenase preparation from *K. pneumoniae.* Their data indicated that the rate of production of the intermediate that produced N_2H_4 was proportional to the rate of production of NH_3 by the preparation. Chatt (46) had earlier given evidence supporting his speculation that a hydride probably was an intermediate in the fixation of N_2.

Control of Nitrogenase

N_2 fixation is an energy-demanding process; under normal conditions about 20–30 MgATP are required for fixation of $1N_2$ to $2NH_3$, so the N_2-fixing organisms have devised methods to turn off N_2 fixation when utilizable N compounds, such as NH_3, are available. Various mechanisms have evolved, but one of the most interesting depends upon covalent modification of dinitrogenase reductase by ADP-ribosylation to inactivate the nitrogenase system.

After Ludden and Burris (47) observed that there was an activating enzyme that restored activity to cell-free preparations from *Rhizobium rubrum*, the hunt was on for both the activating and inactivating enzyme systems. Inactivation occurs when an enzyme designated DRAT (dinitrogenase reductase ADP-ribosyltransferase) in response to darkness or the presence of NH_4^+ adds an ADP-ribosyl group to arginine 101 of *R. rubrum* dinitrogenase reductase (arginine 100 is modified in certain other organisms). This process, referred to as "switch-off" in response to covalent modification of dinitrogenase reductase, can be reversed by the action of DRAG (dinitrogenase reductase-activating glycohydrolase). This enzyme removes the ADP-ribosyl group from the modified arginine to restore activity to the system. Both DRAT and DRAG have been purified, and the genetics of their production has been studied (48).

Intermediates in N_2 Fixation and Assimilation

The Michaelis constant for N_2 fixation varies among organisms, but is in the reported range of 5–15 kilopascals of N_2. Thus, the ambient level of about 79 kilopascals of N_2 in air approaches saturation of the enzyme. With $^{15}N_2$ as a tracer it was possible to show that NH_4^+ was an early product of N_2 fixation and the product that was assimilated by the N_2-fixing organisms (49). As indicated, Thorneley *et al.* (45) have recorded evidence that a diimide precedes formation of NH_4^+ in N_2-fixing systems. Assimilation occurs primarily via the GSH-GOGAT (glutamine synthetase-glutamine oxoglutarate aminotransferase) system. Thus, NH_4^+ is assimilated into glutamine and the glutamine in turn reacts with oxoglutarate (α-ketoglutarate) to form two glutamates.

Alternative Nitrogenases

It was long accepted that molybdenum was indispensable at the active site of dinitrogenase, although there had been some evidence of growth of the organisms on N_2 in the apparent absence of molybdenum. When Bishop *et al.* (50) reported "Evidence for an alternative nitrogen fixation system in *Azotobacter vinelandii*," their paper was greeted with marked skepticism. However, it now is accepted that there probably are three distinct nitrogenases, the MoFe nitrogenase, a vanadium nitrogenase, and an iron nitrogenase that apparently is devoid of both molybdenum and vanadium. So these alternative nitrogenases currently are attracting considerable research attention, and the genetics of the systems are being explored. In general they are less active than the MoFe nitrogenases, and they differ in substrate specificity.

The nucleotide sequence for dinitrogenase reductase has been established for each of the structural genes, *nif*H, *vnf*H and *anf*H. A major difference is that both the alternative vanadium and iron nitrogenases carry a third subunit, δ, that is coded for by *vnf*G and *anf*G, rather than only two subunits characteristic of other dinitrogenases.

Cyanobacteria

The cyanobacteria group (blue-green algae) has long been recognized to include species capable of fixing N_2, usually filamentous, heterocystous organisms. Certain nonheterocystous cyanobacteria are capable of fixing N_2, but they must have some way to avoid inactivation by O_2 produced photosynthetically.

In heterocystous, filamentous cyanobacteria, up to 10% of the cells in the filament may differentiate into heterocysts. These cells have heavy walls that limit influx of O_2 and other gases, and in differentiating from vegetative cells they lose photosystem II that generates O_2. The vegetative cells carry on active photosynthesis and pass photosynthate to the heterocysts. The heterocysts in turn use this photosynthate to fix N_2, and they export fixed nitrogen to the vegetative cells.

An interesting symbiosis exists between certain water ferns and cyanobacteria. In this association (*e.g. Azolla Anabaena azollae*), up to a third of the cyanobacteria may differentiate into heterocysts. The photosynthesis of the cyanobacteria is supplemented by that of the water fern. This highly effective symbiotic system has been used in southeast Asia for hundreds of years to add fixed nitrogen to rice fields; it may fix over 100 kg of nitrogen/hectare per season, and as the water fern and cyanobacteria decompose they release fixed nitrogen to the rice.

Certain nonheterocystous cyanobacteria exploit a diurnal cycle. During light periods they photosynthesize, generate O_2, and turn off the nitrogenase system. During darkness, they turn on the N_2-fixing system and utilize photosynthate stored during the day.

The array of genes in the cyanobacteria is similar to that of other N_2-fixing organisms, but there is a sequence of nucleic acids that separates genetic components and prevents them from functioning. Upon conversion of vegetative cells to heterocysts, a segment of the DNA apparently is excised; the genetic elements now assume an array comparable with that of other N_2-fixing organisms, and N_2 fixation is initiated in the heterocysts (51).

Conclusions

Biological N_2 fixation poses interesting problems to the biochemist, because reduction of N_2 requires the collaboration of two enzymes, dinitrogenase and dinitrogenase reductase, involves an unusually heavy investment of energy, and operates on a very stable substrate, N_2. It attracts attention, because as is true for photosynthesis, it catalyzes a reaction that is a key to survival of life forms on the earth. The biochemistry of the process was the center of attention for research during three decades, but in the last decade emphasis has turned to the genetics of biological N_2 fixation, an attention which in turn has done much to clarify details of the biochemistry of the process.

REFERENCES

1. Postgate, J. R. (1982) *The Fundamentals of Nitrogen Fixation*, Cambridge University Press, London
2. Emerich, D. W., and Burris, R. H. (1978) *J. Bacteriol.* **134,** 936–943
3. Burris, R. H. (1966) *Annu. Rev. Plant Physiol.* **17,** 155–184
4. Carnahan, J. E., Mortenson, L. E., Mower, H. F., and Castle, J. E. (1960) *Biochim. Biophys. Acta* **38,** 188–189
5. Carnahan, J. E., Mortenson, L. E., Mower, H. F., and Castle, J. E. (1960) *Biochim. Biophys. Acta* **44,** 520–535
6. Dua, R. D., and Burris, R. H. (1963) *Proc. Natl. Acad. Sci. U. S. A.* **50,** 169–175
7. Mortenson, L. E. (1965) in *Non-Heme Iron Proteins: Role in Energy Conversion* (San Pietro, A., ed) pp. 243–259, Antioch Press, Yellow Springs, OH
8. Mortenson, L. E. (1964) *Biochim. Biophys. Acta* **81,** 473–478
9. Shah, V. K., Stacey, G., and Brill, W. J. (1983) *J. Biol. Chem.* **258,** 12064–12068
10. Bulen, W. A., Burns, R. C., and LeComte, J. R. (1965) *Proc. Natl. Acad. Sci. U. S. A.* **53,** 532–539
11. McNary, J. E., and Burris, R. H. (1962) *J. Bacteriol.* **84,** 598–599
12. Winter, H. C., and Burris, R. H. (1968) *J. Biol. Chem.* **243,** 940–944
13. Orme-Johnson, W. H., Hamilton, W. D., Ljones, T., Tso, M.-Y. W., Burris, R. H., Shah, V. K., and Brill, W. J. (1972) *Proc. Natl. Acad. Sci. U. S. A.* **69,** 3142–3145
14. Zumft, W. G., and Mortenson, L. E. (1975) *Biochim. Biophys. Acta* **416,** 1–52
15. Thorneley, R. N. F., Ashby, G., Horvath, J. V., Millar, N. L., and Gutfreund, H. (1989) *Biochem. J.* **264,** 657–661
16. Thorneley, R. N. F., and Lowe, D. J. (1983) *Biochem. J.* **215,** 393–403
17. Shah, V. K., and Brill, W. J. (1977) *Proc. Natl. Acad. Sci. U. S. A.* **74,** 3249–3253
18. Shah, V. K., Madden, M. S., and Ludden, P. W. (1990) in *Nitrogen Fixation: Achievements and Objectives* (Gresshoff, P. M., Roth, L. E., Stacey, G., and Newton, W. E., eds) pp. 87–93, Chapman and Hall, New York
19. Imperial, J., Ugalde, R. A., Shah, V. K. and Brill, W. J. (1984) *J. Bacteriol.* **158,** 187–194
20. Roberts, G. P., MacNeil, T., MacNeil, D., and Brill, W. J. (1978) *J. Bacteriol.* **136,** 267–279
21. McLean, P. A., Smith, B. E., and Dixon, R. A. (1983) *Biochem. J.* **211,** 589–597
22. Ugalde, R. A., Imperial, J., Shah, V. K., and Brill, W. J. (1984) *J. Bacteriol.* **159,** 888–893
23. Liang, J., and Burris, R. H. (1989) *J. Bacteriol.* **171,** 3176–3180
24. Liang, J., Madden, M., Shah, V. K., and Burris, R. H. (1990) *Biochemistry* **29,** 8577–8581
25. Hoover, T. R., Robertson, A. D., Cerny, R. L., Hayes, R. N., Imperial, J., Shah, V. K., and Ludden, P. W. (1987) *Nature* **329,** 855–857
26. Madden, M. S., Kindon, N. D., Ludden, P. W., and Shah, V. K. (1990) *Proc. Natl. Acad. Sci. U. S. A.* **87,** 6517–6521
27. Mozen, M. M., Burris, R. H. (1954) *Biochim. Biophys. Acta* **14,** 577–578
28. Rivera-Ortiz, J. M., and Burris, R. H. (1975) *J. Bacteriol.* **123,** 537–545
29. Rasmussen, L. J., Peters, G. K., and Burris, R. H. (1989) *Phykos* **28,** 64–79
30. Kohl, D. H., Bryan, B. A., and Shearer, G. (1983) *Plant Physiol.* **73,** 514–516
31. Yoneyama, T., Yamada, N., Kojima, H., and Yazaki, J. (1984) *Plant Cell Physiol.* **25,** 1561–1565
32. Simpson, F. B., and Burris, R. H. (1984) *Science* **224,** 1095–1097
33. Hoch, G. E., Schneider, K. C., and Burris, R. H. (1960) *Biochim. Biophys. Acta* **37,** 273–279
34. Burgess, B. K., Wherland, S., Newton, W. E., and Stiefel, E. I. (1981) *Biochemistry* **20,** 5140–5146
35. Li, J.-L., and Burris, R. H. (1983) *Biochemistry* **22,** 4472–4480
36. Guth, J. H., and Burris, R. H. (1983) *Biochemistry* **22,** 5111–5122
37. Liang, J., and Burris, R. H. (1988) *Proc. Natl. Acad. Sci. U. S. A.* **85,** 9446–9450
38. Evans, H. J., Ruiz-Argueso, T., and Russell, S. A. (1978) *Basic Life Sci.* **10,** 209–222
39. Gallon, J. R., and Chaplin, A. E. (1987) *An Introduction to Nitrogen Fixation* pp. 179–232, Cassell Educational, Ltd., London
40. Detroy, R. W., Witz, D. F., Parejko, R. A., and Wilson, P. W. (1967) *Science* **158,** 526–527
41. Emerich, D. W., and Burris, R. H. (1976) *Proc. Natl. Acad. Sci. U. S. A.* **73,** 4369–4373
42. Lowe, D. J., and Thorneley, R. N. F. (1984) *Biochem. J.* **224,** 877–886
43. Thorneley, R. N. F., and Lowe, D. J. (1984) *Biochem. J.* **224,** 887–894
44. Ashby, G. A., and Thorneley, R. N. F. (1987) *Biochem. J.* **246,** 455–465
45. Thorneley, R. N. F., Eady, R. R., and Lowe, D. J. (1978) *Nature* **272,** 557–558
46. Chatt, J. (1976) in *Biological Aspects of Inorganic Chemistry* (Addison, A. W., Cullen, W. R., Dolphin, D., and James, B. R., eds) pp. 229–243, John Wiley & Sons, Inc., New York
47. Ludden, P. W., and Burris, R. H. (1976) *Science* **194,** 424–426
48. Ludden, P. W., and Roberts, G. P. (1989) *Curr. Top. Cell. Regul.* **30,** 23–56
49. Wilson, P. W., and Burris, R. H. (1947) *Bacteriol. Rev.* **11,** 41–73
50. Bishop, P. E., Jarlenski, D. M. L., and Hetherington, D. R. (1980) *Proc. Natl. Acad. Sci. U. S. A.* **77,** 7342–7346
51. Golden, J. W., Carrasco, C. D., Mulligan, M. E., Schneider, G. J., and Haselkorn, R. (1988) *J. Bacteriol.* **170,** 5034–5041

Minireview

THE JOURNAL OF BIOLOGICAL CHEMISTRY
Vol. 266, No. 16, Issue of June 5, pp. 10019–10022, 1991

Printed in U.S.A.

Reactions and Significance of Cytochrome P-450 Enzymes*

F. Peter Guengerich

From the Department of Biochemistry and Center in Molecular Toxicology, Vanderbilt University School of Medicine, Nashville, Tennessee 37232-0146

Cytochrome P-450 enzymes (EC 1.14.14.1, nonspecific monooxygenase) catalyze the oxidations of many chemicals (1). The mass of the substrates ranges from that of ethylene (M_r 28) to that of cyclosporin A (M_r 1201). The classification of a hemoprotein as a P-450[1] is defined by its absorption spectrum; the Fe(II)-CO complex has a characteristic absorption maximum (Soret band) near 450 nm due to axial ligation with a cysteine thiolate of the protein (with or without substrate present). This cysteine residue is present in a relatively well conserved region, ~80% into the protein from the N terminus. Collectively there are thousands of potential substrates for the P-450s; each of the P-450s may have a rather strict limitation of catalytic specificity (*e.g.* P-450s involved in steroid anabolism) or be a catalyst for the oxidation of many substrates (*e.g.* some of the inducible P-450s utilized in xenobiotic oxidation). There are many different P-450 enzymes; we now realize that there can be >30 P-450 (or so-called "CYP") genes expressing their products in a single organism, and many of these are concurrently produced in a single tissue. These genes have been classified on the basis of their coding sequences (2).

Functional Roles of Cytochrome P-450 Enzymes

There is broad interest in the P-450s because of the significance of these enzymes in a wide variety of disciplines, ranging from medical genetics to inorganic chemistry. Historically the P-450 gene family has been considered to be a very large one, with at least 38 genes identified in the rat to date (2). However, numerous other cases of large gene families now exist, such as the steroid receptors, interferons, and the glutathione *S*- and UDP-glucuronosyl transferases involved in drug metabolism (3, 4). There have been two major views on the function of P-450 enzymes: (i) the enzymes have critical and specific roles in the metabolism of endogeneous chemicals and (ii) the enzymes process the burden of natural products (and, in today's world, chemicals such as drugs and other xenobiotics, *i.e.* "foreign" chemicals) in a relatively non-selective manner (5). There is certainly ample evidence for the latter view, and a special case can occur when microorganisms are selected for growth on a particular chemical (or for growth in the presence of a toxic chemical). Thus, Gunsalus and his associates (6, 7) were able to develop a useful bacterial model system with a pseudomonad that harbors a plasmid coding the redox system containing P-450 101 (P-450_{cam}), an enzyme which allows growth of the organism on the terpene camphor as the carbon source. More recently, Gunsalus and his colleagues (8) have also isolated the distantly related P-450_{lin} from bacteria selected for growth on another terpene, linalool. As has been pointed out elsewhere (5, 9), mammals consume a considerable daily burden of natural products such as terpenes, steroids, and alkaloids, and P-450 enzymes are involved in much of the clearance. In the same way, the P-450s encounter numerous new drugs and pollutants each year. The lack of complete selectivity of some of the P-450s and the number of the individual forms are thus an advantage. However, in some cases the oxidation of these chemicals can generate dangerous electrophiles that are detrimental to the host and may cause toxicity or cancer (10).

The other major hypothesis, that of the importance of P-450s in normal metabolism, also has validity. Even in yeasts, lanosterol 14α-demethylation is a key P-450 activity (11). In mammals many steroidogenic tissues contain critical P-450s. The lack of functional P-450 21A2 (catalyzing 21-hydroxylation of progesterone and pregnenolone) constitutes a serious congenital disease (12), and P-450s are also known to play important roles in vitamin D homeostasis (13). Most of the eukaryotic P-450s have been considered to be located in the endoplasmic reticulum (*i.e.* microsomal), except for the two P-450 11 products in steroidogenic tisuses (14). These mitochondrial P-450s and their bacterial counterparts utilize electron transport systems consisting of a flavoprotein and an iron-sulfur protein (adrenodoxin), instead of the single flavoprotein (containing FMN and FAD) in the endoplasmic reticulum. Further, the mitochondrial and bacterial P-450s are usually much more selective in terms of the range of substrates each will oxidize. In recent years Avadhani and others (13, 15) have found evidence that some hepatic P-450s are found in mitochondria and have significant catalytic activities toward xenobiotics as well as steroids (13, 15).

What do we know about the significance of the P-450s involved in the processing of normally endogenous chemicals? While it is clear that some of the P-450s such as P-450 21A2 are critical for humans, the lack of other P-450s such as P-450 2D6 does not generally appear to be a detriment to health (16, 17). Sometimes particular P-450s can be targets for drugs that function by mechanism-based inactivation, such as the estrogen-forming aromatase P-450 19 in breast cancer (18, 19). ω-Hydroxylation of fatty acids was the assay that Lu and Coon (20) first used in the isolation and reconstitution of hepatic microsomal P-450s, but the physiological contribution of this activity does not appear to be dramatic in the case of short-chain fatty acids; with longer fatty acids and eicosanoids the processes of ω-, ω-1-, and ω-2-hydroxylation may be more important (21). While it is accepted that some steroid hydroxylations are extremely critical (22), the situation with the "nonspecific" hepatic microsomal enzymes that also oxidize xenobiotics is unclear; while many can hydroxylate androgens and other steroids regio- and stereoselectively (23), it is not obvious that these reactions are particularly critical (24). A low K_m for a particular reaction is not strong evidence of inherent selectivity or function, and a low K_d can actually work against catalysis by increasing the activation energy (25).

Recently Spector and others (26) have identified metabolic pathways in mammals leading to the *de novo* synthesis of the analgesic morphine, and it is highly likely that at least several of these steps are catalyzed by P-450s, as might be expected from the known roles of P-450s in oxidizing these (27) and other nitrogen-containing drugs and alkaloids (28).

There is considerable interest in the function of eicosanoids and other long-chain fatty acid derivatives as messengers and the function of P-450s in relation to the metabolism of these compounds. Arachidonic acid is converted to alcohols and to epoxides by P-450s (21, 29, 30), and these epoxides can even be incorporated into phospholipids (31). There is significant evidence that the epoxides of arachidonic acid can exert important physiological responses at low concentrations. Recently Kauser *et al.* (32) have shown that P-450 inhibitors can attenuate the myogenic response of dog renal arteries *in vivo*, and such eicosanoid-linked pathways may be involved. Such arachidonic acid epoxides have been implicated in pregnancy-induced hypertension (33). Although some of the less selective hepatic P-450s have been shown to account for much of the epoxidation of arachidonic acid in microsomal systems (34), the contributions of individual P-450s to *in vivo* function remain to be defined. Iwai and Inagami (35) have found that genetically hypertensive rats fail to express

* Research in this laboratory was supported by United States Public Health Service Grants CA 44353, ES 00267, ES 01590, and ES 02205.

[1] The abbreviation used is: P-450, cytochrome P-450.

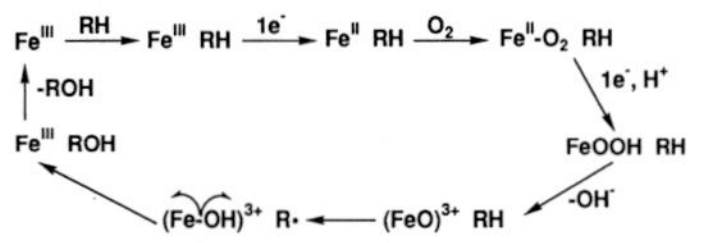

SCHEME 1. **Steps in the oxygen activation and substrate oxidation by most P-450s.** The Fe in the heme prosthetic group is featured; the substrate (*RH*) is bound near the distal ligand (site of oxygen binding). The starting point for the cycle is Fe^{III}, followed by substrate binding, reduction, and oxygen binding. The three intermediates following Fe^{II}-O_2 *RH* are putative and deduced primarily from model work and other inferences; the *superscript 3+* refers to the overall charge on the entity, the precise localization of which is yet undetermined. The role of a transient radical $R^{\cdot}$ is best accepted in carbon hydroxylation. At the end of the sequence the product ROH is released.

mRNA coding for P-450 4A2; the importance and mechanism of this defect also require further study.

Ullrich and his associates (36) have characterized the important enzymes prostacyclin synthase and thromboxane synthase as P-450s which operate via specialized mechanisms involving rearrangement of oxidized chemicals as opposed to oxygen activation *per se.* These enzymes are critical in influencing many functions. Recently Song and Brash (37, 38)[2] have characterized a flaxseed allene oxide synthase as a P-450 (*vide infra*). The hydroperoxide substrates are found in high amounts in plants, where the enzyme catalyzes a key step in the biosynthesis of the plant growth hormone jasmonic acid. This activity has also been demonstrated in lower animals, including coral and starfish oocytes. These are but a few examples of situations where P-450 enzymes may be considerably important.

Mixed Function Oxidation Reaction Chemistry

Most P-450 reactions proceed with the stoichiometry characteristic of monooxygenases (39).

$$NAD(P)H + O_2 + RH \rightarrow NAD(P)^+ + H_2O + ROH$$

In many cases the product does not appear to be a simple alcohol because rearrangement has occurred. P-450 has not been shown to act as a lipoxygenase or other type of dioxygenase, *e.g.* with a reaction of the type

$$RH \rightarrow R^{\cdot} \xrightarrow{O_2} RO_2^{\cdot} \rightarrow ROOH,$$

except perhaps in the case of unusually labile compounds. However, P-450s often have the capability to utilize hydroperoxides in various modes (*vide infra*).

The catalytic mechanism of monooxygenation may be considered in two parts, oxygen activation and substrate oxidation (Scheme 1). Current views of the details of both aspects have been discussed in detail elsewhere (39–43). The Fe(II)-O_2/substrate complex is unstable but has been characterized both with the bacterial P-450 101 (P-450_{cam}) and mammalian P-450s (7, 44, 45). Some evidence for the other oxygenated complexes has been seen with the P-450 enzymes (46, 47), but much of our inference is based upon studies of diagnostic rearrangements catalyzed by the enzymes and biomimetic metalloporphyrin models (39, 48, 49). The overall oxygenation reactions include such processes as hydroxylation at carbon and the heteroatoms N, S, and I, dealkylation of amines and ethers, and epoxidation (39, 40). These reactions can all be rationalized in terms of two steps, abstraction of a hydrogen atom (or electron) and oxygen rebound (radical recombination).

$$(FeO)^{3+} + RH \rightarrow (FeOH)^{3+}R^{\cdot} \rightarrow Fe^{3+} + ROH$$

The chemistry in the different reactions is thought to be rather invariant, and the key influence on catalytic specificity is the apoprotein (40). Further, the different P-450 enzymes should not individually be considered as strictly being epoxidases, *N*-demethylases, etc.; the reaction effected is a function of the fit of the substrate (or, more properly, its transition state) with the protein. A single protein can catalyze all of the types of reactions, depending upon the substrate presented (39). Moreover, seemingly unusual reactions such as dehydrogenation (42, 50), ester cleavage (51), ring expansion, *N*-hydroxylation, exchange of protons with solvent (52), and modification of the heme prosthetic group (53, 54) may be understood in these paradigms (39, 40, 42).

Other Mechanistic Modes

P-450 enzymes can catalyze reduction reactions as well as oxidations. Notable examples include carbon tetrachloride (55) and azo dyes (56). Other reported reductions (*e.g.* of epoxides) (57) are poorly understood (1). P-450 enzymes can form H_2O_2 when the oxidation of substrates is not tightly coupled to electron flow. Evidence has been accumulated that most of this H_2O_2 is formed from the nonenzymatic dismutation of superoxide anion, $O_2^{\overline{\cdot}}$, which is a breakdown product of the Fe(II)-O_2 complex (58).

$$Fe(II)\text{-}O_2 \rightarrow Fe(III) + O_2^{\overline{\cdot}} \xrightarrow{H^+} \tfrac{1}{2}O_2 + \tfrac{1}{2}H_2O_2$$

However, in some cases superoxide anion has not been detected in the decomposition of P-450 Fe(II)-O_2 complexes (45). The complete reduction of O_2 to H_2O by P-450 has also been documented (59).

$$2NADPH + 2H^+ + O_2 \rightarrow 2H_2O$$

This reaction, analogous to that of the mitochondrial cytochrome oxidase, appears to be exacerbated in the presence of "uncoupling agents" such as perfluoroalkanes whose C–F bonds are not easily broken (58). The *in vivo* significance of these nonproductive reactions is still vague; in liver microsomes and with some purified P-450s a large fraction of the reducing equivalents of NADPH is used nonproductively, and it is not clear that such a stress on the reduced pyridine nucleotide pool is incurred in intact tissues.

P-450s can also utilize H_2O_2, hydroperoxides, and peroxides, and in this regard they have some relationship with the peroxidases, which also utilize formal Fe(IV) porphyrin radical cation intermediates (FeO^{3+}) (43, 60). The fungal enzyme chloroperoxidase is by spectral definition a P-450; it catalyzes the following reaction (61).

$$H_2O_2 + HCl + RH \rightarrow RCl + 2H_2O$$

The turnover number for chlorination of dimedone is on the order of 10^5 min^{-1}.[3] The enzyme can also carry out some of the oxidations diagnostic of the more typical P-450s (drug *N*-demethylation, oxygenation, etc.) in the presence of H_2O_2 (62, 63). Oxidations by the hepatic P-450s can also be supported by artificial "oxygen surrogates" such as iodosylbenzene (64)

$$RIO + R' \rightarrow R'O + RI$$

and hydroperoxides. The hydroperoxide-supported reactions are complex because P-450s can apparently cleave hydroperoxides either homolytically or heterolytically, depending upon the particular protein,

$$Fe^{3+} + ROOH \rightarrow RO^- + (FeOH)^{3+}$$

$$Fe^{3+} + ROOH \rightarrow RO^{\cdot} + (FeOH)^{2+}$$

and in the latter case the alkoxide anion may propagate radical reactions (43, 49). P-450s can also catalyze "reductive" β-scission of hydroperoxides to yield alkanes and carbonyl products (65). The flaxseed P-450 isolated by Song and Brash[2] uses a linolenic hydroperoxide to generate an allene oxide in what appears to be an internal rearrangement (Scheme 2). Other interesting rearrangements of prostaglandin H_2 are catalyzed by the important P-450s thromboxane synthase and prostacyclin synthase (Scheme 3).

What Regulates Rates of Catalysis?

Most of the P-450 reactions are relatively slow, and rates of ~1 nmol of product formed/nmol of P-450/min (or min^{-1}) are common for many substrates. Rates this slow can explain *in vivo* drug disposition in many cases. Rates with some other substrates

[2] W-C. Song, and A. R. Brash, submitted for publication.

[3] All rates are expressed as turnover numbers, in units of min^{-1} (*i.e.* nmol of product formed/nmol of P-450 or metalloporphyrin/min, usually under optimal conditions).

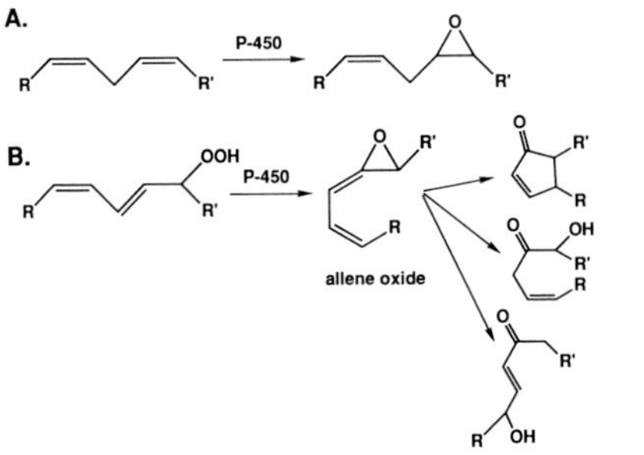

SCHEME 2. **P-450-catalyzed (*A*) epoxidation of arachidonate and (*B*) rearrangement of an allylic hydroperoxide to an allene oxide and subsequent products (37, 38).**

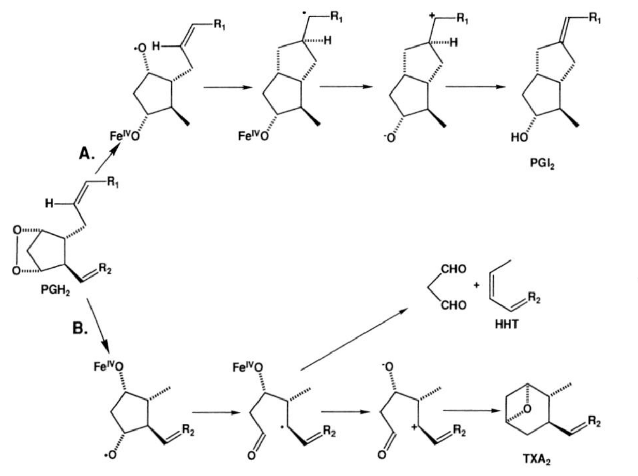

SCHEME 3. **Proposed mechanism for P-450-catalyzed rearrangements of prostaglandin H_2 to thromboxane (*TXA_2*) and prostacyclin (*PGI_2*) (36).** *HHT*, 12(*S*)-hydroxy-5,8,10-(*Z,E,E*)-heptadecatrienoic acid).

such as vitamin D are much slower (13). As mentioned above, chloroperoxidase appears to form product at a rate of ~10^5 min^{-1}, and the flaxseed P-450 of Song and Brash[2] converts the linoleic hydroperoxide to allene oxide at a rate of ~80,000 min^{-1}.[2] Thromboxane synthase catalyzes the rearrangement of prostaglandin H_2 at ~2500 min^{-1}. Among the P-450 reactions where electron transfer is actually used in oxygen activation, the catalytically fastest enzymes are those in bacteria. P-450 101 (P-450_{cam}) and P-450_{lin} form products at rates >10^3 min^{-1} and so does an unusual *Bacillus subtilis* P-450 which contains its accessory flavoprotein covalently attached to its N terminus (66). Among the hepatic P-450 enzymes, the use of an oxygen surrogate (*e.g.* iodosylbenzene) instead of NAD(P)H and the reductase can increase rates of *N*-demethylation by nearly an order of magnitude, to ~800 min^{-1} (67). For comparison, Traylor has developed a model system consisting of an iron porphyrin and pentafluoroisodosyl benzene which can carry out oxidations with a turnover number of 18,000 min^{-1} (68). A more complete biomimetic system consisting of a manganese porphyrin, *N*-methylimidazole, O_2, flavin, and *N*-methylnicotinamide (as the electron source) catalyzed the oxidation of nerol at a rate of 9 min^{-1} with a 33% yield of product based upon reducing equivalents (69).

Much of the early literature considered the nature of *the* rate-limiting step in P-450 catalysis. What has emerged is the view that the rate-limiting step probably varies depending upon the particular P-450 enzyme and substrate. In some cases it is apparent that the rate of introduction of the first or second electron is limiting (Scheme 1), and cytochrome b_5 can play an important role in the transfer of the second electron in some cases (70). In other situations the existence of large intermolecular kinetic deuterium isotope effects argues that hydrogen atom abstraction is limiting (51). An argument for rate-limiting product release has been reported (71), and actually in some cases (*e.g.* conversion of testosterone to estradiol) a single P-450 can catalyze sequential reactions on a substrate (19); the extent of equilibration of the product with the medium is unknown. In the liver, the oxygen supply can actually limit the rates of P-450 reactions (72).

Ultimately all of these features may be understood at a molecular level. The three-dimensional structures of several kinds of bacterial P-450_{cam} (P-450 101) crystals are known at high resolution (73, 74). However, none of the intrinsic membrane-bound P-450s has been crystallized to date. Extrapolation of the domains of the soluble bacterial protein to the other enzymes may be useful in illuminating several features of function but will probably not reveal high resolution details of catalytic specificity, which can be highly sensitive to small amino acid replacements (75). The structure of the bacterial protein, however, has revealed a general domain structure and that the iron spin state is controlled by the accessibility of an H_2O residue (or OH^-) as a distal iron ligand (76). The same principle probably holds in the other P-450s, but the correlation between spin state, Fe^{3+}/Fe^{2+} oxidation-reduction potential, and catalysis is clearly not universal among P-450s (77). Negishi and his associates (78) have also shown that minor variations in a single residue (position 209 of a mouse P-450) can produce dramatic changes in the catalytic specificity toward a pair of substrates, coumarin and testosterone, and in the iron spin state, but there is no obvious correlation. Site-directed mutagenesis experiments have also been used to implicate the basic amino acids (Lys and Arg) clustered in two regions of rat P-450 1A2 (in the vicinity of residues 100 and 450) in electron transfer from NADPH-P-450 reductase (79). Ishimura's laboratory (80) has used site-directed mutagenesis to show that (in bacterial P-450 101) the Thr-252 residue is near the distal heme ligand region and is critical for appropriate heterolytic scission of the oxygenated iron complex.

$$\underset{\text{Cys}}{\underset{|}{Fe^{2+}\text{—}O_2}} \xrightarrow{1e^-\ \ H^+} \overset{\text{Thr-252}}{\underset{\text{Cys}}{\underset{|}{Fe\text{—}O\text{—}OH}}} \rightarrow \underset{\text{Cys}}{\underset{|}{(FeO)^{3+}}} + OH^-$$

In the absence of this residue the enzyme is "uncoupled," that is it acts as an oxidase.

$$NADH + H^+ + O_2 \rightarrow NAD^+ + H_2O_2$$

This finding is surprising in that the appropriate heterolytic cleavage was thought to be largely a function of the axial Cys thiolate ligand (73). The question can then be raised as to what the function of the universal Cys ligand really is in catalysis. Does it have a function as such? The biomimetic models seem to work at least somewhat effectively without it, although the presence of imidazole often improves these systems (49). However, the ability of a Thr hydroxyl to assist in such a protonation has been questioned by Raag and Poulos, who feel that the role of Thr-252 is better understood in the context of a network of residues near the distal face of the heme (73, 74).

Other Aspects of P-450 and Prospects for the Future

In this brief review it is really not possible to deal with all aspects of P-450 biochemistry or to adequately acknowledge all studies that have led to our current views. The reader is referred to lead reviews and articles on P-450 protein structure (73), nomenclature and sequence similarity (2), biomimetic and other chemical models (49), catalytic mechanism (39–42), molecular biology studies (16) and regulation (81), features of structure/activity relationship (74, 82), and roles of P-450s in steroidogenesis (14), drug metabolism (83, 84), and carcinogenesis (10).

Great progress has been made in the understanding of these complex enzymes through the efforts of many individuals and the availability of new technologies. Some areas in which research needs remain include the following. (i) Analysis of more protein structures and their interactions with substrates, in three dimensions, is needed. The P-450s are too large for the application of global NMR methods, and a need exists for crystallization of the membrane proteins. (ii) Further details of catalytic mechanisms remain to be elucidated. For instance, some points regarding *N*-oxidation and epoxidation remain controversial. (iii) The regulation of levels of P-450 enzymes is complex, and much remains to be learned. The individual P-450s appear to be regulated in different manners, with some elements of similarity among certain genes. While much attention has been focused on transcription, there are other points of control as well. (iv) There are still important reactions for which the P-450s remain to be isolated and characterized. (v) Finally, *in vivo* and *in vitro* systems are needed to determine the significance of the individual P-450s in influencing humans to risk and disease from xenobiotics and also endogenous chemicals.

Acknowledgments—I thank Drs. J. H. Capdevila, L. J. Marnett, M. J. Coon, T. D. Porter, T. Shimada, and A. Brash for their comments on the manuscript. The reader is referred to a coming minireview on P-450 by Drs. M. J. Coon and T. D. Porter in this journal and to an issue of *FASEB J.* on P-450s to appear in late 1991.

REFERENCES

1. Wislocki, P. G., Miwa, G. T., and Lu, A. Y. H. (1980) in *Enzymatic Basis of Detoxication* (Jakoby, W. B., ed) Vol. 1, pp. 135–182, Academic Press, New York
2. Nebert, D. W., Nelson, D. R., Coon, M. J., Estabrook, R. W., Feyereisen, R., Fujii-Kuriyama, Y., Gonzalez, F. J., Guengerich, F. P., Gunsalus, I. C., Johnson, E. F., Loper, J. C., Sato, R., Waterman, M. R., and Waxman, D. J. (1991) *DNA Cell Biol.* **10,** 1–14
3. Coles, B., and Ketterer, B. (1990) *Crit. Rev. Biochem. Mol. Biol.* **25,** 47–70
4. Tephly, T., Green, M., Puig, J., and Irshaid, Y. (1988) *Xenobiotica* **18,** 1201–1210
5. Jakoby, W. B. (1980) in *Enzymatic Basis of Detoxication* (Jakoby, W. B., ed) Vol. 1, pp. 1–6, Academic Press, New York
6. Gunsalus, I. C., Bhattacharyya, P. K., and Suhara, K. (1985) *Curr. Top. Cell. Regul.* **26,** 295–309
7. Tyson, C. A., Lipscomb, J. D., and Gunsalus, I. C. (1972) *J. Biol. Chem.* **247,** 5777–5784
8. Ulah, A. J. H., Murray, R. I., Bhattacharyya, P. K., Wagner, G. C., and Gunsalus, I. C. (1990) *J. Biol. Chem.* **265,** 1345–1351
9. Ames, B. N., Profet, M., and Gold, L. S. (1990) *Proc. Natl. Acad. Sci. U. S. A.* **87,** 7782–7786
10. Guengerich, F. P. (1988) *Cancer Res.* **48,** 2946–2954
11. Aoyama, Y., Yoshida, Y., Nishino, T., Katsuki, H., Maitra, U. S., Mohan, V. P., and Sprinson, D. B. (1987) *J. Biol. Chem.* **262,** 14260–14264
12. Higashi, Y., Hiromasa, T., Tanae, A., Miki, T., Nakura, J., Kondo, T., Ohura, T., Ogawa, E., Nakayama, K., and Fujii-Kuriyama, Y. (1991) *J. Biochem.,* in press
13. Hollis, B. W. (1990) *Proc. Natl. Acad. Sci. U. S. A.* **87,** 6009–6013
14. Waterman, M. R., and Simpson, E. R. (1989) *Recent Prog. Horm. Res.* **45,** 533–566
15. Niranjan, B. G., Raza, H., Shayiq, R. M., Jefcoate, C. R., and Avadhani, N. G. (1988) *J. Biol. Chem.* **263,** 575–580
16. Gonzalez, F. J. (1989) *Pharmacol. Rev.* **40,** 243–288
17. Gonzalez, F. J., and Nebert, D. W. (1990) *Trends Genet.* **66,** 164–168
18. Brodie, A. M. H. (1985) *Biochem. Pharmacol.* **34,** 3213–3219
19. Corbin, C. J., Graham-Lorence, S., McPhaul, M., Mason, J. I., Mendelson, C. R., and Simpson, E. R. (1988) *Proc. Natl. Acad. Sci. U. S. A.* **85,** 8948–8952
20. Lu, A. Y. H., and Coon, M. J. (1968) *J. Biol. Chem.* **243,** 1331–1332
21. Falck, J. R., Lumin, S., Blair, I. A., Waxman, D. J., Dishman, E., Martin, M. V., Guengerich, F. P., and Capdevila, J. H. (1990) *J. Biol. Chem.* **265,** 10244–10249
22. Andersson, S., Davis, D. L., Dahlback, H., Jornvall, H., and Russell, D. W. (1989) *J. Biol. Chem.* **264,** 8222–8229
23. Wood, A. W., Ryan, D. E., Thomas, P. E., and Levin, W. (1983) *J. Biol. Chem.* **258,** 8839–8847
24. Fishman, J., Bradlow, H. L., Schneider, J., Anderson, K. E., and Kappas, A. (1980) *Proc. Natl. Acad. Sci. U. S. A.* **77,** 4957–4960
25. Kraut, J. (1988) *Science* **242,** 533–539
26. Kodaira, H., and Spector, S. (1988) *Proc. Natl. Acad. Sci. U. S. A.* **85,** 1267–1271
27. Dayer, P., Desmeules, J., Leemann, T., and Striberni, R. (1988) *Biochem. Biophys. Res. Commun.* **152,** 411–416
28. Guengerich, F. P., Muller-Enoch, D., and Blair, I. A. (1986) *Mol. Pharmacol.* **30,** 287–295
29. Oliw, E. H., Guengerich, F. P., and Oates, J. A. (1982) *J. Biol. Chem.* **257,** 3771–3781
30. Capdevila, J., Mamett, L. J., Chacos, N., Prough, R. A., and Estabrook, R. W. (1982) *Proc. Natl. Acad. Sci. U. S. A.* **79,** 767–770
31. Karara, A., Dishman, E., Falck, J. R., and Capdevila, J. H. (1991) *J. Biol. Chem.* **266,** 7561–7569
32. Kauser, K., Clark, J. E., Masters, B. S., Ortiz de Montellano, P. R., Ma, Y-H., Harder, D. R., and Roman, R. J. (1991) *Circ. Res.,* in press
33. Catella, F., Lawson, J. A., Fitzgerald, D. J., and FitzGerald, G. A. (1990) *Proc. Natl. Acad. Sci. U. S. A.* **87,** 5893–5897
34. Capdevila, J. H., Karara, A., Falck, J. R., Martin, M. V., and Guengerich, F. P. (1990) *J. Biol. Chem.* **265,** 10865–10871
35. Iwai, N., and Inagami, T. (1991) *Hypertension* **17,** 161–169
36. Hecker, M., and Ullrich, V. (1989) *J. Biol. Chem.* **264,** 141–150
37. Brash, A. R., Baertschi, S. W., Ingram, C. D., and Harris, T. M. (1988) *Proc. Natl. Acad. Sci. U. S. A.* **85,** 3382–3386
38. Brash, A. R., Baertschi, S. W., and Harris, T. M. (1990) *J. Biol. Chem.* **265,** 6705–6712
39. Guengerich, F. P. (1990) *Crit. Rev. Biochem. Mol. Biol.* **25,** 97–153
40. Guengerich, F. P., and Macdonald, T. L. (1990) *FASEB J.* **4,** 2453–2459
41. Ortiz de Montellano, P. R. (1986) in *Cytochrome P-450* (Ortiz de Montellano, P. R., ed) pp. 217–271, Plenum Press, New York
42. Ortiz de Montellano, P. R. (1989) *Trends Pharmacol. Sci.* **10,** 354–359
43. Marnett, L. J., Weller, P., and Battista, J. R. (1986) in *Cytochrome P-450* (Ortiz de Montellano, P. R., ed) pp. 29–76, Plenum Press, New York
44. Lipscomb, J. D., Sligar, S. G., Namtvedt, M. J., and Gunsalus, I. C. (1976) *J. Biol. Chem.* **251,** 1116–1124
45. Oprian, D. D., Gorsky, L. D., and Coon, M. J. (1983) *J. Biol. Chem.* **258,** 8684–8691
46. Blake, R. C., II, and Coon, M. J. (1989) *J. Biol. Chem.* **264,** 3694–3701
47. Larroque, C., Lange, R., Maurin, L., Bienvenue, A., and van Lier, J. E. (1990) *Arch. Biochem. Biophys.* **282,** 198–201
48. Bondon, A., Macdonald, T. L., Harris, T. M., and Guengerich, F. P. (1989) *J. Biol. Chem.* **264,** 1988–1997
49. Mansuy, D., Battioni, P., and Battioni, J. P. (1989) *Eur. J. Biochem.* **184,** 267–285
50. Guengerich, F. P., and Kim, D-H. (1991) *Chem. Res. Toxicol.* **4,** in press
51. Guengerich, F. P., Peterson, L. A., and Bocker, R. H. (1988) *J. Biol. Chem.* **263,** 8176–8183
52. Groves, J. T., Avaria-Neisser, G. E., Fish, K. M., Imachi, M., and Kuczkowski, R. L. (1986) *J. Am. Chem. Soc.* **108,** 3837–3838
53. Guengerich, F. P. (1986) *Biochem. Biophys. Res. Commun.* **138,** 193–198
54. Osawa, Y., and Pohl, L. R. (1989) *Chem. Res. Toxicol.* **2,** 131–141
55. Mico, B. A., and Pohl, L. R. (1983) *Arch. Biochem. Biophys.* **225,** 596–609
56. Hernandez, P. H., Mazel, P., and Gillette, J. R. (1967) *Biochem. Pharmacol.* **16,** 1877–1888
57. Kato, R., Iwasaki, K., Shiraga, T., and Noguchi, H. (1976) *Biochem. Biophys. Res. Commun.* **70,** 681–687
58. Kuthan, H., and Ullrich, V. (1982) *Eur. J. Biochem.* **126,** 583–588
59. Gorsky, L. D., Koop, D. R., and Coon, M. J. (1984) *J. Biol. Chem.* **259,** 6812–6817
60. Ortiz de Montellano, P. R. (1987) *Acc. Chem. Res.* **20,** 289–294
61. Nakajima, R., Yamazaki, I., and Griffin, B. W. (1985) *Biochem. Biophys. Res. Commun.* **128,** 1–6
62. Ortiz de Montellano, P. R., Choe, Y. S., DePillis, G., and Cataiano, C. E. (1987) *J. Biol. Chem.* **262,** 11641–11646
63. Colonna, S., Gaggero, N., Manfredi, A., Casella, L., Gulloti, M., Carrea, G., and Pasta, P. (1990) *Biochemistry* **29,** 10465–10468
64. Burka, L. T., Thorsen, A., and Guengerich, F. P. (1980) *J. Am. Chem. Soc.* **102,** 7615–7616
65. Vaz, A. D. N., Roberts, E. S., and Coon, M. J. (1990) *Proc. Natl. Acad. Sci. U. S. A.* **87,** 5499–5503
66. Wen, L. P., and Fulco, A. J. (1987) *J. Biol. Chem.* **262,** 6676–6682
67. Macdonald, T. L., Gutheim, W. G., Martin, R. B., and Guengerich, F. P. (1989) *Biochemistry* **28,** 2071–2077
68. Traylor, T. G., Marsters, J. C., Jr., Nakano, T., and Dunlap, B. E. (1985) *J. Am. Chem. Soc.* **107,** 5537–5539
69. Tabushi, I., and Kodera, M. (1986) *J. Am. Chem. Soc.* **108,** 1101–1103
70. Pompon, D., and Coon, M. J. (1984) *J. Biol. Chem.* **259,**15377–15385
71. Ling, K. H. J., and Hanzlik, R. P. (1989) *Biochem. Biophys. Res. Commun.* **160,** 844–849
72. Wu, Y-R., Kauffman, F. C., Qu, W., Ganey, P., and Thurman, R. G. (1990) *Mol. Pharmacol.* **38,** 128–133
73. Poulos, T. P. (1988) *Pharm. Res. (NY)* **5,** 67–75
74. Raag, R., and Poulos, T. L. (1991) *Biochemistry* **30,** 2674–2684
75. Lindberg, R. L. P., and Negishi, M. (1989) *Nature* **339,** 632–634
76. Raag, R., and Poulos, T. L. (1989) *Biochemistry* **28,** 917–922
77. Guengerich, F. P. (1983) *Biochemistry* **22,** 2811–2820
78. Iwasaki, M., Juvonen, R., Lindberg, R., and Negishi, M. (1991) *J. Biol. Chem.* **266,** 3380–3382
79. Shimizu, T., Tateishi, T., Hatano, M., and Fujii-Kuriyama, Y. (1991) *J. Biol. Chem.* **266,** 3372–3375
80. Imai, M., Shimada, H., Watanabe, Y., Matsushima-Hibiya, Y., Makino, R., Koga, H., Horiuchi, T., and Ishimura, Y. (1989) *Proc. Natl. Acad. Sci. U. S. A.* **86,** 7823–7827
81. Whitlock, J. P., Jr. (1990) *Annu. Rev. Pharmacol. Toxicol.* **30,** 251–277
82. Stayton, P. S., and Sligar, S. G. (1990) *Biochemistry* **29,** 7381–7386
83. Eichelbaum, M. (1984) *Fed. Proc.* **43,** 2298–2302
84. Idle, J. R., and Smith, R. L. (1979) *Drug Metab. Rev.* **9,** 301–317

THE JOURNAL OF BIOLOGICAL CHEMISTRY
Vol. 266, No. 17, Issue of June 15, pp. 10711–10714, 1991

Minireview

Visual Excitation and Recovery*

Lubert Stryer

From the Department of Cell Biology, Sherman Fairchild Center, Stanford University School of Medicine, Stanford, California 94305

Retinal rod cells are exquisitely sensitive detectors. Psychophysical studies carried out a half century ago by Selig Hecht and his co-workers (1) revealed that a rod cell can be excited by a single photon. Biochemists and physiologists in the ensuing decades have been challenged to discover how a rod cell achieves this ultimate sensitivity. The answer is now known as a result of studies carried out by numerous investigators around the world. The flow of information in visual excitation is summarized in Fig. 1. In essence, light triggers a nerve signal by activating a cascade that closes cation-specific channels in the plasma membrane of the rod outer segment (for reviews, see Refs. 2–9). The photoisomerization of the 11-*cis*-retinal chromophore of rhodopsin (R)[1] to the all-*trans* form generates photoexcited rhodopsin (R*). R* then activates transducin (T), a member of the G-protein family, by catalyzing the exchange of GTP for bound GDP. The GTP form of transducin (specifically, T_α-GTP, the activated α subunit) in turn switches on a potent phosphodiesterase (PDE) that rapidly hydrolyzes cGMP. In the dark, Na^+ and Ca^{2+} enter the outer segment through cation-specific channels, which are kept open by cGMP. The light-induced decrease in the level of cGMP closes these channels, which hyperpolarizes the plasma membrane and generates the neural signal. Recovery of the dark state is mediated by deactivation of PDE and activation of guanylate cyclase. The light-induced lowering of the cytosolic calcium level (Ca_i) is detected by recoverin, a new member of the EF-hand family. Stimulation of guanylate cyclase by recoverin elevates the cGMP level, which reopens channels and restores the dark state. *The capacity of rods to detect single photons and to operate over a wide range of background light level is conferred by multiple stages of positive and negative feedback that are coordinated and precisely timed.* This cascade also displays in vivid form many recurring motifs in sensory and hormonal transduction.

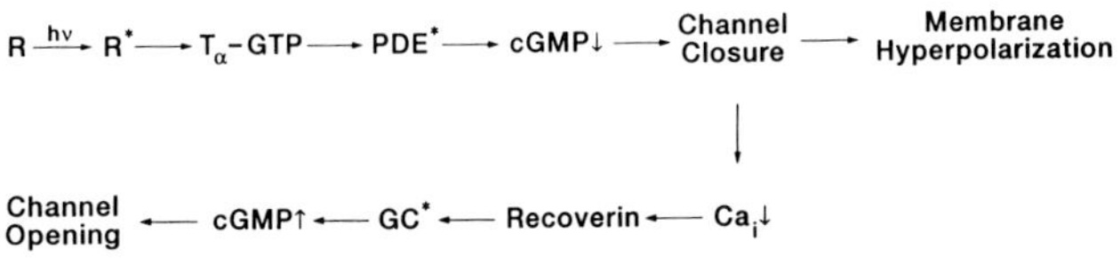

FIG. 1. **Flow of information in visual excitation and recovery.** Photoisomerization of rhodopsin triggers a cascade leading to cGMP hydrolysis and the closure of membrane channels, which generates a nerve signal. Channel closure also induces a drop in the cytosolic calcium level, which leads to the activation of guanylate cyclase and the reopening of channels. *GC**, activated guanylate cyclase.

Photoexcited Rhodopsin Catalyzes the Activation of Transducin

Rhodopsin, a 40-kDa protein with seven transmembrane helices (10), contains an 11-*cis*-retinal chromophore which is covalently linked to a lysine side chain by a protonated Schiff base linkage. Photoisomerization of the 11-*cis*-retinal chromophore (2) is both efficient (quantum yield = 0.6) and rapid. Picosecond resonance Raman studies have shown that much of the isomerization process takes place in less than 20 ps (11). In contrast, the spontaneous isomerization of retinal occurs about once in a thousand years (12). The capacity of rods to detect single photons is critically dependent on this exceptional signal-to-noise ratio. The key intermediate produced by photoisomerization is metarhodopsin II (R*), which contains an unprotonated Schiff base of all-*trans*-retinal. This conformationally activated intermediate stores 27 kcal/mol of the energy of the incident photon (13), enabling it to be a highly effective and reliable trigger.

The next step in visual excitation is the activation of transducin by R*. Transducin is a multisubunit peripheral membrane protein consisting of three chains: α (39 kDa), β (36 kDa), and γ (8 kDa). The α subunit contains a binding site for GTP or GDP and a catalytic site for the hydrolysis of bound GTP. The β and γ chains form a $T_{\beta\gamma}$ subunit. T_α is associated with $T_{\beta\gamma}$ when GDP is bound (the inactive dark state), whereas they are separate when GTP is bound (the light-activated state). The GTPase activity of transducin is essential for bringing the system back to the dark state.

Reconstitution studies involving rhodopsin, transducin, and phosphodiesterase (14, 15) have led to the elucidation of the light-triggered amplification cycle shown in Fig. 2. In the dark, nearly all of the transducin is in the inactive T-GDP state. Following illumination, T-GDP encounters R* in the plane of the disc membrane. R* induces the release of GDP from transducin and allows GTP to enter. R*-T-GTP then dissociates into T_α-GTP, $T_{\beta\gamma}$, and R*. T_α-GTP carries the excitation signal to PDE, and R* is free again to catalyze another round of GTP-GDP exchange. *The activation of hundreds of molecules of transducin by a single R* is the primary stage of amplification in vision* (15). Indeed, the only known role of R* is to activate transducin. Activation is rapid; a molecule of R* triggers the formation of a molecule of T_α-GTP in 1 ms (16). In contrast, spontaneous GTP-GDP exchange in the absence of R* takes several hours. R* stabilizes the transition state for nucleotide exchange by some 10 kcal/mol and thereby accelerates the activation of transducin by a factor of 10^7. The transducin cycle is powered by the phosphoryl potential of GTP rather than by the energy of the absorbed photon. The −12 kcal/mol provided by GTP is neatly partitioned among the intermediates so that the transducin cycle flows unidirectionally once triggered by R* (17). Both the formation of T_α-GTP and its hydrolysis are essentially irreversible.

Transducin Activates the Phosphodiesterase by Removing Its Inhibitory Subunits

The phosphodiesterase consists of three kinds of polypeptide chains: α (88 kDa), β (85 kDa), and γ (9 kDa) (18). The subunit structure of this peripheral membrane protein is $\alpha\beta\gamma_2$ (19, 20). The α and β catalytic subunits of PDE are kept

* This research was supported by National Eye Institute Grant EY-02005 and National Institute of General Medical Sciences Grant GM-24032.

[1] The abbreviations used are: R, rhodopsin; R*, photoexcited rhodopsin; T, transducin; T_α-GTP, activated transducin; Ca_i, cytosolic calcium level; PDE, phosphodiesterase.

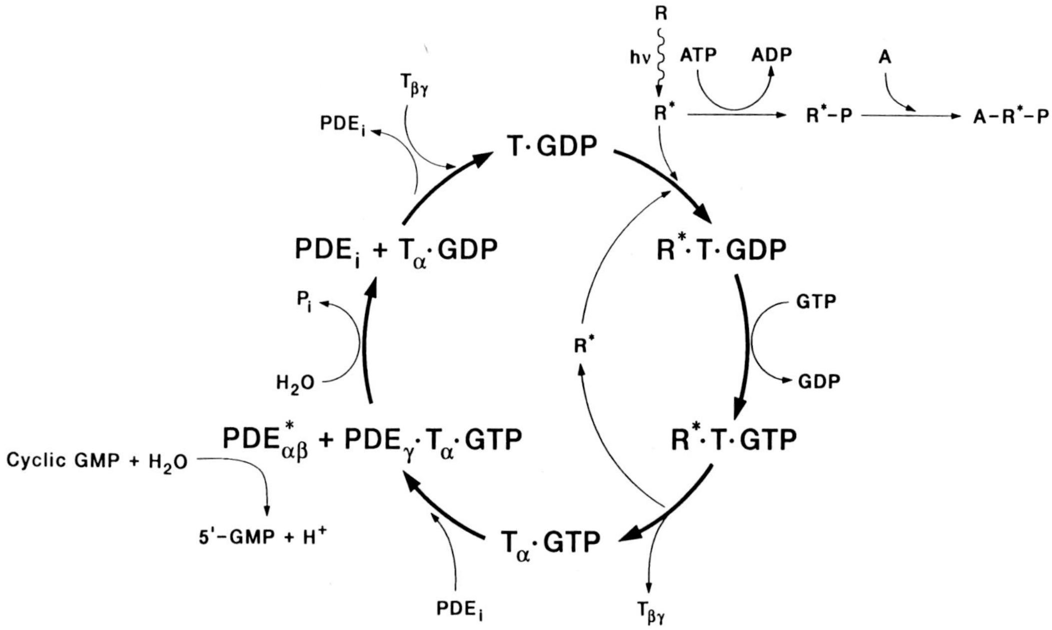

FIG. 2. **Light-activated transducin cycle.** Excitation is mediated by the amplified formation of the active form of transducin (*T_α-GTP*), which stimulates the phosphodiesterase. Transducin is deactivated by hydrolysis of bound GTP. Photoexcited rhodopsin is switched off by phosphorylation and capping by arrestin. *PDE_i*, inhibited phosphodiesterase ($\alpha\beta\gamma_2$); *PDE**, activated phosphodiesterase; *A*, arrestin.

inhibited in the dark by the γ subunits, which bind very tightly (K_i = 10 pM) (21). The maximal activity obtained by activating PDE with transducin is nearly the same as that achieved by proteolytic destruction of its γ subunits (14). This identity suggested that T_α-GTP activates PDE by overcoming the inhibitory constraint imposed by γ. Indeed, fluorescence and biochemical studies have shown that *transducin activates PDE by carrying away its inhibitory subunits* (22–25). T_α-GTP interacts first with the $\alpha\beta\gamma_2$ holoenzyme and carries away one of its γ subunits to form a partially active $\alpha\beta\gamma$ complex. The remaining γ subunit is then carried away by another T_α-GTP to form fully active $\alpha\beta$. The removal of the γ subunits stimulates PDE more than 1500-fold. The catalytic power of the fully activated enzyme is impressive; its K_{cat}/K_m ratio of 6×10^7 M^{-1} s^{-1} is near the limit set by the diffusion-controlled encounter of enzyme and substrate (21). Under physiological conditions, fully activated PDE hydrolyzes cGMP to $1/e$ of its initial concentration in only 0.6 ms. The rate-limiting step in vision in dim light is the activation of hundreds of transducins rather than the subsequent switching of PDE or the hydrolysis of cGMP by activated PDE.

Site-specific mutagenesis studies of the γ subunit are providing insight into its interactions with its alternative partners, the catalytic $\alpha\beta$ subunit and T_α-GTP. A gene encoding the γ chain of bovine PDE with codons optimized for expression in *Escherichia coli* and conveniently spaced restriction sites was chemically synthesized and cloned into a vector that uses the P_L promoter of λ phage (26). *E. coli* was transformed with this vector, which encodes a fusion protein consisting of the first 31 residues of the λ cII protein, a 7-residue joining sequence that can be specifically cleaved at its COOH terminus by clotting protease X_a and the 87-residue γ subunit. The fusion protein is as effective as native γ in inhibiting PDE. Moreover, its inhibitory action can be reversed by T_α-GTP. Thus, the amino terminus of γ is not essential for interaction with the catalytic moiety or with transducin. In contrast, the carboxyl-terminal region of γ is critical for inhibition. A mutant lacking the last 5 residues of native γ activates rather than inhibits PDE activity. The simplest interpretation is that the carboxyl-terminal tail of native γ blocks the entry of cGMP into the active site. A region important for the interaction of γ and T_α-GTP has also been identified.[2] Activation by transducin of a mutant containing glutamine in place of lysine at residues 41, 44, and 45 is markedly impaired.

cGMP Hydrolysis Leads Directly to Channel Closure

How does the light-triggered hydrolysis of cGMP lead to a nerve signal? The answer came from patch-clamp studies of excised pieces of plasma membrane from the outer segment. cGMP applied to the cytosolic side of the membrane directly opens cation-specific channels (27). Channel opening is highly cooperative (the Hill coefficient is 3), which makes the channel highly responsive to small changes in the cGMP level. The channel, an oligomer of 80-kDa subunits, has been functionally reconstituted into liposomes and planar bilayer membranes (28, 29). Under physiological conditions, the channel opens and closes in times of milliseconds in response to changes in the cGMP level, as shown by photolysis of a caged analog of cGMP and voltage-jump conductance studies (30, 31). The introduction of 8-bromo-cGMP, a hydrolysis-resistant analog of cGMP, into retinal rod cells blocks the normal light-induced channel closure, strongly suggesting that the channel is gated solely by cGMP (32). Thus, *the plasma membrane of the rod outer segment is in essence a cGMP electrode. The channel responds to the instantaneous level of cGMP, which it samples each millisecond.*

Transducin GTPase and Rhodopsin Kinase Terminate Excitation

Channels in amphibian rods close within a second after a very dim light flash and reopen about 2 s later. Channel reopening depends on the restoration of the cGMP level, which requires activation of guanylate cyclase and inhibition of the phosphodiesterase. Deactivation of PDE* requires hy-

[2] R. L. Brown and L. Stryer, manuscript in preparation.

drolysis of T_α-GTP to T_α-GDP (Fig. 2). Thus, the GTPase rate of transducin determines the duration of the excitation phase. Recent studies indicate that hydrolysis of T_α-GTP under physiological conditions occurs in less than a second (33–36), commensurate with the kinetics of the electrophysiological response (37). Deactivation of transducin and hence of PDE is necessary but not sufficient for the restoration of the dark state. Photoexcited rhodopsin too must be quenched. The isomerization of all-*trans*-retinal back to the 11-*cis* form takes many minutes. A much faster shutoff is achieved by the phosphorylation of multiple serines and threonines in the carboxyl-terminal region of R* by rhodopsin kinase, a 68-kDa cytosolic protein (38, 39). Arrestin, a 48-kDa cytosolic protein, then caps multiply phosphorylated R* to prevent it from interacting with transducin. Rhodopsin is regenerated many minutes later by insertion of 11-*cis*-retinal, release of arrestin, and removal of the COOH-terminal phosphates by protein phosphatase 2A (40, 41).

Recoverin Activates Guanylate Cyclase to Restore the Dark State

Restoration of the dark state also requires the stimulation of guanylate cyclase. What is the signal for activation of cGMP synthesis? The clue came from electrophysiological studies, which revealed a negative feedback loop between cGMP and the cytosolic calcium level (Ca_i) (42). In the dark, the entry of Ca^{2+} through the cGMP-gated channel is matched by its efflux through a 220-kDa exchanger (43) that is driven by the influx of three Na^+ and the efflux of one K^+ (44). Following illumination, the influx of Ca^{2+} through the cGMP-gated channel ceases, but its export by the exchanger continues until Ca_i drops markedly, from about 500 to 50 nM. A lowering of Ca_i in this range stimulates guanylate cyclase about 5-fold in bovine rod outer segment suspensions (45). The steep dependence of activity on Ca_i indicates that calcium acts cooperatively. Stimulation at low Ca_i was shown to be mediated by a protein that can be eluted from outer segment membranes with a low ionic strength buffer. Thus, *guanylate cyclase, like the phosphodiesterase, consists of separable regulatory and catalytic subunits.*

This calcium-sensitive stimulatory protein, named recoverin, has recently been purified (46–48). The amino sequence of recoverin (46), a 23-kDa polypeptide, exhibits three potential calcium-binding sites characteristic of the EF-hand superfamily (49). A ^{45}Ca blot shows that recoverin in fact binds Ca^{2+} in the submicromolar range (46). Furthermore, recoverin activates guanylate cyclase in a calcium-dependent manner in membranes stripped of the endogenous regulatory protein (Fig. 3). The activity of guanylate cyclase in this reconstituted system is half-maximal at 240 nM Ca^{2+}. Anti-recoverin antibody blocks the calcium-sensitive stimulation of guanylate cyclase in native rod outer segment membranes, indicating that recoverin is likely to be the endogenous activator.

A simple scheme for the action of recoverin is depicted in Fig. 4. In the dark, the calcium-bound form of recoverin is dissociated from guanylate cyclase, which has a low basal activity. Following illumination, the calcium level drops in the submicromolar range, resulting in the formation of the calcium-free form of recoverin, which binds to guanylate cyclase and markedly stimulates it. This model (46) accounts for the finding that guanylate cyclase has low activity in the absence of recoverin and that the enzyme is stimulated when the calcium level is lowered. Thus, recoverin appears to differ from Ca^{2+}-dependent activators such as calmodulin and troponin C, which require Ca^{2+} to activate their targets. Does the Ca^{2+} bound form of recoverin have a distinct function?

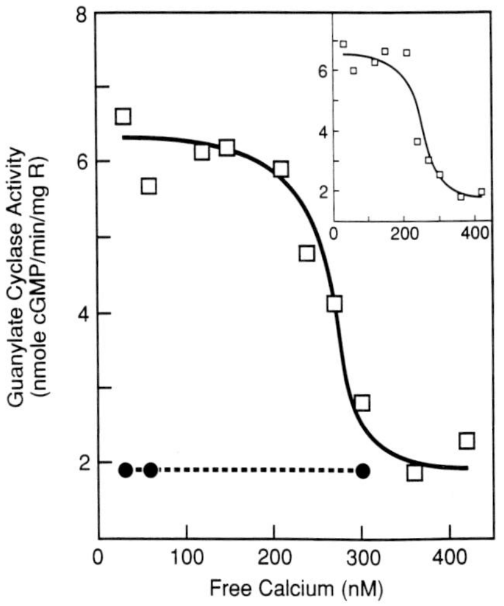

FIG. 3. **Recoverin activates guanylate cyclase at low calcium levels.** *Filled circles* show the cyclase activity of stripped rod outer segment membranes, and *open circles* show the activity of the same membranes in the presence of recoverin. The *inset* shows the calcium dependence of cyclase activity in native membranes (from Ref. 46).

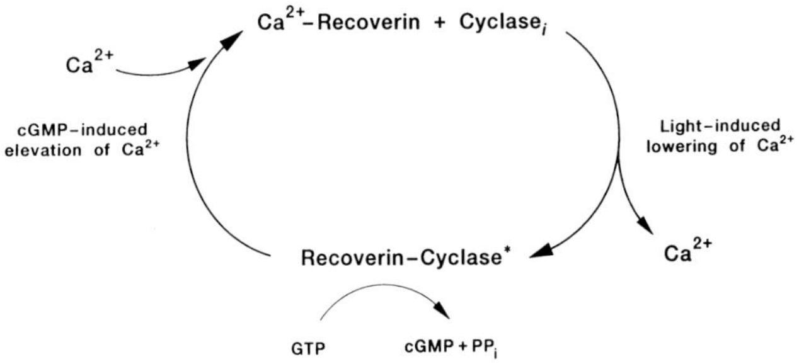

FIG. 4. **Proposed mode of action of recoverin.** The basal and activated forms of guanylate cyclase are denoted by *i* and *, respectively.

The recent finding that micromolar calcium accelerates the activation of phosphodiesterase in dim light (50) raises the intriguing possibility that this effect is mediated by the interaction of Ca^{2+}-recoverin with another protein in the cascade. It will also be interesting to learn whether the calcium-free form of calmodulin has regulatory roles. The cloning of the cDNA for recoverin and its expression in *E. coli* opens the way to a deeper understanding of this new calcium switch and its relation to other members of the EF-hand superfamily.[3] The challenge now is to elucidate the molecular mechanism of adaptation. The sensitivity of a rod cell is continuously adjusted according to the background light level, enabling it to detect incremental stimuli over a 10^5-fold range of ambient light intensity. The light-induced lowering of the calcium level is important for both recovery and adaptation (51, 52). In effect, *calcium ion is the remembrance of photons past.* Recoverin may be the calcium sensor in adaptation as well as recovery.

Recurring Motifs

The transduction mechanism of vertebrate cones, which mediate color vision, is very much like that of rods. The three human visual pigments mediating color vision also use 11-*cis*-retinal as the chromophore and have the same architectural plan as rhodopsin (53). The transducin, phosphodiesterase,

[3] S. Ray, G. Niemi, D. Brolley, J. B. Hurley, and L. Stryer, manuscript in preparation.

and cGMP-gated channel in cones also closely resemble their counterparts in rods. Visinin (54) may be the cone equivalent of recoverin. Thus, the transduction mechanisms of rods and cones are variations on the same fundamental theme. Moreover, the initial events in visual excitation in invertebrates closely resemble those of vertebrates. Octopus rhodopsin can trigger mammalian transducin, showing that complementary sites on these proteins have been conserved over 700 million years of evolution. Olfaction is mediated by a similar transduction process (55). Odorants trigger a cAMP cascade involving a G-protein (56, 57) and a cyclic nucleotide-gated channel (58, 59).

The seven-helix rhodopsin motif is found in many other signal transduction systems, as exemplified by adrenergic receptors (60) and muscarinic receptors (61). This motif is an early eukaryotic invention, as illustrated by yeast mating factor receptors (62) and the *Dictyostelium* cAMP chemotaxis receptor (63). Furthermore, each of these receptors is deactivated by phosphorylation of their COOH-terminal tail by a kinase that recognizes the activated receptor (64). Yet another similarity is the capping of phosphorylated β-adrenergic receptor by a homolog of arrestin (65). Transducin is the best understood member of a large family of signal-coupling proteins, the G proteins, that play key roles in sensory transduction, hormone action, and growth control (66, 67). Recoverin and visinin may be members of a new branch of the EF-hand superfamily (49) that act in diverse cells as switches at submicromolar calcium levels.

Many different disciplines—biochemistry, biophysics, molecular genetics, and electrophysiology—have come together, each enriching the other in revealing the molecular mechanism of vision and in providing glimpses of how this beautiful sensory transduction process came into being.

REFERENCES

1. Hecht, S., Shlaer, S., and Pirenne, M. H. (1942) *J. Gen. Physiol.* **25,** 819–840
2. Wald, G. (1968) *Nature* **219,** 800–807
3. Stryer, L. (1988) *Cold Spring Harbor Symp. Quant. Biol.* **53,** 283–294
4. Baylor, D. A. (1987) *Invest. Ophthalmol. & Visual Sci.* **28,** 34–49
5. Ho, Y. K., Hingorani, V. N., Navon, S. E., and Fung, B. K. (1989) *Curr. Top. Cell. Regul.* **30,** 171–202
6. Liebman, P. A., Parker, K. R., and Dratz, E. A. (1987) *Annu. Rev. Physiol.* **49,** 765–791
7. Chabre, M., and Deterre, P. (1989) *Eur. J. Biochem.* **179,** 255–266
8. Hurley, J. B. (1987) *Annu. Rev. Physiol.* **49,** 793–812
9. Miller, W. H. (1990) *Invest. Ophthalmol. & Visual Sci.* **31,** 1664–1673
10. Dratz, E. A., and Hargrave, P. A. (1983) *Trends Biochem. Sci.* **8,** 128–132
11. Hayward, G., Carlsen, W., Siegman, A., and Stryer, L. (1981) *Science* **211,** 942–944
12. Yau, K. W., Matthews, G., and Baylor, D. A. (1979) *Nature* **279,** 806–807
13. Cooper, A. (1981) *FEBS Lett.* **123,** 324–326
14. Fung, B. K., Hurley, J. B., and Stryer, L. (1981) *Proc. Natl. Acad. Sci. U. S. A.* **78,** 152–156
15. Fung, B. K.-K., and Stryer, L. (1980) *Proc. Natl. Acad. Sci. U. S. A.* **77,** 2500–2504
16. Vuong, T. M., Chabre, M., and Stryer, L. (1984) *Nature* **311,** 659–661
17. Stryer, L. (1987) *Chem. Scr.* **27B,** 161–171
18. Baehr, W., Devlin, M. J., and Applebury, M. L. (1979) *J. Biol. Chem.* **254,** 11669–11677
19. Deterre, P., Bigay, J., Robert, M., Pfister, C., Kuhn, H., and Chabre, M. (1986) *Proteins Struct. Funct. Genet.* **1,** 188–193
20. Fung, B. K., Young, J. H., Yamane, H. K., and Griswold-Prenner, I. (1990) *Biochemistry* **29,** 2657–2664
21. Hurley, J. B., and Stryer, L. (1982) *J. Biol. Chem.* **257,** 11094–11099
22. Wensel, T. G., and Stryer, L. (1990) *Biochemistry* **29,** 2155–2161
23. Wensel, T. G., and Stryer, L. (1986) *Proteins Struct. Funct. Genet.* **1,** 90–99
24. Fung, B. K., and Griswold-Prenner, I. (1989) *Biochemistry* **28,** 3133–3137
25. Yamazaki, A., Hayashi, F., Tatsumi, M., Bitensky, M. W., and George, J. S. (1990) *J. Biol. Chem.* **265,** 11539–11548
26. Brown, R. L., and Stryer, L. (1989) *Proc. Natl. Acad. Sci. U. S. A.* **86,** 4922–4926
27. Fesenko, E. E., Kolesnikov, S. S., and Lyubarsky, A. L. (1985) *Nature* **313,** 310–313
28. Kaupp, U. B., Hanke, W., Simmoteit, R., and Luhring, H. (1988) *Cold Spring Harbor Symp. Quant. Biol.* **53,** 407–415
29. Kaupp, U. B., Niidome, T., Tanabe, T., Terade, S., Bonigk, W., Stuhmer, W., Cook, N. J., Kangawa, K., Matsuo, H., Hirose, T., Miyata, T., and Numa, S. (1989) *Nature* **342,** 762–766
30. Karpen, J. W., Zimmerman, A. L., Stryer, L., and Baylor, D. A. (1988) *Proc. Natl. Acad. Sci. U. S. A.* **85,** 1287–1291
31. Karpen, J. W., Zimmerman, A. L., Stryer, L., and Baylor, D. A. (1988) *Cold Spring Harbor Symp. Quant. Biol.* **53,** 325–332
32. Zimmerman, A. L., Yamanaka, G., Eckstein, F., Baylor, D. A., and Stryer, L. (1985) *Proc. Natl. Acad. Sci. U. S. A.* **82,** 8813–8817
33. Dratz, E. A., Lewis, J. W., Schaechter, L. E., Parker, K. R., and Kliger, D. S. (1987) *Biochem. Biophys. Res. Commun.* **146,** 379–386
34. Vuong, T. M., and Chabre, M. (1990) *Nature* **346,** 71–74
35. Vuong, T. M., and Chabre, M. (1991) *Biophys. J.* **59,** 348a
36. Arshavsky, V. Y., Gray-Keller, M. P., and Bownds, M. D. (1991) *Biophys. J.* **59,** 407a
37. Baylor, D. A., Lamb, T. D., and Yau, K. W. (1979) *J. Physiol. (Lond.)* **288,** 613–634
38. Wilden, U., Hall, S. W., and Kuhn, H. (1986) *Proc. Natl. Acad. Sci. U. S. A.* **83,** 1174–1178
39. Miller, J. L., Fox, D. A., and Litman, B. J. (1986) *Biochemistry* **25,** 4983–4988
40. Fowles, C., Akhtar, M., and Cohen, P. (1989) *Biochemistry* **28,** 9385–9391
41. Palczewski, K., Hargrave, P. A., McDowell, J. H., and Ingebritsen, T. S. (1989) *Biochemistry* **28,** 415–419
42. Yau, K. W., and Nakatani, K. (1985) *Nature* **313,** 579–582
43. Cook, N. J., and Kaupp, U. B. (1988) *J. Biol. Chem.* **263,** 11382–11388
44. Cervetto, L., Lagnado, L., Perry, R. J., Robinson, D. W., and McNaughton, P. A. (1989) *Nature* **337,** 740–743
45. Koch, K. W., and Stryer, L. (1988) *Nature* **334,** 64–66
46. Dizhoor, A. M., Ray, S., Kumar, S., Niemi, G., Spencer, M., Brolley, D., Walsh, K. A., Philipov, P. P., Hurley, J. B., and Stryer, L. (1991) *Science* **251,** 915–918
47. Lambrecht, H. G., and Koch, K. W. (1991) *EMBO J.* **10,** 793–798
48. Polans, A. S., Bucylko, J., Crabb, J., and Palczewski, K. (1991) *J. Cell Biol.* **112,** 981–989
49. Moncrief, N. D., Kretsinger, R. H., and Goodman, M. (1990) *J. Mol. Evol.* **30,** 522–562
50. Kawamura, S., and Murakami, M. (1991) *Nature* **349,** 420–423
51. Nakatani, K., and Yau, K. W. (1988) *Nature* **334,** 69–71
52. Fain, G. L., Lamb, T. D., Matthews, H. R., and Murphy, R. L. (1989) *J. Physiol. (Lond.)* **416,** 215–243
53. Nathans, J., Thomas, D., and Hogness, D. S. (1986) *Science* **232,** 193–202
54. Yamagata, K., Goto, K., Kuo, C. H., Kondo, H., and Miki, N. (1990) *Neuron* **4,** 469–476
55. Lancet, D., Lazard, D., Heldman, J., Khen, M., and Nef, P. (1988) *Cold Spring Harbor Symp. Quant. Biol.* **53,** 343–348
56. Bakalyar, H. A., and Reed, R. R. (1990) *Science* **250,** 1403–1406
57. Jones, D. T., and Reed, R. R. (1989) *Science* **244,** 790–795
58. Dhallan, R. S., Yau, K. W., Schrader, K. A., and Reed, R. R. (1990) *Nature* **347,** 184–187
59. Ludwig, J., Margalit, T., Eismann, E., Lancet, D., and Kaupp, U. B. (1990) *FEBS Lett.* **270,** 24–29
60. Lefkowitz, R. J., Kobilka, B. K., Benovic, J. L., Bouvier, M., Cotecchia, S., Hausdorff, W. P., Dohlman, H. G., Regan, J. W., and Caron, M. G. (1988) *Cold Spring Harbor Symp. Quant. Biol.* **53,** 507–514
61. Numa, S., Fukuda, K., Kubo, T., Maeda, A., Akiba, I., Bujo, H., Nakai, J., Mishina, M., and Higashida, H. (1988) *Cold Spring Harbor Symp. Quant. Biol.* **53,** 295–301
62. Blumer, K. J., Reneke, J. E., Courchesne, W. E., and Thorner, J. (1988) *Cold Spring Harbor Symp. Quant. Biol.* **53,** 591–603
63. Pupillo, M., Klein, P., Vaughan, R., Pitt, G., Lilly, P., Sun, T., Devreotes, P., Kumagai, A., and Firtel, R. (1988) *Cold Spring Harbor Symp. Quant. Biol.* **53,** 657–665
64. Benovic, J. L., DeBlasi, A., Stone, W. C., Caron, M. G., and Lefkowitz, R. J. (1989) *Science* **246,** 235–240
65. Lohse, M. J., Benovic, J. L., Codina, J., Caron, M. G., and Lefkowitz, R. J. (1990) *Science* **248,** 1547–1550
66. Casey, P. J., Graziano, M. P., Freissmuth, M., and Gilman, A. G. (1988) *Cold Spring Harbor Symp. Quant. Biol.* **53,** 203–208
67. Bourne, H. R., Sanders, D. A., and McCormick, F. (1990) *Nature* **348,** 125–132

Minireview

THE JOURNAL OF BIOLOGICAL CHEMISTRY
Vol. 266, No. 18, Issue of June 25, pp. 11405–11408, 1991

Printed in U.S.A.

Time-resolved Fluorescence Spectroscopy

APPLICATIONS TO CALMODULIN

Sonia R. Anderson

From the Department of Biochemistry and Biophysics, Oregon State University, Corvallis, Oregon 97331-6503

Fluorescence spectroscopy is a distinctively useful method characterized by high sensitivity and responsiveness to molecular processes that occur in the picosecond to nanosecond time range. Static fluorescence measurements provide *time-averaged* information on diverse phenomena (1). These include both the internal flexibility and the overall rotational diffusion of macromolecules, membrane fluidity, solvent relaxation, excited state reactions, and resonance energy transfer. *Time-resolved* fluorescence analyzes both excited state lifetimes and the individual processes occurring therein. Impulse response and harmonic response techniques (referred to as pulse fluorometry and frequency-domain or phase fluorometry, respectively) are independent methods that give equivalent information on the dynamics of fluorescence. The purpose of this article is to summarize the principles of pulse and frequency-domain fluorometry, to present two recent applications to calmodulin, and to provide a bibliography.

Frequency-domain Fluorometry

Elements of the design (2) and theory (3) of the phase fluorometer have been known for 60 years. Spencer and Weber (4) pointed out that each single fluorescent species, for which the response to an infinitely narrow pulse of light decays in time according to $I(t) = I(0)e^{-t/\tau}$, is an exponential detector. When a population of these detectors is excited by light that is sinusoidally modulated in intensity, the emitted light is sinusoidally modulated at the same frequency (f), with no attenuation or change of phase, provided that $2\pi f << \tau^{-1}$. (The product $2\pi f$ equals the angular frequency of modulation ω.) As $\omega \rightarrow \tau^{-1}$, finite persistence of the excited state leads to a phase delay (δ) and attenuation or demodulation of the fluorescence relative to the exciting light (3).

$$\delta = \tan^{-1}\omega\tau \quad (1)$$

When $\omega = \tau^{-1}$ the phase delay is 45°. When $\omega >> \tau^{-1}$, the phase lag approaches 90° and the attenuation becomes complete. Values of δ between 20 and 70° are favorable for determinations of τ (5). Fig. 1 contains a schematic representation of the separate excitation and emission wave forms. The inset shows the values of δ and the demodulation ratio M as a function of the product $\omega\tau$. Note that the modulation is the ratio of the signal amplitude to the average signal at that frequency (the AC/DC ratio) while the *demodulation ratio M* relates the modulation of the emitted light to that of the exciting light (3).

$$M = (AC_{em}/DC_{em})/(AC_{ex}/DC_{ex}) \quad (2)$$

$$M = (1 + (\omega\tau)^2)^{-1/2} \quad (3)$$

Modulation in the cross-correlation phase fluorometer of Spencer and Weber (4) was achieved through the generation of an ultrasonic standing wave perpendicular to the light wave. Either of two modulation frequencies (14.2 and 28.4 MHz) could be selected. A second feature was the transformation of high-frequency signals to low-frequency signals that contain the same information as the original. This technique, *cross-correlation,* is used in all modern frequency-domain fluorometers.

Examples of single exponential decay proved to be the exception (6). Complex decay results in discrepancies between the values of τ calculated from δ and M, with τ^P being smaller than ω^M (4). Weber (7) later derived the mathematical solution for the determination of N exponential lifetime components and their fractional intensities from data collected at N modulation frequencies. Nonlinear least squares fitting routines are used to determine the individual lifetimes and their fractional intensity contributions (5, 8, 9). In a hypothetical case of double exponential decay, Gratton *et al.* (10) estimated that the minimum ratio of lifetimes allowing resolution is ~1.6. This assumed errors of ± 0.2° and ±0.004 in the values of δ and M, respectively.

The development of a continuously variable frequency (1–160 MHz) cross-correlation phase fluorometer by Gratton and Limkeman (11) extended the range of excited state lifetimes that can be determined and made possible the analysis of complex decay by the harmonic response technique. Berndt *et al.* (12) recently described a 4-GHz frequency-domain fluorometer with internal microchannel plate photomultiplier cross-correlation. This instrument contains a synchronously pumped dye laser system, in which an argon-ion laser optically pumps the dye laser. Frequency doubling then generates the wavelengths needed for excitation. Light modulation is achieved through the actions of a mode locker, cavity dumper, and frequency synthesizer. Several of these components are employed in both frequency-domain and pulse fluorometers. The theory of operation of synchronously pumped dye lasers and microchannel plate photomultipliers and their use in pulse fluorometry have been reviewed by Small (13).

Phase fluorometry is inherently a differential method. In determinations of τ, parallel measurements are made with the fluorescent sample and a light scattering suspension in order to correct for phase shifts in the electronics and optics (4). However, other types of differential measurements are also made. By equipping the instrument with polarizers, one can determine the difference between the phase angles ($\Delta\delta$) of the parallel and perpendicular components of the emitted light. Weber (14) showed that in the case of a single spherical rotator in an isotropic medium,

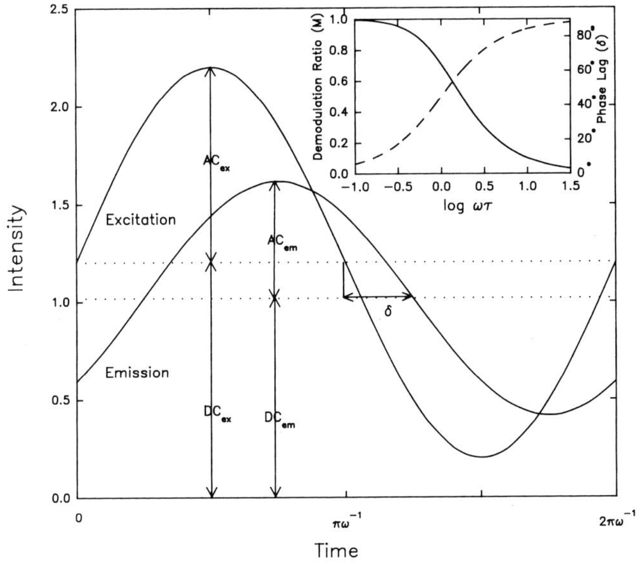

FIG. 1. Wave forms of excitation and emission as monitored in frequency-domain fluorometry. The time scale is given in terms of the period ($2\pi\omega^{-1}$), which is the time interval between harmonics. The intensity is expressed in arbitrary units. In this example, $\omega^{-1} = \tau$. Hence the phase lag $\delta = 45°$ and $M = 0.707$. δ can be calculated from the time interval (Δt) between corresponding points of the two wave forms. $\delta = \omega\Delta t\ (360°/2\pi)$. The *inset* shows the dependence of the demodulation ratio M (*solid line*) and phase lag δ (*dashed line*) on the angular modulation frequency ω for a single fluorescent species with excited state lifetime τ.

$\Delta\delta$ is related to D^1 (rotational diffusion coefficient), to r_0 (limiting anisotropy), to $k = \tau^{-1}$, and to ω.

$$\Delta\delta = \tan^{-1}\left(\frac{18\ \omega r_0 D}{(k^2 + \omega^2)(1 + r_0 - 2r_0^2) + 6D(6D + 2k + kr_0)}\right) \quad (4)$$

Plots of $\Delta\delta$ *versus* log $\omega\tau$ are *bell-shaped.* The value of D is directly related to the value of ω giving the *maximum value* of $\Delta\delta$.

$$D = ((2r_0 + 1)(1 - r_0)(k^2 + \omega^2))^{1/2}/3 \quad (5)$$

For more complex cases, multiple peaks and shoulders may be evident in the plots of $\Delta\delta$ *versus* log $\omega\tau$. The modulated (frequency-dependent) anisotropy $r(\omega)$ can be calculated from the modulation ratios (15).

$$r(\omega) = (m_\parallel - m_\perp)/(m_\parallel + 2m_\perp) \quad (6)$$

Application of Frequency-domain Fluorometry to the Intrinsic Fluorescence of Calmodulin

The intrinsic fluorescence of calmodulin, which undergoes an ~2.5-fold increase in quantum yield on calcium binding (16), is due entirely to two tyrosine residues occurring in the third (Tyr^{99}) and fourth (Tyr^{138}) calcium binding domains (17). Preliminary studies suggested both the occurrence of resonance energy transfer between the two tyrosines and the existence of segmental mobility within the calmodulin molecule (18). The intensity decay measurements shown in Fig. 2 were performed by Gryczynski *et al.* (19) with a 2-GHz fluorometer, in which the maximum modulation frequency is 2×10^9/s. The modulated exciting light consists of a train of 8-ps pulses from a cavity-dumped rhodamine 6G dye laser (frequency doubled to 285 nm). Polarization bias, recruiting from rotational diffusion of fluorescent molecules, was eliminated by placement of a vertically oriented polarizer next to the sample in the excitation beam and of a second polarizer, oriented at 54.7° to the vertical, on the emission side (1).

Estimations based on the frequency corresponding to a phase shift of 45° give values for τ^P and τ^M of ~2.3 and ~2.6 ns in the presence of calcium and of ~1.0 and ~2.0 ns in the absence of calcium. Rigorous data analysis was carried out for exponential decay, employing transforms of the frequency-domain equations. The intensity decay was analyzed as a sum of exponentials according to the equation,

$$I(t) = \sum f_i\, e^{-t/\tau_i} \quad (7)$$

where f_i and τ_i are the amplitude and decay time of the ith decay mode. The goodness of fit over the entire frequency range was determined by a nonlinear least squares algorithm (8, 9).

Table I summarizes decay times of the fluorescence intensity determined in a three-component analysis, which gave a better fit than either a one- or two-component analysis. Since the thrombic fragments (each containing a single tyrosine) also show complex decays, individual lifetimes cannot be assigned to specific tyrosine residues. The results may reflect a continuous distribution of excited state lifetimes resulting from fluctuations in the microenvironment rather than three discrete components. The former, popularized by Alcala *et al.* (20), is difficult to distinguish from the average decay obtained with several distinct lifetimes.

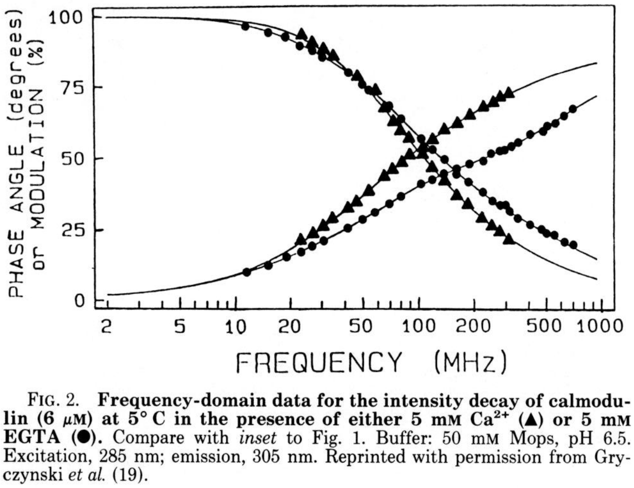

FIG. 2. **Frequency-domain data for the intensity decay of calmodulin (6 μM) at 5° C in the presence of either 5 mM Ca^{2+} (▲) or 5 mM EGTA (●).** Compare with *inset* to Fig. 1. Buffer: 50 mM Mops, pH 6.5. Excitation, 285 nm; emission, 305 nm. Reprinted with permission from Gryczynski *et al.* (19).

TABLE I

Frequency-domain measurements of the intensity decay of calmodulin and its thrombic fragments

Data are compiled from Gryczynski *et al.* (19). TM1, residues 1–106 of calmodulin (CaM); TM2, residues 107–148. Temperature, 5 °C. Refer to Fig. 2 for other conditions.

	CaM-EGTA	CaM-Ca^{2+}	TM1-Ca^{2+}	TM2-Ca^{2+}
$\bar{\tau}$ (ns)	2.66	2.70	3.81	2.58
α_1	0.674	0.318	0.486	0.516
τ_1 (ns)	0.352	0.716	0.269	0.372
α_2	0.299	0.628	0.397	0.143
τ_2 (ns)	2.09	2.58	2.58	2.73
α_3	0.028	0.054	0.123	0.341
τ_3 (ns)	7.18	5.08	6.11	2.94
χ^2_R	1.8	0.5	1.2	1.1

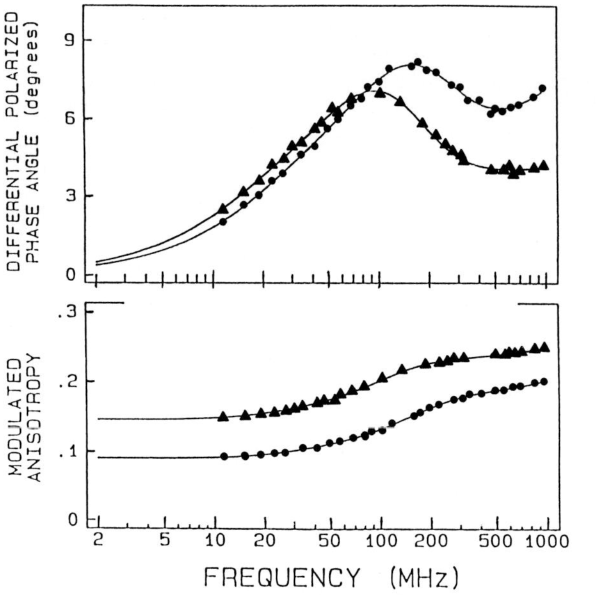

FIG. 3. **Frequency-domain anisotropy data for calcium-free calmodulin (5 mM EGTA) collected at 5 °C (▲) and 37 °C (●).** The *solid lines* show the best two-correlation time fits: 5 °C, 0.089 and 2.88 ns; 37 °C, 0.067 and 1.35 ns. Refer to Fig. 2 for other conditions. Reprinted with permission from Gryczynski *et al.* (19).

The data in Fig. 3 were obtained in differential polarized frequency-domain experiments with calcium-free calmodulin. Analysis for exponential anisotropy decay utilized the equation,

$$r(t) = \sum_i \beta_i e^{-t/\phi_i} \quad (8)$$

where β_i and ϕ_i[1] represent the amplitude and correlation time, respectively, of the ith decay mode. The amplitudes are related to the apparent limiting anisotropy (r_0) at zero time.

$$r_0 = \sum_i \beta_i$$

Both calmodulin and the thrombic fragments demonstrate two resolvable components, one with a rotational correlation time in the sub-nanosecond range (Table II). The addition of calcium significantly affects both rotations. However, none of the correlation times is large enough to represent the overall global rotation of calmodulin (see the next section).

The sums of the amplitudes exceed the limiting anisotropies obtained in steady state experiments, 0.243 (0 Ca^{2+}) and 0.211 (+Ca^{2+}). Frequency-domain measurements at −60 °C and high

[1] The Debye rotational relaxation time (ρ) of a sphere equals $1/2D$ while the rotational correlation time (ϕ) equals $1/6D$.

TABLE II

Frequency-domain anisotropy measurements on calmodulin and its thrombic fragments

Data compiled from Gryczynski *et al.* (19). TM1, residues 1–106 of calmodulin (CaM). TM2, residues 107–148. Value of ϕ were corrected to standard conditions at 20 °C through multiplication by factors of 0.628 (5 °C) and 1.53 (37 °C). Refer to Figs. 2 and 3 for other conditions.

Sample	T	β_1	ϕ_1^{20}	β_2	ϕ_2^{20}
	°C		*ns*		*ns*
CaM-EGTA	5	0.047	0.06	0.238	1.81
CaM-EGTA	37	0.115	0.11	0.179	2.07
CaM-Ca^{2+}	5	0.071	0.60	0.175	5.28
CaM-Ca^{2+}	37	0.077	0.38	0.201	5.52
TM1-EGTA	37	0.253	0.09	0.111	0.84
TM1-Ca^{2+}	37	0.074	0.14	0.217	4.14
TM2-EGTA	37	0.157	0.23	0.192	0.34
TM2-CA^{2+}	37	0.17	0.31	0.144	3.78

viscosity demonstrated a fast component (ϕ = 1 ns, β = 0.049) believed to reflect energy transfer (19, 21). They gave recovered anisotropies (0.288) approaching the values obtained with proteins and peptides containing only one tyrosine.

Pulse Fluorometry

Frequency-domain measurements are performed under stationary conditions, in which all of the emissions from $t = 0$ to $t = \infty$ contribute to the values of δ and M (7). Pulse or time-domain fluorometry utilizes flash lamps operated at repetition frequencies such that the decays initiated by one flash do not overlap the next. There are two general types of pulse fluorometer: 1) time-correlated monophoton counting instruments, in which one photon per pulse is monitored and collected in one of a number of channels corresponding to varying periods of time after excitation; and 2) instruments measuring many photons per pulse (23).

The experiments to be described were performed with a monophoton decay fluorometer based on the original design of Small *et al.* (24). Frequency-doubled pulses from the cavity-dumped output of a synchronously pumped dye laser constitute the excitation source. The system includes a mode-locked Nd:YAG laser and an experimental version Hamamatsu R2287U triple microchannel plate photomultiplier. The narrow half-width of the measured excitation pulse (80 ps) approaches a Δ function, simplifying the analysis of many experiments. Adjustment of the exciting light intensity together with *energy windowing* facilitate the recording of only monophoton events (25, 26).

Anisotropy Decay of the Fluorescent Derivative of Calmodulin Obtained Upon Photoactivated Cross-linking of Tyr[99] and Tyr[138]

UV irradiation of bovine calmodulin results in calcium-dependent changes in its intrinsic tyrosine fluorescence and the appearance of a new emission maximum at 400 nm. Characterization of the 400-nm emitting species revealed the presence of *dityrosine,* a product of the intramolecular coupling of Tyr[99] and Tyr[138] (27). The occurrence of this reaction is consistent with the locations of the two tyrosines within the three-dimensional structure of calmodulin (17, 22) and with the localized mobilities discussed in the preceding section (19, 21). Steady state fluorescence intensity and anisotropy determinations, monitoring the fluorescence of the ionized dityrosine chromophore (λ_{ex} = 320 nm), revealed a generally weakened interaction with calcium occurring in two stages (24).

The two-point anchoring of the dityrosine chromophore eliminated localized side chain rotation. Small and Anderson (28) applied a mathematical model for the rotational diffusion of dumbbell-shaped molecules to the cross-linked calmodulin. The fluorescence anisotropy of a rigid molecule with cylindrical symmetry decays essentially as a sum of three exponentials (5, 29, 30).

$$r(t) = \beta_1 e^{-(4D_1+2D_2)t} + \beta_2 e^{-(D_1+5D_2)t} + \beta_3 e^{-(6D_2)t} \tag{9}$$

The rotational diffusion coefficient D_1 describes the relatively fast rotation of the molecule about its long symmetry axis while

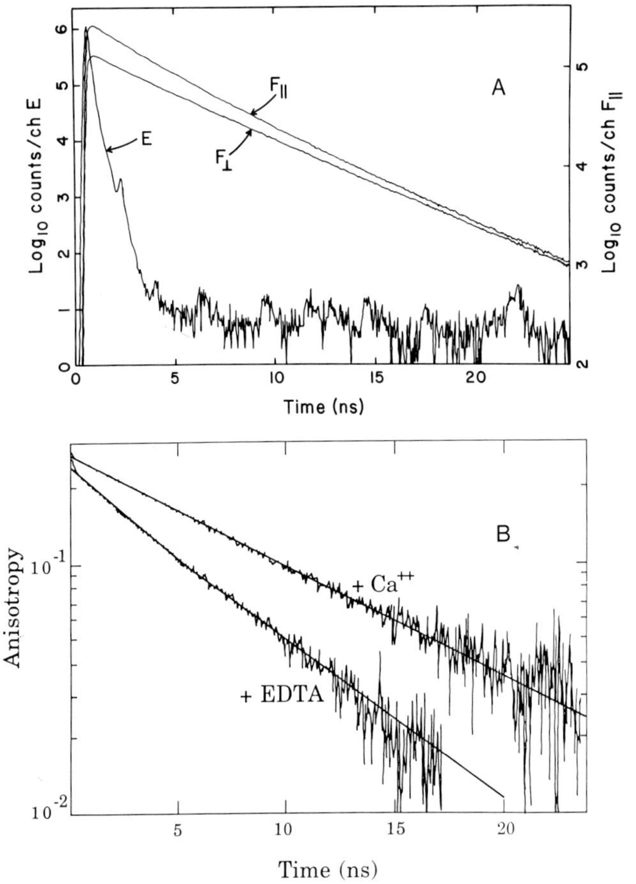

FIG. 4. *A,* logarithmic plots of the excitation (E) and the polarized fluorescence intensity decays ($F_\parallel$ and $F_\perp$) for dityrosyl calmodulin (1.0 μM). Conditions: 20 mM calcium acetate, 50 mM Mops, pH 7.5 (20.0 °C). Excitation, 310 nM; emission, Corning CSO-52 cut-off filter. *B,* anisotropy decay of dityrosyl calmodulin (1.0 μM) in the presence of 0.0 and 20 mM calcium acetate. Calculated curves corresponding to the values of β and ϕ given in Table III are superimposed on the data. Reprinted with permission from Small and Anderson (28).

D_2 applies to the slower rotation about an axis perpendicular to the long axis.

Equations for the computation of the rotational diffusion coefficients of variously shaped molecules (31) allowed estimation of the values of D_1 and D_2 for the calcium-calmodulin complex. The 64-Å-long dumbbell shape determined in x-ray crystallography (17) was approximately by a pair of spheres that are 28.5 Å in diameter and separated from each other by 7 Å. The molecular volume (25,200 $Å^3$) was calculated from the molecular weight and partial specific volume of calmodulin together with an assumed hydration of 0.2 g of H_2O/g of protein. The amplitudes (β) are functions of the inclinations and orientations of the absorption and emission dipoles, quantities that are seldom both known.

Fig. 4*A* illustrates time-domain determinations of the parallel and perpendicular components of the fluorescence of dityrosyl calmodulin in the presence of excess calcium. Also shown is the excitation pulse, measured with a dilute scattering sample. The output of the rhodamine 575 dye laser was frequency-doubled to 310 nm for excitation. The data were collected in 1024 channels, representing 0.0464 ns/channel. The counting rate was near 20 kHz, with each curve representing approximately 2.4 × 10^7 counts. The anisotropy is calculated from $r(t) = (F_\parallel - SF_\perp)/(F_\parallel + 2SF_\perp)$, where S is a factor correcting for both unequal transmission of the two components and unequal counting rates. The average excited state lifetimes[2] (3.59 ns (0 Ca^{2+}) and 3.81 ns (+Ca^{2+})) allowed accurate monitoring of ~90% of the total anisotropy decay. The existence of more than one lifetime has no

[2] Method of moments calculations performed on total intensity decays gave the following values of τ and relative intensities: 1. 0 Ca^{2+}, 4.3 ns (80%), 2.4 ns (11.7%), 8.8 ns (5%), and 0.9 ns (3.3%); 2. 20 mM Ca^{2+}, 4.0 ns (73.8%), 6.3 ns (21.3%), 1.3 ns (4.6%), and 0.4 ns (0.4%). The data also are compatible with a continuous distribution of lifetimes (20, 28).

TABLE III

Anisotropy decay of dityrosyl calmodulin determined by pulse fluorometry

Conditions are given in the legends to Fig. 4. Data are compiled from Small and Anderson (28).

$[Ca^{2+}]$	β_1	ϕ_1^{20}	β_2	ϕ_2^{20}
		ns		*ns*
0	0.213	6.84	0.025	2.1
20 μM	0.220	8.27	0.035	2.0
200 μM	0.235	7.84	0.019	~1.1
2 mM	0.257	9.49	0.0026	~1.6
20 mM	0.265	9.91		

effect on the latter unless the individual values are nonrandomly associated with specific values of ϕ (30).

The anisotropy decay of dityrosyl calmodulin in the presence of 20 mM calcium is apparently first order while that determined in its absence is more complex (Fig. 4*B*). The data were analyzed by the method of moments, which was advanced by Isenberg and associates (32–34) for the study of multiexponential decays. Nonlinear least squares iterative reconvolution has been more commonly used in pulse fluorometry (35). Small (34) considers that the method of moments is preferable in cases of nonrandom instrument error or difficult-to-resolve lifetimes while nonlinear least squares iterative reconvolution is advantageous in the fitting of complex models. Interrelated experiments sufficient to constitute a "multidimensional data surface" allow a comprehensive approach known as *global analysis,* which emphasizes physical model evaluation. Programs are available for applications to both pulse and frequency-domain fluorometry (30).

Table III contains analyses of the anisotropy decays of dityrosyl calmodulin determined at varying calcium concentrations selected from the results of steady state measurements (27). At zero calcium, a moderate rotational correlation time of ϕ_1 = 6.84 ns predominates (β_1 = 0.213). This value is close to that calculated for a sphere of molecular volume of 25,200 Å^3 (6.25 ns). A shorter rotational correlation time of ~2 ns (β_2 = 0.025), probably a reflection of segmental flexibility, also is evident at the lower calcium concentrations. There may be additional unresolvable decays (χ^2 = 1.056). Increases in ϕ_1 occur on the addition of calcium, leading to a clearly resolved (reduced χ^2 = 0.995) single correlation time of 9.9 ns (β = 0.265).

The theoretical calculations of D_1 and D_2 predicted three major rotational correlation times for the calcium-calmodulin complex: $(4D_1 + 2D_2)^{-1}$ = 7.2 ns, $(D_1 + 5D_2)^{-1}$ = 9.4 ns, and $(6D_2)^{-1}$ = 10.4 ns. The figure of 9.9 ns apparently corresponds to one of the two larger values or an unresolved average thereof. Consideration of the effects of varied length and hydration shows that the data are consistent with a dumbbell ranging in length from about 62 to 66 Å. Inability to detect the shorter component probably reflects the orientations of the absorption and emission dipoles relative to the molecular axes of calmodulin.

Other Applications and Summary

This review has emphasized two related applications of pulse and frequency-domain fluorometry. Frequency-domain measurements demonstrate the considerable mobility of the two tyrosine residues of calmodulin and the existence of resonance energy transfer between them (19). Time-domain studies of the fluorescent derivative of calmodulin obtained upon the photoactivated cross-linking of the two tyrosines reveal a molecule which undergoes elongation upon calcium binding. The rotational diffusion of the calcium-saturated species monitored in the measurements agrees with that predicted for the global rotation of native calmodulin, with no localized flexibility detected (28).

Segmental flexibility in proteins appears to be a common phenomenon. Anisotropy decay studies employing genetically engineered immunoglobulins show that the degree of segmental flexibility is controlled by both the heavy chain constant region (C_H1) and the hinge (36). Resonance energy transfer monitored through donor fluorescence decay reveals flexibility in rabbit skeletal troponin, with the distance between labels attached to Cys[98] of troponin C and Cys[133] of troponin I being dependent on calcium and magnesium binding (37). Pulse fluorometry shows that the two redox centers of NADPH-cytochrome P-450 reductase are separated by a relatively large average distance (~20 Å), even though electrons are transferred between them during reaction (38).

A time-correlated single photon-counting nanosecond microscope, allows observation of fluorescent molecules within single cells (39). A pilot study demonstrates the binding of a major fraction of fura-2, the calcium indicator, to macromolecular components in rat basophilic leukemia cells.

Acknowledgments—I thank the following individuals for providing reprints, manuscripts, and suggestions: Enrico Gratton, David M. Jameson, Susan M. Keating, Gregory D. Reinhart, Enoch W. Small, Robert F. Steiner, Lubert Stryer, Terence Tao, and Antonie J. W. G. Visser. In addition, I am grateful to colleague Enoch W. Small for reading this article and to Gregorio Weber for introducing me to fluorescence spectroscopy.

REFERENCES

1. Lakowicz, J. R. (1983) *Principles of Fluorescence Spectroscopy,* pp. 1–428, Plenum Press, New York
2. Gaviola, E. (1927) *Z. Physik* **42,** 852–859
3. Dushinsky, F. (1933) *Z. Physik* **81,** 7–12
4. Spencer, R. D. & Weber, G. (1969) *Ann. N. Y. Acad. Sci.* **158,** 361–376
5. Jameson, D. M. & Hazlett, T. L. (1991) in *Biophysical and Biochemical Aspects of Fluorescence Spectroscopy* (Dewey, G., ed) pp. 105–133, Plenum Press, New York
6. Rayner, D. M. & Szabo, A. G. (1978) *Can. J. Chem.* **56,** 743–745
7. Weber, G. (1981) *J. Phys. Chem.* **85,** 949–953
8. Jameson, D. M. & Gratton, E. (1983) in *New Directions in Molecular Luminescence* (Eastwood, D., ed) ASTM STP 4822, p. 67, American Society for Testing and Materials, Philadelphia
9. Jameson, D. M., Gratton, E. & Hall, R. D. (1984) *Appl. Spectrosc. Rev.* **20,** 55–106
10. Gratton, E., Jameson, D. M. & Hall, R. D. (1984) *Annu. Rev. Biophys. Bioeng.* **13,** 105–124
11. Gratton, E. & Limkeman, M. (1983) *Biophys. J.* **44,** 315–324
12. Berndt, K. W., Gryczynski, I. & Lakowicz, J. R. (1991) *Anal. Biochem.* **192,** 131–137
13. Small, E. W. (1991) in *Modern Fluorescence Spectroscopy. Vol. 1: Techniques* (Lakowicz, J. R., ed) Plenum Press, New York, in press
14. Weber, G. (1977) *J. Chem. Phys.* **66,** 4081–4091
15. Gratton, E., Alcala, J. R. & Barbieri, B. (1990) in *Handbook of Luminescence Techniques in Chemical and Biochemical Analysis,* pp. 47–71, Marcel Dekker, Inc., New York
16. Kilhoffer, M.-C., Demaille, J. G. & Gerard, D. (1981) *Biochemistry* **20,** 4407–4414
17. Babu, Y. S., Sack, J. S., Greenhough, T. J., Bugg, C. E., Means, A. R. & Cook, W. J. (1985) *Nature* **315,** 37–40
18. Gryczynski, I., Lakowicz, J. R. & Steiner, R. F. (1988) *Biophys. Chem.* **30,** 45–59
19. Gryczynski, I., Steiner, R. F. & Lakowicz, J. R. (1991) *Biophys. Chem.* **39,** 69–78
20. Alcala, J. R., Gratton, E., and Prendergast, F. G. (1987) *Biophys. J.* **51,** 597–604
21. Steiner, R. F., Albaugh, S. & Kilhoffer, M.-C. (1991) *J. Fluorescence Spectrosc.* **1,** in press
22. Babu, Y. S., Bugg, C. E. & Cook, W. J. (1988) *J. Mol. Biol.* **204,** 191–204
23. Badea, M. G. & Brand, L. (1979) *Methods Enzymol.* **61,** 378–425
24. Small, E. W., Libertini, L. J. & Isenberg, I. (1984) *Rev. Sci. Instrum.* **55,** 879–885
25. Schuyler, R. & Isenberg, I. (1971) *Rev. Sci. Instrum.* **42,** 813–817
26. Hutchings, J. J. & Small, E. W. (1990) in *Time-Resolved Laser Spectroscopy in Biochemistry II* (Lakowicz, J. R., ed) Proceedings of the SPIE 1204, pp. 184–191, Bellingham, WA
27. Malencik, D. A. & Anderson, S. R. (1987) *Biochemistry* **26,** 695–704
28. Small, E. W. & Anderson, S. R. (1988) *Biochemistry* **27,** 419–428
29. Weber, G. (1972) *Proc. Natl. Acad. Sci. U. S. A.* **69,** 1392–1393
30. Beechem, J. M., Gratton, E., Ameloot, M., Knutson, J. R. & Brand, L. (1991) in *Fluorescence Spectroscopy. Vol. II: Principles* (Lakowicz, J. R. ed) Plenum Press, New York, in press
31. de la Torre, J. G. & Bloamfield, V. A. (1977) *Biopolymers* **16,** 1765–1778
32. Isenberg, I. & Dyson, R. D. (1969) *Biophys. J.* **9,** 1340–1350
33. Libertini, L. J. & Small, E. W. (1989) *Biophys. Chem.* **34,** 269–282
34. Small, E. W. (1991) *Methods Enzymol.,* in press
35. Johnson, M. L. & Frasier, S. G. (1985) *Methods Enzymol.* **117,** 301–342
36. Schneider, W. P., Wensel, T. G., Stryer, L. & Oi, V. T. (1988) *Proc. Natl. Acad. Sci. U. S. A.* **85,** 2509–2513
37. Tao, T., Gowell, E., Strasburg, G. M., Gergely, J. & Leavis, P. C. (1989) *Biochemistry* **28,** 5902–5908
38. Bastiaens, P. I. H., Bonauts, P. J. M., Muller, F. & Visser, A. J. W. G. (1989) *Biochemistry* **28,** 8416–8425
39. Keating, S. M. & Wensel, T. G. (1991) *Biophys. J.,* in press

Minireview

THE JOURNAL OF BIOLOGICAL CHEMISTRY
Vol. 266, No. 19, Issue of July 5, pp. 12115–12118, 1991
Printed in U.S.A.

Caldesmon, a Novel Regulatory Protein in Smooth Muscle and Nonmuscle Actomyosin Systems

Kenji Sobue‡§ and James R. Sellers¶

From the ‡Department of Neurochemistry and Neuropharmacology, Biomedical Research Center, Osaka University Medical School, Osaka 530, Japan and the ¶Laboratory of Molecular Cardiology, National Heart, Lung, and Blood Institute, National Institutes of Health, Bethesda, Maryland 20892

Caldesmon is a major calmodulin- and actin-binding protein found in smooth muscle and nonmuscle cells (Ref. 1, and reviewed in Ref. 2). Current studies suggest a vital role for this protein in the regulation of smooth muscle and nonmuscle contraction. The actomyosin system, which converts the chemical energy of ATP into mechanical force, is the molecular basis for contraction of smooth muscle and various contractile and motile events in nonmuscle cells such as cytokinesis, capping, and phagocytosis (3–5). In smooth muscle and nonmuscle cells actomyosin interactions are controlled in part by phosphorylation of the 20-kDa regulatory light chain subunit of myosin by myosin light chain kinase and dephosphorylation by myosin phosphatase (3–5). In addition, thin filament-based regulation involving caldesmon and calmodulin is also possible. The *in vitro* evidence for this is compelling. Caldesmon inhibits superprecipitation and actomyosin Mg^{2+}ATPase activity *in vitro* (6–9). It also inhibits movement of actin filaments by myosin in an *in vitro* motility assay,[1] and exogenously added caldesmon relaxes permeabilized smooth muscle fiber preparations (10, 11). Direct experimental evidence for a regulatory role *in vivo* has not yet been provided, however. Here we will overview the structure and function of caldesmon.

Characterization of Caldesmon Isoforms

Two isoforms of caldesmon can be discerned by mobility upon SDS-PAGE[2]: *h*-caldesmon (high M_r of 120,000–150,000) and *l*-caldesmon (low M_r of 70,000–80,000) (1, 2, 12–25). However, the molecular weights of the two isoforms calculated from the deduced amino acid sequence of their respective full-length cDNA from chicken tissues are in the range of 87,000–89,000 and 59,000–60,000, respectively (26–29).

The basic properties of caldesmon isoforms are summarized in Table I. Both isoforms are heat-stable (12, 22). They both bind calmodulin, tropomyosin, and actin and are inhibitors of the actin-activated Mg^{2+}ATPase activity of myosin (2, 6–9, 17, 21, 30). In addition, both undergo what was termed "flip-flop" binding (31) to either actin filaments or calmodulin depending upon the free Ca^{2+} concentration since the Ca^{2+}/calmodulin-caldesmon complex binds very weakly to actin (1, 12, 13, 15, 17, 19, 32).

The tissue and cell distributions of the two isoforms are distinctively different; *h*-caldesmon is predominantly expressed in smooth muscle, whereas *l*-caldesmon is primarily found in nonmuscle tissue and cells (19, 20). Neither caldesmon isoform is detected in skeletal and cardiac muscle (33). Immunohistochemistry of smooth muscle at the light and electron microscope levels has revealed that *h*-caldesmon is localized within the subset of actin filaments that are also co-localized with myosin (13). In cultured nonmuscle cells, *l*-caldesmon distributes along stress fibers co-locally with tropomyosin (16, 18, 19, 23). See Fig. 1.

Primary Structures of Caldesmon Isoforms

The structural organization of caldesmon and the relationship between the two caldesmon isoforms were originally investigated using protein chemical and immunological techniques. These results suggest that the amino- and carboxyl-terminal domains of the two isoforms are composed of similar or identical sequences, while the central domain of *h*-caldesmon is missing in the smaller *l*-caldesmon molecule (34). The amino acid sequences of the two isoforms deduced by cloning and sequencing two caldesmons of each have led to a clarification of the above suggestion (26–29, 35).

Fig. 2 shows a diagrammatic comparison of the predicted amino acid sequences between chick embryo gizzard *h*-caldesmon and chick brain *l*-caldesmon based on the assignments of Hiyashi *et al.* (27). We will use these assignments for the remainder of this discussion. There is heterogeneity within *h*-caldesmon isoforms from the same species. The amino acid sequence predicted from an independent cloning of chick gizzard *h*-caldesmon (28) is identical to that of Hiyashi *et al.* (26) except for a deletion of residues 319–333. This central region of *h*-caldesmon (residues 201–477) contains 8–10 repeats where each repeating unit is based on a glutamic acid-rich sequence of 13–15 amino acids. It is one of these repeating units that is missing in the cDNA clone of Bryan *et al.* (28). Heterogenous *h*-caldesmons have also been detected by SDS-PAGE or isoelectric focusing (1, 12, 36). Proteolytic studies demonstrate that this heterogeneity exists in the central repeating domain (36).

The central region of *h*-caldesmon (residues 201–447) is deleted in the smaller *l*-caldesmon molecule (27, 29). Amino acid residues 25–200 and 448–771 for gizzard *h*-caldesmon are completely identical with residues 17–192 and 193–517 for chick brain *l*-caldesmon except for the insertion of alanine at residue 508 of *l*-caldesmon (27). There are several differences in individual amino acids in the amino-terminal regions of the two isoforms (residues 1–24 for *h*-caldesmon and residues 1–16 for *l*-caldesmon). A cDNA clone for *l*-caldesmon from chicken gizzard is identical to that of *h*-caldesmon with the exception of the deletion of 232 residues in the central region (29), which given the sequence of *l*-caldesmon from chicken brain suggests that there is also heterogeneity in *l*-caldesmons from a given species. In support of this finding two species of *l*-caldesmon cDNAs have been isolated from HeLa cells.[3]

Northern blot analysis has revealed two species of mRNA with different molecular weights probably coding for *h*- and *l*-caldesmons, respectively (27, 28). Southern blot analysis of genomic DNA suggests that the two species of caldesmon mRNA are probably generated from a single gene via alternative splicing and are not the products of different genes (27).

Domain Mapping of Caldesmon

Hydrodynamic measurements and direct electron microscopic observations show that both caldesmon isoforms have an elongated shape with lengths of about 74 and 53 nm for the high and low isoforms, respectively (37–40). Studies utilizing limited proteolysis, chemical cleavage, and bacterial expression of genetically engineered caldesmon fragments have led to the identification of functional domains within the caldesmon molecule. A carboxyl-terminal 35-kDa chymotryptic fragment of *h*-caldesmon (residues 466–771) binds calmodulin, tropomyosin, and actin and inhibits the actin-activated Mg^{2+}ATPase activity of myosin (41–46). An homologous 32-kDa domain of adrenal medullary *l*-caldesmon exhibits similar behavior (46). Further studies have shown that a 10-kDa carboxyl-terminal fragment (residues 674–771) of *h*-caldesmon prepared by CNBr cleavage at methionine residues also

§ To whom reprint requests should be addressed.
[1] J. R. Sellers and V. P. Shirinski, manuscript in preparation.
[2] The abbreviations used are: SDS, sodium dodecyl sulfate; PAGE, polyacrylamide gel electrophoresis; HMM, heavy meromyosin; S-1, subfragment-1.
[3] K. Hayashi, I. Kato, and K. Sobue, manuscript in preparation.

TABLE I

Characterization of h- and l-caldesmons

	h-Caldesmon	*l*-Caldesmon
M_r		
From SDS-PAGE	120–150 kDa	70–80 kDa
From primary structure	89,000	59,000
Molecular length	74 nm	53 nm
Calmodulin binding	+	+
Tropomyosin binding	+	+
Actin binding	+	+
Flip-flop binding	+	+
Heat stability	+	+
Inhibition of actomyosin ATPase	+	+
Immunological cross-reactivity		
h-Caldesmon antibody	+	+
l-Caldesmon antibody	+	+
Tissue distribution	Smooth muscle	Nonmuscle
Intracellular localization	Thin filaments	Actin filaments or stress fibers

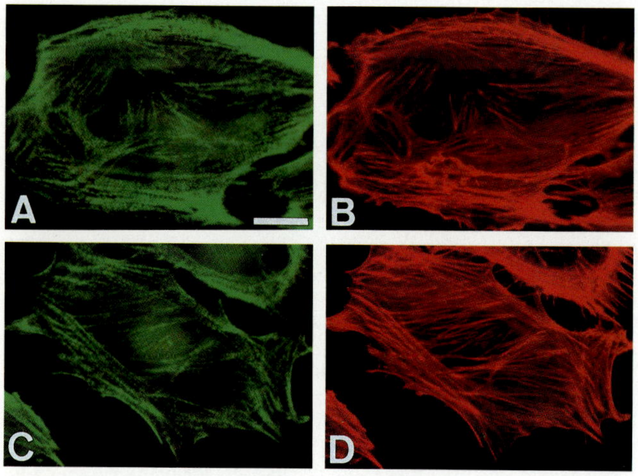

FIG. 1. **Double-fluorescent light micrographs of rat 3YI cells stained with antibodies against caldesmon (*A*) or tropomyosin (*C*) using fluorescein isothiocyanate-conjugated second antibodies and rhodamine-labeled phalloidin for actin filaments (*B* and *D*) (J. Tanaka, T. Watanabe, N. Nakamura, and K. Sobue, manuscript in preparation).** Note that caldesmon and tropomyosin are periodically distributed along stress fibers (bundles of actin filament). *Bar*, 10 μm.

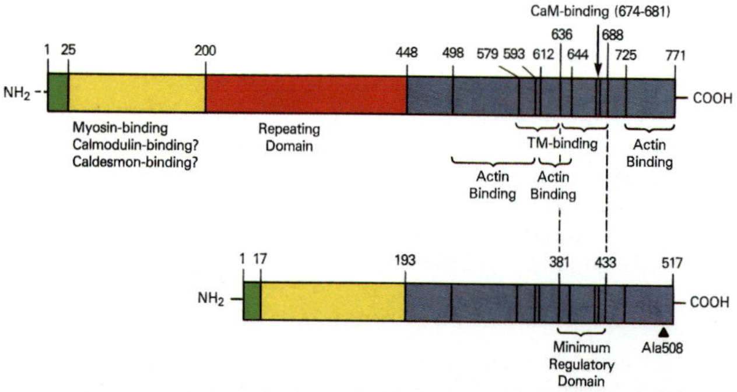

FIG. 2. **Comparison of primary structures and domain mappings between *h*- and *l*-caldesmons.** Residues 25–200 (*yellow*) and 448–771 (*blue*) for *h*-caldesmon are completely identical with residues 17–192 (*yellow*) and 193–517 (*blue*) for *l*-caldesmon except for the insertion of Ala-508 of *l*-caldesmon. The short NH_2 termini (residues 1–24 for *h*-caldesmon and residues 1–16 for *l*-caldesmon) show individually unique sequences (*green*). Residues 251–392 for *h*-caldesmon in the *red box* are the central repeating sequence. Each binding site for calmodulin, actin, tropomyosin, and myosin in the caldesmon molecules are indicated in this figure. Amino acid residues of each caldesmon are *numbered.*

binds actin and calmodulin and partially inhibits the actin-activated Mg^{2+}ATPase of myosin subfragment 1 (48).

Domain mapping studies using deletion mutants of caldesmon are in good agreement with the proteolytic and chemical studies described above and have provided a more precise localization of the actin, tropomyosin, and calmodulin binding regions (Fig. 2). There is evidence for three actin-binding sites distributed throughout the 35-kDa region (27, 49, 50). A strong actin-binding site is located between residues 612 and 644, with a weaker binding site located between 725 and 771 (50). Another weak actin-binding site appears to be localized between residues 498 and 593 as determined by studies with a 15-kDa peptide (49). Studies using numerous different truncated constructs and the 10-kDa CNBr fragment restrict the calmodulin-binding domain to 7 amino acids, WEKGNVF, starting with residue 674 (27, 48, 50). This location would explain the enhancement of tryptophan fluorescence observed upon binding of calmodulin to caldesmon (51). The tropomyosin-binding sites lie between residues 579–635 and 636–771 for *h*-caldesmon (residues 324–380 and 381–571 for *l*-caldesmon) (27). A region with homology to the tropomyosin-binding sites of troponin T (residues 523–580 for *h*-caldesmon and 268–325 for *l*-caldesmon) (28, 42, 49, 52) appears to be nonfunctional based on the expression experiments with truncated caldesmon molecules (27). While constructs containing another region with troponin T homology (residues 622–632 for *h*-caldesmon and residues 367–377 for *l*-caldesmon) bind tropomyosin, it seems unlikely that these short sequences are essential for the regulation of actin-myosin interactions since some constructs that do not contain this region inhibit the actin-activated Mg^{2+}ATPase of myosin (27). Instead, a new tropomyosin-binding site, found within residues 636–688 for *h*-caldesmon (residues 381–443 for *l*-caldesmon), might be important for such regulations (27). An amino-terminal construct in which the carboxyl-terminal 178 amino acids are deleted did not inhibit actin-activated Mg^{2+}ATPase activity and bound poorly to tropomyosin (53).

In summary, actin-, calmodulin-, and tropomyosin-binding regions all appear to lie within a short region of the sequence spanning residues 636–688 for *h*-caldesmon and 381–433 for *l*-caldesmon. All expressed caldesmon constructs or proteolytic fragments having this region are able to inhibit the actin-activated Mg^{2+}ATPase activity of myosin and bind to calmodulin.

The function of the central repeating region of *h*-caldesmon that is missing in *l*-caldesmon is not known. It has been postulated to be highly α-helical based on secondary structure prediction programs (26, 28) and electron microscopic observations (37). Its deletion is probably responsible for the shorter length of *l*-caldesmon (38).

The amino-terminal domain of caldesmon has been postulated to be involved in the dimerization of caldesmon molecules (54), and a second calmodulin-binding site in this region has been suggested on the basis of cross-linking experiments and directly visualized by rotary shadowing of cross-linked caldesmon-calmodulin complexes (40, 55). Other studies of calmodulin binding to caldesmon, however, suggest a single calmodulin-binding site (51, 56–58). Interestingly, the amino-terminal domain of caldesmon also binds myosin (47, 58–60). A K_b of about 1×10^6 M^{-1} was determined for the binding constant of intact caldesmon to myosin filaments (60). The subfragment 2 region of myosin appears to be most directly involved in caldesmon binding (59–61). The affinity of caldesmon to myosin filaments is weakened in the presence of Ca^{2+}/calmodulin (60).

Phosphorylation of Caldesmon

h-Caldesmon is phosphorylated by at least four protein kinases: protein kinase C, Ca^{2+}/calmodulin-dependent protein kinase II, casein kinase II, and the cdc2 protein kinase. Protein kinase C incorporates more than 3 mol of phosphate/mol of *h*-caldesmon into the carboxyl-terminal 35-kDa domain (62–64). This phosphorylation reduces the affinity of caldesmon or its fragment to both calmodulin and F-actin, and subsequently the inhibition of the actin-activated Mg^{2+}ATPase activity of myosin by caldesmon is reduced (64). The Ca^{2+}/calmodulin-dependent kinase II phosphorylates several sites on *h*-caldesmon scattered throughout the molecule with the primary sites lying in the amino-terminal domain (62, 65). Slower phosphorylation occurs in the carboxyl-terminal region of the molecule. This phosphorylation weakens the binding of caldesmon to myosin (60). One report suggested that phosphorylation of caldesmon to an extent of about 1 mol of phosphate/mol by this kinase reversed the inhibition of the actin-activated Mg^{2+}ATPase activity of myosin (59). This finding is difficult to reconcile with the known sequence of phosphorylation sites by the kinase which suggests that at this level of phosphorylation only amino-terminal sites should be phosphorylated (65). Others find little effect on the inhibition of Mg^{2+}ATPase activity following phosphorylation of caldesmon with Ca^{2+}/calmodulin-dependent kinase (66). Casein kinase II phosphorylates *h*-caldesmon within the amino-terminal domain, but neither the amino acid residue(s) phosphorylated nor the functional significance of this phosphorylation has been explored (62). Recently, human platelet *l*-caldesmon has been shown to be phosphorylated by cAMP-dependent protein kinase (67).

Caldesmon is also phosphorylated *in vivo*. Treatment of intact platelets with tumor-promoting phorbol ester results in an increased incorporation of phosphate into *l*-caldesmon (22). Treatment of platelets with prostacyclin results in phosphorylation of *l*-caldesmon at sites phosphorylated *in vitro* by cAMP-dependent protein kinase (67). Recently, *l*-caldesmon has been shown to be a substrate for the cdc2 kinase which is activated during mitosis with phosphorylation occurring at several sites probably located in the carboxyl-terminal domain of the molecule (35, 68, 69). During mitosis of HeLa cells, caldesmon is phosphorylated at sites similar to those phosphorylated *in vitro* by the purified cdc2 kinase (35). This phosphorylation reduces the affinity of caldesmon for actin filaments (68, 69). *h*-Caldesmon is phosphorylated when porcine carotid arteries are stimulated with various agonists, but the kinase responsible for this phosphorylation is not known (70).

Regulatory Mode of Caldesmon in the Actomyosin System

The maximum inhibition of the actomyosin Mg^{2+}ATPase activity occurs at caldesmon to actin molar ratios ranging from 1:6 to 1:26 depending on the report (6–9, 17, 66, 71). The reason for this variability is not known but may be related to differing assay conditions and the use of proteins (actin, myosin, tropomyosin, etc.) from different species and/or tissues. The molar ratio of caldesmon to actin in isolated gizzard smooth muscle thin filaments is about 1:16, assuming a molecular weight of 89,000 for *h*-caldesmon (72). This is higher than the caldesmon:actin ratio in gizzard tissue which is 1:25 to 1:35 (20, 72). The difference is likely due to the differential compartmentalization of actin with its various actin-binding proteins (13).

Direct measurements of the binding of caldesmon to actin have also given a variety of affinities and stoichiometries. The binding constant for binding of caldesmon to actin is typically in the range of 10^6–10^7 M^{-1} (32, 57). One study found a single class of binding sites which saturates at a caldesmon:actin ratio of 1:7 to 1:10 (32), while another found two classes of binding sites: a high affinity site which saturates at about 1:21 and another lower affinity site which saturates at about 1:5 assuming a molecular weight of 89,000 for *h*-caldesmon (57). Tropomyosin does not affect the stoichiometry of caldesmon to actin but does increase the affinity about 2–4-fold (32, 57). This effect may be due to the direct interaction of tropomyosin with *h*-caldesmon (42, 43, 46, 73, 74). Several models have been presented to explain how caldesmon could be such an effective inhibitor of the Mg^{2+}ATPase activity at molar ratios of less than 1 caldesmon:1 tropomyosin:7 actins. These usually involve one elongated caldesmon molecule spanning two tropomyosin molecules (38, 72).

h-Caldesmon inhibits the actin-activated Mg^{2+}ATPase of intact smooth or skeletal muscle myosin when assayed in the presence, but not the absence, of tropomyosin (6, 8, 9, 57, 75). In contrast, *h*-caldesmon dramatically suppresses the actin-activated Mg^{2+}ATPase activity of both heavy meromyosin (HMM) or subfragment-1 (S-1) even when no tropomyosin is present (32, 58, 66, 76, 77). These discrepant results might be generated from different assay conditions (for example, ionic strength, pH, and divalent cations, etc.) and/or different myosin preparations used.[4]

Calmodulin reverses the caldesmon-induced inhibition of myosin's Mg^{2+}ATPase activity in a Ca^{2+}-dependent manner. The affinity of *h*-caldesmon for calmodulin (K_b less than 10^6 M^{-1}), however, is weak compared with many other calmodulin-binding proteins (45, 51, 56, 57). Ca^{2+}/calmodulin also reduces the affinity of caldesmon to actin (1, 12, 13, 15, 19, 21); however, recent findings suggest that the reversal of caldesmon-induced inhibition by Ca^{2+}/calmodulin does not closely correlate with the dissociation of caldesmon. For example, near physiological ionic strength conditions, Ca^{2+}/calmodulin relieves the *h*-caldesmon-induced inhibition while the *h*-caldesmon-Ca^{2+}/calmodulin complex remains bound to the actin-tropomyosin complex[4] (57, 78). Furthermore, it is noteworthy that troponin C, which also interacts with *h*-caldesmon, can overcome the *h*-caldesmon-induced inhibition even when it remains on the *h*-caldesmon-actin-tropomyosin complex.[4]

Studies of the mechanism of inhibition of the actin-myosin Mg^{2+}ATPase activity by caldesmon are complicated by the fact that the amino-terminal region of caldesmon binds myosin in addition to actin (47, 58, 59, 66). Caldesmon increases the binding of gizzard smooth muscle HMM to actin while inhibiting its actin-activated Mg^{2+}ATPase activity (66, 77). Subsequently, this increase in binding was determined to be due to "tethering" of HMM probably via the subfragment-2 binding site on caldesmon (60, 61). The carboxyl-terminal fragments of caldesmon which do not bind myosin also inhibit Mg^{2+}ATPase activity (46, 58, 60, 79). Therefore, it is clear that the tethering of myosin to actin cannot fully explain this inhibition. Studies with S-1 which binds caldesmon only weakly have shown the mode of caldesmon inhibition is probably due to a direct competition between myosin and caldesmon for binding sites on actin (58). This is supported by studies using the carboxyl-terminal fragments of caldesmon which also appear to directly compete with S-1 for binding to actin (58, 79). In this regard, it is interesting that both myosin and caldesmon appear to bind to the amino-terminal region of actin (80–82).

The ability of caldesmon to bind to both myosin and actin has led to speculation about its function in smooth muscle cells. A model (59) has been proposed where simultaneous binding of caldesmon to both actin and myosin filaments may support tension in smooth muscle fibers and may be partly responsible for the "latch" state (83). In order for a caldesmon molecule to bind to both actin and the S-2 region of myosin in a filament in the muscle, this model would require that caldesmon molecules project some distance out from the thin filaments. Recent structural studies on native smooth muscle thin filaments suggest that caldesmon is bound lengthwise along the actin filament with little tendency to project outward (84).

Role of Caldesmon in Nonmuscle Cells

As described above *l*-caldesmon is associated with the stress fibers in nonmuscle cells. There are several instances of *l*-caldesmon redistribution during physiological changes in cells. Lymphocyte caldesmon, tropomyosin, and actin are found to accumulate into subcaps during ligand-induced receptor capping (85, 86). *l*-Caldesmon also redistributes at the periphery of chromaffin cells during stimulant-induced catecholamine secretion (24). This finding suggests the involvement of caldesmon in control of

[4] Y. Fujio and K. Sobue, manuscript in preparation.

reorganization of actin filaments at the cell periphery following an increase in intracellular free Ca^{2+} during secretion. In Rous sarcoma virus-transformed cells, there is a 3-fold decrease in caldesmon levels present in cells, and the intracellular distribution of *l*-caldesmon changes to a diffuse and blurred appearance (16). This change in *l*-caldesmon may correlate with a loss of Ca^{2+} regulation in the transformed cells. Finally, caldesmon appears to dissociate from thin filaments during mitosis as a result of phosphorylation by cdc2 kinase (35, 68).

In addition to the obvious role in regulating actomyosin interactions, caldesmon may also be involved in directly regulating cytoskeletal structure. The binding of low molecular weight tropomyosin subunits to actin is potentiated by *l*-caldesmon (87). Caldesmon potentiates the tropomyosin inhibition of the severing of actin filaments by gelsolin (88) and can enhance the reannealing of gelsolin-severed actin filaments (89).

Conclusions—While the available *in vitro* data suggest an important role for caldesmon in regulation of smooth muscle contraction and of control of contractile events and the cytoskeleton in nonmuscle cells, questions remain as to the function of this protein and its stoichiometry and arrangement on thin filaments. Further studies are required for these questions. The presence of caldesmon, however, may have important implications in contractility where it is clear that a simple on/off mechanism based on myosin phosphorylation is not sufficient to account for the physiological function.

REFERENCES

1. Sobue, K., Muramoto, Y., Fujita, M., and Kakiuchi, S. (1981) *Proc. Natl. Acad. Sci. U. S. A.* **78,** 5652–5655
2. Sobue, K., Kanda, K., Tanaka, T., and Ueki, N. (1988) *J. Cell. Biochem.* **37,** 317–325
3. Korn, E. D., and Hammer, H. A., III (1988) *Annu. Rev. Biophys. Biophys. Chem.* **17,** 23–45
4. Hartshorne, D. J. (1987) in *Physiology of the Gastrointestinal Tract* (Johnson, L. R., ed) pp. 423–482, Raven Press, New York
5. Sellers, J. R. (1991) *Curr. Opin. Cell Biol.* **3,** 98–104
6. Sobue, K., Morimoto, K., Inui, M., Kanda, K., and Kakiuchi, S. (1982) *Biomed. Res.* **3,** 188–196
7. Ngai, P. K., and Walsh, M. P. (1984) *J. Biol Chem.* **259,** 13656–13659
8. Sobue, K., Takahashi, K., and Wakabayashi, I. (1985) *Biochem. Biophys. Res. Commun.* **132,** 645–651
9. Marston, S. B., and Lehman, W. (1985) *Biochem. J.* **231,** 517–522
10. Szpacenko, A., Wagner, J., Dabrowska, R., and Rüegg, J. C. (1985) *FEBS Lett.* **192,** 9–12
11. Taggart, M. J., and Marston, S. B. (1988) *FEBS Lett.* **242,** 171–174
12. Bretscher, A (1984) *J. Biol. Chem.* **259,** 12873–12880
13. Furst, D. O., Cross, R. A., Mey, J. D., and Small, J. V. (1986) *EMBO J.* **5,** 251–257
14. Clark, T., Ngai, P. K., Sutherland, C., Gröschel-Stewart, U., and Walsh, M. P. (1986) *J. Biol. Chem.* **261,** 8028–8035
15. Yamazaki, K., Itoh, K., Sobue, K., Mori, T., and Shibata, N. (1987) *J. Biochem. (Tokyo)* **101,** 1–9
16. Owada, M. K., Hakura, A., Iida, K., Yahara, I., Sobue, K., and Kakiuchi, S. (1984) *Proc. Natl. Acad. Sci. U. S. A.* **81,** 3133–3137
17. Sobue, K., Tanaka, T., Kanda, K., Ashino, N., and Kakiuchi, S. (1985) *Proc. Natl. Acad. Sci. U. S. A.* **82,** 5025–5029
18. Bretscher, A., and Lynch, W. (1985) *J. Cell Biol.* **100,** 1656–1663
19. Dingus, J., Hwo, S., and Bryan, J. (1986) *J. Cell Biol.* **102,** 1748–1757
20. Ueki, N., Sobue, K., Kanda, K., Hada, T., and Higashino, K. (1987) *Proc. Natl. Acad. Sci. U. S. A.* **84,** 9049–9053
21. Onji, T., Takagi, M., and Shibata, N. (1987) *Biochem. Biophys. Res. Commun.* **143,** 475–481
22. Litchfield, D. W., and Ball, E. H. (1987) *J. Biol. Chem.* **262,** 8056–8060
23. Yamashiro-Matsumura, S., and Matsumura, F. (1988) *J. Cell Biol.* **106,** 1973–1983
24. Burgoyne, R. D., Cheek, T. R., and Norman, K-M. (1986) *Nature* **319,** 68–70
25. Terrossian, E. D., Deprette, C., and Cassoly, R. (1989) *Biochem. Biophys. Res. Commun.* **159,** 395–401
26. Hayashi, K., Kanda, K., Kimizuka, F., Kato, I., and Sobue, K. (1989) *Biochem. Biophys. Res. Commun.* **164,** 503–511
27. Hayashi, K., Fujio, Y., Kato, I., and Sobue, K. (1991) *J. Biol. Chem.* **266,** 355–361
28. Bryan, J., Imai, M., Lee, R., Moore, P. L., Cook, R. G., and Lin, W-G. (1989) *J. Biol. Chem.* **264,** 13873–13879
29. Bryan, J., and Lee, R. (1991) *J. Muscle Res. Cell Motil.,* in press
30. Sobue, K., Tanaka, T., Kanda, K., Takahashi, K., Itoh, K., and Kakiuchi, S. (1985) *Biomed. Res.* **6,** 93–102
31. Kakiuchi, S., and Sobue, K. (1983) *Trends Biochem. Sci.* **8,** 59–62
32. Velaz, L., Hemric, M. E., Benson, C. E., and Chalovich, J. M. (1989) *J. Biol. Chem.* **264,** 9602–9610
33. Ban, T., Ishimura, K., Fujita, H., Sobue, K., and Kakiuchi, S. (1984) *Acta Histochem. Cytochem.* **17,** 331–338
34. Ball, E. H., and Kovala, T. (1988) *Biochemistry* **27,** 6093–6098
35. Yamashiro, S., Yamakita, Y., Hosoya, H., and Matsumura, F. (1991) *Nature* **349,** 169–172
36. Riseman, V. M., Lynch, W. P., Nefsky, B., and Bretscher, A. (1989) *J. Biol. Chem.* **264,** 2869–2875
37. Wang, C.-L. A., Chalovich, J. M., Graceffa, P., Lu, R. C., Mabuchi, K., and Stafford, W. F. (1991) *J. Biol. Chem.* **266,** in press
38. Graceffa, P., Wang, C.-L. A., and Stafford, W. F. (1988) *J. Biol. Chem.* **263,** 14196–14202
39. Stafford, W. F., III, Jansco, I., and Graceffa, P. (1990) *Arch. Biochem. Biophys.* **281,** 66–69
40. Mabuchi, K., and Wang, C.-L. A. (1991) *J. Muscle Res. Cell Motil.* **12,** 145–151
41. Szpacenko, A., and Dabrowska, R. (1986) *FEBS Lett.* **202,** 182–186
42. Hayashi, K., Yamada, S., Kanda, K., Kimizuka, F., Kato, I., and Sobue, K. (1989) *Biochem. Biophys. Res. Commun.* **161,** 38–45
43. Fujii, T., Ozawa, J., Ogoma, Y., and Kondo, Y. (1988) *J. Biochem. (Tokyo)* **104,** 734–737
44. Fujii, T., Imai, M., Rosenfeld, G. C., and Bryan, J. (1987) *J. Biol. Chem.* **262,** 2757–2763
45. Yazawa, M., Yagi, K., and Sobue, K. (1987) *J. Biochem. (Tokyo)* **102,** 1065–1073
46. Sobue, K., and Fujio, Y. (1989) *Adv. Exp. Med. Biol.* **255,** 325–335
47. Katayama, E., Horiuchi, K. Y., and Chacko, S. (1989) *Biochem. Biophys. Res. Commun.* **160,** 1316–1322
48. Bartegi, A., Fattoum, A., Derancourt, J., and Kassab, R. (1990) *J. Biol. Chem.* **265,** 15231–15238
49. Mornet, D., Audemard, E., and Derancourt, J. (1988) *Biochem. Biophys. Res. Commun.* **154,** 564–571
50. Wang, C.-L. A., Wang, L.-W. C., Xu, S., Lu, R. C., Saavedra-Alanis, V., and Bryan, J. (1991) *J. Biol. Chem.* **266,** 9166–9172
51. Shirinsky, V. P., Bushueva, T. L., and Frolova, S. I. (1988) *Biochem. J.* **255,** 203–208
52. Leszyk, J., Mornet, D., Audemard, E., and Collins, J. H. (1989) *Biochem. Biophys. Res. Commun.* **160,** 1371–1378
53. Redwood, C. S., Marston, S. B., Bryan, J., Cross, R. A., and Kendrick-Jones, J. (1990) *FEBS Lett.* **270,** 53–56
54. Cross, R. A., Cross, K. E., and Small, J. V. (1987) *FEBS Lett.* **219,** 306–310
55. Wang, C.-L. A., Wang, L.-W. C., and Lu, R. C. (1989) *Biochem. Biophys. Res. Commun.* **162,** 746–752
56. Sobue, K., Muramoto, Y., Fujita, M., and Kakiuchi, S. (1981) *Biochem. Int.* **2,** 469–476
57. Smith, C. W. J., Pritchard, K., and Marston, S. B. (1987) *J. Biol. Chem.* **262,** 116–122
58. Velaz, L., Ingraham, R. H., and Chalovich, J. M. (1990) *J. Biol. Chem.* **265,** 2929–2934
59. Sutherland, C., and Walsh, M. P. (1989) *J. Biol. Chem.* **264,** 578–583
60. Hemric, M. E., and Chalovich, J. M. (1990) *J. Biol. Chem.* **265,** 19672–19678
61. Ikebe, M., and Reardon, S. (1988) *J. Biol. Chem.* **263,** 3055–3058
62. Vorotnikov, A. V., Shirinsky, V. P., and Gusev, N. B. (1988) *FEBS Lett.* **236,** 321–324
63. Umekawa, H., and Hidaka, H. (1985) *Biochem. Biophys. Res. Commun.* **132,** 56–62
64. Tanaka, T., Ohta, H., Kanda, K., Tanaka, T., Hidaka, H., and Sobue, K. (1990) *Eur. J. Biochem.* **188,** 495–500
65. Ikebe, M., and Reardon, S. (1990) *J. Biol. Chem.* **262,** 17607–17612
66. Lash, J. A., Sellers, J. R., and Hathaway, D. R. (1986) *J. Biol. Chem.* **261,** 16155–16160
67. Hettasch, J. M., and Sellers, J. R. (1991) *J. Biol. Chem.* **266,** 11876–11881
68. Yamashiro, S., Yamakita, Y., Ishikawa, R., and Matsumura, F. (1990) *Nature* **344,** 675–678
69. Mak, A. S., Watson, M. H., Litwin, C. M. E., and Wang, J. H. (1991) *J. Biol. Chem.* **266,** 6678–6681
70. Adam, L. P., Haeberle, J. R., and Hathaway, D. R. (1989) *J. Biol. Chem.* **264,** 7698–7703
71. Marston, S. B. (1986) *Biochem. J.* **237,** 605–607
72. Lehman, W., Craig, R., Lui, J., and Moody, C. (1989) *J. Muscle Res. Cell Motil.* **10,** 101–112
73. Graceffa, P. (1987) *FEBS Lett.* **218,** 139–142
74. Watson, M. H., Kuhn, A. E., and Mark, A. S. (1990) *Biochim. Biophys. Acta* **1054,** 103–113
75. Dabrowska, R., Goch, A., Galazkiewicz, B., and Osinska, H. (1985) *Biochim. Biophys. Acta* **842,** 70–75
76. Chalovich, J. M., Cornelius, P., and Benson, C. E. (1987) *J. Biol. Chem.* **262,** 5711–5716
77. Hemric, M. E., and Chalovich, J. M. (1988) *J. Biol. Chem.* **263,** 1878–1885
78. Pritchard, K., and Marston, S. B. (1989) *Biochem. J.* **257,** 839–843
79. Horiuchi, K. Y., and Chacko, S. (1989) *Biochemistry* **22,** 470–476
80. Adams, S., Das Gupta, G., Chalovich, J. M., and Reisler, E. (1990) *J. Biol. Chem.* **265,** 19652–19657
81. Bartegi, A., Fattoum, A., and Kassab, R. (1990) *J. Biol. Chem.* **265,** 2231–2237
82. Levine, B. A., Moir, A. J. G., Audemard, E., Mornet, D., Patchell, V. B., and Perry, S. V. (1990) *Eur. J. Biochem.* **195,** 687–696
83. Hai, C. M., and Murphy, R. A. (1988) *Am. J. Physiol.* **254,** C99–C106
84. Moody, C., Lehman, W., and Craig, R. (1990) *J. Muscle Res. Cell Motil.* **11,** 176–185
85. Mizushima, Y., Kanda, K., Hamaoka, T., Fujiwara, H., and Sobue, K. (1987) *Biomed. Res.* **8,** 73–78
86. Walker, G., Kerrick, G. L., and Bourguignon, L. Y. W. (1989) *J. Biol. Chem.* **264,** 496–500
87. Yamashiro-Matsumura, S., and Matsumura, F. (1988) *J. Cell Biol.* **106,** 1973–1983
88. Ishikara, R., Yamashira, S., and Matsumura, F. (1989) *J. Biol. Chem.* **264,** 7490–7497
89. Ishikawa, R., Yamashiro, S., and Matsumura, F. (1989) *J. Biol. Chem.* **264,** 16764–16770

Minireview

THE JOURNAL OF BIOLOGICAL CHEMISTRY
Vol. 266, No. 20, Issue of July 15, pp. 12809-12812, 1991
Printed in U.S.A.

Adhesive Recognition Sequences

Kenneth M. Yamada

From the Laboratory of Developmental Biology, National Institute of Dental Research, National Institutes of Health, Bethesda, Maryland 20892

Specific cellular adhesion and migration of cells are recurring themes in embryonic development, tumor cell metastasis, and wound healing. Recent advances in our understanding of the molecular basis of cell adhesive and migratory interactions with extracellular matrix molecules have converged on the concept that many cell interactions are dependent on specific adhesive recognition sequences (reviewed in Refs. 1–8). As will be described in detail below, certain short peptide sequences in adhesion proteins are thought to serve as sites for recognition and binding by specific plasma membrane receptors (Table I).

A number of specific cell surface receptors, particularly integrins, mediate the adhesion of cells to fibronectin, laminin, or collagen by recognizing different, specific peptide sequences in each (Fig. 1). Moreover, some receptors can recognize the same specific sequence in several different proteins (1–5, 8). Conversely, a single adhesion protein can contain several different sequences that are recognized by distinct receptors (1–8). The existence of so many potential combinations of binding activities provides considerable complexity to the repertoire of interactions possible for an individual cell.

Fibronectin, a Prototype Cell Adhesion System

The adhesive glycoprotein fibronectin is involved in a variety of biological processes, particularly in mediating cell attachment and cell migration (reviewed in Refs. 1–5). Fibronectin is bound by several cell surface receptors, including the "classical" fibronectin receptor $\alpha_5\beta_1$ and several other integrin receptors (2–5, 8). As reviewed below, short peptide sequences appear to be key determinants in the recognition of fibronectin by these receptors, although contributions from other polypeptide sequences are also quite important (Fig. 2). Most concepts in this review were established initially for fibronectin, as were most of the current tests of biological relevance summarized in Table II. Consequently, this review will focus in detail on fibronectin and then will compare adhesive recognition sequences in other molecules.

Adhesion and Competition Assays—The first peptide sequence in fibronectin to be identified as an adhesive recognition sequence was Arg-Gly-Asp (RGD) (9–11). Subsequently other cell adhesion sites of fibronectin have been identified, including a Leu-Asp-Val (LDV)-containing sequence that displays cell-type specificity and alternative splicing (12–15). Synthetic peptides based on the RGD or LDV motifs can mimic activity of the intact protein at least partially. They can mediate cell attachment when adsorbed to a substrate or when conjugated to a carrier, *e.g.* to albumin, IgG, beads, or a substrate (9, 14, 16). Conjugation to a carrier may avoid substrate adsorption artifacts (*cf.* Refs. 13 and 14).

An equally important property of such peptides is their capacity for competitive inhibition of adhesion or of other processes involving the native protein, *e.g.* to compete for the ligand in cell attachment and spreading assays or to compete for binding of radiolabeled ligand to a cell surface receptor (9–11, 14, 15). Both adhesion and competitive inhibition assays demonstrate specificity for such peptides when comparing a series of peptide substitutions. For example, RGDS-containing peptides are active in the binding of the central cell-binding domain of fibronectin by $\alpha_5\beta_1$, $\alpha_v\beta_3$, and $\alpha_3\beta_1$, whereas RG$\underline{E}$S-containing peptides are much less active (10). In contrast, LD$\overline{V}$-containing peptides are active for the $\alpha_4\beta_1$ receptor, whereas L$\underline{E}$V-containing peptides have minimal activity (15).

Analysis by Mutagenesis or Natural Mutation—The most direct test for function of a putative peptide recognition sequence is mutation. Deletion or mutation of RGD to RGE in fibronectin cDNA expressed as a fusion protein in bacteria leads to a loss of activity (17). Interspecies comparisons of evolutionarily mutated sequences can also be enlightening; although the LDV sequence is present in fibronectins from a variety of species, the REDV adhesion sequence is missing from chicken fibronectin (15, 18). Surprisingly, even though the RGD sequence is present and potentially functional in chicken tenascin and mouse laminin, it is missing from mouse tenascin and moved in position in human laminin (19–22); the RGD sequences in tenascin and laminin may thus be fortuitous or used variably. The CS1 alternatively spliced sequence appears to contribute 40% or much less of the activity of intact fibronectin depending on the cell type and the system for analyzing its activity (23–26).

Loss of Avidity and Specificity in Synthetic Peptides—Although peptide sequences such as RGD and LDV appear to be critical minimal sequences, they do not function alone. Competitive inhibition assays and avidity determinations show that these sequences by themselves are substantially less active than either the native protein or fully active fragments (9, 11, 27). The RGD sequence has about 100-fold less activity than intact fibronectin according to binding assays (27), and the LDV sequence is up to 25-fold less active than the complete CS1 sequence of 25 residues (15). Since the RGD sequence can also inhibit the function of receptors that do not normally bind to fibronectin (1–4), this minimal adhesion sequence alone also appears to lack information determining receptor specificity.

Synergistic or Helper Sequences—Recent studies indicate considerable complexity in the effects of other regions of the molecule on the function of minimal adhesion sequences (15, 17, 24–33). The simplest case so far appears to be that of the LDV sequence in the alternatively spliced CS1 region of fibronectin. The full-length CS1 peptide displays an impressive 40% of the full activity of fibronectin (14). Truncation of the 25-mer CS1 sequence leads to a gradual loss of activity, until reaching the LDV sequence which shows a 20–25-fold loss of activity (15); structural determinations and amino acid substitution studies should help elucidate why activity is lost during truncation.

The well studied RGD sequence is dependent on additional polypeptide information for full function, but studies to date can be interpreted as supporting roles for such sequences in either conformational stabilization or function as a second binding region (1–5). Cyclizing the RGD sequence can dramatically increase its effectiveness for interacting *in vitro* with vitronectin receptors, but not with fibronectin receptors (34). Moreover, antibodies can recognize different conformations or environments of the RGD sequence in different proteins (35, 36). These studies support the importance of conformation of the RGD sequence.

On the other hand, mutagenesis studies have located key functional polypeptide regions considerable distances away from the RGD site, as far as 14,000 and 28,000 daltons toward the amino terminus (17, 29–32). Deletions show 100–200-fold less activity than the native protein, consistent with a synergistic function (Refs. 17 and 32, but compare Ref. 30). Weak synergism can even be found between such sequences and the RGD sequence when each is located in separate polypeptides (17). Moreover, a monoclonal antibody binding to an epitope mapping about 15,000 daltons away from RGD blocks cell adhesive functions, whereas another monoclonal binding closer to the RGD site does not inhibit, ruling out steric inhibition (31).

Interestingly, the absolute distance between at least one of these sequences and the RGD sequence may also be important,

TABLE I

Adhesive recognition sequences

Name and number[a]	Recognition sequence[b]	Receptor(s)[c]	Ref.
Fibronectin			
Cell-binding determinant (1)	Gly-**Arg-Gly-Asp**-Ser	$\alpha_5\beta_1$, $\alpha_{IIb}\beta_3$, $\alpha_v\beta_3$, $\alpha_3\beta_1$, $\alpha_v\beta_1$	1, 9–11
IIICS: CS1 site (4)	Asp-Glu-Leu-Pro-Gln-Leu-Pro-His-Pro-Asn-Leu-His-Gly-Pro-Glu-Ile-**Leu-Asp-Val**-Pro-Ser	$\alpha_4\beta_1$	12–15
IIICS: REDV site (5)	**Arg-Glu-Asp-Val**	$\alpha_4\beta_1$	13
Peptide I (6)	Tyr-Glu-Lys-Pro-Gly-Ser-Pro-Pro-Arg-Glu-Val-Val-Pro-Arg-Pro-Arg-Pro-Gly-Val	?	38, 39
Peptide II (6)	Lys-Asn-Asn-Gln-Lys-Ser-Glu-Pro-Leu-Ile-Gly-Arg-Lys-Lys-Thr	?	38
Laminin			
YIGSR (1)	**Tyr-Ile-Gly-Ser-Arg**-Cys	67-kDa binding protein	40
PDSGR (2)	**Pro-Asp-Ser-Gly-Arg**	?	46
F9 (RYVVLPR) (3)	**Arg-Tyr-Val-Val-Leu-Pro-Arg**-Pro-Val-Cys-Phe-Glu-Lys-Gly-Lys-Gly-Met-Asn-Tyr-Val-Arg	?	42
LGTIPG (4)	**Leu-Gly-Thr-Ile-Pro-Gly**	67-kDa binding protein	47
RGD site (5)	**Arg-Gly-Asp**-Asn	?	21, 44
p20 (6)	Arg-Asn-Ile-Ala-Glu-Ile-Ile-Lys-Asp-Ile	?	48
PA22-2 (IKVAV) (7)	Ser-Arg-Ala-Arg-Lys-Gln-Ala-Ala-Ser-**Ile-Lys-Val-Ala-Val**-Ser-Ala-Asp-Arg	110 kDa, other?	41
LRE site (8)	**Leu-Arg-Glu**	?	43
Vitronectin	**Arg-Gly-Asp**-Val	$\alpha_v\beta_3$, $\alpha_v\beta_5$, $\alpha_{IIb}\beta_3$	2, 49
Fibrinogen			
RGD site	**Arg-Gly-Asp**-Ser	$\alpha_v\beta_3$, $\alpha_{IIb}\beta_3$	1, 2, 50
RGD site	**Arg-Gly-Asp**-Phe	$\alpha_{IIb}\beta_3$	1, 2, 50
γ chain peptide	His-His-Leu-Gly-Gly-Ala-Lys-Gln-Ala-Gly-Asp-Val	$\alpha_{IIb}\beta_3$	50
von Willebrand factor			
RGD site	**Arg-Gly-Asp**-Ser	$\alpha_{IIb}\beta_3$	1, 2
GPIb site	Cys-Gln-Glu-Pro-Gly-Gly-Leu-Val-Val-Pro-Pro-Thr-Asp-Ala-Pro *plus* Leu-Cys-Asp-Leu-Ala-Pro-Glu-Ala-Pro-Pro-Pro-Thr-Leu-Pro-Pro	Glycoprotein Ib	52
Entactin	Ser-Ile-Gly-Phe-**Arg-Gly-Asp**-Thr-Cys	?	51
Circumsporozoite protein	**Val-Thr-Cys-Gly**	?	54
Thrombospondin	**Val-Thr-*X*-Gly**	?	53
Collagen type I			
RGD site	**Arg-Gly-Asp**-Thr-Pro	30, 70, and 250 kDa	56
RGD site	Ser-**Arg-Gly-Arg**-Thr-Gly	?	57
DGEA site	**Asp-Gly-Glu-Ala**	$\alpha_2\beta_1$	55
Collagen type IV			
IV-H1	Gly-Val-Lys-Gly-Asp-Lys-Gly-Asn-Pro-Gly-Trp-Pro-Gly-Ala-Pro	?	58
Hep III	Gly-Glu-Phe-Tyr-Phe-Asp-Leu-Arg-Leu-Lys-Gly-Asp-Lys	?	59
Amyloid P component	**Phe-Thr-Leu-Cys-Phe-Arg**	?	60

[a] Numbers in parentheses refer to *numerals* in *triangles* in Figs. 2 or 3.
[b] Bold letters indicate putative minimal active sequences.
[c] Greek letters indicate the specific α and β subunit of each integrin receptor heterodimer (see Fig. 1).

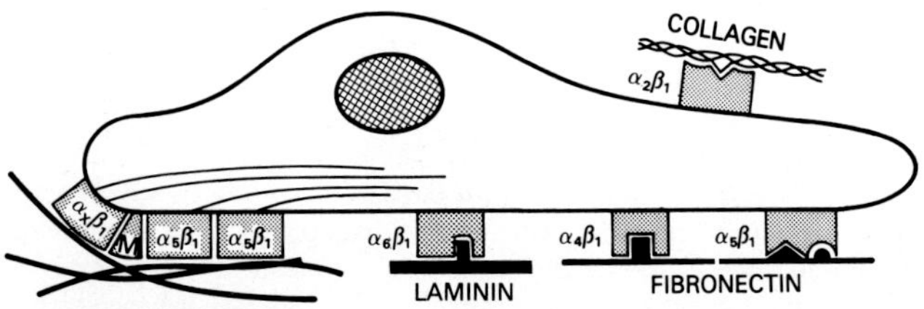

FIG. 1. **Schematic model of several cell interactions with extracellular glycoproteins via specific recognition sequences.** Cells can use integrin receptors (as depicted here) or other types of receptors to interact with specific peptide regions of ligands. Integrin receptors consist of heterodimers with one α and one β subunit, each of which can determine ligand specificity. For example, $\alpha_2\beta_1$ can bind collagen, while $\alpha_5\beta_1$ binds fibronectin; at least 17 different vertebrate integrins have been described so far. Some of their cell adhesion ligands such as fibronectin can contain multiple sites involved in recognition by several different integrin receptors. These receptors can participate in a number of biological processes, including adhesion, migration, and assembly of a fibronectin matrix. Matrix assembly may involve other molecules (*M*) and is accompanied by clustering of receptors and assembly of intracellular microfilament bundles (*thin lines*).

since moving a putative synergistic sequence away from the RGD sequence causes a major loss of biological activity (30). Taken together, these and other studies suggest that both RGD conformation and external sequence information contribute to interactions of the RGD sequence with the $\alpha_5\beta_1$ fibronectin receptor. Function of such "synergistic" sequences is also important for cell migration mediated by a fibronectin substrate (24, 31) and for assembly of fibronectin into extracellular fibrils (31). It is obvious that the three-dimensional structure of fibronectin needs to be determined to complement these functional studies.

Other Recognition Sequences in Fibronectin—Besides the CS1 site, another alternatively spliced region that contains cell adhesion activity is the REDV sequence (13). This sequence, like LDV, is recognized by the $\alpha_4\beta_1$ integrin receptor (8). It has been suggested that the presence of a common D (Asp) residue in RGD, LDV, and REDV may reflect its use as part of hypothesized cation-binding structures found in a number of integrins (37).

The high affinity heparin-binding domain of fibronectin also contains sequences that show adhesive activity, either when adsorbed to substrates or when used as competitive inhibitors of cell attachment (38, 39) (Table I). These sequences may function independently of the CS1 sequence, and they bind to heparin.

Laminin: A Surprising Number of Active Peptides

The glycoprotein laminin is a prominent constituent of basement membranes and can serve as an adhesion protein for a variety of cell types, especially epithelial and neuronal cells (reviewed in Refs. 6 and 7). A number of peptides with attachment activity has been derived from this protein (Table I, Fig. 3). The first sequence reported from this protein was the YIGSR sequence

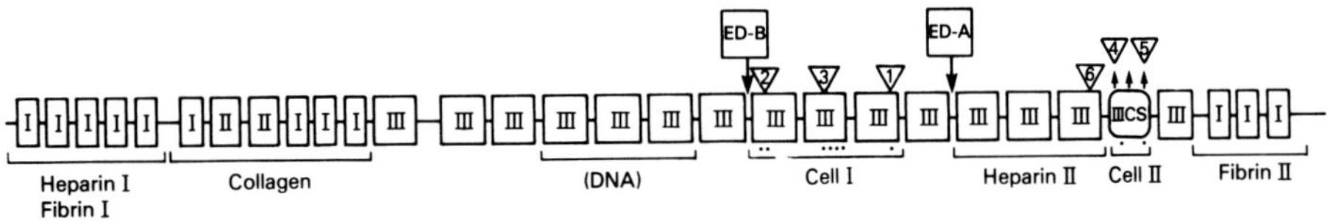

FIG. 2. **Structural organization of fibronectin, indicating putative adhesive recognition sites and regions.** The *diagram* depicts a subunit of fibronectin, which generally exists as a dimer linked by two COOH-terminal interchain disulfide bonds (the COOH terminus is at the *far right*). Fibronectin contains three types of repeating motif, termed types I, II, and III. These motifs are organized into the functional domains listed along the *bottom* of the figure; these domains are linearly arranged along the molecule and can be separated by proteolytic cleavage into fragments that can bind to heparin, fibrin, collagen, and weakly to DNA. The *arrows* indicate sites of alternative splicing, in which entire type III units (*ED-A* or *ED-B*) or parts of the IIICS region can be inserted into the mature sequence. The *numerals* in *triangles* represent putative adhesion sites or regions as identified using synthetic peptides or site-directed mutagenesis. Site 1 is the RGD sequence, with which regions 2 and 3 must synergize for full activity. Adhesive sites 4 and 5 are alternatively spliced from either end of the IIICS region. Site 6 contains at least two distinct peptide sequences that can mediate adhesion and binding to heparin. See Table I and the text for details.

TABLE II

Tests for biological relevance of a putative adhesive recognition sequence

Synthetic peptides containing the sequence display activity after conjugation to a carrier, even if inactive when adsorbed directly on substrates
Synthetic peptides competitively inhibit function of the intact protein
Biological activity is lost after site-directed mutagenesis of the protein and after amino acid substitutions in synthetic peptides
Similar activities are present in the intact protein and are retained in progressively truncated proteolytic fragments or recombinant proteins
Anti-peptide antibodies inhibit function of the native protein
The same plasma membrane receptor is used for the intact protein as for synthetic peptides

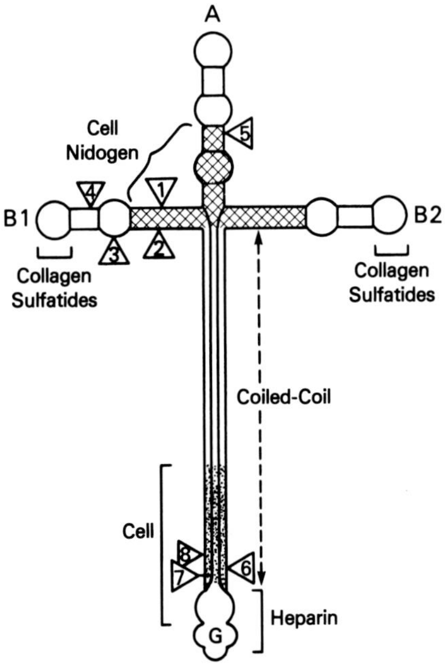

FIG. 3. **Peptide recognition sites on laminin.** Laminin consists of various disulfide-bonded subunits, *e.g.* in the form depicted, it contains an A chain linked to B1 and B2 chains by disulfide bonds and forming a triple-helical coiled-coil structure comprising the long arm of a distinctive cross-shaped molecule; the short arms consist of the NH_2-terminal portions of each subunit. Laminin also contains globular regions, as well as a large, multilobular globule (*G*) at the carboxyl-terminal end of the A chain involved in binding to heparin. Two major cell-binding regions of the protein identified using proteolytic fragmentation are shown *cross-hatched* or *stippled.* Eight putative peptide adhesive recognition sites identified using synthetic peptides are indicated by the *triangles* (see Table I and the text for discussion). Laminin can also bind to sulfated lipids (sulfatides) in some cells. Besides binding to cell surfaces, laminin interacts tightly with the basement membrane protein nidogen (entactin), as well as binding to type IV collagen and to heparin (or heparan sulfate) via other domains; these interactions help organize basement membranes.

(40), followed by others such as IKVAV (41) and RYVVLPR (42). In general, these peptides appear to have considerably lower avidities of binding to cells than native laminin, as suggested by the higher molar amounts needed for assays (40–48).

Some laminin peptides show adhesive activity for only highly specialized cell types, *e.g.* LRE for only ciliary ganglion cells (43). At least one of the putative laminin recognition sites, the RGD-containing sequence in the A chain of mouse laminin (site 5 in Fig. 3), is reported to be cryptic in the intact molecule (44). Additionally, the IKVAV sequence of laminin appears to require additional chain-chain polypeptide interactions for efficient adhesion of certain neuronal cells (45).

The rather large number of proposed recognition sequences reported to date in laminin is puzzling (at least 8 have been reported so far, Table I). It will be more reassuring when more of the criteria listed in Table II are applied to each of these sites in laminin (and fibronectin) to establish the biological relevance of each. The activities of different sites for different cell types also need further evaluation. Finally, it will also be important to learn which receptor(s) recognize each sequence and the biological sequelae of binding.

Other Adhesive Extracellular Matrix Proteins

A number of other proteins can mediate cell adhesion or serve as cell migration substrates. Many of them contain the RGD sequence, *e.g.* vitronectin, fibrinogen, von Willebrand factor, entactin, thrombospondin, and collagen (Table I). Since these proteins are bound by specific integrin receptors, the roles of other sequences in determining specificity requires elucidation. A variety of novel adhesive recognition sequences are also present in these and other proteins (Table I); the relative importance of each *in vivo* remains to be established.

Sequences in Receptors and Cell-Cell Adhesion Molecules

Surprisingly, a short peptide from the $\alpha_{IIb}\beta_3$ integrin retains the capacity to bind directly to fibrinogen, while retaining RGD specificity; this Thr-Asp-Val-Asn-Gly-Asp-Gly-Arg-His-Asp-Leu peptide is highly conserved among integrins (61). Besides their involvement in cell-substrate interactions, adhesive recognition sequences may also be involved in recognition events involving certain cell-cell adhesion molecules. The octapeptide Tyr-Lys-Leu-Asn-Val-Asn-Asp-Ser inhibits aggregation of *Dictyostelium* mediated by the glycoprotein 80 adhesion molecule and can mimic its homophilic binding *in vitro* (62). The tripeptide sequence His-Ala-Val present in a variety of cadherin molecules inhibits mouse embryo compaction (63). These promising findings suggest that cell-cell adhesion mechanisms may also have a component involving recognition of a short peptide sequence.

Synthetic Peptides as Probes of Biological Functions

Since synthetic peptides can specifically inhibit the function of adhesive recognition sites, they can be used to test the roles of these sites (and of the protein as a whole) in living animals. Roles for RGD-dependent processes have been identified for gastrulation, neural crest cell migration, and experimental metastasis (Ref. 64, reviewed in Refs. 1–5). Similarly, the YIGSR peptide has been reported to inhibit experimental metastasis and migration of neural crest cells (reviewed in Refs. 1 and 7). Platelet functions such as attachment to extracellular matrix proteins and aggregation in suspension are inhibited by RGD peptides, which block interactions with the $\alpha_{IIb}\beta_3$ (glycoprotein IIb-IIIa) receptor (reviewed in Refs. 1–4 and 8).

Signaling Activities of Synthetic Peptides

Besides serving as competitive inhibitors of cell adhesion and migration, synthetic peptides containing adhesive recognition sequences appear to signal certain metabolic responses directly. RGD-containing peptides can induce cell-cell adhesion of segmental plate cells *in vitro* and promote somite formation *in vivo* (65), perhaps by increasing expression of cell-cell adhesion molecules. RGD-containing peptides can also dramatically signal the secretion of certain proteases such as collagenase and stromelysin from cultured cells (66).

Certain peptides from laminin containing the IKVAV sequence can display strong biological activities, some of which may be distinct from those of the intact molecule. The PA-22-2 peptide can promote experimental metastasis and angiogenesis (7, 67). It may function by directly stimulating tissue plasminogen activator activity, which may in turn activate latent type IV collagenase and promote the metastatic process via increased cell invasion (68). It is conceivable that this sequence may function in some cases only after proteolytic degradation of laminin, *e.g.* in promoting angiogenesis after tissue destruction.

Summary and Perspective

The importance of short peptide recognition sequences in binding to cell surface receptors such as the integrins during cell adhesion now appears to be well established. Less clear, however, is the manner in which additional polypeptide sequences in some proteins such as fibronectin function to enhance or synergize with such sequences and to provide receptor specificity. Such contributions are important to elucidate, since they can account for 100–200-fold differences in biological activity. Although a number of short adhesive recognition sequences has been proposed, their overall patterns of cell-type specificity remain to be determined. Some of these sites may be cryptic or otherwise inactive in the native protein and may therefore require proteolysis of the molecule for function. Some putative recognition sequences may ultimately fail the functional tests listed in Table II.

The receptors for most of these sequences still remain to be determined, and the relationship of receptor expression to cell-type specificity also remains to be clarified. The use of these peptides as probes of biological functions in living animals has established the general importance of certain sequences in processes such as cell migration, but more protein-specific analyses are needed, *e.g.* by including specificity-determining synergistic sequences along with the short peptide recognition sequence.

These studies of adhesive recognition sequences should eventually lead to an understanding of the molecular basis of the cell adhesive and migratory mechanisms and of the binding specificities involved in embryonic development, tumor cell invasion, and wound healing. Obvious potential practical applications include the development of specific peptide inhibitors of thrombosis and tumor cell invasion, and possibly the rational design of molecules for promoting aspects of wound healing.

REFERENCES

1. Hay, E. D. (ed) (1991) *Cell Biology of Extracellular Matrix,* 2nd Ed., Plenum Press, New York, in press
2. Ruoslahti, E., and Pierschbacher, M. D. (1987) *Science* **238,** 491–497
3. Mosher, D. F. (ed) (1989) *Fibronectin,* Academic Press, New York
4. Hynes, R. O. (1990) *Fibronectins,* Springer-Verlag, New York
5. Yamada, K. M. (1990) *Curr. Opin. Cell Biol.* **1,** 956–963
6. Timpl, R. (1989) *Eur. J. Biochem.* **180,** 487–502
7. Kleinman, H. K., and Weeks, B. S. (1989) *Curr. Opin. Cell Biol.* **1,** 964–967
8. Mecham, R. P., and McDonald, J. A. (eds) (1991) *Receptors for Extracellular Matrix Proteins,* Academic Press, New York, in press
9. Pierschbacher, M. D., and Ruoslahti, E. (1984) *Nature* **309,** 30–33
10. Pierschbacher, M. D., and Ruoslahti, E. (1984) *Proc. Natl. Acad. Sci. U. S. A.* **81,** 5985–5988
11. Yamada, K. M., and Kennedy, D. W. (1984) *J. Cell Biol.* **99,** 29–36
12. McCarthy, J. B., Hagen, S. T., and Furcht, L. T. (1986) *J. Cell Biol.* **102,** 179–188
13. Humphries, M. J., Akiyama, S. K., Komoriya, K., Olden, K., and Yamada, K. M. (1986) *J. Cell Biol.* **103,** 2637–2647
14. Humphries, M. J., Komoriya, A., Akiyama, S. K., Olden, K., and Yamada, K. M. (1987) *J. Biol. Chem.* **262,** 6886–6892
15. Komoriya, A., Green, L. J., Mervic, M., Yamada, S. S., Yamada, K. M., and Humphries, M. J. (1991) *J. Biol. Chem.* **266,** in press
16. Brandley B. K., and Schnaar, R. L. (1989) *Dev. Biol.* **135,** 74–86
17. Obara, M., Kang, M. S., and Yamada, K. M. (1988) *Cell* **53,** 649–657
18. Norton, P. A., and Hynes, R. O. (1987) *Mol. Cell. Biol.* **7,** 4297–4307
19. Bourdon, M. A., and Ruoslahti, E. (1989) *J. Cell Biol.* **108,** 1149–1155
20. Weller, A., Beck, S., and Ekblom, P. (1991) *J. Cell Biol.* **112,** 355–362
21. Grant, D. S., Tashiro, K., Segui-Real, B., Yamada, Y., Martin, G. R., and Kleinman, H. K. (1989) *Cell* **58,** 933–943
22. Olsen, D., Nagayoshi, T., Fazio, M., Peltonen, J., Jaakkola, S., Sanborn, D., Sasaki, T., Kuivaniemi, H., Chu, M. L., Deutzmann, R., Timpl, R., and Uitto, U. (1989) *Lab Invest.* **60,** 772–782
23. Humphries, M. J., Akiyama, S. K., Komoriya, A., Olden, K., and Yamada, K. M. (1988) *J. Cell Biol.* **106,** 1289–1297
24. Dufour, S., Duband, J-L., Humphries, M. J., Obara, M., Yamada, K. M., and Thiery, J. P. (1988) *EMBO J.* **7,** 2661–2671
25. Guan, J. L., Trevithick, J. E., and Hynes, R. O. (1990) *J. Cell Biol.* **110,** 833–847
26. Kocher, O., Kennedy, S. P., and Madri, J. A. (1990) *Am. J. Pathol.* **137,** 1509–1524
27. Akiyama, S. K., and Yamada, K. M. (1985) *J. Biol. Chem.* **260,** 10402
28. Streeter, H. B., and Rees, D. A. (1987) *J. Cell Biol.* **105,** 507–515
29. Bowditch, R. D., Halloran, C. E., Obara, M., Aota, S., Plow, E. F., Yamada, K. M., and Ginsberg, M. H. (1990) *J. Cell Biol.* **111,** 403 (abstr.)
30. Kimizuka, F., Ohdate, Y., Kawase, Y., Shimojo, T., Taguchi, Y., Hashino, K., Goto, S., Hashi, H., Kato, I., Sekiguchi, K., and Titani, K. (1991) *J. Biol. Chem.* **266,** 3045–3051
31. Nagai, T., Yamakawa, N., Aota, S., Yamada, S. S., Akiyama, S. K., Olden, K., and Yamada, K. M. (1991) *J. Cell Biol.,* in press
32. Aota, S., Nagai, T., and Yamada, K. M. (1991) *J. Biol. Chem.* **266,** in press
33. Mugnai, G., Lewandowska, K., Carnemolla, B., Zardi, L., and Culp, L. A. (1988) *J. Cell Biol.* **106,** 931–943
34. Pierschbacher, M. D., and Ruoslahti, E. (1987) *J. Biol. Chem.* **262,** 17294
35. Cierniewski, C. S., Swiatkowska, M., Poniatowski, J., and Niewiarowska, J. (1988) *Eur. J. Biochem.* **177,** 109–115
36. Berliner, S., Niiya, K., Roberts, J. R., Houghten, R. A., and Ruggeri, Z. M. (1988) *J. Biol. Chem.* **263,** 7500–7505
37. Loftus, J. C., O'Toole, T. E., Plow, E. F., Glass, A., Frelinger, A. L., and Ginsberg, M. H. (1990) *Science* **249,** 915–918
38. McCarthy, J. B., Chelberg, M. K., Mickelson, D. J., and Furcht, L. T. (1988) *Biochemistry* **27,** 1380–1388
39. McCarthy, J. B., Skubitz, A. P., Qi, Z., Yi, X. Y., Mickelson, D. J., Klein, D. J., and Furcht, L. T. (1990) *J. Cell Biol.* **110,** 777–787
40. Graf, J., Iwamoto, Y., Sasaki, M., Martin, G. R., Kleinman, H. K., Robey, F. A., and Yamada, Y. (1987) *Cell* **48,** 989–996
41. Tashiro, K., Sephel, G. C., Weeks, B., Sasaki, M., Martin, G. R., Kleinman, H. K., and Yamada, Y. (1989) *J. Biol. Chem.* **264,** 16174–16182
42. Skubitz, A. P. N., McCarthy, J. B., Zhao, Q., Yi, X., and Furcht, L. T. (1990) *Cancer Res.* **50,** 7612–7622
43. Hunter, D. D., Porter, B. E., Bulock, J. W., Adams, S. P., Merlie, J. P., and Sanes, J. R. (1989) *Cell* **59,** 905–913
44. Aumailley, M., Gerl, M., Sonnenberg, A., Deutzmann, R., and Timpl, R. (1990) *FEBS Lett.* **262,** 82–86
45. Deutzmann, R., Aumailley, M., Wiedemann, H., Pysny, W., Timpl, R., and Edgar, D. (1990) *Eur. J. Biochem.* **191,** 513–522
46. Kleinman, H. K., Graf, J., Iwamoto, Y., Sasaki, M., Schasteen, C. S., Yamada, Y., Martin, G. R., and Robey, F. A. (1989) *Arch. Biochem. Biophys.* **272,** 39–45
47. Mecham, R. P., Hinek, A., Griffin, G. L., Senior, R. M., and Liotta, L. A. (1989) *J. Biol. Chem.* **264,** 16652–16657
48. Liesi, P., Narvanen, A., Soos, J., Sariola, H., and Snounou, G. (1989) *FEBS Lett.* **244,** 141–148
49. Tomasini, B. R., and Mosher, D. F. (1990) *Prog. Hemostasis Thromb.* **10,** 269–305
50. Smith, J. W., Ruggeri, Z. M., Kunicki, T. J., and Cheresh, D. A. (1990) *J. Biol. Chem.* **265,** 12267–12271
51. Chakravarti, S., Tam, M. F., and Chung, A. E. (1990) *J. Biol. Chem.* **265,** 10597–10603
52. Mohri, H., Fujimura, Y., Shima, M., Yoshioka, A., Houghten, R. A., Ruggeri, Z. M., and Zimmerman, T. S. (1988) *J. Biol. Chem.* **263,** 17901–17904
53. Prater, C. A., Plotkin, J., Jaye, D., and Frazier, W. A. (1991) *J. Cell Biol.* **112,** 1031–1040
54. Rich, K. A., George, F. W., Law, J. L., and Martin, W. J. (1990) *Science* **249,** 1574–1577
55. Staatz, W. D., Fok, K. F., Zutter, M. M., Adams, S. P., Rodriguez, B. A., and Santoro, S. A. (1991) *J. Biol. Chem.* **266,** 7363–7367
56. Dedhar, S., Ruoslahti, E., and Pierschbacher, M. D. (1987) *J. Cell Biol.* **104,** 585–593
57. Pignatelli, M., and Bodmer, W. F. (1988) *Proc. Natl. Acad. Sci. U. S. A.* **85,** 5561–5565
58. Chelberg, M. K., McCarthy, J. B., Skubitz, A. P., Furcht, L. T., and Tsilibary, E. C. (1990) *J. Cell Biol.* **111,** 261–270
59. Wilke, M. S., and Furcht, L. T. (1990) *J. Invest. Dermatol.* **95,** 264–270
60. Dhawan, S., Fields, R. L., and Robey, F. A. (1990) *Biochem. Biophys. Res. Commun.* **171,** 1284–1290
61. D'Souza, S. E., Ginsberg, M. H., Matsueda, G. R., and Plow, E. F. (1991) *Nature* **350,** 66–68
62. Kamboj, R. K., Gariepy, J., and Siu, C. H. (1989) *Cell* **59,** 615–625
63. Blaschuk, O. W., Sullivan, R., David, S., and Pouliot, Y. (1990) *Dev. Biol.* **139,** 227–229
64. Boucaut, J.-C., Darribere, T., Poole, T. J., Aoyama, H., Yamada, K. M., and Thiery, J. P. (1984) *J. Cell Biol.* **99,** 1822–1830
65. Lash, J. W., Linask, K. K., and Yamada, K. M. (1987) *Dev. Biol.* **123,** 411
66. Werb, Z., Tremble, P. M., Behrendtsen, O., Crowley, E., and Damsky, C. H. (1989) *J. Cell Biol.* **109,** 877–889
67. Kanemoto, T., Reich, R., Royce, L., Greatorex, D., Adler, S. H., Shiraishi, N., Martin, G. R., Yamada, Y., and Kleinman, H. K. (1990) *Proc. Natl. Acad. Sci. U. S. A.* **87,** 2279–2283
68. Stack, S., Gray, R. D., and Pizzo, S. V. (1991) *Biochemistry* **30,** 2073–2077

Minireview

THE JOURNAL OF BIOLOGICAL CHEMISTRY
Vol. 266, No. 21, Issue of July 25, pp. 13469–13472, 1991

Printed in U.S.A.

Cytochrome P-450

MULTIPLICITY OF ISOFORMS, SUBSTRATES, AND CATALYTIC AND REGULATORY MECHANISMS*

Todd D. Porter and Minor J. Coon

From the Department of Biological Chemistry, Medical School, University of Michigan, Ann Arbor, Michigan 48109

The carbon monoxide-binding pigment of liver microsomes (1, 2) was shown over 25 years ago to be a hemoprotein of the *b* type (3) and, as judged by the photochemical action spectrum, to be involved in the oxidation of drugs and steroids (4). The solubilization and resolution of the components of this enzyme system from microsomal membranes and the reconstitution of an active complex containing cytochrome P-450, NADPH-cytochrome P-450 reductase, and phosphatidylcholine (5, 6) permitted the purification and thorough characterization of these constituents. What is now known to be the cytochrome P-450 gene superfamily encodes numerous enzymes that are remarkable in the variety of chemical reactions catalyzed and in the number of substrates attacked (7–9). Indeed, it is no exaggeration to state that P-450 is the most versatile biological catalyst known. Considering the rapid progress that has been made in recent years in the characterization of over 150 isoforms, it may seem surprising that P-450, a name first used to describe a red pigment having a reduced CO-difference spectrum with a major band at an unusually long wavelength (about 450 nm) (3), has not been replaced by a terminology based on function. Even the term cytochrome is unsuitable, since in most instances P-450 acts as an oxygenase rather than simply as an electron carrier. Since many of the individual P-450s catalyze multiple reactions, the usual method of naming enzymes is inadequate for this group of heme proteins, and a systematic nomenclature has been devised based on structural homology (10).[1] It may be noted in this connection that chloroperoxidase and P-450 exhibit some physicochemical and catalytic similarities (11) but have no antigenic determinants in common (12). The aims of this review are to summarize briefly our current knowledge of the function, structure, mechanism of action, and regulation of this interesting group of catalysts and to describe recent progress in several areas where much still remains to be learned.

Reactions Catalyzed

Our knowledge of the scope of P-450-catalyzed reactions is still incomplete, as this cytochrome is widespread in nature and many isoforms have yet to be fully characterized or even identified. Animals, plants, and microorganisms contain P-450, and in mammals the enzyme system has been found in all tissues examined. P-450 is found predominantly in the endoplasmic reticulum and mitochondria, and in greatest abundance in the liver. As shown in Table I, the substrates for cytochrome P-450 encompass a host of xenobiotics, including substances that occur biologically but are foreign to animals, such as antibiotics and unusual compounds in plants, as well as synthetic organic chemicals, and a variety of steroids and other physiologically occurring lipids (7–9, 13). The number of man-made "environmental chemicals" has been estimated as greater than 200,000, most of which are thought to be potential substrates for P-450; many may also serve as inducers or inhibitors of various isoforms. Given the possibilities for synthetic modification of new and existing drugs and xenobiotics, one cannot put an upper limit on the number of compounds acted on by this enzyme family.

Most of the reactions begin with the transfer of electrons from NAD(P)H to either NADPH-cytochrome P-450 reductase in the microsomal system or a ferredoxin reductase and a nonheme iron protein in the mitochondrial and bacterial systems, and then to cytochrome P-450; this leads to the reductive activation of molecular oxygen followed by the insertion of one oxygen atom into the substrate. The reactions that have been demonstrated include hydroxylation, epoxidation, peroxygenation, deamination, desulfuration, and dehalogenation, as well as reduction. Most of the substrates are lipophilic, and there is no evidence that charge interactions contribute to binding by the cytochrome. Some of the transformations are essential for life, as with the conversion of cholesterol to corticoid and sex hormones, and others, particularly with xenobiotics, lead to the formation of more polar compounds that are more readily excreted directly or after conjugation with water-soluble agents such as glucuronic acid and glutathione (14). This is usually a detoxication process, but in some instances foreign compounds are converted to products with much greater cytotoxicity, mutagenicity, or carcinogenicity. Some of the P-450s, such as those involved in steroid transformations, are fairly selective in their choice of substrates, whereas other P-450s, particularly those in liver microsomes, have unusually broad and overlapping substrate specificity.

Mechanism of Oxygen and Peroxide Activation

The active site of P-450 contains iron protoporphyrin IX bound in part by hydrophobic forces. The fifth ligand is a thiolate anion provided by a cysteine residue, a feature that contributes to the unusual spectral and catalytic properties of P-450, and the sixth coordination position may be occupied by an exchangeable water molecule. Upon reduction of the iron, O_2 (or, in a competitive fashion, CO) can be bound in the sixth position.

The scheme in Fig. 1, which is modified from an earlier version (15), is in accord with findings in a number of laboratories and with the known stoichiometry of the hydroxylation reaction, where RH represents the substrate.

$$RH + O_2 + NADPH + H^+ \rightarrow ROH + H_2O + NADP^+$$

The first step in the reaction cycle is substrate binding, which perturbs the spin state equilibrium of the cytochrome and facilitates uptake of the first electron. Substrates that undergo reduction rather than oxygenation, such as epoxides, *N*-oxides, nitro and azo compounds, and lipid hydroperoxides, accept two electrons in a stepwise fashion as shown, to give $RH(H)_2$. To initiate the oxidative reactions, O_2 is bound to the ferrous P-450 with coordination to iron *trans* to thiolate. This intermediate can also be written as the resonance form, $Fe^{3+}(O_2^-)$, with substrate still present. Transfer of the second electron then occurs, with the possible involvement of cytochrome b_5 as an additional electron donor in mammalian microsomal systems (16, 17). The next step

* Research in this laboratory was supported by Grant DK-10339 from the National Institutes of Health and Grant AA-06221 from the National Institute on Alcohol Abuse and Alcoholism.

[1] The systematic nomenclature based on structural homology (10) is still evolving as more P-450s are characterized but may be described in general as follows. Those P-450 proteins with 40% or greater sequence identity are included in the same family (designated by an Arabic number), and those with greater than 55% identity are then included in the same subfamily (designated by a capital letter). Presently there are 27 families, of which 10 exist in all mammals. The individual genes are arbitrarily assigned numbers. For example, the major phenobarbital-inducible cytochrome in rabbit liver microsomes, originally called $P\text{-}450_{LM2}$, or form 2, is assigned to family 2 and subfamily B, and the gene and the enzyme are designated *CYP2B4* and CYP2B4, respectively; the enzyme may also be called P-450 2B4. Other examples from rabbit liver are 3-methylcholanthrene-inducible form 4 (now 1A2), ethanol-inducible form 3a (now 2E1), antibiotic-inducible form 3c (now 3A6), and constitutive form 3b (now 2C3). The main advantage of the unified nomenclature is that structurally identical or highly similar P-450s are easily recognizable, regardless of the source (species, tissue, or organelle), the inducer, or the catalytic activity examined.

TABLE I
Substrates for cytochrome P-450

Xenobiotics	Physiologically occurring compounds
Drugs, including antibiotics	Steroids
Carcinogens	Eicosanoids
Antioxidants	Fatty acids
Solvents	Lipid hydroperoxides
Anesthetics	Retinoids
Dyes	Acetone, acetol
Pesticides	
Petroleum products	
Alcohols	
Odorants	

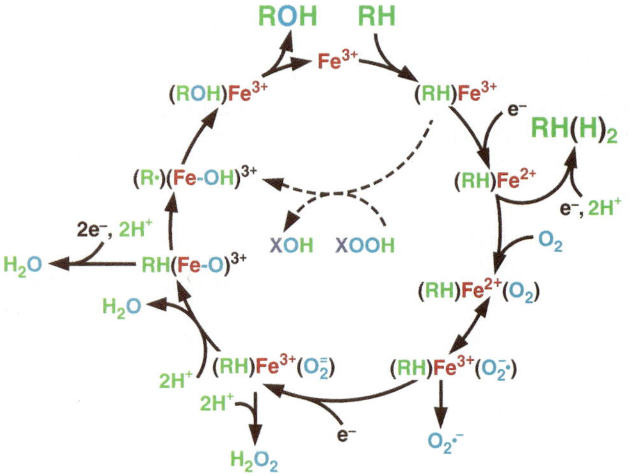

FIG. 1. **Scheme for mechanism of action of P-450.** *Fe* represents the heme iron atom at the active site, *RH* the substrate, *RH(H)₂* a reduction product, *ROH* a monooxygenation product, and *XOOH* a peroxy compound that can serve as an alternative oxygen donor.

is not well understood but involves splitting of the oxygen–oxygen bond with the uptake of two protons at some stage and the generation of an "activated oxygen," perhaps an iron-oxene species, and the release of H_2O. Several resonance forms are possible for the active oxygen intermediate, considering the redox possibilities with the sulfur, iron, and oxygen atoms. Oxygen insertion into the substrate is believed to involve hydrogen abstraction from the substrate and recombination of the resulting transient hydroxyl and carbon radicals to give the product (18). Dissociation of ROH then restores the P-450 to the starting ferric state. Also shown is the way in which a peroxy compound may substitute for O_2 and reducing equivalents in what is termed the peroxide shunt. Homolytic cleavage is envisioned with the formation of an iron-bound hydroxyl radical and an alkoxy radical ($XO^{\cdot}$) capable of hydrogen abstraction from the substrate (15). Thus, oxygenation by O_2 and by peroxy compounds has some common mechanistic features. Although it is not clear what role peroxy compounds play *in vivo*, they have proved useful in generating "activated oxygen" intermediates for spectral (19) and EPR analysis (20).

Much remains to be learned about the factors controlling regio- and stereospecificity in P-450-catalyzed reactions, as well as the identity of the powerful oxidant that is necessary for oxygen insertion into those substrates without activating groups at or near the position attacked, as in fatty acid ω-hydroxylation. Some particularly interesting variations on the scheme shown are the proposal of a cage radical mechanism for the rearrangement of a prostaglandin endoperoxide to a prostacyclin and a thromboxane (21), of radical intermediates in dehydrogenation reactions (22), and of aminium radical intermediates in amine oxidations (23). The unusual catalytic properties of P-450 contribute to the regulation of its own activities, as with competitive inhibition by alternate substrates, mechanism-based inactivation, and the effects of a variety of effectors and other lipophilic substances (24).

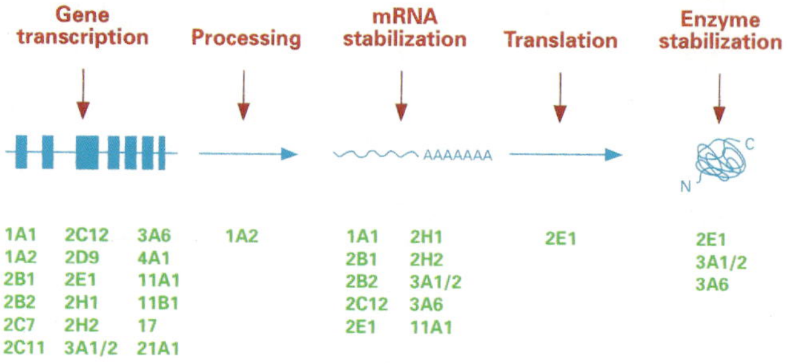

FIG. 2. **Multiplicity in the regulation of cytochrome P-450 expression.** P-450 designations (shown at the *bottom* of the figure) are as described in Nebert *et al.* (10) and are based on selected examples from various mammalian species. mRNA stabilization was inferred from transcriptional and message level studies, rather than measured directly, but may in some instances actually result from enhanced precursor processing, as shown for P-450 1A2 (41).

Regulation of Expression

Consonant with the multiplicity of P-450 cytochromes is the considerable diversity in the mechanisms of regulation of these enzymes, as depicted schematically in Fig. 2. Not surprisingly, the most common means of regulation is transcriptional. Post-transcriptional mechanisms include mRNA stabilization and protein stabilization or degradation that may be mediated through changes in the phosphorylation state of the enzyme. Moreover, many P-450s are subject to tissue-specific patterns of expression, with resulting differences in isoform compositions and activities in various tissues. Several recent, in-depth reviews of these topics are available (25, 26).

The most extensively characterized P-450 with regard to regulation is P-450 1A1 (27), a member of the PAH[2]-inducible gene family. This is the only P-450 for which a receptor-mediated mechanism of induction has been clearly demonstrated, via the *Ah* or TCDD receptor (28). The 5′-flanking region of the gene for this P-450 contains several short sequence motifs, termed xenobiotic responsive elements, or XREs, that function as transcriptional enhancers when *Ah* receptor ligands, such as TCDD, are added to *Ah*-responsive cells in culture (29). These XREs bear some resemblance to the glucocorticoid-responsive element, an interesting finding in light of the apparent similarity between the *Ah* and glucocorticoid receptors (30, 31). Evidence that these sequences bind a *trans*-acting factor, presumably the ligand-bound *Ah* receptor, has been provided by a variety of *in vitro* studies (29, 31–33). In addition to these enhancer elements, several transcription factor recognition sequences have been defined in the promoter region of *CYP1A1* (34), as well as a possible repressor binding site (35), and, surprisingly, a glucocorticoid-responsive element in the first intron (36).

In contrast to P-450 1A1, much less is known about the mechanism of regulation of P-450 1A2, a closely related isoform. Although the latter P-450 is strongly induced by PAHs, suggesting a role for the *Ah* receptor, the 1A2 gene lacks XREs in the proximal 5′-flanking region.[3] Indeed, the 20–40-fold increase in the level of 1A2 mRNA found after treatment of cells or animals with various PAHs has been shown to be mediated for the most part post-transcriptionally (38–40), apparently through enhanced stability and intranuclear processing of the 1A2 mRNA precursor (41). Interestingly, the *Ah* receptor appears to be involved in this post-transcriptional induction.

mRNA stabilization as a mechanism of induction is found with several other P-450s, including the alcohol-inducible cytochrome, P-450 2E1. In fact, this isoform is subject to multiple modes of regulation. Although diabetes and fasting each produce up to a 10-fold elevation in 2E1 mRNA, the increase with diabetes results from mRNA stabilization (42), whereas the increase following

[2] The abbreviations used are: PAH, polycyclic aromatic hydrocarbon; TCDD, 2,3,7,8-tetrachlorodibenzo-*p*-dioxin; XRE, xenobiotic-responsive element; GH, growth hormone; ACTH, adrenocorticotropic hormone; CRE, cAMP-responsive element.

[3] Gene transfection studies have revealed a modest transcriptional response to *Ah* receptor ligands, suggesting a more remote location for these elements (up to 3 kilobases upstream of the transcription start site) (37).

fasting (43, 44) results from an increase in gene transcription (45). Furthermore, chemical inducers of this cytochrome act predominantly to stabilize the protein, as has been demonstrated with acetone (46). The mechanism of this stabilization appears to be through ligand-mediated protection from phosphorylation, which otherwise leads to denaturation and degradation (47). The presence of phosphorylation sites on other P-450s and the ability of substrates to stabilize other P-450s suggest that this may be an important mechanism of P-450 regulation.

In rodents, but less so in man, the expression of a number of P-450s is sexually determined, by neonatal imprinting and by hormonal regulation in mature animals. These constitutively expressed P-450s generally are not responsive to xenobiotic induction. Neonatal castration has been shown to abolish the expression in adult animals of several male-specific steroid hydroxylases and to diminish the expression of a female-specific P-450 (48–52). Early administration of androgens to neonatally castrated male or female animals imprints the male pattern of P-450 expression; this steroidal programming in immature animals is thought to be mediated through pituitary growth hormone (GH), with neonatal androgens imprinting a pulsatile pattern of secretion that is characteristic of males. Indeed, intermittent administration of GH to hypophysectomized animals results in the expression of male-specific P-450s, and a more continuous administration of GH, obtained in castrated males and characteristic of females, results in the female pattern of P-450 expression (51, 53, 54). Notably, the administration of sex steroids to hypophysectomized animals is generally without effect on P-450 expression, consistent with these hormones acting through the hypothalamopituitary axis (51, 52, 55, 56). It should be noted that not all gender-determined P-450s are regulated as described above; several appear to be suppressed by GH regardless of the mode of its administration (51, 52).

The P-450 cytochromes responsible for the biosynthesis of steroid hormones are regulated in part by ACTH, which acts intracellularly through cAMP to increase gene transcription (57, 58). However, several of the steroidogenic P-450 genes do not contain the canonical cAMP-responsive element (CRE) found upstream of many other genes regulated by this nucleotide. Recent studies on the 17α-hydroxylase gene have identified the cAMP-responsive region and have shown that it binds a 47-kDa protein that may be a member of the proposed CRE-binding protein family (59). Surprisingly, none of the other known steroidogenic P-450 genes contain this cAMP-responsive sequence element, suggesting that they may bind one or more unique CRE transcription factors. Indeed, each steroidogenic P-450 appears to have its own CRE and perhaps its own subset of CRE-binding proteins (58).

Structure-Function Relationships

Studies on the relationship of structure to function in the P-450s have been limited by the lack of a three-dimensional structure for a mammalian isoform. Currently, all mammalian models are based on the known structure of P-450cam (60), a cytosolic cytochrome from *Pseudomonas putida* that shares only 10–20% sequence identity with the mammalian forms (7, 61). The validity of using P-450cam as a model has not yet been clearly established, although most structural studies on mammalian P-450s have been supportive, as discussed below. The crystal structure of P-450cam resembles a triangular prism, with 45% α-helix and 15% antiparallel β-structure. Although the helices are distributed throughout the polypeptide chain, the tertiary structure reveals an asymmetric arrangement, with the helices clustered on one side of the protein and β-structure located predominantly on the opposite side. The heme is positioned between two helices and is held in place by hydrophobic interactions, by hydrogen-bonding interactions between the heme propionates and Arg and His residues, and by a cysteine-thiolate ligand to the iron. In common with other hemoproteins such as catalase and cytochrome *c* peroxidase, and in contrast to typical cytochromes, the heme of P-450cam is not directly accessible from the surface of the protein; the closest approach is to what is termed the proximal surface, a distance of about 8 Å. Site-directed mutagenesis studies have demonstrated that several basic amino acids on this surface are involved in the electrostatic interaction of P-450cam with its redox partner, putidaredoxin (62). The lack of surface accessibility to the heme makes it likely that one or more amino acid side chains are involved in the conduction of electrons from putidaredoxin to the heme, but the identity of these residues is not yet known.

Crystallographic and site-directed mutagenesis studies have identified 2 residues critical to both substrate and O_2 binding at the active site of P-450cam. The substrate-binding pocket is lined with hydrophobic residues and is buried within the protein; access is gained via a small, dynamic solvent channel leading to the distal face of the heme. The substrate, camphor, is bound by van der Waals contacts and a single sterically important hydrogen bond with Tyr-96 (60, 63). The O_2-binding pocket is centered on Thr-252, which forms a hydrogen bond with Gly-248 to produce a local deformation, or kink, in the distal helix. When this Thr is replaced with Ala or Val, the monooxygenase reaction is uncoupled, such that O_2 and NADPH consumption is funneled into the production of H_2O_2 (64). These results point to a role for the threonyl hydroxyl group in the cleavage of the dioxygen bond during catalysis. Notably, the substitution of Ser for Thr at this position does not alter monooxygenase activity appreciably.

Cys-357 of P-450cam provides the axial heme ligand that dictates many of the spectral and functional characteristics of cytochrome P-450 (60). The peptide containing this invariant residue is the single most highly conserved P-450 segment and can be readily recognized in P-450s from organisms as diverse as bacteria and man (7, 25, 65). Substitutions of amino acids in this segment by oligonucleotide-directed mutagenesis produce a variety of effects on the spectral and heme-binding properties of the resultant cytochrome. Substitution of His or Tyr for the axial Cys of rat P-450 1A2 apparently prevents heme incorporation into the mutant proteins, as indicated by the minimal ferrous-CO Soret absorbance at either 448 or 420 nm (66). This finding, coupled with the concurrent loss of catalytic activity in the mutant proteins, demonstrates the importance of the cysteine-thiolate ligand to heme binding. Substitutions of more polar for hydrophobic amino acids in this segment also generally results in decreased Soret absorbance and in complete or partial loss of activity toward some, but not all, substrates (66–68).

Interestingly, substitutions in several mammalian P-450s at the highly conserved Thr that is thought to help form the O_2-binding pocket (corresponding to Thr-252 of P-450cam) also affected substrate selectivity and binding (69, 70). Most notably, and in contrast to the studies with P-450cam noted above, the substitution of Ser at this position decreased the rate of some reactions but left others unaffected or even slightly enhanced. Moreover, substrate-binding affinities were significantly reduced with this mutant. As might be expected, His and Ile substitutions at this position yielded inactive enzymes; however, enzymes with Val or Asn at this position retained limited activity toward some substrates. Evidently, the hydroxyl moiety is not essential to function in the two mammalian P-450s examined; perhaps the substrate-binding pocket is sufficiently large or flexible that with some substrates it can accommodate O_2 binding and catalysis in the absence of the bend in the distal helix otherwise provided by the hydroxyl group hydrogen bond.

Most of the studies on substrate binding to the mammalian P-450 cytochromes have used chimeric constructs between related isoforms to identify functional segments and residues. Based on these studies, the middle third of the P-450 sequence was assigned to substrate binding. More recent studies have identified specific residues that are responsible for substrate recognition. Kronbach *et al.* (71) utilized two structurally similar cytochromes that exhibit a 10-fold difference in the K_m of progesterone to identify 3 residues that are critical to the binding of this steroid. The segment containing these closely spaced residues corresponds to the region of P-450cam that contains Tyr-96, a residue involved in substrate binding in this cytochrome, as noted above. Similarly, Aoyama *et al.* (72) identified two amino acids in the NH_2-terminal third of the P-450 2B cytochromes that are necessary for testosterone 16β-hydroxylation; a variant with substitutions at these positions lacked this activity. A noteworthy study by Lindberg and Negishi (73) on two highly similar P-450s that catalyze coumarin and testosterone hydroxylation similarly narrowed substrate recognition to 3 residues, one of which conveys greater than 80% of the specificity for testosterone. In all of these studies, a most surprising aspect is the conservative nature of the amino acid substitutions that confer changes in substrate specificity.

REFERENCES[4]

1. Klingenberg, M. (1958) *Arch. Biochem. Biophys.* **75,** 376–386
2. Garfinkel, D. (1958) *Arch. Biochem. Biophys.* **77,** 493–509
3. Omura, T., and Sato, R. (1964) *J. Biol. Chem.* **239,** 2370–2378
4. Omura, T., Sato, R., Cooper, D. Y., Rosenthal, O., and Estabrook, R. W. (1965) *Fed. Proc.* **24,** 1181–1189
5. Lu, A. Y. H., and Coon, M. J. (1968) *J. Biol. Chem.* **243,** 1331–1332
6. Strobel, H. W., Lu, A. Y. H., Heidema, J., and Coon, M. J. (1970) *J. Biol. Chem.* **245,** 4851–4854
7. Black, S. D., and Coon, M. J. (1987) *Adv. Enzymol. Relat. Areas Mol. Biol.* **60,** 35–87
8. Ryan, D. E., and Levin, W. (1990) *Pharmacol. & Ther.* **45,** 153–239
9. Guengerich, F. P. (1991) *J. Biol. Chem.* **266,** 10019–10022
10. Nebert, D. W., Nelson, D. R., Coon, M. J., Estabrook, R. W., Feyereisen, R., Fujii-Kuriyama, Y., Gonzalez, F. J., Guengerich, F. P., Gunsalus, I. C., Johnson, E. F., Loper, J. C., Sato, R., Waterman, M. R., and Waxman, D. J. (1991) *DNA Cell Biol.* **10,** 1–14
11. Dawson, J. H. (1988) *Science* **240,** 433–439
12. Pandey, R. N., Kuemmerle, S. C., and Hollenberg, P. F. (1987) *Drug Metab. Dispos.* **15,** 518–523
13. Coon, M. J., and Koop, D. R. (1983) in *The Enzymes* (Boyer, P. D., ed) Vol. XVI, pp. 645–677, Academic Press, New York
14. Jakoby, W. B., and Ziegler, D. M. (1990) *J. Biol. Chem.* **265,** 20715–20718
15. White, R. E., and Coon, M. J. (1980) *Annu. Rev. Biochem.* **49,** 315–356
16. Schenkman, J. B., Jansson, I., and Robie-Suh, K. M. (1976) *Life Sci.* **19,** 611–624
17. Pompon, D., and Coon, M. J. (1984) *J. Biol. Chem.* **259,** 15377–15385
18. Groves, J. T., McClusky, G. A., White, R. E., and Coon, M. J. (1978) *Biochem. Biophys. Res. Commun.* **81,** 154–160
19. Blake, R. C., II, and Coon, M. J. (1989) *J. Biol. Chem.* **264,** 3694–3701
20. Larroque, C., Lange, R., Maurin, L., Bienvenue, A., and van Lier, J. E. (1990) *Arch. Biochem. Biophys.* **282,** 198–201
21. Hecker, M., and Ullrich, V. (1989) *J. Biol. Chem.* **264,** 141–150
22. Ortiz de Montellano, P. R. (1989) *Trends Pharmacol. Sci.* **10,** 354–359
23. Bondon, A., Macdonald, T. L., Harris, T. M., and Guengerich, F. P. (1989) *J. Biol. Chem.* **264,** 1988–1997
24. Ortiz de Montellano, P. R., and Reich, N. O. (1986) in *Cytochrome P-450 Structure, Mechanism, and Biochemistry* (Ortiz de Montellano, P. R., ed) pp. 273–314, Plenum Publishing Corp., New York
25. Gonzalez, F. J. (1988) *Pharmacol. Rev.* **40,** 243–288
26. Okey, A. S. (1990) *Pharmacol. & Ther.* **45,** 241–298
27. Nebert, D. W., and Jones, J. E. (1989) *Int. J. Biochem.* **21,** 243–252
28. Poland, A., Glover, E., and Kende, A. S. (1976) *J. Biol. Chem.* **251,** 4936–4946
29. Fujisawa-Sehara, A., Sogawa, K., Yamane, M., and Fujii-Kuriyama, Y. (1987) *Nucleic Acids Res.* **15,** 4179–4191
30. Wilhelmsson, A., Wikström, A.-C., and Poellinger, L. (1986) *J. Biol. Chem.* **261,** 13456–13463
31. Hapgood, J., Cuthill, S., Denis, M., Poellinger, L., and Gustafsson, J. Å. (1989) *Proc. Natl. Acad. Sci. U. S. A.* **86,** 60–64
32. Denison, M. S., Fisher, J. M., and Whitlock, J. P., Jr. (1988) *J. Biol. Chem.* **263,** 17221–17224
33. Saatcioglu, F., Perry, D. J., Pasco, D. S., and Fagan, J. B. (1990) *J. Biol. Chem.* **265,** 9251–9258
34. Jones, K. W., and Whitlock, J. P., Jr. (1990) *Mol. Cell. Biol.* **10,** 5098–5105
35. Gonzalez, F. J., and Nebert, D. W. (1985) *Nucleic Acids Res.* **13,** 7269–7288
36. Mathis, J. M., Houser, W. H., Bresnick, E., Cidlowski, J. A., Hines, R. N., Prough, R. A., and Simpson, E. R. (1989) *Arch. Biochem. Biophys.* **269,** 93–105
37. Quattrochi, L. C., and Tukey, R. H. (1989) *Mol. Pharmacol.* **36,** 66–71
38. Kimura, S., Gonzalez, F. J., and Nebert, D. W. (1986) *Mol. Cell. Biol.* **6,** 1471–1477
39. Pasco, D. S., Boyum, K. W., Merchant, S. N., Chalberg, S. C., and Fagan, J. B. (1988) *J. Biol. Chem.* **263,** 8671–8676
40. Silver, G., and Krauter, K. S. (1988) *J. Biol. Chem.* **263,** 11802–11807
41. Silver, G., and Krauter, K. S. (1990) *Mol. Cell. Biol.* **10,** 6765–6768
42. Song, B. J., Matsunaga, T., Hardwick, J. P., Park, S. S., Veech, R. L., Yang, C. S., Gelboin, H. V., and Gonzalez, F. J. (1987) *Mol. Endocrinol.* **1,** 542–547
43. Hong, J., Pan, J., Gonzalez, F. J., Gelboin, H. V., and Yang, C. S. (1987) *Biochem. Biophys. Res. Commun.* **142,** 1077–1083
44. Porter, T. D., Khani, S. C., and Coon, M. J. (1989) *Mol. Pharmacol.* **36,** 61–65
45. Johansson, I., Lindros, K. O., Eriksson, H., and Ingelman-Sundberg, M. (1990) *Biochem. Biophys. Res. Commun.* **173,** 331–338
46. Song, B.-J., Veech, R. L., Park, S. S., Gelboin, H. V., and Gonzalez, F. J. (1989) *J. Biol. Chem.* **264,** 3568–3572
47. Eliasson, E., Johansson, I., and Ingelman-Sundberg, M. (1990) *Proc. Natl. Acad. Sci. U. S. A.* **87,** 3225–3229
48. Chao, H., and Chung, L. W. K. (1982) *Mol. Pharmacol.* **21,** 744–752
49. Waxman, D. J., Dannan, G. A., and Guengerich, F. P. (1985) *Biochemistry* **24,** 4409–4417
50. Wong, G., Kawajiri, K., and Negishi, M. (1987) *Biochemistry* **26,** 8683–8690
51. Waxman, D. J., LeBlanc, G. A., Morrissey, J. J., Staunton, J., and Lapenson, D. P. (1988) *J. Biol. Chem.* **263,** 11396–11406
52. McClellan-Green, P. D., Linko, P., Yeowell, H. N., and Goldstein, J. A. (1989) *J. Biol. Chem.* **264,** 18960–18965
53. Morgan, E. T., MacGeoch, C., and Gustafsson, J.-Å. (1985) *J. Biol. Chem.* **260,** 11895–11898
54. Zaphiropoulos, P. G., Mode, A., Ström, A., Möller, C., Fernandez, C., and Gustafsson, J.-Å. (1988) *Proc. Natl. Acad. Sci. U. S. A.* **85,** 4214–4217
55. Colby, H. D., Gaskin, J. H., and Kitay, J. I. (1973) *Endocrinology* **92,** 769–774
56. Kamataki, T., Shimada, M., Maeda, K., and Kato, R. (1985) *Biochem. Biophys. Res. Commun.* **130,** 1247–1253
57. Simpson, E. R., and Waterman, M. R. (1988) *Annu. Rev. Physiol.* **50,** 427–440
58. Simpson, E. R., Lund, J., Ahlgren, R., and Waterman, M. R. (1990) *Mol. Cell. Endocrinol.* **70,** C25–C28
59. Lund, J., Ahlgren, R., Wu, D., Kagimoto, M., Simpson, E. R., and Waterman, M. R. (1990) *J. Biol. Chem.* **265,** 3304–3312
60. Poulos, T. L., Finzel, B. C., Gunsalus, I. C., Wagner, G. C., and Kraut, J. (1985) *J. Biol. Chem.* **260,** 16122–16130
61. Haniu, M., Armes, L. G., Yasunobu, K. T., Shastry, B. A., and Gunsalus, I. C. (1982) *J. Biol. Chem.* **257,** 12664–12671
62. Stayton, P. S., and Sligar, S. G. (1990) *Biochemistry* **29,** 7381–7386
63. Atkins, W. M., and Sligar, S. G. (1988) *J. Biol. Chem.* **263,** 18842–18849
64. Imai, M., Shimada, H., Watanabe, Y., Matsushima-Hibiya, Y., Makino, R., Koga, H., Horiuchi, T., and Ishimura, Y. (1989) *Proc. Natl. Acad. Sci. U. S. A.* **86,** 7823–7827
65. Kalb, V. F., and Loper, J. C. (1988) *Proc. Natl. Acad. Sci. U. S. A.* **85,** 7221–7225
66. Shimizu, T., Hirano, K., Takahashi, M., Hatano, M., and Fujii-Kuriyama, Y. (1988) *Biochemistry* **27,** 4138–4141
67. Furuya, H., Shimizu, T., Hatano, M., and Fujii-Kuriyama, Y. (1989) *Biochem. Biophys. Res. Commun.* **160,** 669–676
68. Furuya, H., Shimizu, T., Hirano, K., Hatano, M., Fujii-Kuriyama, Y., Raag, R., and Poulos, T. L. (1989) *Biochemistry* **28,** 6848–6857
69. Imai, Y., and Nakamura, M. (1988) *FEBS Lett.* **234,** 313–315
70. Imai, Y., and Nakamura, M. (1989) *Biochem. Biophys. Res. Commun.* **158,** 717–722
71. Kronbach, T., Larabee, T. M., and Johnson, E. F. (1989) *Proc. Natl. Acad. Sci. U. S. A.* **86,** 8262–8265
72. Aoyama, T., Korzekwa, K., Nagata, K., Adesnik, M., Reiss, A., Lapenson, D. P., Gillette, J., Gelboin, H. V., Waxman, D. J., and Gonzalez, F. J. (1989) *J. Biol. Chem.* **264,** 21327–21333
73. Lindberg, R. L. P., and Negishi, M. (1989) *Nature* **339,** 632–634

[4] We regret that, due to space limitations, many important contributions to this field could not be noted here.

Minireview

THE JOURNAL OF BIOLOGICAL CHEMISTRY
Vol. 266, No. 22, Issue of August 5, pp. 14139-14142, 1991

Printed in U.S.A.

Multisite and Hierarchal Protein Phosphorylation*

Peter J. Roach

From the Department of Biochemistry and Molecular Biology, Indiana University School of Medicine, Indianapolis, Indiana 46202-5122

Phosphorylation of proteins at Ser, Thr, and Tyr residues is one of the most frequent forms of posttranslational modification in eukaryotic cells and is linked to the control of a multitude of cellular functions (1, 2). The historical prototype for proteins controlled by phosphorylation is mammalian glycogen phosphorylase, an enzyme activated by modification of a single Ser per subunit by a single protein kinase (3). As the number of examples of phosphorylated proteins has escalated in recent years, it has become apparent that the majority of phosphoproteins contain multiple sites. This review seeks to survey the phenomenon of multisite[1] phosphorylation and to evaluate its significance.

Which Proteins Are Multiply Phosphorylated?

In assessing any specific role of multiple, as opposed to single site, phosphorylation, an immediate question is whether the proteins so modified fall into any particular classes. Almost certainly, only a fraction of multiply phosphorylated proteins have yet been identified, and efforts to survey multiply phosphorylated proteins, as in Fig. 1, inevitably involve a restricted data base. In compiling Fig. 1, the guiding principle was to identify proteins for which the location of phosphorylations was either known precisely or at least localized to some specific domain of the protein.[2] The best response to the question posed at the outset of this paragraph is that the occurrence of multiple phosphorylation appears no more restricted than the occurrence of phosphorylation in general.

The survey of Fig. 1 does reveal some interesting features. First, the number of phosphorylation sites observed in proteins varies from 1 to over 100 (Fig. 1) and no special numerology emerges. The only distinction that can perhaps be made is between proteins that contain relatively few sites and those that are heavily phosphorylated. The egg yolk protein phosvitin, neurofilaments, and the COOH-terminal tails of the large subunit of eukaryotic RNA polymerases are good examples of the latter phenomenon. Second, Ser(P), Thr(P), and Tyr(P) can be found in the same protein. Third, multiple phosphorylation sites tend not to be randomly distributed and are usually concentrated in relatively short segments of the polypeptide chain. Often, these phosphorylated regions are located at the extreme NH_2 or COOH termini of proteins. Since phosphorylation requires interaction with protein kinase(s), regions of phosphorylation will be defined in part by their accessibility, such as on the surface of a globular protein. In addition, their location must also be related to function which will vary from protein to protein. These restrictions alone, however, are unlikely to explain the occurrence of clustered sites. Another perspective on the clustering of sites in short segments is an evolutionary one. An interesting comparison is between the α and β regulatory subunits of phosphorylase kinase (6). These polypeptides have significant overall sequence homology, but 9 out of the 10 phosphorylation sites are in sequences specific to one subunit or the other. The α-subunit contains an ~100-residue insertion that harbors some seven sites while the β-subunit has an NH_2-terminal extension that contains two sites. In this regard, it is relevant that related proteins in different species or tissues often differ most at their termini. Glycogen synthase is an enzyme with its phosphorylation sites localized to the termini of the polypeptide (7). In comparing mammalian isoforms and the yeast enzymes, the greatest divergence is seen in precisely these regions. The yeast versions lack entirely the NH_2-terminal regulatory domain, and the other major differences are in the region of COOH-terminal phosphorylation sites. Two messages for acetyl-CoA carboxylase have been detected that differ as to the presence or absence of sequences encoding an 8-amino acid segment just upstream of an important phosphorylation site (8). The insertion disrupts the ability of cAMP-dependent protein kinase to phosphorylate this site (Ser-1200 in the shorter message).

How Are Proteins Multiply Phosphorylated?

Any discussion of mechanism leads to consideration of the specificity of protein kinases. These enzymes range from being highly specific, designed to phosphorylate even a single substrate, to having very broad substrate specificity. The latter class of protein kinase is faced with recognizing specific determinants that can be duplicated in several or many protein substrates. Studies of the specificity of protein kinases, based both on surveys of natural sites and analysis of synthetic peptide substrates, have defined, in several cases, local sequence motifs that appear to accomplish this goal (4, 5, 9). Many protein kinases have requirements for positively or negatively charged groups in the vicinity of the modified residue and have been termed basotropic or acidotropic accordingly (10).

There are two fundamental mechanisms of multisite protein phosphorylation (Fig. 2). The simplest is one in which the substrate protein has multiple copies of the recognition determinants for one or more protein kinases. The density at which multiple phosphates can be introduced in this way is obviously linked to the amount of sequence information involved in the recognition, enzymes with very simple requirements being able to introduce more phosphate groups within a given stretch of the polypeptide chain. Protein kinases of the cdc2 family, which appear to recognize simple -S/T-P- or -K-S/T-P- motifs, may be examples here (9). It is interesting that some very heavily phosphorylated proteins are modified in regions of sequence repeats. Examples are the neurofilaments NF-H and NF-M which contain multiple copies of a -K-S-P- sequence (11, 12) and the RNA polymerase II large subunit which, depending on species, can have as many as 52 repeats of a heptameric sequence unit (13, 14). Microtubule-associated protein 2 has been reported to contain more than 30 phosphates per polypeptide (15) and is a substrate *in vitro* for numerous protein kinases (see Ref. 16). Identification of these sites is incomplete, but the multiplicity of sites in this case is not related to any sequence repeat but rather to the presence of numerous Ser and Thr residues in a generally polar protein. Many other proteins (Fig. 1) contain more moderate numbers of sites whose modification can be explained by the independent action of one or more protein kinases and do not involve global sequence features like sequence repeats.

A mechanism unique to multiply phosphorylated proteins is when the introduction of phosphate groups influences the subsequent phosphorylation reaction(s) (Fig. 2). The first example worked out in molecular terms was that of the phosphorylation

* Work from my laboratory was supported by National Institutes of Health Grants DK27221 and DK42576.

[1] For this article, multisite phosphorylation is defined as the occurrence of non-identical phosphorylation sites in a protein complex, usually but not necessarily in a single polypeptide. Phosphorylation of the same site in an oligomer composed of identical subunits is not considered multisite.

[2] Space limitations prohibit a totally comprehensive citation of all the information contained in Fig. 1. Refs. 4 and 5 contain listings of many of the individual phosphorylation sites.

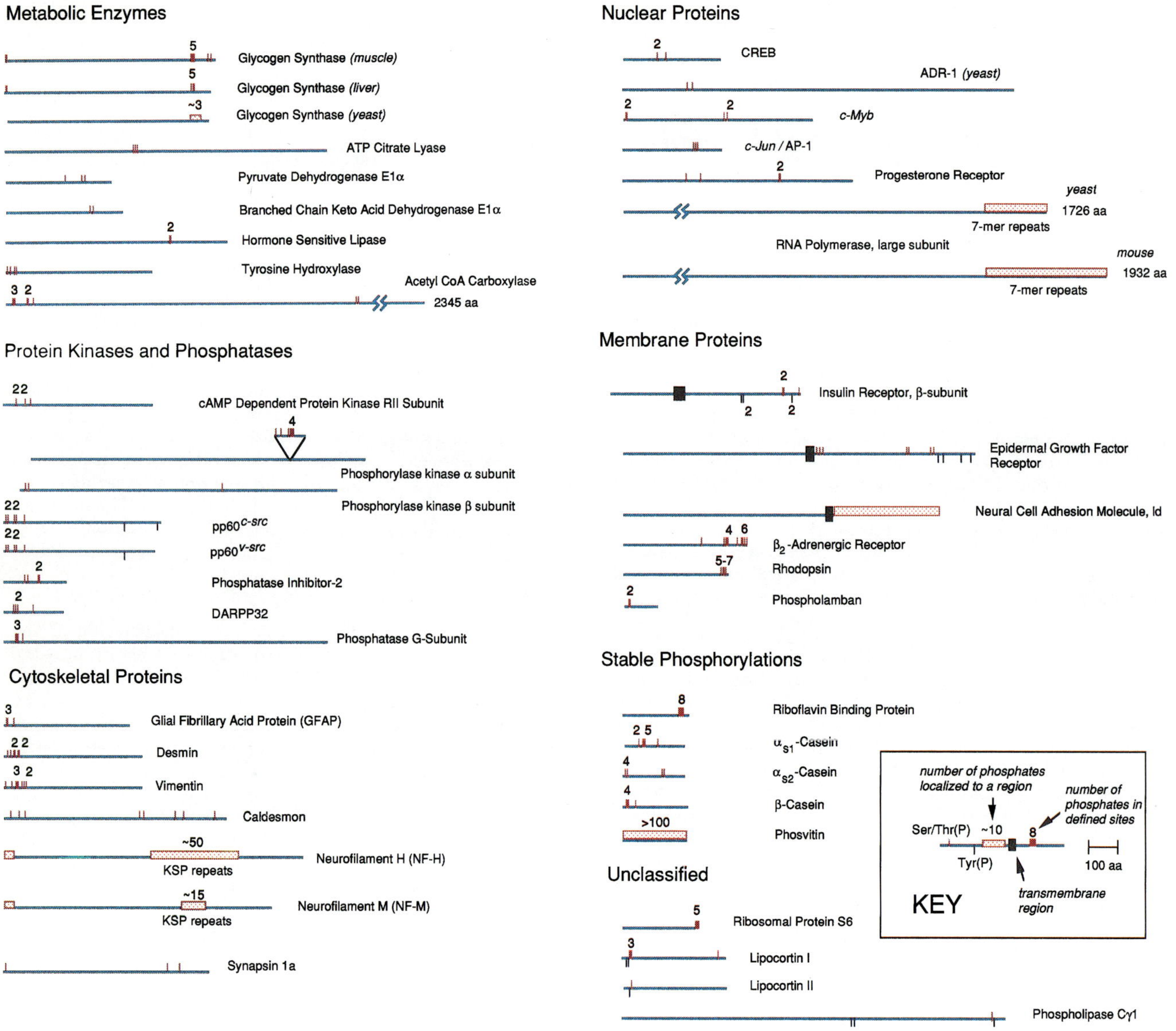

FIG. 1. **Some multiply phosphorylated proteins.** In compiling this figure, it became apparent that anything close to a comprehensive listing was impossible, and there are many omissions. Multiple entries of related proteins were curtailed unless a specific point was to be made. For example, many receptors of the β-adrenergic/rhodopsin family, more than 30 members characterized by seven membrane-spanning segments, are likely to have multiply phosphorylated cytosolic tails. Some phosphorylations not yet proven unequivocally to occur *in vivo* are also included. Phosphorylation sites are indicated by a *vertical line*, above the *horizontal* for Ser/Thr or below for Tyr. Where no number is shown, a single site is implied.

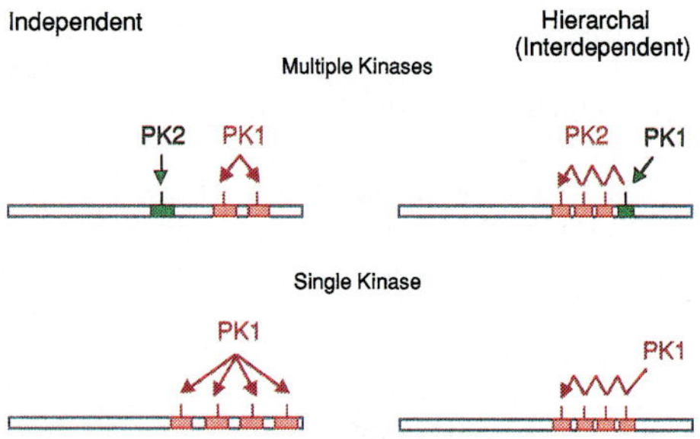

FIG. 2. **Mechanisms of multisite phosphorylation.** Some phosphorylations involve independent recognition of multiple sites in the substrate by one or more protein kinases (*left*). Interdependent phosphorylations occur when one phosphorylation event influences another (*right*). In these hierarchal schemes, one can distinguish primary phosphorylations which affect the course of subsequent secondary phosphorylations. Usually, different protein kinases are involved (*upper right*), but it is formally possible for a single protein kinase to act in both a primary and secondary way (*lower right*), as may be exemplified by casein kinase II (see text). The *shadings* represent different recognition determinants in the substrate.

of glycogen synthase by glycogen synthase kinase-3 (GSK-3)[3] and casein kinase II (7). Totally dephosphorylated glycogen synthase is not a substrate for GSK-3. Upon introduction of phosphate at one specific site by casein kinase II, the protein becomes a substrate for GSK-3, which sequentially modifies 4 Ser residues. The results are explained if GSK-3 recognizes sites in the motif -S-*X*-*X*-*X*-S(P)- (17). In this example, another feature of the reaction is that GSK-3 introduces multiple phosphates, due to the presence of adjacent repeats of the -S-*X*-*X*-*X*-S- motif. As one phosphate is introduced, a new GSK-3 site is generated. At the time of writing, seven examples of hierarchal phosphorylation involving more than a dozen GSK-3 sites have been recorded (Table I). In three instances the primary protein kinase is cAMP-dependent protein kinase and in three cases casein kinase II. The important feature for recognition by GSK-3 is the presence of a phosphate in an appropriate site and not the kinase that introduced it. Another enzyme that can act as a secondary protein kinase in the sense above is casein kinase I which has a particular selectivity for sites in the motif -S(P)-*X*-*X*-S- (10). Again, sites in glycogen synthase provide the best examples to date. The Golgi

[3] The abbreviation used is: GSK-3, glycogen synthase kinase-3.

TABLE I

Hierarchal phosphorylation systems

"Hierarchal," as explained further in the text, refers to multiple phosphorylations in which the introduction of one phosphate group influences the introduction of subsequent ones.

Primary protein kinase	Secondary protein kinase	Ref.	Substrate
Synergistic			
CK II[a]	GSK-3	7	Glycogen synthase
CK II	GSK-3	18	cAMP-dependent protein kinase RII subunit
CK II	GSK-3	19	Inhibitor-2 of type 1 phosphatase
cAMP PK	GSK-3	20, 21	G-subunit of type 1 phosphatase
cAMP PK	GSK-3	22	ATP-citrate lyase
cAMP PK	GSK-3	23	CREB
?	GSK-3	24	N-CAM
cAMP PK[b]	CK I	25	Glycogen synthase
?	CK I	24	N-CAM
CK II	cAMP PK	26	DARPP-32
Antagonistic			
AMP PK	cAMP PK	27	Hormone-sensitive lipase
cAMP PK	AMP PK	27	Hormone-sensitive lipase
cAMP PK	AMP PK	28	Acetyl-CoA carboxylase

[a] CK, casein kinase; CREB, cAMP-responsive element binding protein; N-CAM, neural cell adhesion molecule; DARPP-32, dopamine- and cAMP-regulated phosphoprotein; cAMP PK, cAMP-dependent protein kinase; AMP AK, AMP-activated protein kinase.

[b] Other protein kinases can also be the primary kinase in this example.

TABLE II

Recognition motifs for acidotropic protein kinases

Note that these sequences indicate only minimal requirements and the most effective sites may have other determinants also. In most cases, Thr can also be phosphorylated; Thr(P) can sometimes substitute Ser(P); Asp can usually replace Glu. However, not all of the combinations have been tested, notably for β-adrenergic receptor kinase (βARK) (32).

Enzyme	Recognition motif
Casein kinase I	S(P)-*X*-*X*-S
Casein kinase II	S-*X*-*X*-E/S(P)
Mammary gland casein kinase	S-*X*-E/S(P)
GSK-3	S-*X*-*X*-*X*-S(P)
βARK	E-*X*-S

casein kinase (9, 29) and casein kinase II (30, 31) are acidotropic kinases that phosphorylate sites in motifs containing either Glu, Asp, or Ser(P) (Table II). Thus, these enzymes can act either as primary or as secondary protein kinases and have the potential to phosphorylate initially on the basis of existing Asp or Glu residues and subsequently on the basis of the covalent phosphate introduced. For example, the sequence -S(P)-*X*-S(P)-S(P)-S(P)-E-E-, found in caseins (33) and riboflavin-binding protein (34), can be explained by the ordered action of an enzyme recognizing the motif -S-*X*-E/S(P)-.

One possibility raised by the occurrence of hierarchal phosphorylation is for "cross-talk" between protein Ser/Thr kinases and protein Tyr kinases. The protein Tyr kinases appear often, although not always, to be acidotropic, and one could thus ask whether Ser(P)/Thr(P) could be involved in recognition. Likewise, could Tyr(P) be recognized by protein Ser/Thr kinases? No physiological example of either is known, but Pinna and colleagues (35) have shown that Tyr(P) in a synthetic peptide can act as a recognition determinant for casein kinase II. A potential example of protein tyrosine kinases in a hierarchal scheme is the autophosphorylation of the insulin receptor in which the modification of 3 tyrosine residues involved in control of kinase activity is reported to be ordered (36).

Most of the known interdependent phosphorylations involve a positive role for the initial phosphorylation though *a priori* there is no reason why phosphorylation might not impair the action of a second kinase. For example, hormone-sensitive lipase (27) has two phosphorylation sites, Ser-563 (site 1) and Ser-565 (site 2). Phosphorylation of Ser-563 by cAMP-dependent protein kinase activates the enzyme whereas Ser-565 modification is without effect. However, phosphorylation of the two sites is mutually exclusive so that modification of Ser-565 could control activity indirectly by reducing phosphorylation at Ser-563. A similar situation holds for acetyl-CoA carboxylase. Phosphorylation of Ser-77 and Ser-1200 by cAMP-dependent protein kinase prevents phosphorylation of Ser-79 by the AMP-dependent protein kinase (28).

The exact role of the phosphate in hierarchal phosphorylation schemes has not been defined. One possibility is that the phosphoserine itself is involved in recognition contacts at the active site of the secondary kinase. This idea fits with the fact that most enzymes so far identified as acting in a secondary manner are acidotropic, their recognition motifs characterized by acidic residue(s) close to the modified amino acid (Table II). In the case of GSK-3, it should be noted that -S-*X*-*X*-*X*-S(P)- is the shortest unit recognition sequence and in two substrates, phosphatase inhibitor 2 and cAMP-dependent protein kinase RII subunit, the phosphates are more distant in the linear sequence. One can speculate that the folding of the protein might bring the target and the recognition serine phosphate into similar juxtaposition as in a -S-*X*-*X*-*X*-S(P)- unit. Two multiply phosphorylated GSK-3 substrates, c-Jun and c-Myb, do not fit the recognition criteria noted above and are reported to be phosphorylated without prior phosphorylation (37).

A second possibility is that the introduction of a phosphate group changes the conformation of the substrate into one that is recognized by a secondary protein kinase without the phosphate group itself participating in kinase-substrate contacts. One potential example is the enhanced phosphorylation of Thr-34 in DARPP32 by cAMP-dependent protein kinase once Ser-45 and/or Ser-102 has first been phosphorylated by casein kinase II (26). The cAMP-dependent protein kinase typically recognizes basic motifs such as -R-R-*X*-S-, and a phosphate group would not be expected to participate directly in recognition. Hierarchal phosphorylation mediated by conformational changes in the substrate would not require the secondary kinase to be acidotropic.

Why Are Proteins Multiply Phosphorylated?

There are two related but distinct perspectives on the possible role of multisite phosphorylation. One is at the level of the structural changes elicited by the introduction of phosphate groups and the subsequent effects on protein function. Multiple phosphorylations could correlate with the generation of a variety of protein forms, in which one or more properties are altered. Different phosphorylations could thus be linked to distinct protein functions or graded effects on a single function. Alternatively, multiple phosphates might be necessary to cause one critical conformational change. The other perspective is at the level of the regulation of the different phosphorylations. Obviously, if the actions of more than one kinase influence the functional status of a target protein, more complex regulation could be exerted relative to the action of a single kinase. Examples below are selected to illustrate some of the features that may be exclusive to multiple phosphorylations.

Hormones regulate the phosphorylation of glycogen synthase, and the multisite phosphorylation is in part linked to the occurrence of the hierarchal mechanism described above (7). An important feature is that the primary phosphorylations, such as mediated by casein kinase II and cAMP-dependent protein kinase, at best have moderate effects on activity; effective inactivation requires the occurrence of the secondary phosphorylations catalyzed by GSK-3 or casein kinase I. Thus, some phosphorylations function to alter activity whereas others influence kinase recognition. Not all the details linking hormone action to the control of phosphorylation have been worked out, but an important aspect of glycogen synthase control is that, mechanistically, the dozen or more phosphorylations do not occur totally independently and a smaller number of multiply phosphorylated units can be defined.

Acetyl-CoA carboxylase is another metabolic enzyme whose multiple phosphorylation is regulated by hormones (28, 38). In this protein, there is evidence that different phosphorylation sites have different influences on the kinetic properties of the enzyme. From site-directed mutagenesis studies, it appears that phosphorylation of Ser-77 and Ser-79 is linked primarily to decreases in V_{max} whereas modification of Ser-1200 increases the K_a for the allosteric activator citrate.[4] It has also been reported that phosphorylation of the enzyme by casein kinase II at site 6 (Ser-29), itself without effect on activity, may influence dephosphorylation of other sites that do control activity (39).

The tyrosine protein kinase pp60$^{c\text{-}src}$ provides an example of a protein in which different tyrosine phosphorylations can potentially modulate protein function (40). The autophosphorylation site, Tyr-416, may activate the kinase, whereas Tyr-527 is an inactivating site that lies, interestingly, in precisely the region of the molecule missing in the retroviral transforming gene product pp60$^{v\text{-}src}$. The viral protein therefore lacks the negative control, and the unconstrained activity of pp60$^{v\text{-}src}$ is thought to be linked to transformation by Rous sarcoma virus. pp60$^{c\text{-}src}$ is also phosphorylated at several Ser and Thr residues close to the NH_2 terminus, including sites for cAMP-dependent protein kinase and protein kinase C that may also be linked to activation of the kinase. The activity of pp60$^{c\text{-}src}$ may thus depend on the phosphorylation of multiple sites.

The β-adrenergic receptor undergoes phosphorylation in two sets of functionally distinguishable sites (41), modified by cAMP-dependent protein kinase and the β-adrenergic receptor kinase, respectively. Receptor activation promotes its own phosphorylation by cAMP-dependent protein kinase which is thought to mediate short-term desensitization of the receptor by low levels of agonist, so-called heterologous desensitization. Exposure to high agonist levels results additionally in phosphorylation in the extreme COOH terminus at sites for β-adrenergic receptor kinase. The β-adrenergic receptor kinase only recognizes receptor occupied by agonist, leading to what has been termed homologous desensitization. The precise mechanisms of desensitization have still to be elucidated but potentially involve altered interactions with the G-protein G_s and/or the accessory molecule arrestin. Both types of phosphorylation represent a feedback control on receptor function, but the phosphorylations are distinguishable both in the feedback circuit utilized and possibly the exact functional effect of the phosphorylation.

Some phosphorylations are irreversible in the sense that they occur once in the lifetime of a protein, and the phosphate groups themselves must be required for protein function (33). For example, caseins are proteins in which the phosphate is important for the structure of the casein micelles that maintain otherwise insoluble ions, notably calcium, in suspension. Phosvitin and riboflavin-binding protein are synthesized in the liver of laying hens and are transported to the egg. In phosvitin, the most highly phosphorylated protein known with over 100 phosphates per polypeptide, the covalent phosphorylation serves as a physical source of phosphate and perhaps also to bind and transport cations. There is also evidence for both proteins that phosphorylation is needed for their proper deposition in egg yolk and the phosphate could be involved in receptor recognition (42).

Other situations in which it is tempting to speculate a requirement for multiple phosphorylations are when large numbers of phosphates are clustered in a segment of a protein (Fig. 1). Some of the more heavily phosphorylated proteins come from the ranks of the intermediate filaments. Neurofilaments NF-L and NH-M, for example, are phosphorylated in both their NH-terminal heads (43, 44) and more densely in the COOH-terminal tails (11, 12). Phosphorylation has been implicated in determining the structures of intermediate filaments as well as interactions with other cytoskeletal elements. Though a precise need for multisite phosphorylation has not yet been demonstrated, one could surmise that the high degree of phosphorylation correlates with major protein structural changes which a single phosphorylation might not be sufficient to invoke. When massive phosphorylation occurs, one could also ask whether a precise set of sites is involved or whether there is a degree of randomness, the evolved goal being simply to introduce a sufficient number of phosphates.

[4] K-H. Kim, personal communication.

Summary

Multisite phosphorylation is a prevalent form of protein modification whose full implications are just beginning to be understood. Multiple protein modifications expand the repertoire of structural changes that can be elicited in proteins and permit more intricate regulatory circuits to operate.

Acknowledgments—Special thanks go to Peter Kennelly and Edwin Krebs, and to Bruce Kemp and Richard Pearson for allowing me access to their reviews and compilations of phosphorylation sites prior to publication. Robert Swift of Eli Lilly provided invaluable help with the figures.

REFERENCES

1. Hunter, T. (1987) *Cell* **50,** 823–829
2. Edelman, A. M., Blumenthal, D. K., and Krebs, E. G. (1987) *Annu. Rev. Biochem.* **56,** 567–613
3. Sprang, S. R., Acharya, K. R., Goldsmith, E. J., Stuart, D. I., Varvill, K., Fletterick, R. J., Madsen, N. B., and Johnson, L. N. (1988) *Nature* **336,** 215–221
4. Krebs, E. G., and Kennelly, P. J. (1991) *J. Biol. Chem.*, in press
5. Pearson, R. B., and Kemp, B. E. (1991) *Methods Enzymol.* **200,** in press
6. Kilimann, M. W., Zander, N. F., Kuhn, C. C., Crabb, J. W., Meyer, H. E., and Heilmeyer, L. M. G., Jr. (1988) *Proc. Natl. Acad. Sci. U. S. A.* **85,** 9381–9385
7. Roach, P. J. (1990) *FASEB J.* **4,** 2961–2968
8. Kong, I-S., Lopez-Casillas, F., and Kim, K-H. (1990) *J. Biol. Chem.* **265,** 13695–13701
9. Kemp, B. E., and Pearson, R. B. (1990) *Trends Biochem. Sci.* **15,** 342–346
10. Flotow, H., Graves, P. R., Wang, A., Fiol, C. J., Roeske, R. W., and Roach, P. J. (1990) *J. Biol. Chem.* **265,** 14264–14269
11. Lee, V. M.-Y., Otvos, L., Jr., Carden, M. J., Hollosi, M., Dietzschold, B., and Lazzarini, R. A. (1988) *Proc. Natl. Acad. Sci. U. S. A.* **85,** 1998–2002
12. Lees, J. F., Shneidman, P. S., Skuntz, S. F., Carden, M. J., and Lazzarini, R. A. (1988) *EMBO J.* **7,** 1947–1955
13. Allison, L. A., Moyle, M., Shales, M., and Ingles, C. J. (1985) *Cell* **42,** 599–610
14. Ahearn, J. M., Jr., Bartolomei, M. S., West, M. L., Cisek, L. J., and Corden, J. L. J. (1987) *J. Biol. Chem.* **262,** 10695–10705
15. Tsuyama, S., Bramblett, G. T., Huang, K.-P., and Flavin, M. (1986) *J. Biol. Chem.* **261,** 4110–4116
16. Jefferson, A. B., and Schulman, K. (1991) *J. Biol. Chem.* **266,** 346–354
17. Fiol, C. J., Mahrenholz, A. M., Wang, T., Roeske, R. W., and Roach, P. J. (1987) *J. Biol. Chem.* **262,** 14042–14048
18. Hemmings, B. A., Aitken, A., Cohen, P., Rymond, M., and Hofmann, F. (1982) *Eur. J. Biochem.* **127,** 473–481
19. DePaoli-Roach, A. A. (1984) *J. Biol. Chem.* **259,** 12144–12152
20. Fiol, C. J., Haseman, J. H., Wang, Y., Roach, P. J., Roeske, R. W., Kowulczuk, M., and DePaoli-Roach, A. A. (1988) *Arch. Biochem. Biophys.* **267,** 797–802
21. Dent, P., Campbell, D. G., Hubbard, M. J., and Cohen, P. (1989) *FEBS Lett.* **248,** 67–72
22. Ramakrishna, S., D'Angelo, G., and Benjamin, W. B. (1990) *Biochemistry* **29,** 7617–7624
23. Fiol, C. J., Andrisani, O. M., Flotow, H., Hrubey, T. W., Corbett, C. A., Tian, Z., Roeske, R. W., Dixon, J. E., and Roach, P. J. (1991) *FASEB J.* **5,** A1167
24. Mackie, K., Sorkin, B. C., Nairn, A. C., Greengard, P., Edelman, G. M., and Cunningham, B. A. (1989) *J. Neurosci.* **9,** 1883–1896
25. Flotow, H., and Roach, P. J. (1989) *J. Biol. Chem.* **264,** 9126–9128
26. Girault, J.-A., Hemmings, H. C., Jr., Williams, K. R., Nairn, A. C., and Greengard, P. (1989) *J. Biol. Chem.* **264,** 21748–21759
27. Yeaman, S. J. (1990) *Biochim. Biophys. Acta* **1052,** 128–132
28. Haystead, T. A., Moore, F., Cohen, P., and Hardie, D. G. (1990) *Eur. J. Biochem.* **187,** 199–205
29. Ribadeau-Dumas, B., Grosclaude, F., and Mercier, J. C. (1970) *Eur. J. Biochem.* **14,** 451–459
30. Kuenzel, E. A., Mulligan, J. A., Sommercorn, J., and Krebs, E. G. (1987) *J. Biol. Chem.* **262,** 9136–9140
31. Pinna, L. A., Meggio, F., and Marchiori, F. (1990) in *Peptides and Protein Phosphorylation* (Kemp, B. E., ed) pp. 145–169, CRC Press, Inc., Boca Raton, FL
32. Onorato, J. J., Palczewski, K., Regan, J. W., Caron, M. G., Lefkowitz, R. J., and Benovic, J. L. (1991) *Biochemistry*, in press
33. Weller, M. (1979) *Protein Phosphorylation*, Pion Limited, London
34. Fenselau, C., Heller, D. H., Miller, M. S., and White, H. B., III (1985) *Anal. Biochem.* **150,** 309–314
35. Meggio, F., Perich, J. W., Reynolds, E. C., and Pinna, L. A. (1991) *FEBS Lett.* **279,** 307–309
36. White, M. F., Shoelson, S. E., Keutmann, H., and Kahn, C. R. (1988) *J. Biol. Chem.* **263,** 2969–2980
37. Boyle, W. J., Smeal, T., Defize, L. H. K., Angel, P., Woodgett, J. R., Karin, M., and Hunter, T. (1991) *Cell* **64,** 573–584
38. Kim, K.-H., Lopez-Casillas, F., Bai, D. H., Luo, X., and Pape, M. E. (1989) *FASEB J.* **3,** 2250–2256
39. Sommercorn, J., McNall, S. J., Fischer, E. H., and Krebs, E. G. (1987) *Fed. Proc.* **46,** 2003
40. Cantley, L. C., Auger, K. R., Carpenter, C., Duckworth, B., Graziani, A., Kapeller, R., and Soltoff, S. (1991) *Cell* **64,** 281–302
41. Hausdorff, W. P., Caron, M. G., and Lefkowitz, R. J. (1990) *FASEB J.* **4,** 2881–2889
42. Miller, M. S., Benore-Parsons, M., and White, H. B., III (1982) *J. Biol. Chem.* **257,** 6818–6824
43. Sihag, R. K., and Nixon, R. A. (1989) *J. Biol. Chem.* **264,** 457–464
44. Sihag, R. K., and Nixon, R. A. (1990) *J. Biol. Chem.* **265,** 4166–4171

THE JOURNAL OF BIOLOGICAL CHEMISTRY
Vol. 266, No. 23, Issue of August 15, pp. 14831–14834, 1991

Extracellular Proteins That Modulate Cell-Matrix Interactions

SPARC, TENASCIN, AND THROMBOSPONDIN*

E. Helene Sage‡ and Paul Bornstein§¶

From the Departments of ‡Biological Structure and §Biochemistry, University of Washington, Seattle, Washington 98195

It is almost axiomatic in cell biology to accept the postulate that cells normally adherent *in vivo* require attachment and spreading *in vitro* to proliferate and to express their differentiated properties. Experimental verification of this hypothesis focused attention on the nature of adhesive macromolecules in the extracellular matrix (ECM)[1] and the characteristics and consequences of their interactions with cell-surface receptors (1–8). The major adhesive macromolecules for epithelial and mesenchymal cells include fibronectin (FN), vitronectin, laminin, entactin, the fibrillar collagens (types I, II, III, V, and XI), and the collagen in basement membranes (type IV); these components interact with a variety of integrins, heparan sulfate proteoglycans, and other cell-surface receptors. However, it is apparent that the interaction of a cell with its environment must be a dynamic one, since cells that ordinarily remain attached *in vivo* must also be capable of diminishing adhesive forces for rounding and division. In addition, it seems probable that, for many cells, differentiation requires a number of cell divisions, and cell movement requires the making and breaking of interactions with an ECM.

We describe in this review a group of secreted glycoproteins that presently includes SPARC, tenascin (TN), and thrombospondin (TSP) but that is likely to include other macromolecules with related functions. These proteins do not function primarily and generally as cell adhesion factors, in the sense that they foster both attachment and spreading in most cells. Instead, for many cells, these proteins exert an "anti-adhesive" function that leads to cell rounding and partial detachment from a substratum. SPARC, TN, and TSP possess certain common attributes. Although these proteins are secreted and retained in the local environment where they can function in an autocrine and paracrine fashion, they do not usually accumulate or function as structural components in the normal adult organism. SPARC, TN, and TSP appear to play dynamic roles in embryogenesis and morphogenesis. *In vitro*, there is a positive association between synthesis and cell proliferation, and all three proteins have been shown to promote changes in cell shape, with attendant consequences for cell behavior.

Cells that produce an ECM are also influenced by it, a concept that we and others have termed "dynamic reciprocity." Since there are no major structural similarities among SPARC, TN, and TSP, it is likely that these proteins exert their proximal effects, *i.e.* an interference with focal adhesions leading to changes in cell shape (9–12), by different mechanisms. Possibilities include activation of intracellular signaling pathways (13), occlusion or interference with cell attachment to proteins like FN (3, 14, 15), an influence on Ca^{2+} ion flux, presentation or monopolization of growth factors and cytokines (16, 17), and modulation of extracellular protease activity (18, 19). We consider, in this review, the manner in which SPARC, TN, and TSP influence interactions between a cell and its ECM and the consequences for cell function.

SPARC

SPARC (also termed osteonectin) is an acidic, cysteine-rich component of the extracellular milieu that displays a high degree of interspecies sequence conservation (20–27). Clues to its function were first obtained from studies that showed a restricted expression in embryonic and adult tissues. The location of SPARC in fetal growth plates and in zones of mineralization, as well as its demonstrated affinity for type I collagen, hydroxyapatite, and Ca^{2+}, indicated a role for the protein in the mineralization of bone and cartilage (20, 28, 29). High levels of SPARC mRNA and protein were found in somites, limb buds, and invasive cells of extraembryonic tissue (30–32), whereas SPARC in the adult was limited to tissues exhibiting high rates of cellular proliferation and remodeling, as well as to nondividing steroidogenic cells (30, 32, 33). The temporal and spatial restriction of SPARC to certain active cellular populations raised the possibility that the protein modulated cell proliferation and/or facilitated acquisition of a differentiated phenotype (30, 32).

That SPARC might participate in cell-matrix interactions during remodeling, development, or tissue response to injury was suggested by its ability to function as an inhibitor of cell spreading *in vitro* (34). Endothelial cells, smooth muscle cells, and fibroblasts attached to plastic or collagen-coated surfaces but failed to spread when treated with exogenous SPARC; this activity has been attributed to two distinct Ca^{2+}-binding domains (34, 35). From these studies we proposed that SPARC interacted with proteins at the cell surface to facilitate changes in cell shape and release of cells from their ECM (32, 34). This hypothesis is supported by the behavior of teratocarcinoma cells stably transfected with sense and antisense SPARC cDNA. Cell lines that overexpressed SPARC were rounded and clumped, whereas underexpressors were attached and spread (36). SPARC therefore appears to interfere with cell-matrix interactions in a manner that produces subtle to overt changes in cell morphology.

Mason *et al.* (23) reported a 20-fold increase in SPARC mRNA in teratocarcinoma cells after their differentiation to a phenotype resembling parietal endoderm. During implantation, these extraembryonic cells withdraw from the cell cycle, synthesize a basement membrane, and facilitate embryonic invasion of the uterine stroma. It is therefore interesting that SPARC was recently identified as an inhibitor of the $G_1 \rightarrow S$ progression in BAE cells (37). SPARC might affect cells by retarding progression through the cell cycle or by maintaining a rounded phenotype in cells that have already divided. Cells consequently arrested in G_1 could conceivably exercise one of several options (*e.g.* migration or adoption of a postmitotic phenotype) that are dependent on cessation of cell division.

Tenascin

A large, multisubunit glycoprotein with both adhesive and anti-adhesive properties (38, 39), TN (cytotactin, J1, hexabrachion) is a mosaic of structural units that include EGF-like repeats, an integrin-specific cell-binding sequence RRGDM, and Ca^{2+}-binding domains (3, 14). Various approaches have been taken to elucidate the roles that TN might play in regulating cell behavior. Addition of soluble TN to cultured tumor cells resulted in partial detachment and loss of intercellular contacts, as well as an inhibition of cell migration and spreading on basal lamina (14). Experiments with TN-coated substrates have shown that TN retards the attachment and/or spreading of several cell types on

* The research performed in our laboratories is supported by National Institutes of Health Grants HL 18645, GM 40711, DE 08229, and HL 03174.
¶ To whom correspondence should be addressed.
[1] The abbreviations used are: ECM, extracellular matrix; BAE, bovine aortic endothelial; bFGF, basic fibroblast growth factor; EGF, epidermal growth factor; FN, fibronectin; PDGF, platelet-derived growth factor; SPARC, secreted protein acidic and rich in cysteine; TGF-β, transforming growth factor β; TN, tenascin; TSP, thrombospondin.

FN, laminin, the peptide GRGDS, and basal lamina (3, 15, 40). Other assays provided evidence for adhesive domains in TN that interacted with a chondroitin sulfate proteoglycan or with an RGD-sensitive integrin (3). It nevertheless appears that TN supports only weak if any attachment of cells to ECM, a conclusion supported by the low affinity of TN for FN, collagens, and laminin (15), and by the identification of an anti-spreading sequence in the EGF-like domain (38). Several investigators have in fact presented a strong case for a functional antagonism between TN and FN (40, 41), a relationship nicely illustrated by the nearly ubiquitous distribution of the latter protein and the restricted expression of TN during morphogenesis and remodeling. Moreover, the ability of FN to promote complete cell-substrate adhesion contrasts markedly with the tendency of TN to promote cell rounding and detachment (41). These results are compatible with a role for TN as an "avoidance molecule" (39) that could act as a steric blocker of cellular receptors for other ECM components (41).

The purported interactions facilitated by TN were suggested in part by the intriguing patterns of expression reported for this protein *in vivo*. Although expression in the normal adult is both restricted and minimal, high but transient expression of TN has been found coincident with actively migrating or proliferating cells during embryonic morphogenesis, wound healing, tissue repair, and oncogenesis (3, 14, 43, 44). Studies *in vitro* support a role for substrate-associated TN in both the migration of neural crest cells and the outgrowth of neurites from spinal cord explants (14, 45). It has also been proposed that TN facilitates epithelial-mesenchymal interactions during organogenesis, since it appears in the mesenchyme adjacent to growing epithelial structures (14, 46).

An interesting example of a physiologic role for TN in destabilizing cell-matrix adhesions may be found in the shedding of epithelial cells from intestinal villi. In embryonic mice, TN is concentrated at the junction between epithelium and mesenchyme in an increasing gradient from the bases of crypts to the tips of villi; a reverse gradient was observed for FN (47). In crypt regions containing minimal TN, epithelial cells were in close, uninterrupted opposition to the basal lamina, whereas numerous microvilli extended from the basal surfaces of the cells at the villus tip. When an epithelial cell line was studied in attachment assays, little or no binding was observed to TN, and TN interfered with binding to FN but not to collagen or laminin (47). Since the net effect of TN is anti-adhesive for intestinal epithelial cells, epithelial migration and shedding could result from an interplay of adhesive forces that vary from the core to the tip of intestinal villi.

Thrombospondin

TSP participates in the aggregation of activated platelets (48) but also functions as a secreted product of most mesenchymal and epithelial cells. The protein contains globular N- and C-terminal domains, which have been implicated in binding to heparin and to Ca^{2+} and platelets, respectively, and central stalk-like regions (49, 50). The latter regions interact with collagen, FN, fibrinogen, laminin, and plasminogen (48, 50, 51). The nature and consequences of the interaction of TSP with cells are complex and controversial. TSP has been found to bind to heparan sulfate proteoglycans, integrins, glycoprotein IV, and sulfated glycolipids, but not all of these interactions occur in all cells (see Ref. 52 for a recent discussion), and cooperative interactions between receptor systems might occur (53). TSP has been described as an adhesive protein, and for keratinocytes, melanoma cells, and platelets, its predominant effect might indeed be adhesive (48, 54, 55). On the other hand, TSP has been shown to reduce focal adhesions in endothelial cells and fibroblasts (10), which attach but do not spread on TSP-treated substrates (56). TSP has also been shown to induce chemotaxis in neutrophils and endothelial cells (57, 58) but to inhibit the migration of endothelial cells toward angiogenic factors such as bFGF (59). The interpretation of some of these experiments may be complicated by the ability of TSP to bind TGF-β (17). We have also identified a second TSP gene in the mouse which encodes a protein that is likely to have distinctive properties (60) and could contribute to the observed effects of TSP on cells.

There are a number of considerations, in addition to its effect on focal adhesions, which lead us to suggest that TSP may modulate cell-matrix interactions and influence both cell shape and cell function. bFGF and PDGF rapidly induce the synthesis and secretion of TSP in smooth muscle and 3T3 cells (51, 61); this stimulation does not extend to other cell surface-associated and matrix proteins such as collagen and FN. TSP also acts synergistically with EGF to stimulate mitogenesis in smooth muscle cells (59, 62), and monoclonal antibodies to TSP inhibit smooth muscle cell proliferation and reduce cell surface-associated TSP (63). Heparin, a growth inhibitor for smooth muscle cells, has varied effects, among them a reduction in cell-associated TSP (51). Finally, the level of synthesis of TSP is positively correlated with growth for several types of cells *in vitro* (64). For example, in the mouse embryo, TSP is expressed in areas of active organogenesis and tissue morphogenesis during postimplantation development (65).

Roles in Morphogenetic Processes

It is perhaps not surprising that SPARC, TN, and/or TSP are found consistently but transiently in tissues that are undergoing rapid change. Processes such as wound healing, angiogenesis, bone remodeling, reinnervation, and embryonic morphogenesis are characterized, in part, by cell-matrix interactions that are continually changing to facilitate tissue growth and repair. Thus, a striking coincidence of expression of these proteins has been noted in osteogenic differentiation (32, 65, 66) and in remodeling basement membranes (31, 43, 66). Table I is a compilation of experimental data that support active roles for these proteins in both embryonic and adult vertebrates. As an example, we examine the contribution of SPARC and TSP to the process of angiogenesis.

Angiogenesis, defined as the formation of new capillaries from preexisting vessels, is a multistep process that minimally requires endothelial cell migration, proliferation, alterations in gene expression, and changes in shape. Cultured endothelial cells under a variety of conditions form cords and tubes with patent lumina, a process referred to as angiogenesis *in vitro* (73). We have recently shown that both SPARC and TSP are produced by BAE cells that spontaneously form cords and tubes *in vitro* (67, 68). Whereas SPARC mRNA was increased in cord-forming cultures of both macrovascular and microvascular endothelial cells (74), TSP mRNA was decreased significantly in the cord-forming cultures (67). Immunostaining revealed an association of both proteins with sprouting BAE cells and endothelial cords, although SPARC appeared to be predominantly intracellular whereas TSP was distributed extracellularly in fibrillar arrays that delimited the cords (Fig. 1, *A* and *C*) (67, 68, 74).

Although the net effect of SPARC on capillary growth is not known, TSP has been characterized as an inhibitor of angiogenesis *in vivo* (59, 70) and *in vitro* (67). Points at which SPARC and TSP might participate during a sequence of endothelial cord/tube formation *in vitro* are shown in Fig. 1, *B* and *D*. For example, sprouting BAE cells that proliferate above or below the normally quiescent monolayer secrete SPARC and TSP (68), and both proteins diminish focal adhesions (10, 11). Moreover, TSP interferes with attachment (75) and promotes chemotaxis in BAE cells (58), and SPARC additionally affects cell shape by preventing spreading (34). Migration would be facilitated by the ability of SPARC to decrease the expression of FN and to increase plasminogen activator inhibitor-1, an antiprotease with known effects on angiogenesis *in vitro* (19, 76, 77). The organization of cords and tubes is further accompanied by the production of an altered ECM and a cessation of cell division (68). Given the affinities of SPARC and TSP for certain types of collagens and proteoglycans (34, 50, 51), it is reasonable to expect that some of their effects on cell behavior are mediated through direct binding to insoluble components of the ECM.

An important component of angiogenesis is the regulation of capillary progression and dissolution. *In vitro*, the assembly and disassembly of cords and tubes are dynamic processes accom-

TABLE I

Morphogenetic processes that involve modulation of cell-matrix adhesion by SPARC, TN, and/or TSP

Process	Protein	Action[a]	Refs.
Angiogenesis	TSP	Inhibition	59, 67
	SPARC	Sprout formation*	68
Epithelial shedding	TN	Stimulation	47
Reinnervation	TN	Axon growth and regeneration of neuromuscular junction*	42, 44
Thrombosis	TSP	Platelet aggregation	48
Tumor growth	TN	Enhancement in stroma of epithelial neoplasms*	69
	TSP	Inhibition	70
	SPARC	Basement membrane remodeling*	31
Wound healing	TN	Transient expression*	43
	TSP	Organization of perivascular ECM*	71
Chondrogenesis, osteogenesis	TN	Promotion of differentiation*	66
	SPARC	Promotion of mineralization*	28, 29
Neural crest development	TN	Directed cell migration; control of differentiation*	72; 3, 14
	TSP	Migration and aggregation*	65
Peripheral nerve development	TN	Stimulation of neurite outgrowth *in vitro*; axon outgrowth and fasciculation*	45; 44
	TSP	Growth and process formation*	65
Renal development	TN	Epithelial-mesenchymal interaction	46

[a] In some instances, indicated by *, roles have been inferred from immunolocalization and/or *in situ* hybridization.

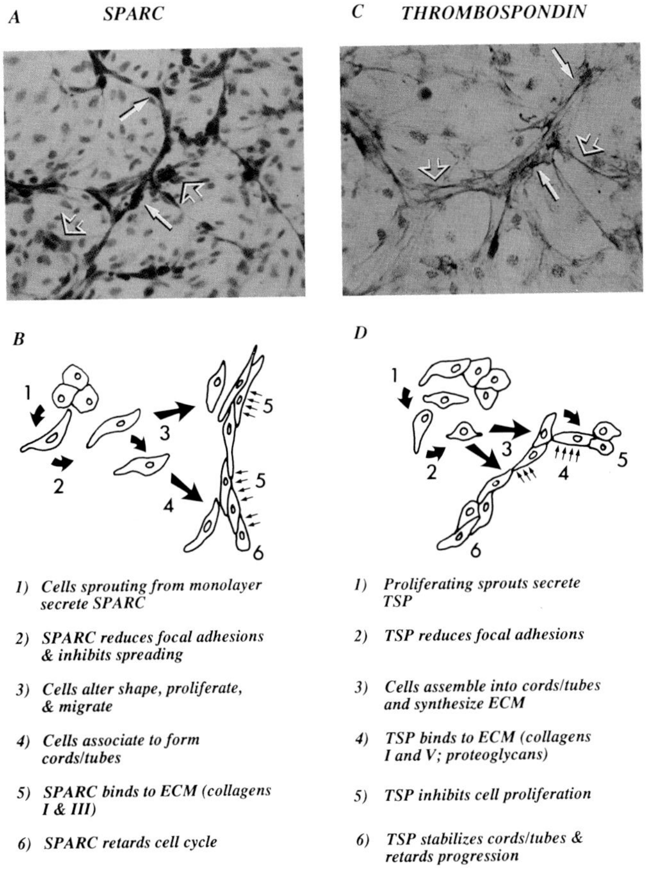

A SPARC
C THROMBOSPONDIN
B
D
1) Cells sprouting from monolayer secrete SPARC
2) SPARC reduces focal adhesions & inhibits spreading
3) Cells alter shape, proliferate, & migrate
4) Cells associate to form cords/tubes
5) SPARC binds to ECM (collagens I & III)
6) SPARC retards cell cycle
1) Proliferating sprouts secrete TSP
2) TSP reduces focal adhesions
3) Cells assemble into cords/tubes and synthesize ECM
4) TSP binds to ECM (collagens I and V; proteoglycans)
5) TSP inhibits cell proliferation
6) TSP stabilizes cords/tubes & retards progression

FIG. 1. **Proposed roles for SPARC and TSP during angiogenesis *in vitro*.** *A* and *C*, immunolocalization of SPARC (*A*) and TSP (*C*) in cultures of BAE cells displaying extensive cord/tube formation. Cells were fixed and exposed to affinity-purified anti-SPARC IgG (*A*) or anti-TSP IgG (*C*); reaction product was visualized by an avidin-biotin-peroxidase technique, and cells were counterstained with toluidine blue. SPARC was preferentially associated with cords/tubes (*solid arrows*) and with sprouting cells in the proximity of these structures (*open arrows*). Although the staining is mainly intracellular, a filamentous pattern is occasionally apparent on cords or tubes. In contrast, TSP is found primarily on mature cords and tubes and is predominantly extracellular. In *panel C*, *open arrows* denote fibrillar arrays surrounding tube-like structures (*solid arrows*); nuclei are labeled with [^{3}H]thymidine. *B* and *D*, diagrams of BAE cells in the process of cord formation. Accompanying text describes functions ascribed to SPARC (*B*) and TSP (*D*) in the context of angiogenesis *in vitro*. Numbered steps in the angiogenic process correspond to functions listed at the bottom (see text for details). Micrographs were provided by M. L. Iruela-Arispe.

panied by changes in mitotic index and gene expression. It is noteworthy that, in BAE cells, SPARC retards cell cycle progression (37) and TSP abrogates the mitogenic effects of serum and bFGF (58). Thus, the temporary withdrawal of endothelial cells from the cell cycle mediated by SPARC might potentiate a migratory response to other factors or might initiate changes in gene expression requisite to the formation of a capillary tube. In contrast, the inhibitory effect of TSP on angiogenesis is viewed as a later event in the process. The synthesis of TSP by sprouts might facilitate migration and recruitment of cells at the site of a developing cord or tube, but further progression of these structures could then be limited by the association of TSP with the newly secreted ECM.

Concluding Remarks: Anti-adhesive Molecules

Alterations in cell shape *in vitro* are related directly to the number and distribution of focal adhesions on the cell surface. These structures link proteins and proteoglycans of the ECM to the cytoskeleton and its associated components via several of the integrins (2, 4, 5, 12). It is therefore interesting that SPARC, TN, and TSP have been shown to diminish the number of focal adhesions in BAE cells (10, 11). Proteins that disrupt focal adhesions by displacing ligands from their receptors would provide a highly regulated mechanism for the disengagement of cells from the ECM. Since SPARC, TN, and TSP promote reorganization of the actin cytoskeleton through modulation of adhesive contacts, cellular receptivity to motility factors (78) or disintegrins (79) might be dependent in part on the "priming" of cells by these anti-adhesive extracellular proteins.

Anti-adhesive mechanisms involving repulsive interactions among cells and their ECM may be important for the directional inhibition of axon growth during neural development (39, 80). Growth cones located at the tips of elongating axons project filopodia for exploration of extracellular terrain during axon advancement; avoidance of an area is thought to be mediated by glycoproteins that contribute to changes in shape and motility of the growth cones. Keynes and Cook (80) have proposed a spectrum of molecules ranging from the highly adhesive *N*-CAM and *N*-cadherin to inhibitory glycoproteins that mediate cellular repulsion. Candidates for the last category are TN, which acts as a repulsive substrate for neurons of the central nervous system *in vitro* (39), and a neural protein, SC-1, which contains an extended sequence homologous to SPARC and is associated with synapse formation, motility, and/or extension of filopodia (81).

Proteoglycans also modulate cell behavior by mediating the interaction of cells with ECM. Both TN and TSP form complexes with proteoglycans that could modulate their respective adhesive properties (3, 7), *e.g.* the inhibitory effect of TN on neural crest cell movement and neurite extension has been attributed in part to its affinity for chondroitin sulfate. Another role proposed for proteoglycans is the sequestration of extracellular growth factors and the modulation of their activity (7). The concept of cell-surface or extracellular components as molecular sinks for morphogenetic factors can be extended to include SPARC and TSP, which have been shown to bind PDGF and TGF-β, respectively (16, 17). Furthermore, a SPARC·PDGF complex effectively inhibited the binding of the growth factor to its cognate receptor (16).

A recurrent theme underlying cellular responses to injury and developmental signals is the necessity for changing cellular morphology. Acquisition of a terminally differentiated phenotype, induction of migration, stimulation or inhibition of the cell cycle, and modulation of the expression of certain gene products, for example, are associated with or contingent upon changes in cell shape. Many of the properties ascribed collectively but not uniquely to SPARC, TN, and TSP in this review are related to the ability of these secreted glycoproteins to modulate cell-cell and cell-matrix interactions. At the molecular level we do not understand the specific associations that take place between these proteins and their target sequences, nor do we understand the mechanisms that promote the dynamic organization of cells and ECM into unique tissues. Studies with embryos carrying experimentally altered genes for these anti-adhesive proteins should stimulate some unifying hypotheses and elucidate specific morphoregulatory mechanisms.

Acknowledgments—We thank co-workers in our laboratories for data and stimulating discussions, and colleagues who read drafts prior to publication. We are grateful to Brenda Wood for assistance with the manuscript. Since limitations of space preclude extensive citation, we have endeavored to describe recent, representative studies and to direct the reader to comprehensive reviews in which other relevant literature has been cited.

REFERENCES

1. Watt, F. (1986) *Trends Biochem. Sci.* **11,** 482–485
2. Burridge, K., Fath, K., Kelly, T., Nuckolls, G., and Turner, C. (1988) *Annu. Rev. Cell Biol.* **4,** 487–526
3. Erickson, H. P., and Bourdon, M. A. (1989) *Annu. Rev. Cell Biol.* **5,** 71–92
4. Albelda, S. M., and Buck, C. A. (1990) *FASEB J.* **4,** 2868–2880
5. Humphries, M. J. (1990) *J. Cell Sci.* **97,** 585–592
6. Beck, K., Hunter, I., and Engel, J. (1990) *FASEB J.* **4,** 148–160
7. Bernfield, M., and Sanderson, R. D. (1990) *Philos. Trans. R. Soc. Lond. B Biol. Sci.* **327,** 171–186
8. Ruoslahti, E. (1989) *J. Biol. Chem.* **264,** 13369–13372
9. Mosher, D. F. (1990) *Annu. Rev. Med.* **41,** 85–97
10. Murphy-Ullrich, J. E., and Höök, M. (1989) *J. Cell Biol.* **109,** 1309–1319
11. Murphy-Ullrich, J. E., Lightner, V. A., Erickson, H. P., and Höök, M. (1990) *J. Cell Biol.* **111,** 144a
12. Woods, A., and Couchman, J. R. (1988) *Collagen Relat. Res.* **8,** 155–182
13. Jaken, S., Leach, K., and Klauck, T. (1989) *J. Cell Biol.* **109,** 697–704
14. Chiquet-Ehrismann, R. (1990) *FASEB J.* **4,** 2598–2604
15. Lightner, V. A., and Erickson, H. P. (1990) *J. Cell Sci.* **95,** 263–277
16. Lane, T. F., Iruela-Arispe, L., Ross, R., Sage, H., and Raines, E. (1991) *J. Cell. Biochem.* **15F,** 182
17. Murphy-Ullrich, J. E., Schultz-Cherry, S., and Höök, M. (1990) *J. Cell Biol.* **111,** 148a
18. Silverstein, R. L., Leung, L. L. K., Harpel, P. C., and Nachman, R. L. (1985) *J. Biol. Chem.* **260,** 10346–10352
19. Hasselaar, P., Loskutoff, D. J., Sawdey, M., and Sage, H. (1991) *J. Biol. Chem.* **266,** 13178–13184
20. Termine, J. D., Kleinman, H. K., Whitson, S. W., Conn, K. M., McGarvey, M. L., and Martin, G. R. (1981) *Cell* **26,** 99–105
21. Sage, H., Johnson, C., and Bornstein, P. (1984) *J. Biol. Chem.* **259,** 3993–4007
22. Bolander, M. F., Young, M. F., Fisher, L. W., Yamada, Y., and Termine, J. D. (1988) *Proc. Natl. Acad. Sci. U. S. A.* **85,** 2919–2923
23. Mason, I. J., Taylor, A., Williams, J. G., Sage, H., and Hogan, B. L. M. (1986) *EMBO J.* **5,** 1465–1472
24. Dziadek, M., Paulsson, M., Aumailley, M., and Timpl, R. (1986) *Eur. J. Biochem.* **161,** 455–464
25. Domenicucci, C., Goldberg, H. A., Hofmann, T., Isenman, D., Wasi, S., and Sodek, J. (1988) *Biochem J.* **253,** 139–151
26. Engel, J., Taylor, W., Paulsson, M., Sage, H., and Hogan, B. (1987) *Biochemistry* **26,** 6958–6965
27. Villarreal, X. C., Mann, K. G., and Long, G. L. (1989) *Biochemistry* **28,** 6483–6491
28. Metsäranta, M., Young, M. F., Sandberg, M., Termine, J., and Vuorio, E. (1989) *Calcif. Tissue Int.* **45,** 146–152
29. Pacifici, M., Oshima, O., Fisher, L. W., Young, M. F., Shapiro, I. M., and Leboy, P. S. (1990) *Calcif. Tissue Int.* **47,** 51–61
30. Holland, P., Harper, S., McVey, J., and Hogan, B. L. M. (1987) *J. Cell Biol.* **105,** 473–482
31. Wewer, U. M., Albrechtsen, R., Fisher, L. W., Young, M. F., and Termine, J. D. (1988) *Am. J. Pathol.* **132,** 345–355
32. Sage, H., Vernon, R., Decker, J., Funk, S., and Iruela-Arispe, M.-L. (1989) *J. Histochem. Cytochem.* **37,** 819–829
33. Vernon, R. B., and Sage, E. H. (1989) *Biol. Reprod.* **40,** 1329–1340
34. Sage, H., Vernon, R., Funk, S., Everitt, E., and Angello, J. (1989) *J. Cell Biol.* **109,** 341–356
35. Lane, T. F., and Sage, E. H. (1990) *J. Cell Biol.* **111,** 3065–3076
36. Everitt, E. A., and Sage, H. (1990) *J. Cell. Biochem.* **14E,** 121
37. Funk, S. E., and Sage, E. H. (1991) *Proc. Natl. Acad. Sci. U. S. A.* **88,** 2648–2652
38. Spring, J., Beck, K., and Chiquet-Ehrismann, R. (1989) *Cell* **59,** 325–334
39. Faissner, A., and Kruse, J. (1990) *Neuron* **5,** 627–637
40. Chiquet-Ehrismann, R., Kalla, P., Pearson, C. A., Beck, K., and Chiquet, M. (1988) *Cell* **53,** 383–390
41. Lotz, M. M., Burdsal, C. A., Erickson, H. P., and McClay, D. R. (1989) *J. Cell Biol.* **109,** 1795–1805
42. Gatchalian, C. L., Schachner, M., and Sanes, J. R. (1989) *J. Cell Biol.* **108,** 1873–1890
43. Mackie, E. J., Halfter, W., and Liverani, D. (1988) *J. Cell Biol.* **107,** 2757–2767
44. Daniloff, J. K., Crossin, K. L., Pincon-Raymond, M., Murawsky, M., Rieger, F., and Edelman, G. M. (1989) *J. Cell Biol.* **108,** 625–635
45. Wehrle, B., and Chiquet, M. (1990) *Development* **110,** 401–415
46. Aufderheide, E., Chiquet-Ehrismann, R., and Ekblom, P. (1987) *J. Cell Biol.* **105,** 599–608
47. Probstmeier, R., Martini, R., and Schachner, M. (1990) *Development* **109,** 313–321
48. Asch, A. S., and Nachman, R. L. (1989) *Prog. Hemostasis Thromb.* **9,** 157–176
49. Lawler, J., and Hynes, R. O. (1987) *Semin. Thromb. Hemostasis* **13,** 245–254
50. Frazier, W. A. (1987) *J. Cell Biol.* **105,** 625–632
51. Majack, R. A., and Bornstein, P. (1987) in *Cell Membranes: Methods and Reviews* (Elson, E., Frazier, W., and Glaser, L., eds) pp. 55–77, Plenum Publishing Co., New York
52. Prater, C. A., Plotkin, J., Jaye, D., and Frazier, W. A. (1991) *J. Cell Biol.* **112,** 1031–1040
53. Asch, A. S., Tepler, J., Silbiger, S., and Nachman, R. L., (1991) *J. Biol. Chem.* **266,** 1740–1745
54. Roberts, D. D., Sherwood, J. A., and Ginsburg, V. (1987) *J. Cell Biol.* **104,** 131–139
55. Varani, J., Nickoloff, B. J., Riser, B. L., Mitra, R. S., O'Rourke, K., and Dixit, V. M. (1988) *J. Clin. Invest.* **81,** 1537–1544
56. Lawler, J., Weinstein, R., and Hynes, R. O. (1988) *J. Cell Biol.* **107,** 2351–2361
57. Mansfield, P. J., Boxer, L. A., and Suchard, S. J. (1990) *J. Cell Biol.* **111,** 3077–3086
58. Taraboletti, G., Roberts, D., Liotta, L. A., and Giavazzi, R. (1990) *J. Cell Biol.* **111,** 765–772
59. Good, D. J., Polverini, P. J., Rastinejad, F., Le Beau, M. M., Lemons, R. S., Frazier, W. A., and Bouck, N. P. (1990) *Proc. Natl. Acad. Sci U. S. A.* **87,** 6624–6628
60. Bornstein, P., O'Rourke, K., Wikstrom, K., Wolf, F. W., Katz, R., Li, P., and Dixit, V. M. (1991) *J. Biol. Chem.* **266,** 12821–12824
61. Donoviel, D. B., Amacher, S. L., Judge, K. W., and Bornstein, P. (1990) *J. Cell. Physiol.* **145,** 16–23
62. Majack, R. A., Cook, S. C., and Bornstein, P. (1986) *Proc. Natl. Acad. Sci. U. S. A.* **83,** 9050–9054
63. Majack, R. A., Goodman, L. V., and Dixit, V. M. (1988) *J. Cell Biol.* **106,** 415–422
64. Mumby, S. M., Abbott-Brown, D., Raugi, G. J., and Bornstein, P. (1984) *J. Cell. Physiol.* **120,** 280–288
65. O'Shea, K. S., and Dixit, V. M. (1988) *J. Cell Biol.* **107,** 2737–2748
66. Mackie, E. J., Thesleff, I., and Chiquet-Ehrismann, R. (1987) *J. Cell Biol.* **105,** 2569–2579
67. Iruela-Arispe, M.-L., Bornstein, P., and Sage H. (1991) *Proc. Natl. Acad. Sci. U. S. A.* **88,** 5026–5030
68. Iruela-Arispe, M.-L., Hasselaar, P., and Sage, H. (1991) *Lab. Invest.* **64,** 174–186
69. Mackie, E. J., Chiquet-Ehrismann, R., Pearson, C. A., Inaguma, Y., Taya, K., Kawarada, Y., and Sakakura, T. (1987) *Proc. Natl. Acad. Sci. U. S. A.* **84,** 4621–4625
70. Rastinejad, F., Polverini, P. J., and Bouck, N. P. (1989) *Cell* **56,** 345–355
71. Raugi, G. J., Olerud, J. E., and Gown, A. M. (1987) *J. Invest. Dermatol.* **89,** 551–554
72. Bronner-Fraser, M. (1988) *J. Neurosci. Res.* **21,** 135–147
73. Ingber, D. E., and Folkman, J. (1989) *Cell* **58,** 803–805
74. Iruela-Arispe, M.-L., Diglio, C. A., and Sage, H. (1991) *Arterioscler. Thromb.*, in press
75. Murphy-Ullrich, J. E., and Mosher, D. F. (1987) *Semin. Thromb. Hemostasis* **13,** 343–351
76. Sage, H., Lane, T., and Iruela-Arispe, M. L. (1990) *J. Cell. Biochem.* **14E,** 127
77. Montesano, R., Pepper, M. S., Möhle-Steinlein, U., Risau, W., Wagner, E. F., and Orci, L. (1990) *Cell* **62,** 435–445
78. Rosen, E. M., Maromsky, L., Goldberg, I., Bhargava, M., and Setter, E. (1990) *J. Cell Sci.* **96,** 639–649
79. Gould, R. J., Polokoff, M. A., Friedman, P. A., Huang, T.-F., Holt, J. C., Cook, J. J., and Niewiarowski, S. (1990) *Proc. Soc. Exp. Biol. Med.* **195,** 168–171
80. Keynes, R., and Cook, G. (1990) *Cell* **62,** 609–610
81. Johnston, I. G., Paladino, T., Gurd, J. W., and Brown, I. R. (1990) *Neuron* **4,** 165–176

THE JOURNAL OF BIOLOGICAL CHEMISTRY
Vol. 266, No. 24, Issue of August 25, pp. 15555–15558, 1991

Consensus Sequences as Substrate Specificity Determinants for Protein Kinases and Protein Phosphatases

Peter J. Kennelly‡§ and Edwin G. Krebs¶

From the ‡Department of Biochemistry and Nutrition, Virginia Polytechnic Institute and State University, Blacksburg, Virginia 24061-0308 and the ¶Department of Pharmacology, University of Washington School of Medicine, Seattle, Washington 98195

Protein phosphorylation plays a pivotal role in the execution and regulation of many cellular functions. Consequently, phosphoproteins and the enzymes that catalyze their phosphorylation/dephosphorylation have been intensely studied. Central to our understanding of protein phosphorylation is the question of how these enzymes recognize their diverse substrate proteins. Since detailed information on substrate recognition is currently confined to those enzymes that phosphorylate/dephosphorylate serine and threonine, the remarks that follow will be confined to this group.

In 1964, Nolan and co-workers (1) observed that a chymotryptic peptide from glycogen phosphorylase could be phosphorylated by phosphorylase kinase. Later, Daile and Carnegie (2) reported that proteolytic fragments of myelin basic protein, a substrate for cAMP-dependent protein kinase (cAMP-PK),[1] could also serve as substrates for cAMP-PK. Bylund and Krebs (3) observed that denaturation actually transformed some proteins into substrates for cAMP-PK, while Humble and co-workers (4) demonstrated that cAMP-PK phosphorylated denatured pyruvate kinase as well as a cyanogen bromide peptide from the enzyme. The implications of these findings were that protein kinases recognized the local structure surrounding the phosphoacceptor group and that distant regions of the polypeptide chain or higher ordered protein structures play little role in determining substrate specificity. The concept rapidly evolved that the sites phosphorylated by a particular protein kinase shared a set of common sequence elements, its "consensus sequence," whose existence was necessary and sufficient for recognition by that enzyme.

In this short review we will attempt to summarize our current understanding of the role of consensus sequences in substrate recognition by protein kinases and phosphatases. While this has been the subject of several recent reviews (5–7), these works have largely focused upon the features that render synthetic peptides optimal substrates for these enzymes. Although such peptides represent powerful investigative tools, their small size and random conformation significantly limit their abilities to mimic the proteins they are intended to model. Therefore, we have focused upon the sequence features surrounding the phosphorylation sites on protein substrates. This was done both to discern patterns indicative of the existence and influence of consensus sequences in substrate recognition and to assess the degree to which the consensus sequence paradigm reliably predicts the behavior of protein kinases toward substrate proteins, both known and potential. Synthetic peptide data have been used to illuminate the significant features of those patterns revealed through the comparison of protein substrates. The limitations of space preclude citing many of the individuals who have contributed to our understanding of protein kinase and protein phosphatase substrate specificity. We apologize for this unfortunate circumstance.

Definition

The term consensus sequence refers to those sequence elements immediately surrounding the site(s) phosphorylated by a given protein kinase that are considered essential for its recognition and phosphorylation by that kinase. It generally takes the form of a short linear sequence of amino acids indicating the identity of the minimum set of amino acids comprising such a site and their position relative to the phosphoacceptor residue. The following assumptions are implicit in the formulation of such consensus sequences: 1) The presence of a consensus sequence on a protein is necessary and sufficient for its recognition as a substrate by a particular protein kinase. 2) The specificity-determining features of the phosphorylation site are contained in a contiguous sequence of amino acids around the phosphoacceptor. It does not include elements from different polypeptide chains or from widely scattered portions of a single polypeptide chain. 3) Not all sequence positions surrounding the phosphoacceptor group, regardless of their proximity thereto, carry equal weight in determining the recognition code.

Consensus sequences are typically represented as in the following example for $p34^{cdc2}$ (8): S*/T*-P-*X*-R/K. The phosphoacceptor group is denoted by an asterisk or by the letter P in parentheses. We will use the former convention and reserve the use of a P in parentheses to denote pre-existing phosphoamino acids. Where two amino acids function interchangeably, both are listed with a slash (/) separating them. Sequence positions judged to be recognition neutral are denoted by an *X*. This does not guarantee that all possible substitutions at these positions, or combinations thereof, will be without effect upon the properties of the resulting substrates.

Applications and Limitations

Consensus sequences have many useful applications. As models of critical substrate recognition determinants they presumably form reflected images of the corresponding substrate binding domains. They have been used to identify autoinhibitory domains involved in the regulation of a number of protein kinases and phosphatases. They also have served as guides for the design of synthetic peptide substrates of great utility.

Much of the usefulness of the consensus sequence model lies in its simplicity. Summarizing the complexities of the substrate recognition process as sets of short recognition sequences has facilitated the evaluation and application of a large body of observations. However, it must be borne in mind that in practice the model's assumption that local primary sequence alone controls recognition represents an oversimplification. Factors such as secondary/tertiary structure or distant secondary recognition sites can and do play significant roles in substrate recognition, sometimes completely overshadowing primary sequence considerations. This can apply especially to intramolecular autophosphorylation, where the sheer physical proximity of the phosphoacceptor may drive its phosphorylation. The existence of often unknown or ill-defined negative determinants contributes further complexity to the model's application. Thus, the existence of an apparent consensus sequence does not assure that a protein can be phosphorylated, nor is it a foolproof indicator of the protein kinase responsible if phosphorylation does occur. Consensus sequence information functions best as a guide whose implications must be confirmed or refuted experimentally. Despite these limitations, the tremendous success engendered by the application of consensus sequences attests both to their practical usefulness and their genuine importance in substrate recognition.

§ To whom correspondence should be addressed.
[1] The abbreviations used are: cAMP-PK, cAMP-dependent protein kinase; cGMP-PK, cGMP-dependent protein kinase; PKC, protein kinase C; AMP-PK, AMP-activated protein kinase; CaM, calmodulin; MLCK, myosin light chain kinase; sm, smooth muscle; sk, skeletal muscle; MHCK, myosin heavy chain kinase; GSK-3, glycogen synthase kinase-3; CK, casein kinase.

Specific Consensus Sequences

Below are discussed those protein kinases for which sufficient data exists concerning the sites they phosphorylate to provide a reasonable expectation that patterns of common features indicative of consensus sequences should be apparent, if recognition proceeds through such a mechanism. A much abbreviated summary of this information is given in Table I. Exclusion of a particular kinase does not imply that it has no consensus sequence. Unless indicated, no attempt has been made to weigh information gained from proteins known to be phosphorylated *in vivo* more heavily than that obtained *in vitro*. This has been done to secure the largest possible data base from which to make comparisons. Autophosphorylation sites were omitted for the reasons outlined earlier. To simplify discussion, the phosphoacceptor residue will be considered to be at the zero position and the adjacent N-terminal and C-terminal amino acids will be designated by the numbers NH_2- . . . ,−3, −2, −1, 0, +1, +2, +3, . . . -COOH, etc.

cAMP-dependent Protein Kinase—The presence of basic amino acids, particularly arginine, N-terminal to the phosphoacceptor serine or threonine (as with most protein kinases, phosphorylation of serine is generally preferred over threonine) is a key factor in substrate recognition by cAMP-PK (9, 10). Of 93 phosphorylation sites on 52 proteins, 88 possess at least one arginine at the −2 or, more frequently (65 *versus* 54 cases), the −3 position. However, the optimal sequence for peptide substrates of R-R-*X*-S*/T* (10, 11) only describes 30 of these sites, the major deviations being the substitution of lysine for arginine at position −2 (14 cases) or the presence of only a single arginine in the −1 to −3 area (33 cases). It appears that *in vitro* cAMP-PK will phosphorylate sites possessing one arginine as frequently, if perhaps not as efficiently, as those with two. Surveying probable physiological targets, *i.e.* those thought to be phosphorylated by the enzyme in living cells, substrates having a pair of basic residues in the −1 to −3 region were favored over those with a single basic residue by nearly 2:1, indicating that cAMP-PK may be more discriminating in nature than in the laboratory, where several factors, especially the use of supraphysiological enzyme concentrations, come into play. A consensus sequence for cAMP-PK might therefore be R-R/K-*X*-S*/T* > R-X_2-S*/T* = R-X-S*/T*, which describes 95% of the sequences surveyed.

cGMP-dependent Protein Kinase—A survey of 16 sites phosphorylated by the cGMP-dependent protein kinase (cGMP-PK) on 10 proteins indicates the universal presence of at least one N-terminal (−1 to −4) arginine. In most cases (13/16) there are 2 or 3 basic residues present. Thus, it appears that cGMP-PK possesses a more stringent requirement for multiple basic residues than does its cAMP-regulated cousin. A possible consensus sequence for cGMP-PK might therefore be R/K_{2-3}-*X*-S*/T*, which describes 75% of the sites surveyed. Work with synthetic peptides has led to the suggestion that a +1 arginine forms the key specificity determinant for cGMP-PK (7). However, this position is so occupied in only one of the sites surveyed. Moreover, the only physiological substrates identified to date, G-substrate (12) and the cGMP-binding phosphodiesterase (13), follow the pattern R/K_{2-3}-*X*-S*-*X*, in which *X* + 1 is neutral.

Protein Kinase C—Protein kinase C (PKC) also requires basic amino acid residues near the phosphoacceptor group. PKC can be influenced by both N- and C-terminal basic residues, and avidly phosphorylates substrates containing both. A survey of 68 sites of phosphorylation on 29 proteins showed that 16 contained at least one arginine or lysine at positions −1 through −3, 19 possessed one or more at positions +1 through +3, and nearly half (31/67) were bracketed by these amino acids in the −1 through −3 and +1 through +3 regions. The positions most frequently occupied by a basic residue were +2 (34/67), −2 (31/67), −3 (25/67), and +3 (24/67). Positions −1 and +1 were so occupied in only nine and five instances, respectively.

Studies with synthetic peptides indicate that although either C- or N-terminal basic residues can serve as determinants for PKC, optimal peptides contained both (14–17). Arginine was found superior to lysine. A survey of eight proteins phosphorylated by PKC *in vivo* shows examples of sites that are double-sided (6/11), that have only C-terminal basic residues (2/11), or that have N-terminal basics only (3/11), a pattern resembling that displayed *in vitro*. A definitive consensus sequence for PKC has yet to be determined, a task complicated by potential substrate specificity differences among its isozymic forms (17). However, a summary of our current understanding might be (R/K_{1-3}, X_{2-0})-S*/T*-(X_{2-0}, R/K_{1-3}) > S*/T*-(X_{2-0}, R/K_{1-3}) ≥ (R/K_{1-3}, X_{2-0})-S*/T*.

AMP-activated Protein Kinase—To date six sites phosphorylated by the AMP-activated protein kinase (AMP-PK) distributed over four proteins have been sequenced (summarized in Ref. 18). Five reside on presumed physiological targets. While no clear consensus has emerged, they do share some common characteristics including a propensity for hydrophobic amino acids at positions −5 (5/6), −1 (5/6), +2 (4/6), +4 (5/6), and +5 (5/6). All possess two and often (3/6) three hydrophilic amino acids at positions −4 through −2, one of which is basic (arginine or histidine) with position −3 usually (4/6) so occupied. No acidic amino acids appear in region −1 to −6. Hydrophilic residues are generally present at positions +3 (5/6) and +6 (4/6).

CaM Kinase II—Exempting its behavior towards caldesmon, the multifunctional calcium/calmodulin-dependent protein kinase (CaM kinase II) is a close adherent to the consensus sequence paradigm. A survey of 16 sites phosphorylated by CaM kinase II on 12 proteins indicates that 14 possess the sequence R-*X*-*X*-S*/T*, a consensus originally suggested by Payne *et al.* (19) and later confirmed using synthetic peptides (20). However, when one considers caldesmon, this trend shifts markedly. CaM kinase II from smooth muscle phosphorylates eight sites on caldesmon *in vitro*, only one of which conforms to the R-*X*-*X*-S*/T* paradigm (21). The implications of this "aberrant" behavior have yet to be resolved.

p34^{cdc2}—The best characterized of the emerging family of protein kinases involved in cell cycle control is p34^{cdc2}. Moreno and

TABLE I

Summary of consensus sequences most frequently recognized by protein kinases

This table represents a simplified version of the information in the text to allow rapid comparisons to be made. The reader is advised to refer to the text to gain more detailed information on the origin and predictive reliability of the sequences presented.

Protein kinase	Consensus sequence
cAMP-PK	R-R/K-*X*-S*/T* > R-X_2-S*/T* = R-*X*-S*/T*
cGMP-PK	$(R/K)_{2-3}$-*X*-S*/T*
PKC	(R/K_{1-3}, X_{2-0})-S*/T*-(X_{2-0}, R/K_{1-3}) > S*/T*-(X_{2-0}, R/K_{1-3}) ≥ (R/K_{1-3}, X_{2-0})-S*/T*
AMP-PK	ND[a]
CaM kinase II	R-*X*-*X*-S*/T*
p34^{cdc2}	S*/T*-P-*X*-R/K[b]
Phosphorylase kinase	ND
smMLCK	(K/R_2, *X*)-X_{1-2}-K/R_3-X_{2-3}-R-X_2-S*-N-V-F
skMLCK	(K/R_2, *X*)-X_{1-2}-K/R_3-X_{2-3}-R-X_2-S*-N-V-F > (K/R_2, *X*)-X_{1-2}-K/R_3-X_{2-3}-E-X_2-S*-N-V-F
MHCK I	R/K_{1-2}-X_{-1-2}-S*/T*-*X*-Y[c]
MHCK II	R-G-*X*-S*-*X*-R
GSK-3	S*-X_3-S(P)
CK I	S(P)-X_{1-3}-S*/T* ≫ (D/E_{2-4}, X_{2-0})-*X*-S*/T*
CK II	S*/T*-(D/E/$S(P)_{1-3}$, X_{2-0})

[a] ND, unable to predict consensus sequence due to lack of sufficient information and/or influence of other factors in influencing substrate specificity.
[b] Where *X* is polar.
[c] Where *X* +1 or −1 is a hydroxyl amino acid.

Nurse (8) surveyed eight sites phosphorylated by $p34^{cdc2}$ on six proteins and observed that seven conformed to the pattern S*/T*-P-*X*-R/K, with *X* being a polar amino acid. They noted that at least four of the six proteins become phosphorylated at the identical sites *in vivo* during mitosis.

Phosphorylase Kinase—Factors additional to primary sequence weigh heavily in substrate recognition by phosphorylase kinase. The evidence for this is 2-fold. First, although phosphorylase kinase will phosphorylate small peptides, such peptides fall well short of being phosphorylated with the same rate or affinity as substrate proteins. Second, cAMP-PK will phosphorylate denatured, but not native, glycogen phosphorylase at the same site as does phosphorylase kinase (3), implying that it possesses a "special" conformation.

The sites phosphorylated by phosphorylase kinase on all tissue/species forms of its major physiological target, glycogen phosphorylase, conform to the pattern K/R-R-K/R-Q-I-S*-V/I-R-G-L (22). Synthetic peptide work indicates that the basic residues beginning at position −3 are essential for recognition (reviewed in Ref. 23). The arginine at position +2 acts as an enhancer; its removal markedly decreased the efficacy of peptide substrates. Thus, a potential consensus sequence might contain one or more basic residues between positions −3 and −5 and possibly a +2 arginine. However, other features, including the conformation of the phosphorylation site, play important roles.

The Myosin Light Chain Kinases—The myosin light chain kinases (MLCKs) possess an absolute specificity for myosin light chains. Comparison of the phosphorylation sites of a number of myosin light chains indicates that they conform to the following sequence pattern (reviewed in Ref. 24): (K/R_2, *X*)-$X_{1\text{-}2}$-K/R_3-$X_{2\text{-}3}$-R(smooth)/E(skeletal, cardiac)-X_2-S*-N-V-F. The key difference between smooth muscle light chains *versus* those from striated (cardiac and skeletal) muscle is at position −3, where the former has arginine and the latter glutamate. Smooth muscle MLCK (smMLCK) recognizes only smooth muscle myosin light chains. Studies with synthetic peptides indicate that the three clusters of basic amino acids at −3, −6 to −10, and −11 to −14 are important recognition determinants for smMLCK, their relative influence increasing with their proximity to the phosphoacceptor serine (25). skMLCK will phosphorylate smooth, cardiac, or skeletal muscle light chains. *In vitro*, skMLCK prefers the same consensus sequence, with a −3 arginine instead of glutamate, as does smMLCK (26). This suggests that the most distant basic cluster may be an important specificity determinant under physiological conditions, even though it is not absolutely essential *in vitro*.

Myosin I Heavy Chain Kinase—The myosin I heavy chain kinase (MHCK I) of *Acanthamoeba castellanii* phosphorylates a single threonine or serine in the heavy chain of myosin I. The sites on three myosin I isozymes conform to the pattern R/$K_{1\text{-}2}$-$X_{1\text{-}2}$-S*/T*-*X*-Y, with either position +1 or −1 occupied by a hydroxyl amino acid (27). Peptide studies indicate that one N-terminal basic residue and tyrosine +2 are required for recognition, with a second N-terminal basic residue enhancing phosphorylation (28).

Myosin II Heavy Chain Kinase—The myosin II heavy chain kinase (MHCK II) of *A. castellanii* phosphorylates three closely clustered serine residues within the heavy chain of myosin II: T-P-S-S-R-G-G-S*-T-R-G-A-S*-A-R-G-A-S*-V-R (29). The enzyme also readily phosphorylates a peptide with this sequence, indicating that it contains the information essential for recognition. These sites follow the pattern R-G-*X*-S*-X-R, and this has been suggested to form the recognition determinant for MHCK II (29).

Glycogen Synthase Kinase-3 and Synergistic Phosphorylation—Glycogen synthase kinase-3 (GSK-3) phosphorylates glycogen synthase, cAMP-PK type II, protein phosphatase inhibitor 2, and protein phosphatase-1_G, all likely physiological targets. However, phosphorylation only takes place after their prior phosphorylation by another protein kinase, a phenomenon termed "synergistic" or "hierarchical" phosphorylation. These sites possess two common characteristics, a propensity for one or more nearby (−3 to +3) proline residues, and a C-terminal location of the synergistic phosphorylation event. Thus, the recognition determinant for GSK-3 appears to be a C-terminal serine-phosphate group. In synthetic peptides, GSK-3 recognizes the consensus sequence S*-X_3-S(P) (30). Most (7/9) sites phosphorylated by GSK-3 can conform to this model, possessing a serine residue at the +3 (one case) or +4 position that can be phosphorylated either by the "synergistic" kinase or by GSK-3 itself in an ordered mechanism. However, this synergistic phosphoserine occupies position +13 or +27 in the other two cases. Thus, proximity of a serine phosphate moiety in space, rather than along the polypeptide chain, may also confer recognition by GSK-3 (30).

Casein Kinase I—Casein kinase I (CK I) targets sites rich in N-terminal, negatively charged, *i.e.*, acidic or phosphorylated, amino acids. Examination of 13 sites phosphorylated by CK I on five proteins indicates that 12 possess one or more negatively charged amino acids immediately N-terminal (−1 to −5) to the phosphoacceptor group. Position −3 was so occupied 10 of 13 times, with positions −2 (7/13) and −4 (6/13) next in frequency. The residues C-terminal to the phosphoacceptor show no pattern of basic, acidic, or other character. The importance of serine phosphate as a determinant was recognized when Tuazon *et al.* (31) observed that prior dephosphorylation of casein adversely affected its phosphorylation by CK I. Flotow *et al.* (32) have since shown that a single N-terminal (−1, −2, or −3) serine or threonine phosphate residue can serve as the recognition determinant for CK I in peptides. Serine phosphate may be the more potent recognition determinant since peptides with a single serine phosphate have lower K_m values than those with multiple glutamates or aspartates (33). This behavior suggests that phosphorylation by CK I can be regulated through a synergistic mechanism (32). The consensus sequence for CK I might be summarized as S(P)-X_2-S*/T* > S(P)-$X_{1\text{ or }3}$-S*/T* > (D/$E_{2\text{-}4}$, $X_{2\text{-}0}$)-S*/T*.

Casein Kinase II—Casein kinase II (CK II) requires the presence of the acidic amino acid residues glutamate, aspartate, or (occasionally) serine phosphate immediately C-terminal (+1 to +3) to the phosphoacceptor serine/threonine. In synthetic peptides a single C-terminal acidic amino acid is sufficient for recognition, with the optimal position being +3 (34–36). However, in protein substrates, the presence of multiple acidic residues appears to be strongly preferred since of 37 phosphorylation sites surveyed on 19 proteins, 30 had two or more aspartic or glutamic acid residues. The distribution of these residues was 23/37 at position +1, 28/37 at +2, and 26/37 at +3. In 10 cases the close clustering of CK II sites could allow an initial phosphorylation event to introduce an additional +1 to +3 acidic determinant for the second site, three of which would have the effect of increasing the number of acidic determinants from one to two. Thus, the consensus sequence for CK II can be summarized as S*/T*-(D/E/$S(P)_{1\text{-}3}$, $X_{2\text{-}0}$).

Phosphorylation Site Geometry

Early on it was recognized that geometry must help determine substrate specificity if only by denying access to potential phosphoacceptor groups. This negative "veto" role was supported by the observation that denaturation often transformed proteins into or improved them as substrates (3, 4). Small *et al.* (37) applied the Chou-Fasman secondary structure algorithm to 30 phosphorylation sites and found that 80% were predicted to exist within β-turns. However, spectroscopic studies of synthetic peptides (38, 39) and the use of conformationally constrained peptides (40) indicate that they bind cAMP-PK in an extended coil conformation. cGMP-PK preferred extended coil peptides as well (40). An extended conformation would be consistent with the ability to manipulate many consensus sequences as if they were linear in space. The propensity with which phosphorylation sites occur near the N and C termini of proteins is also consistent with the idea that phosphorylation sites have extended (or extendable) conformations. On the other hand, given the wide range of protein kinases and phosphoproteins extant, it seems likely that protein phosphorylation sites must exist in a variety of conformations, not just extended ones. Fiol *et al.* (30) have speculated that the frequency of proline and glycine residues proximal to GSK-3 phosphorylation sites might indicate a requirement for a β-turn structure and the sites phosphorylated by the AMP-PK also occur in regions predicted to possess a high probability for forming β-turns (18). Certainly, for some kinases the existence of specialized, very intricate secondary/tertiary structures will no doubt form an important key to substrate recognition. The more discriminating the kinase, the more likely that the determinants recognized are conformationally complex.

Protein Phosphatases

Since kinases and phosphatases share common protein substrates, a natural question is whether they recognize these proteins by similar mechanisms. Early on it was recognized that the number of serine/threonine-specific protein kinases far outstripped the number of serine/threonine-specific protein phosphatases. Moreover, it was observed that many protein phosphatases acted on phosphoryl groups introduced by a number of different protein kinases, indicating that protein phosphatases recognize specificity determinants different from those of protein kinases.

A frequently employed criterion for establishing that an enzyme recognizes a consensus sequence on its target(s) is determining whether it recognizes smaller fragments of substrate proteins containing the putative consensus sequence. In 1960, Graves *et al.* (41) observed that phosphorylase phosphatase would dephosphorylate, albeit slowly, a hexapeptide derived from phosphorylase *a*. More recently, several phosphatases have been observed to dephosphorylate peptide substrates including pyruvate kinase phosphatase (42), protein phosphatase 2A (43, 44), and calcineurin (44, 45). Oftentimes the peptides that have shown promise as phosphatase substrates were much larger than those typically recognized by protein kinases. Blumenthal *et al.* (45), for example, systematically varied the length of a phosphopeptide modeled after the site on the type II regulatory subunit of cAMP-PK that is dephosphorylated by calcineurin. While a 19-residue peptide was dephosphorylated with kinetics comparable with the intact protein, decreasing the length to 15 residues simultaneously increased K_m and decreased V_{max} severalfold. Nolan *et al.* (1) reported that phosphorylase phosphatase dephosphorylated a tetradecapeptide at a 15-fold greater rate than the hexapeptide of Graves *et al.* (41). Such behavior implies that either the primary sequence "window" scanned by phosphatases is larger than that typically scanned by kinases, or that greater length is required to support the formation of higher order structures required for recognition. Attempts to resolve consensus sequences by testing peptides of varying sequence or by comparing sites on substrate proteins have revealed some general trends but have yet to yield any clearly defined consensus sequence.

In general, control of protein phosphorylation is achieved through the selective activation of individual catalysts, protein kinases, that are specifically targeted toward the appropriate substrate proteins. The requirement for a handful of phosphatases to counterbalance the activity of scores of kinases suggests the likelihood of fundamental differences in the nature of the mechanisms that control of protein phosphorylation and dephosphorylation. Among these is a much greater emphasis on mechanisms such as substrate activation, through effector binding to substrates, etc., or catalyst translocation in the control of dephosphorylation reactions (46). This emerging dichotomy in the control of protein kinase and protein phosphatase action suggests that the search for consensus sequences of the type that help provide the highly specific targeting required of many protein kinases may fail to yield similar results with regards to the protein phosphatases.

Acknowledgments—We would like to thank the following individuals for many helpful discussions and for their critical reading of this manuscript: Kenneth Lerea, David Litchfield, Christine Roush, and Erwin Reimann. We would also like to thank Horst Flotow and Peter Roach, Paul Clarke and Graham Hardie, and Steven Hanks for giving us an early look at manuscripts in press.

REFERENCES

1. Nolan, C., Novoa, W. B., Krebs, E. G., and Fischer, E. H. (1964) *Biochemistry* **3,** 542–551
2. Daile, P., and Carnegie, P. R. (1974) *Biochem. Biophys. Res. Commun.* **61,** 852–858
3. Bylund, D. B., and Krebs, E. G. (1975) *J. Biol. Chem.* **250,** 6355–6361
4. Humble, E., Berglund, L., Titanji, V., Ljungstrom, O., Edlund, B., Zetterquist, O., and Engstrom, L. (1975) *Biochem. Biophys. Res. Commun.* **66,** 614–621
5. Pinna, L. A. (1988) *Adv. Exp. Med. Biol.* **231,** 433–443
6. Kemp, B. E. (ed) (1990) *Peptides and Protein Phosphorylation*, CRC Press, Inc., Boca Raton, FL
7. Kemp, B. E., and Pearson, R. B. (1990) *Trends Biochem. Sci.* **15,** 342–346
8. Moreno, S., and Nurse, P. (1990) *Cell* **61,** 549–551
9. Kemp, B. E., Bylund, D. B., Huang, T., and Krebs, E. G. (1975) *Proc. Natl. Acad. Sci. U. S. A.* **72,** 3448–3452
10. Zetterquist, O., Ragnarsson, U., Humble, E., Berglund, L., and Engstrom, L. (1976) *Biochem. Biophys. Res. Commun.* **70,** 696–703
11. Kemp, B. E., Graves, D. J., Benjamini, E., and Krebs, E. G. (1977) *J. Biol. Chem.* **252,** 4888–4894
12. Aitken, A., Bilham, T., Cohen, P., Aswad, D., and Greengard, P. (1981) *J. Biol. Chem.* **256,** 3501–3506
13. Thomas, M. K., Francis, S. H., and Corbin, J. D. (1990) *J. Biol. Chem.* **265,** 14971–14978
14. Woodgett, J. R., Gould, K. L., and Hunter, T. (1986) *Eur. J. Biochem.* **161,** 177–184
15. Ferrari, S., Marchiori, F., Marin, O., and Pinna, L. A. (1987) *Eur. J. Biochem.* **163,** 481–487
16. House, C., Wettenhall, R. E., and Kemp, B. E. (1987) *J. Biol. Cem.* **262,** 772–777
17. Marais, R. M., Nguyen, O., Woodgett, J. R., and Parker, P. J. (1990) *FEBS Lett.* **277,** 151–155
18. Clarke, P. R., and Hardie, D. G. (1990) *EMBO J.* **9,** 2439–2446
19. Payne, M. E., Schworer, C. M., and Soderling T. R. (1983) *J. Biol. Chem.* **258,** 2376–2382
20. Pearson, R. B., Woodgett, J. R., Cohen, P., and Kemp, B. E. (1985) *J. Biol. Chem.* **260,** 14471–14476
21. Ikebe, M., and Reardon, S. (1990) *J. Biol. Chem.* **265,** 17607–17612
22. Dombradi, V., Willis, A. C., Vereb, G., and Johnson, L. N. (1988) *Comp. Biochem. Physiol. B Comp. Biochem.* **91,** 717–721
23. Graves, D. J. (1983) *Methods Enzymol.* **99,** 268–278
24. Stull, J. T., Nunnally, M. H., and Michnoff, C. H. (1986) in *The Enzymes* (Krebs, E. G., and Boyer, P. D., eds) Vol. 17, pp. 113–166, Academic Press, Orlando, FL
25. Kemp, B. E., and Pearson, R. B. (1985) *J. Biol. Chem.* **260,** 3355–3359
26. Michnoff, C. H., Kemp, B. E., and Stull, J. T. (1986) *J. Biol. Chem.* **261,** 8320–8326
27. Brzeska, H., Lynch, T. J., Martin, B., and Korn, E. D. (1989) *J. Biol. Chem.* **264,** 19340–19348
28. Brzeska, H., Lynch, T. J., Martin, B., Corigliano-Murphy, A., and Korn, E. D. (1990) *J. Biol. Chem.* **265,** 16138–16144
29. Côté, G. P., and Bukiejko, U. (1987) *J. Biol. Chem.* **262,** 1065–1072
30. Fiol, C. J., Wang, A., Roeske, R. W., and Roach, P. J. (1990) *J. Biol. Chem.* **265,** 6061–6065
31. Tuazon, P. T., Bingham, E. W., and Traugh, J. A. (1979) *Eur. J. Biochem.* **94,** 497–504
32. Flotow, H., Graves, P. R., Wang, A., Fiol, C. J., Roeske, R. W., and Roach, P. J. (1990) *J. Biol. Chem.* **265,** 14264–14269
33. Flotow, H., and Roach, P. J. (1991) *J. Biol. Chem.* **266,** 3724–3737
34. Marin, O., Meggio, F., Marchiori, F., Borin, G., and Pinna, L. A. (1986) *Eur. J. Biochem.* **160,** 239–244
35. Kuenzel, E. A., Mulligan, J. A., Sommercorn, J., and Krebs, E. G. (1987) *J. Biol. Chem.* **262,** 9136–9140
36. Litchfield, D. W., Arendt, A., Lozeman, F. J., Krebs, E. G., Hargrave, P. A., and Palczewski, K. (1990) *FEBS Lett.* **261,** 117–120
37. Small, D., Chou, P. Y., and Fasman, G. D. (1977) *Biochem. Biophys. Res. Commun.* **79,** 341–346
38. Mildvan, A. S., Rosevear, P. R., Granot, J., O'Brian, C. A., Bramson, H. N., and Kaiser, E. T. (1983) *Methods Enzymol.* **99,** 93–119
39. Reed, J., Kinzel, V., Kemp, B. E., Chen, H., and Walsh, D. A. (1985) *Biochemistry* **24,** 2967–2973
40. Thomas, N. E., Bramson, H. N., Nairn, A. C., Greengard, P., Fry, D. C., Mildvan, A. S., and Kaiser, E. T. (1987) *Biochemistry* **26,** 4471–4474
41. Graves, D. J., Fischer, E. H., and Krebs, E. G. (1960) *J. Biol. Chem.* **235,** 805–809
42. Mullinax, T. R., Stepp, L. R., Brown, J. R., and Reed, L. J. (1985) *Arch. Biochem. Biophys.* **243,** 655–659
43. Agostinis, P., Goris, J., Pinna, L. A., Marchiori, F., Perich, J. W., Meyer, H. E., and Merlevede, W. (1990) *Eur. J. Biochem.* **189,** 235–241
44. Hemmings, H. C., Nairn, A. C., Elliott, J. I., and Greengard, P. (1990) *J. Biol. Chem.* **265,** 20369–20376
45. Blumenthal, D. K., Takio, K., Hansen, R. S., and Krebs, E. G. (1986) *J. Biol. Chem.* **261,** 8140–8145
46. Stralfors, P., Hiraga, A., and Cohen, P. (1985) *Eur. J. Biochem.* **149,** 295–303

THE JOURNAL OF BIOLOGICAL CHEMISTRY
Vol. 266, No. 25, Issue of September 5, pp. 16257–16260, 1991

Printed in U.S.A.

Biosynthesis and Function of Selenocysteine-containing Enzymes

Thressa C. Stadtman

From the Laboratory of Biochemistry, National Heart, Lung, and Blood Institute, National Institutes of Health, Bethesda, Maryland 20892

In this review emphasis will be on the essential roles of selenium in various enzymes and other macromolecules. Identification of selenocysteine as the selenium moiety present in several selenium-dependent enzymes together with the mechanism of its specific insertion into these proteins as directed by the UGA stop codon will be presented. Available information concerning the catalytic advantage of selenium at the active site of an enzyme as compared with sulfur also will be included.

Early information prior to the middle 1950s concerning biological effects of selenium related entirely to the toxicity of the element, particularly to grazing animals. Selenium ingested in the form of seleniferous plants, adapted to growth in high selenium semi-desert soils, often proved lethal. At sublethal doses, neurological disorders and damage to keratinaceous tissues were observed in the animals. Under laboratory conditions selenium was established to be toxic to animals in the low to middle micomolar range and lethal in high micromolar to millimolar concentrations. In view of this history, selenium dietary supplements for animals or humans were prohibited long after the element was shown to be an essential trace element for mammals, birds, and several bacteria. The fact that the threshold between the required and the toxic level for most organisms is only about 1 order of magnitude is unusual for an essential nutrient, and to date the biochemical basis of this is largely unexplained.

Although the occurrence of selenium as an integral component of a few redox-type enzymes in prokaryotes has been known for several years, until recently the only known example in eukaryotes was glutathione peroxidase (1, 2). The ability of this peroxidase to effectively reduce organic peroxides and thus protect cells from damage due to reactive oxygen species provided an explanation at the biochemical level for the requirement of selenium as an essential trace element in mammals and birds. The discovery during the past year (3, 4) that conversion of thyroxine to the active thyroid hormone is catalyzed by a selenoenzyme, tetraiodothyronine 5′-deiodinase, provides an example of an even more ubiquitous role of the element, namely in growth and developmental processes of diverse animal species including amphibia. A new mammalian selenoprotein of unusual composition, the plasma selenoprotein P (5, 6), which contains at least 8 and presumably 10 selenocysteine residues per subunit, should also prove to be of fundamental physiological importance.

Selenocysteine-containing Proteins

Selenocysteine occurs in highly specific positions in three types of bacterial enzymes: glycine reductase in several clostridia; formate dehydrogenases in *Escherichia coli*, *Salmonella*, *Clostridia*, and a *Methanococcus*; and hydrogenases in a *Methanococcus* and certain other anaerobic bacteria (1). Two types of selenoenzymes in eukaryotic species are the family of glutathione peroxidases and a tetraiodothyronine 5′-deiodinase. Among the glutathione peroxidases are a soluble tetrameric form known to occur in mammals and birds (2) and tentatively identified in a marine diatom (7), a glycosylated form in plasma (8, 9), and a monomeric species that specifically reduces lipid peroxides in membranes (10). These three glutathione peroxidases are products of different genes (11). The selenium-containing 5′-deiodinase is the Type I microsomal enzyme found in thyroid, liver, and kidney (3, 4). The very unusual mammalian selenoprotein, selenoprotein P, of still unidentified function, is a glycoprotein present in plasma (5). Representative members of these selenoprotein groups are listed in Table I.

Other Selenoproteins

Two other bacterial selenoenzymes, nicotinic hydroxylase and xanthine dehydrogenase, present in certain clostridial species, contain selenium in the form of an unidentified dissociable cofactor (1). Derivatization of the cofactor from nicotinic acid hydroxylase with various alkylating agents produced the corresponding dialkylselenides as the sole selenium-containing products (12). From this it can be concluded that the selenium was originally present in a form that was labilized by reduction and alkylation. Among possible candidates would be the selenium analog of an FeS center or Se instead of S as an outer ligand of a molybdopterin cofactor.

Although selenomethionine is detected in numerous proteins of both eukaryotic and prokaryotic origin, particularly in methionine-rich proteins, it apparently occurs only as a random substitute for methionine (13–15). No deleterious effects of such substitutions have been reported. Recently a method for crystallographic structural analyses of proteins based on selenomethionine content instead of heavy atom replacement was developed (16). Using a methionine auxotroph of *E. coli* for expression of the ribonuclease H gene, the 4 methionine residues were substituted with selenomethionine (17). A crystal of the fully active selenoprotein was used for structural analysis at 2-Å resolution by the method of multiwavelength anomalous diffraction (MAD).

Synthetic Selenoproteins

A selenium analog of a *Neurospora crassa* copper metallothionein was synthesized using an automated solid phase peptide synthesizer (18). In this metalloselenonein 7 selenocysteines replaced the 7 cysteine residues present in the native protein. In view of the instability of selenocysteine under a variety of conditions, this is a significant technical achievement. Availability of this synthetic selenoprotein has allowed interesting metal binding studies to be performed.

Conversion of the active site serine in subtilisin to a selenocysteine residue was achieved by reaction of the enzyme with phenylmethylsulfonyl fluoride and cleavage of the sulfonyl ester with HSe^- (19). The resulting selenosubtilisin, which exhibited esterase rather than protease activity, can also function as a glutathione peroxidase (Ref. 20; see below).

TABLE I
Selenium-dependent enzymes that contain selenocysteine

Glycine reductase	*Clostridia*
Formate dehydrogenases	Bacteria (*E. coli*, *Salmonella*)
Hydrogenases	Certain anaerobic bacteria
Glutathione peroxidase	Mammals, birds, marine diatom
Selenoprotein P	Mammalian liver, kidney, plasma
Tetraiodothyronine 5′-deiodinase	Thyroid, liver, kidney

Mechanistic Roles of Selenium in Selenoenzymes

The reaction mechanisms of eukaryotic glutathione peroxidases, and more recently that of iodothyronine 5′-deiodinase, have been studied in detail. In these enzymes the ionized selenol group functions as a nucleophile in the reaction with ROOH, possibly to form *E*-SeOH (2) or with RI to form *E*-SeI (3, 21). Regeneration of both enzymes by thiols (*in vivo* by reduced glutathione) liberates OH^- or I^- and reforms *E*-Se^-. Numerous kinetic studies of these types of enzymes support ping-pong mechanisms (2, 22). The artificial selenoenzyme, selenosubtilisin, serves as a redox catalyst and a mimic of glutathione peroxidase (19, 20). Using ^{77}Se-enriched selenosubtilisin, various oxidation states of the protein were studied by ^{77}Se NMR spectroscopy (23). After reaction with peroxide a species that exhibited a ^{77}Se resonance at 1189 ppm typical of a seleninic acid (*E*Se(O)OH) was obtained. Reduction with 3 equivalents of ArSH converted the enzyme to a species thought to be *E*Se-SAr (resonance at 388.5 ppm). In the normal turnover reactions a selenenic enzyme product (*E*SeOH) has been proposed as the oxidized intermediate (2, 20).

An *E. coli* formate dehydrogenase, a selenoenzyme of the formate-hydrogen lyase complex, oxidizes formate to carbon dioxide and reducing equivalents which are transferred to a hydrogenase and liberated as H_2 (24). Studies with the purified form of this selenoenzyme (25, 26) and with a mutant form containing sulfur in place of selenium (27) indicate that the selenocysteine residue interacts directly with formate. Both the selenoenzyme and the mutant sulfur enzyme are resistant to inactivation by alkylation in the absence of substrate. However, when formate is added, inactivation by iodoacetamide is rapid and the pH optima for alkylation of the two enzymes reflect the differing pK_a values for the ionization of a selenol (5.2) *versus* that of a sulfhydryl (8.3 or above) group. At present, the mechanism of generation of the reactive selenol group in the selenium-dependent formate dehydrogenase by formate is unknown.

A novel role for selenium in a biological catalyst came to light with the discovery (28) that the selenoprotein A component of the glycine reductase complex (29, 30) is converted to an Se-carboxymethyl derivative during the overall reaction sequence. In this case the carboxymethyl group is derived from the carbon skeleton of the substrate, glycine, in Schiff base linkage to a carbonyl group protein, protein B (31). In a subsequent reaction the selenoether form of selenoprotein A is cleaved to give an acetylthiol ester derivative of protein C, a third protein of the complex (32–34). In the presence of orthophosphate the protein C thiol ester is converted to acetylphosphate (33, 34),[1] which serves as a source of ATP (36, 37). Alternatively, in the presence of arsenate, acetate is formed directly (34, 36, 37).[1] The reaction mechanisms involving the ionized selenol group of the single selenocysteine residue in selenoprotein A are not known in detail. However, it is apparent that the reductive cleavage of the carboxymethyl group in selenoether linkage to protein A, coupled to regeneration of the reduced selenoprotein A, serves as the energy-conserving step in the overall glycine reductase reaction. The central role of selenoprotein A in the reductive deamination of glycine is depicted in Scheme I. Although a cysteine residue of selenoprotein A present in the sequence -Cys-Phe-Val-SeCys-[2] (38, 39) is indicated as a possible participant in the reductive cleavage of the carboxymethyl group, there is no direct evidence to support this. Conversion of this cysteine residue to serine by site-directed mutagenesis of a base of the corresponding TGC triplet in the *Clostridium sticklandii* selenoprotein A gene[3] may provide useful information on this point.

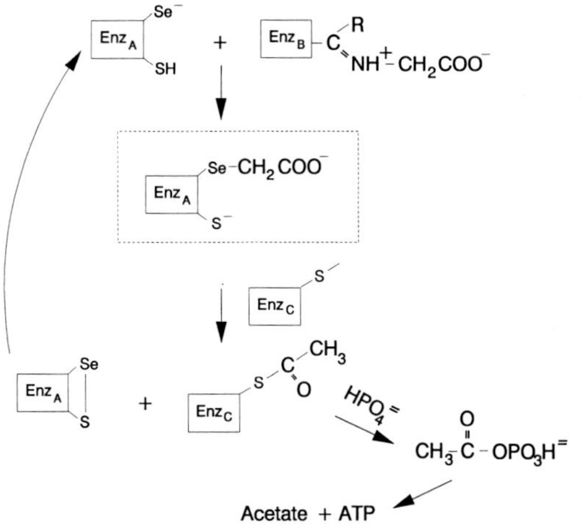

SCHEME I. **Glycine reductase.** The role of selenocysteine in selenoprotein A is shown.

TABLE II
Comparison of selenium versus sulfur in the active site of enzymes

	SeCys enzyme	Cys enzyme
Formate dehydrogenase (*E. coli*)		
K_m	26 mM	5 mM
K_{cat}	2800 s^{-1}	9 s^{-1}
Tetraiodothyronine 5′-deiodinase[a] (K_m)	~5 μM	~2 nM

[a] Values given are for Type I (SeCys enzyme) and Type II (Cys enzyme) deiodinases.

Advantages Conferred by Selenium over Sulfur in Biological Catalysts

A mutant formate dehydrogenase (40) that differs from wild type only by the presence of a sulfur atom instead of selenium at the active site (cysteine replacement directed by conversion of TGA to TGC) was purified and its properties and catalytic activity compared with those of the native selenocysteine-containing enzyme (25–27). The markedly higher rate of catalysis exhibited by the selenoenzyme, even when assayed using an artificial dye as electron acceptor, clearly shows the chemical advantage of selenium in the enzyme (Table II). The fact that both enzymes are extremely oxygen-labile suggests that one of the other redox groups, the molybdopterin or an FeS center, may be the more oxygen-sensitive moiety.

A cysteine mutant of the selenium-containing Type I tetraiodothyronine 5′-deiodinase, also prepared by mutagenesis of a TGA codon to TGT, was reported to exhibit about 10% of the activity of the selenoenzyme (3) and markedly reduced sensitivity to inhibitors such as gold thioglucose (41). Type II iodothyronine 5′-deiodinase from pituitary and brown fat,

[1] T. C. Stadtman and J. N. Davis, manuscript submitted for publication.

[2] The abbreviation used is: SeCys, selenocysteine.
[3] G. E. Garcia, unpublished results.

which apparently contains sulfur instead of selenium, also is much less sensitive to gold thioglucose and more resistant to alkylating agents than is the Type I selenoenzyme. With triiodothyronine as substrate, approximate K_m values for the two enzymes reported in several studies (21, 42–44) differ considerably (Table II). The synthetic metalloselenonein analog of *N. crassa* metallothionein binds 3 mol of CuI/mol of enzyme, whereas the sulfur protein binds 6 mol. It is suggested that in this case the decreased copper binding is due to the larger ionic radius of selenium (18).

Mechanism of Specific Incorporation of Selenocysteine in Proteins

Elucidation of the overall features governing the highly specific insertion of the unusual amino acid selenocysteine into a few selenium-dependent enzymes has opened up an exciting new area in selenium biochemistry. An early important clue came from the observation of Sunde (45) that the carbon skeleton of the selenocysteine residue in glutathione peroxidase was derived directly from serine. Shortly thereafter the gene encoding this enzyme was found to contain a TGA triplet in the position corresponding to selenocysteine in the gene product (46, 47), and at the same time this termination codon also was shown to occur in an *E. coli* formate dehydrogenase gene (48). Elegant studies of Böck and associates (40) proved that the TGA codon (UGA in the message) directed the cotranslational insertion of selenocysteine in formate dehydrogenase and that readthrough of this codon depended on the availability of selenium in the growth medium. Mutation of the TGA codon to cysteine codons allowed the selenium-independent synthesis of enzymes that exhibited greatly decreased catalytic activity (see above).

A series of *E. coli* mutants (49, 50) and a *Salmonella typhimurium* mutant (51) originally detected by monitoring their inability to decompose formate by the formate-hydrogen lyase pathway have been invaluable in elucidating discrete reaction steps required for selenocysteine incorporation into formate dehydrogenase. One of the mutations resulted in the synthesis of a defective form of a tRNA that has an anticodon complementary to UGA (52). The wild type active form of this tRNA, the product of the *selC* gene, is aminoacylated with L-serine by seryl-tRNA ligase (52). This seryl-tRNA reacts with the *selA* gene product, a 600-kDa enzyme composed of 50-kDa subunits each containing a bound pyridoxal phosphate (53). The resulting Schiff base with the α-amino group of the seryl-tRNA is dehydrated, giving a stabilized dehydroalanyl-tRNA protein complex (54). This pyridoxal phosphate (PLP) enzyme has been termed selenocysteine synthase in view of the fact that addition of a reactive form of RSe^- to the 2=3 double bond of the dehydroalanyl-tRNA in the complex converts it to selenocysteyl-$tRNA^{Sec}_{UCA}$. The reactive reduced selenium species can be generated from HSe^- by the *selD* gene product, a 37-kDa protein, in the presence of ATP (54, 55). Thus, the complete sequence of reactions involving the replacement of the hydroxyl group of serine with selenium is accomplished while the amino acid is in ester linkage to a tRNA (Scheme II). Although there are no comparable mutants in eukaryotes that can be used for delineation of the pathway of selenocysteyl-tRNA biosynthesis in these systems, involvement of the *O*-phosphoseryl derivative of a unique seryl-tRNA that can recognize UGA may serve as intermediate in the process (56). Following transcription of the single copy gene, the UCA anticodon of this tRNA is modified to NCA (where N is an unidentified uridine derivative) or to C_mCA (where C_m is *o*-methylcytidine) (57, 58). Both tRNA species are aminoacylated with L-serine and bind to UGA (58, 59). These tRNAs esterified with selenocysteine have been isolated from various eukaryotic sources including *Thalassiosira*, a diatom, and *Tetrahymena*, a ciliate (59, 60). Furthermore, in a selenocysteyl-$tRNA^{Sec}$ population isolated from a rat mammary tumor cell line, molecules containing putative precursor forms of selenocysteyl-tRNA, namely seryl-tRNA and *O*-phosphoseryl-tRNA were present in addition to those esterified with selenocysteine (59). The reaction sequence depicted in Scheme II is suggested from these findings. In eukaryotic systems, as well as in prokaryotes, the identity of the selenium donor (RSe^-) is unknown.

<u>*E. coli*</u>

$$\text{Seryl-tRNA}^{Sec}_{UCA} \xrightarrow[\text{PLP}]{\text{selA}} [\text{PLP-dehydroalanyl-tRNA}^{Sec}_{UCA}] \xrightarrow[\text{selD}]{RSe^-} \text{Selenocysteyl-tRNA}^{Sec}_{UCA}$$

<u>Mammals</u>

$$\text{Seryl-tRNA}^{Sec} \xrightarrow[\text{kinase}]{\text{ATP}} \text{O-phosphoseryl-tRNA}^{Sec} \xrightarrow{RSe^-} \text{Selenocysteyl-tRNA}^{Sec}$$

SCHEME II. **Conversion of seryl-$tRNA^{Sec}$ to selenocysteyl-$tRNA^{Sec}$.**

In *E. coli* and in *Salmonella* the *selD* gene product plays an additional role, namely in the conversion of the 5-methylaminomethyl 2-thiouridine residue in the anticodons of certain tRNAs to 5-methylaminomethyl 2-selenouridine (51, 61, 62). Thus, the selenium donor generated by the *SELD* protein can react with various types of activated receptor molecules involving addition or replacement processes. The *sel*D mutants synthesize free selenocysteine (51, 62), which thus is available as a potential precursor of an activated form of selenium. This may have some analogy to poorly understood processes involving the addition of sulfur to form biotin, the thiazole of thiamine, 4-thiouridine in tRNAs, etc.

Factors Involved in the Insertion of Selenocysteine

The cotranslational insertion of selenocysteine into formate dehydrogenase in *E. coli* requires the participation of a new translation factor, the *selB* gene product (63, 64). This is a 70-kDa protein that specifically recognizes the selenocysteine-charged *selC* tRNA but not the serine-esterified form. As with elongation factor Tu, GTP is required for complex formation and a GTP binding domain of the protein has been identified.

Finally, discrimination between a UGA sense codon to be used to direct selenocysteine insertion and a UGA stop codon that binds release factor 2 to terminate translation is achieved by base sequences in the message that presumably dictate conformational changes during the readthrough process (65, 44). In the *E. coli* formate dehydrogenase message a minimum of 40 bases immediately downstream of the UGA codon have been shown to be essential for selenocysteine incorporation (65), whereas in the rat and human Type I tetraiodothyronine 5′-deiodinase message, a 250-base sequence in the 3′-untranslated region more than 1 kilobase downstream of the UGA is essential (44). Potential stem-loop secondary structures in the mRNAs involving these sequences are postulated.

The recent report (35) that the gene encoding a pheromone produced by the ciliate, *Euplotes octocarinatus*, contains three "in-frame" TGA triplets, yet these are translated as cysteine in the 11-kDa polypeptide product, provides an interesting new exception to the usage of UGA to specify selenocysteine incorporation. Normal cysteine codons, TGT and TGC, are also present in the pheromone gene, and the termination codon is TAA.

In view of the recent increased interest not only in the beneficial effects but also the toxic effects of selenium as well as exciting new developments at the basic biochemical level, rapid expansion of our understanding of the roles of this trace element in biology can be expected.

REFERENCES

1. Stadtman, T. C. (1990) *Annu. Rev. Biochem.* **59,** 111–127
2. Flohé, L. (1989) in *Glutathione: Chemical, Biochemical, and Medical Aspects* (Dolphin, D., Poulson, R., and Avamovic, O., eds), pp. 644–731, John Wiley & Sons, New York
3. Berry, M. J., Banu, L., and Larsen, P. R. (1991) *Nature* **349,** 438–440
4. Behne, D., Kyriakopoulos, A., Meinhold, H., and Kohrle, J. (1990) *Biochem. Biophys. Res. Commun.* **173,** 1143–1149
5. Read, R., Bellow, T., Yang, J.-G., Hill, K. E., Palmer, I. S., and Burk, R. F. (1990) *J. Biol. Chem.* **265,** 17899–17905
6. Hill, K. E., Lloyd, R. S., Yang, J.-G., Read, R., and Burk, R. F. (1991) *J. Biol. Chem.* **266,** 10050–10053
7. Price, N. M., and Harrison, P. J. (1988) *Plant Physiol.* **86,** 192–199
8. Maddipati, K. R., and Marnett, L. J. (1987) *J. Biol. Chem.* **262,** 17398–17403
9. Takahashi, K. T., Avissar, N., Whitin, J., and Cohen, H. (1987) *Arch. Biochem. Biophys.* **256,** 677–686
10. Ursini, F., Maiorino, M., and Gregolin, C. (1985) *Biochim. Biophys. Acta* **839,** 62–70
11. Schuckelt, R., Brigelius-Flohé, R., Maiorino, M., Roveri, A., Reumkens, J., Strasburger, W., Ursini, F., Wolf, B., and Flohé, L. (1991) *Free Radical Res. Commun.*, in press
12. Dilworth, G. L. (1982) *Arch. Biochem. Biophys.* **219,** 30–38
13. Cowie, D. B., and Cohen, G. N. (1957) *Biochim. Biophys. Acta* **26,** 252–261
14. Sliwkowski, M. X., and Stadtman, T. C. (1985) *J. Biol. Chem.* **260,** 3140–3144
15. Frank, P., Light, A., Tullius, T. D., Hodgson, K. O., and Pecht, I. (1985) *J. Biol. Chem.* **260,** 5518–5525
16. Yang, W., Hendrickson, W. A., Crouch, R. J., and Satow, Y. (1990) *Science* **249,** 1398–1405
17. Yang, W., Hendrickson, W. A., Kalman, E. T., and Crouch, R. J. (1990) *J. Biol. Chem.* **265,** 13553–13559
18. Oikawa, T., Esaki, N., Tanaka, H., and Soda, K. (1991) *Proc. Natl. Acad. Sci. U. S. A.*. **88,** 3057–3059
19. Wu, Z.-P. and Hilvert, D. (1989) *J. Am. Chem. Soc.* **111,** 4513–4514
20. Wu, Z.-P., and Hilvert, D. (1990) *J. Am. Chem. Soc.* **112,** 5647–5648
21. Berry, M. J., Kieffer, J. D., Harney, J. W., and Larsen, P. R. (1991) *J. Biol. Chem.* **266,** 14155–14158
22. Leonard, J. L., and Visser, T. J. (1986) in *Thyroid Hormone Metabolism* (Hennemann, G., ed), pp. 189–229, Marcel Dekker, New York
23. House, K. L., Dunlap, R. B., Odom, J. D., Wu, Z.-P., and Hilvert, D. (1991) *FASEB J.* **5,** A1150 (Abstr. 4515)
24. Peck, H. D., Jr., and Guest, H. (1957) *J. Bacteriol.* **73,** 706–721
25. Axley, M. J., Grahame, D. A., and Stadtman, T. C. (1990) *J. Biol. Chem.* **265,** 18213–18218
26. Axley, M. J., and Grahame, D. A. (1991) *J. Biol. Chem.* **266,** 13731–13736
27. Axley, M. J., Böck, A., and Stadtman, T. C. (1991) *Proc. Natl. Acad. Sci. U. S. A.*, in press
28. Arkowitz, R. A., and Abeles, R. H. (1990) *J. Am. Chem. Soc.* **112,** 870–872
29. Turner, D. C., and Stadtman, T. C. (1973) *Arch. Biochem. Biophys.* **154,** 366–381
30. Cone, J. E., Martin del Rio, R., and Stadtman, T. C. (1976) *J. Biol. Chem.* **252,** 5337–5344
31. Tanaka, H., and Stadtman, T. C. (1979) *J. Biol. Chem.* **254,** 447–452
32. Arkowitz, R. A., and Abeles, R. H. (1989) *Biochemistry* **28,** 4639–4644
33. Arkowitz, R. A., and Abeles, R. H. (1991) *Biochemistry* **30,** 4090–4097
34. Stadtman, T. C. (1989) *Proc. Natl. Acad. Sci. U. S. A.* **86,** 7853–7856
35. Meyer, F., Schmidt, H. J., Plumper, E., Haslik, A., Mersmann, G., Meyer, H. E., Engstrom, A., and Heckmann, K. (1991) *Proc. Natl. Acad. Sci. U. S. A.* **88,** 3758–3761
36. Stadtman, T. C., and Elliott, P. (1956) *J. Am. Chem. Soc.* **78,** 2020–2021
37. Stadtman, T. C., Elliott, P., and Tiemann, L. (1958) *J. Biol. Chem.* **231,** 961–973
38. Sliwkowski, M. X., and Stadtman, T. C. (1988) *Proc. Natl. Acad. Sci. U. S. A.* **85,** 368–371
39. Garcia, G. E., and Stadtman, T. C. (1991) *J. Bacteriol.* **173,** 2093–2098
40. Zinoni, F., Birkman, A., Leinfelder, W., and Böck, A. (1987) *Proc. Natl. Acad. Sci. U. S. A.* **84,** 3156–3160
41. Berry, M. J., Kieffer, J. D., and Larsen, P. R. (1991) *Endocrinology*, in press
42. Visser, T. J., Leonard, J. L., Kaplan, M. M., and Larsen, P. R. (1982) *Proc. Natl. Acad. Sci. U. S. A.* **79,** 5080–5084
43. Goswami, A., and Rosenberg, I. N. (1989) *Biochem. Int.* **19,** 361–368
44. Berry, M. J., Banu, L., Chen, Y., Mandel, S. J., Kieffer, J. D., Harney, J. W., and Larsen, P. R. (1991) *Nature*, in press
45. Sunde, R. A., and Evenson, J. K. (1987) *J. Biol. Chem.* **262,** 933–937
46. Chambers, I., Frampton, J., Goldfarb, P., Affara, N., McBain, W., and Harrison, P. P. (1986) *EMBO J.* **5,** 1221–1227
47. Günzler, W. A., Steffens, G. J., Grossmann, A., Kim, S.-M., Otting, F., Wendel, A., and Flohé, L. (1984) *Hoppe-Seyler's Z. Physiol. Chem.* **365,** 195–212
48. Zinoni, F., Birkman, A., Stadtman, T. C., and Böck, A. (1986) *Proc. Natl. Acad. Sci. U. S. A.* **83,** 4650–4654
49. Haddock, B. A., and Mandrand-Berthelot, M.-A. (1982) *Biochem. Soc. Trans.* **10,** 478–480
50. Leinfelder, W., Forchhammer, K., Zinoni, F., Sawyers, G., Mandrand-Berthelot, M. A., and Böck, A. (1988) *J. Bacteriol.* **170,** 540–546
51. Kramer, G. F., and Ames, B. N. (1988) *J. Bacteriol.* **170,** 736–743
52. Leinfelder, W., Zehelin, E., Mandrand-Berthelot, M.-A., and Böck, A. (1988) *Nature* **331,** 723–725
53. Forchhammer, K., Leinfelder, W., Boesmiller, K., Vepreck, B., and Böck, A. (1991) *J. Biol. Chem.* **266,** 6318–6323
54. Forchhammer, K., and Böck, A. (1991) *J. Biol. Chem.* **266,** 6324–6328
55. Leinfelder, W., Forchhammer, K., Veprek, B., Zehelein, E., and Böck, A. (1990) *Proc. Natl. Acad. Sci. U. S. A.* **87,** 543–547
56. Hatfield, D. (1985) *Trends Biochem. Sci.* **10,** 201–204
57. Diamond, A., Dudock, B., and Hatfield, D. (1981) *Cell* **25,** 497–506
58. Hatfield, D., Diamond, A., and Dudock, B. (1982) *Proc. Natl. Acad. Sci. U. S. A.* **79,** 6215–6219
59. Lee, B. J., Worland, P. J., Davis, J. N., Stadtman, T. C., and Hatfield, D. L. (1989) *J. Biol. Chem.* **264,** 9724–9727
60. Hatfield, D. L., Lee, B. J., Price, N. M., and Stadtman, T. C. (1991) *Mol. Microbiol.* **5,** 1183–1186
61. Böck, A., and Stadtman, T. C. (1988) *BioFactors* **1,** 245–250
62. Stadtman, T. C., Davis, J. N., Zehelein, E., and Böck, A. (1989) *BioFactors* **2,** 35–44
63. Forchhammer, K., Leinfelder, W., and Böck, A. (1989) *Nature* **342,** 453–456
64. Forchhammer, K., Rucknagel, K.-P., and Böck, A. (1990) *J. Biol. Chem.* **265,** 9346–9350
65. Zinoni, F., Heider, J., and Böck, A. (1990) *Proc. Natl. Acad. Sci. U. S. A.* **87,** 4660–4664

Minireview

THE JOURNAL OF BIOLOGICAL CHEMISTRY
Vol. 266, No. 26, Issue of September 15, pp. 16965–16968, 1991

Printed in U.S.A.

Structural Relationships and the Classification of Aminoacyl-tRNA Synthetases*

Jonathan J. Burbaum and Paul Schimmel

From the Department of Biology, Massachusetts Institute of Technology, Cambridge, Massachusetts 02139

The aminoacyl-tRNA synthetases are a family of 20 enzymes that catalyze the same two-step chemical reaction,

$$AA + ATP \rightleftarrows AA{\sim}AMP + PP_i + tRNA^{AA} \rightleftarrows AA{\sim}tRNA^{AA} + AMP$$

where AA represents one of the 20 naturally occurring amino acids and $tRNA^{AA}$ represents a transfer RNA specific for that amino acid. In all, there are 20 such enzymes (one for each amino acid) that share the role of assigning amino acids to triplet codons in genetic translation. The synthetases are among the oldest proteins, and in contrast to more recently evolved and functionally related proteins, such as hemoglobin and myoglobin, their functional similarity is not reflected in a uniform structural framework. (This relationship also contrasts with that of enzymes that share structural homology but are mechanistically distinct, *e.g.* mandelate racemase and muconate lactonizing enzyme (1).) The diversity of synthetase structures has been a problem of long standing, whose solution may lead to a more basic understanding of protein structure/function and evolutionary relationships.

The diversity of synthetases is illustrated in Fig. 1, where the relative sizes and subunit compositions of the Class I and Class II synthetases from *Escherichia coli* are summarized.[1] The enzymes range in quaternary structure from monomers (*e.g.* IleRS)[2] to tetramers (AlaRS), in primary structure from 334 amino acids (TrpRS) to 1112 amino acids (PheRS), and in native molecular mass from 51,000 (CysRS) to 384,000 (AlaRS). The Class I enzymes are chiefly monomeric (only the dimeric Tyr and Trp enzymes cannot function as monomers (4, 5); the Met enzyme is converted into a monomer by limited proteolysis (6)), while the Class II enzymes are entirely oligomeric.

Based on a limited number of sequences, structural information (7, 8) and structural modeling, a group of related synthetases that initially included isoleucine, methionine, tyrosine, and glutamine were recognized (9, 10). This group was later expanded to include the Arg, Glu, Leu, Trp, and Val-tRNA synthetases (11–15). Each member of the group contains a "signature sequence" (9, 10), an 11-mer peptide that ends in a characteristic and mostly conserved tetrapeptide, HIGH (13). Additional sequence comparisons identified a second similarity, the pentapeptide KMSKS, in the same enzymes that contain the HIGH region (16). However, all synthetase sequences do not contain the signature sequence and KMSKS peptides. Even with the development of more sophisticated sequence comparison algorithms (17) and secondary structure predictions (18), similarities that are common to all 20 enzymes were not found. Nevertheless, with sequences and the characterization of the quaternary structures of all 21 synthetases from *E. coli*[3] and the information from five crystal structures (7, 8, 37–41), at least two different structural frameworks were established. Based on limited sequence similarities (2) that correlate with these structural frameworks (38, 41), each enzyme can be placed uniquely into one of two classes. The remarkable diversity in size and quaternary structure (Fig. 1), together with an apparently primordial origin of these two classes, explains the early difficulties in identifying sequence alignments within this group of enzymes.

Structural Basis for Limited Sequence Similarities That Define Two Classes

The first three structures to be solved (MetRS, TyrRS, and GlnRS) provided a structural basis for some of the limited sequence similarities. All three structures contain a nucleotide-binding (or Rossmann) fold (42). This motif of alternating α-helices and β-strands is found in enzymes that bind to an adenine nucleotide (*e.g.* the NADH cofactor of the dehydrogenases or the ATP substrate of the kinases). Furthermore, regions that correspond to the characteristic signature sequence and KMSKS motifs in these three structures have high three-dimensional similarity, and together they form part of the ATP binding site (43).

With the determination of the seryl-tRNA synthetase crystal structure and a complete set of sequences for the *E. coli* synthetases, the diversity of the synthetases was extended to their three-dimensional frameworks. The structure of SerRS (37) is unlike that of the Met, Tyr, or Gln synthetases. The dominant structural motif of SerRS is an eight-stranded antiparallel β-sheet that somewhat resembles the NAD-binding motif of the enterotoxins (44), which contain a seven-stranded sheet, but bears no relationship to the Rossmann folds of the earlier structures. Moreover, in those synthetases that lack the sequence motifs mentioned above, three unique regions of degenerate sequence similarity (motifs 1–3; Ref. 2) were identified. The lengths and the sequences of the central parts of these motifs are as follows: motif 1 (18 amino acids), ⊕GΦ*XX*Φ*XX*PΦΦ; motif 2 (23–31 amino acids), ⊕ΦΦ*X*Φ*XXX*FR*X*E()⊕Φ*X*⊖F; motif 3 (29–34 amino acids), ΦGΦGΦGΦERΦΦΦΦ().[4] As with the signature sequence 11-mer, a mostly conserved tetrapeptide can be used to define two of the three motifs; for motif 2, this tetrapeptide is FRNE, and for motif 3, it is GLER. These motifs (and the signature sequence similarities mentioned above) are the basis for dividing the 20 synthetases into two groups of 10 each (Fig. 1).

Further support for this grouping has come from a partial structure of the *S. cerevisiae* aspartyl-tRNA synthetase:$tRNA^{Asp}$ complex (41). This Class II enzyme contains a segment that is similar to the antiparallel β domain of SerRS. Based on the two structures, the active site of the Class II enzymes has been tentatively identified as being formed in part from amino acids that make up motifs 2 and/or 3. Motif 1 in the Class II alignment forms a portion of a subunit interface in both the SerRS and AspRS structures. This motif may be present because the Class II synthetases are predominantly dimers (Fig. 1). An exception is the tetrameric *E. coli* alanine enzyme, which can function as a monomer (45–47) and is naturally monomeric in higher organisms (for example, *Bombyx mori*; Ref. 48).

Subgroups within Class I Enzymes—Within the Class I synthetases, additional diversity is apparent. Each of the three crystal structures has a unique variant of the nucleotide-binding fold framework. In Fig. 2, the different nucleotide-binding fold topologies of the three known structures are aligned according to the position of the signature sequence and KMSKS motifs. In the enzymes for which no structures are yet known, we assumed that the three-dimensional dispositions of these two motifs are fixed

* This work was supported by Grants GM15539 and GM23562 from the National Institutes of Health.

[1] The classes are defined on the basis of sequence comparisons (2), rather than on their occurrence in multienzyme complexes (3).

[2] The abbreviations used are: RS, tRNA synthetase(s); D, dihydrouridine.

[3] For quaternary structure references, see Schimmel (11) and Mirande (3). Sequences: AlaRS (19); ArgRS (12); AsnRS (20); ArgRS (21); CysRS (22, 23); GlnRS (24); GluRS (15); GlyRS (25); HisRS (26); IleRS (9); LeuRS (27); LysRS (S) (28); LysRS (U) (29); MetRS (30); PheRS (31); ProRS (2); SerRS (32); ThrRS (33); TrpRS (34); TyrRS (35); ValRS (36).

[4] The codes for similar amino acids are as follows: Φ, hydrophobic amino acid; ⊕, positively charged amino acid; ⊖, negatively charged amino acid. Gaps in similarity of fixed length are denoted by strings of the letter *X*, and variable length gaps are denoted by ().

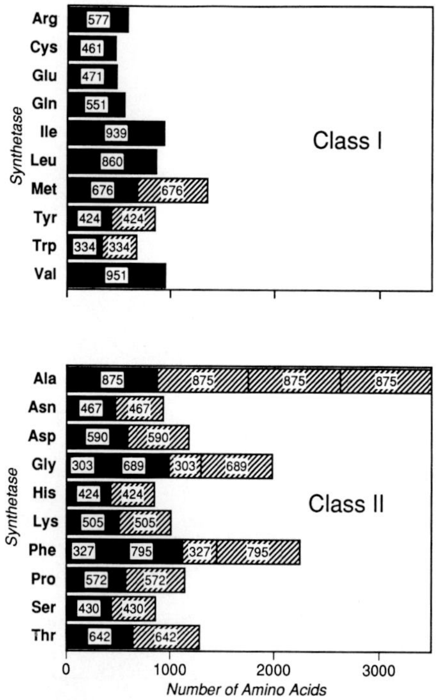

FIG. 1. **Graphic representation of the structural diversity of the aminoacyl-tRNA synthetases.** The synthetases are divided into two classes based on primary sequence comparisons (see text). The total lengths of the *bars* represent the molecular sizes of the enzymes that correspond to the indicated amino acids. Areas shaded in *black* represent the length of non-repetitive protein. *Cross-hatched areas* represent repeating units in the oligomeric synthetases.

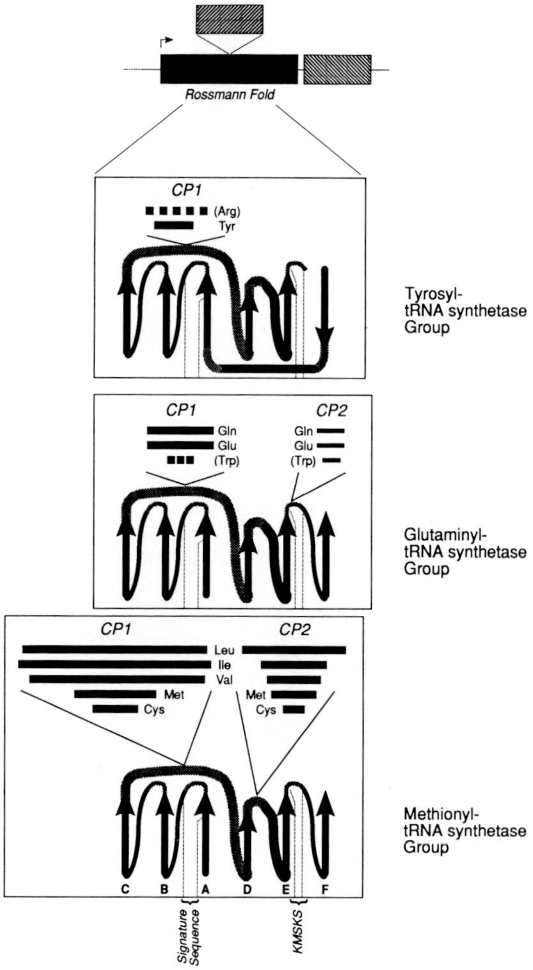

FIG. 2. **Families within the Class I synthetases.** For each of the three solved structures, the synthetases with highest similarity have been grouped. The schematic representation of the Rossmann fold of these enzymes considers a standard, six-stranded β-sheet (*arrows*), flanked by 3–4 helical segments (*solid black rods*). The three known structures have been aligned based on the positions of the signature sequence and KMSKS motifs. In this representation, portions of the polypeptide chain that lie above the plane of the sheet are drawn as *thicker lines*, and those that lie below are drawn as *thinner lines* (*cf.* Refs. 13 and 86). The inserted segments CP1 and CP2 are represented by *solid* or *dashed straight black lines* that are proportional in length to the size of the insertion but are not necessarily drawn to scale. *Dashed lines* represent tentative assignments.

relative to one another and to other secondary structure elements of the fold. Based on this assumption, the seven remaining Class I structures can be tentatively divided into three groups (Fig. 2).

In all three families, the first half of the nucleotide-binding fold contains the signature sequence, and in all but the Arg enzyme, this sequence occurs within the first 70 amino acids. The signature sequence occurs between an inner helix and β-strand within the N-terminal half of the nucleotide-binding fold. The catalytic residues that form the active site include those in a loop after strand E, which contains the KMSKS region. For TyrRS, whose signature sequence is found at residues 40–50, the N-terminal sequence forms the outermost antiparallel strand of the second half of the fold. Because the signature sequence of the Arg enzyme is also displaced (by 122 amino acids) from the N terminus, it is tentatively classified with Tyr rather than with Gln and Glu. Furthermore, the Trp enzyme is not classified with Tyr because its signature sequence is found only 10 residues from the N terminus, not far enough to allow for the extra strand.

Additional differences between the families are found in the connectivity within the nucleotide-binding fold. The largest family corresponds to the methionyl-tRNA synthetase structure and contains synthetases which activate the hydrophobic amino acids Ile, Leu, Met, and Val, as well as Cys (which can also be classified as hydrophobic; Ref. 49). This family has both the largest and the smallest Class I enzymes, with the differences in size concentrated in the insertions that are designated connective polypeptide domains CP1 and CP2 (10, 22). The connectivity differences between families are mainly in the second half of the fold, where the Met family contains a second connective polypeptide (CP2) between the first and second strands (D and E), the Gln family has CP2 between the second and third strands (E and F), and the Tyr family has no identifiable CP2. The alignment and subgroup classification is assisted by a short stretch of similarity that defines the first strand of the second half of the fold (strand D; Ref. 27), which allows the end of CP1 and the start of CP2 to be located.

These specific differences between Class I enzymes should not obscure their close relationship. A recent comparison of the Gln and Met enzyme structures (50) shows that the two enzymes, although in different families, share a rare "left-handed crossover" between strands E and F (where the corresponding strands of TyrRS are not directly connected). This topology places the signature sequence and KMSKS motifs on the same side of the β-sheet. In the GlnRS:tRNA^{Gln} cocrystal, strand F fits into the "armpit" region of the L-shaped tRNA molecule, between the minihelix domain and the D-anticodon domain (Fig. 3). The left-handed crossover, together with the unusually short strand "D," provides part of the docking interaction for the acceptor stem of tRNA.

Motifs for Interaction of Class I Enzymes with tRNA

In the Class I Gln- and Met-tRNA synthetases, two separate domains are used for interactions with the two domains of the L-shaped tRNA structure (51–53; Fig. 3). The acceptor-TΨC minihelix domain of tRNA^{Gln} interacts with the insertions into the nucleotide-binding fold framework of GlnRS (39). This nucleic acid domain ($M_r \sim 12{,}000$) must be placed in position to accept an activated amino acid from a labile adenylate intermediate ($M_r \sim 450$) at the catalytic site of the enzyme. The CP1 insertion occurs at a natural break in the nucleotide-binding fold topology of the catalytic domain, as a connector which joins the two $\beta_3\alpha_2$ halves of the fold (54). In GlnRS, the acceptor helix-binding CP1 structure consists of a five-stranded antiparallel β-sheet flanked

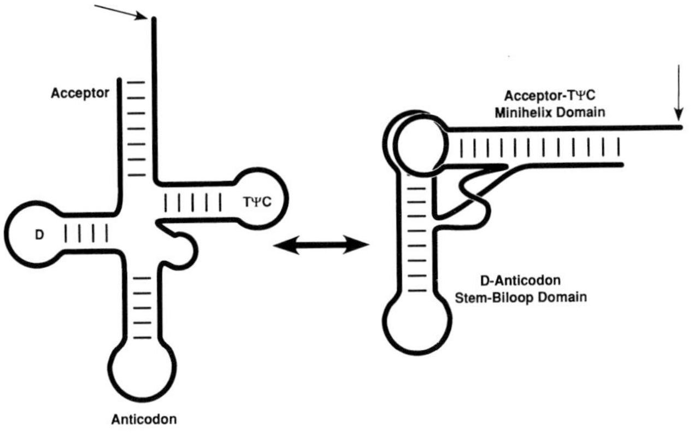

FIG. 3. **Folding of a tRNA molecule.** On the *left*, the familiar "cloverleaf" representation of a tRNA molecule is depicted, and on the *right*, a more accurate two-dimensional L-shaped representation is shown. On the cloverleaf structure, the acceptor stem and the D (dihydrouridine), TΨC, and anticodon stem-loops are indicated. On the L-shaped structure, two domains are indicated. The minihelix domain is comprised of a continuous helix spanning the acceptor stem and TΨC stem-loop. The D-anticodon stem-loop domain is comprised of the D and anticodon stem-loops.

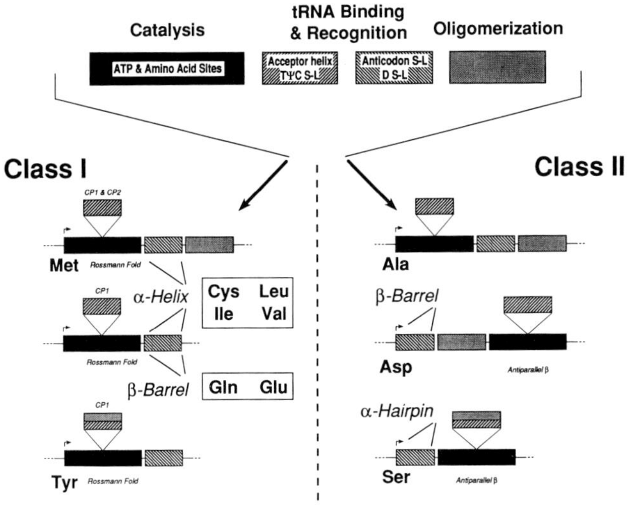

FIG. 4. **Assembly of Class I and Class II aminoacyl-tRNA synthetases.** Three functions of a synthetase polypeptide are illustrated: catalysis, tRNA binding/recognition, and oligomerization. The catalytic domain, which contains sites for ATP binding and amino acid activation, is a Rossmann fold domain in the Class I synthetases and seems to be an antiparallel β-sheet domain in at least two of the Class II synthetases. The tRNA binding/recognition functions of the synthetase are divided into two domains, like the tRNA molecule itself (see text).

by three α-helices, with an antiparallel β-loop that separates the 3′ and 5′ strands of the first base pair of the helix (39). Contacts with the 3:70 base pair are made by Asp^{235} at the N terminus of the helix that follows strand D (Fig. 2), the same point in the nucleotide-binding fold as the CP2 insertion in the methionine family.

A separate C-terminal domain interacts with the distal domain of the tRNA, which is comprised of the dihydrouridine (D) stem-loop and the anticodon stem-loop (Fig. 3). The anticodon-binding domains of the Class I synthetases have at least two divergent structures; in methionyl-tRNA synthetase, this domain is predominantly α-helical, while in GlnRS, it is a β-barrel. By sequence analysis and structural prediction, the α-helical motif of the Met enzyme is also predicted for the other Class I synthetases of the Met family, the Cys, Ile, Leu, and Val enzymes (22). In addition to GlnRS, the Glu synthetase is predicted to have a β-barrel anticodon recognition domain, because it is likely to have been derived from a common ancestor. (In *Bacillus subtilis*, there is a single enzyme which aminoacylates both $tRNA^{Glu}$ and $tRNA^{Gln}$ (Ref. 55), and in *E. coli*, the C-terminal domains of the Gln and Glu enzymes share sequence similarity (Ref. 15).) The Trp enzyme falls into the Gln family based on the predicted nucleotide-binding fold topology, but current structural information (56) is not sufficient to identify a β-barrel in the C-terminal domain.

Assembly of Class I and Class II Enzymes

The assembly of Class I and Class II structures from individual domains is depicted in Fig. 4. The modular arrangement of functional domains common to both classes of synthetase was first suggested by deletion analysis of the Class I *B. stearothermophilus* tyrosyl-tRNA synthetase (57) and of the Class II *E. coli* alanyl-tRNA synthetase (45–47), where domains could individually be isolated and investigated. From genetic and biochemical studies, the Class II alanyl-tRNA synthetase appears to have a domain organization that resembles the Class I enzymes (11). It has an N-terminal amino acid activation domain, with insertions of sequences that contact the acceptor stem of bound tRNA (58, 59). This domain is followed by another motif that contacts tRNA (60), possibly via the D stem-loop and anticodon stem (61), followed by an oligomerization domain (44–46).

Unlike the Class I enzymes and alanyl-tRNA synthetase, the AspRS cocrystal with tRNA (41) demonstrates that contacts outside of the acceptor stem are mediated by an N-terminal domain, followed by a small domain (containing the motif 1 sequence) that is involved with oligomerization. This in turn is followed by an antiparallel β domain that contains an insertion (or insertions) to contact the acceptor stem of the tRNA. As noted above, the topology of the Class II SerRS resembles AspRS, although the oligomerization domain is formally an insertion occurring after the first β-strand of the antiparallel β domain. The Ser enzyme has an unusual antiparallel α-helical coiled coil domain of 100 amino acids that protrudes from the N terminus of the catalytic domain. Because of its unusual shape, this domain is thought to be involved in tRNA binding. In the order of their functional domains from N to C terminus, the Class II enzymes differ both from the Class I enzymes and from one another. Because of these differences, the Class II synthetases seem in general to be more variable in structure.

The reasons for the different quaternary structures within the two classes of synthetases remain obscure. For some synthetases, oligomerization is apparently required for function. In the *B. stearothermophilus* Tyr enzyme, this functional requirement seems to be due to the binding of the tRNA across the dimer interface (62). However, in at least some instances, the quaternary structure can be manipulated. For example, when the genes encoding the α and β subunits of the Class II glycyl-tRNA synthetase are artificially fused, an active enzyme results (63), and, in a different study, the quaternary structure of *E. coli* MetRS was artificially changed from α_2 to $\alpha_2\beta_2$ (64).

Additional diversity within the synthetases is provided by a variable requirement for divalent zinc. At least four enzymes, the Met, Ile, Trp, and Ala synthetases, have been shown to require this metal (65–68), and several other synthetases have sequence motifs that suggest interactions with divalent zinc (68, 69). Alanyl-tRNA synthetase contains a $CX_2CX_6HX_2H$ "Cys-His box" sequence (68) that is found in the *gag* proteins of retroviruses, and is believed to be important for RNA packaging (70–73).

Additional Functions, Relationships to Other Proteins, and Evolution

Other functions of synthetases include roles in RNA splicing for the mitochondrial TyrRS in *Neurospora crassa* (74) and yeast mitochondrial LeuRS (75), transcriptional (76) and translational control (77–79). In one case, the novel function has been attributed to an additional sequence located at the N terminus of the nucleotide-binding fold (80).

Weak similarities of synthetases with other proteins have been observed in at least three cases. First, the sequence of the CP2 region of *E. coli* LeuRS bears a significant similarity to the leucine-specific binding protein from the same organism (81), suggesting a role for this insertion in amino acid recognition. Second, the 180-kDa GCN2 protein, which regulates amino acid biosynthesis in *S. cerevisiae*, has an extensive (60 kDa) segment that is similar to histidyl-tRNA synthetase (82). Finally, the putative catalytic domain of aspartyl-tRNA synthetase has significant sequence similarity to the ammonia-dependent asparagine synthetase of *E. coli* (83). One region of similarity includes motif 3. Because both enzymes proceed via an adenylate intermediate, this similarity supports the assignment of motif 3 to the active site.

The establishment of two broad classes of synthetases may have occurred early. There is no example of a "class switch" of a synthetase in evolution. The eubacterial (*B. stearothermophilus*, Ref. 84; *Bacillus caldotenax*, Ref. 85; *E. coli*, Ref. 35) and eukaryotic (*S. cerevisiae*[5]; *N. crassa*, Ref. 87) tyrosyl-tRNA synthetase sequences established them to be evolutionarily related Class I enzymes. Similarly, the recently sequenced methionyl-tRNA synthetase from the thermophilic bacterium *Thermus thermophilus* (69) shows that it is a Class I enzyme, as are the methionine enzymes from a mesophilic bacterium (*E. coli*; Ref. 30) and a eukaryote (*S. cerevisiae*; Ref. 88). The isoleucyl-tRNA synthetase sequence from an archaebacterium, *Methanobacterium thermoautotrophicum* (89), shows it to be a Class I synthetase, as are its eubacterial (*E. coli*, Ref. 9) and eukaryotic (*S. cerevisiae*; Ref. 90) counterparts. Additional examples of conservation of the classification are evident from comparisons with four human sequences (91–94). Before a generalization is possible, however, more examples are needed.

Acknowledgments—We thank Prof. Dino Moras (Strasbourg) for providing us with results prior to publication and Drs. D. Moras and S. Cusack for helpful comments on Fig. 2. We also thank Drs. Susan Martinis, W. Todd Miller, and Kiyotaka Shiba for helpful comments on the manuscript.

REFERENCES

1. Neidhart, D. J., Kenyon, G. L., Gerlt, J. A. & Petsko, G. A. (1990) *Nature* **347,** 692–694
2. Eriani, G., Delarue, M., Poch, O., Gangloff, J. & Moras, D. (1990) *Nature* **347,** 203–206
3. Mirande, M. (1991) *Prog. Nucleic Acid Res. Mol. Biol.* **40,** 95–142
4. Jones, D. H., McMillan, A. J., Fersht, A. R. & Winter, G. (1985) *Biochemistry* **24,** 5852–5857
5. Iborra, F., Dorizzi, M. & Labouesse, J. (1973) *Eur. J. Biochem.* **39,** 275–282
6. Cassio, D. & Waller, J. P. (1971) *Eur. J. Biochem.* **20,** 283–300
7. Risler, J.-L., Zelwer, C. & Brunie, S. (1981) *Nature* **292,** 384–386
8. Bhat, T. N., Blow, D. M., Brick, P. & Nyborg, J. (1982) *J. Mol. Biol.* **158,** 699–709
9. Webster, T., Tsai, H., Kula, M., Mackie, G. A. & Schimmel, P. (1984) *Science* **226,** 1315–1317
10. Starzyk, R. M., Webster, T. A. & Schimmel, P. (1987) *Science* **237,** 1614–1618
11. Schimmel, P. (1987) *Annu. Rev. Biochem.* **56,** 125–158
12. Eriani, G., Dirheimer, G. & Gangloff, J. (1989) *Nucleic Acids Res.* **17,** 5725–5736
13. Burbaum, J. J., Starzyk, R. M. & Schimmel, P. (1990) *Proteins: Struct. Funct. Genet.* **7,** 99–111
14. Heck, J. D. & Hatfield, G. W. (1988) *J. Biol. Chem.* **263,** 868–877
15. Breton, R., Sanfacọn, H., Papayannopoulos, I., Biemann, K. & Lapointe, J. (1986) *J. Biol. Chem.* **261,** 10610–10617
16. Hountondji, C., Dessen, P. & Blanquet, S. (1986) *Biochimie* **68,** 1071–1078
17. Gribskov, M., McLachlan, A. D. & Eisenberg, D. (1987) *Proc. Natl. Acad. Sci. U. S. A.* **84,** 4355–4358
18. Webster, T. A., Lathrop, R. H. & Smith, T. F. (1987) *Biochemistry* **26,** 6950–6957
19. Putney, S. D., Royal, N. J., De Vegvar, H. N., Herlihy, W. C., Biemann, K. & Schimmel, P. (1981) *Science* **213,** 1497–1501
20. Anselme, J. & Härtlein, M. (1989) *Gene (Amst.)* **84,** 481–485
21. Eriani, G., Dirheimer, G. & Gangloff, J. (1990) *Nucleic Acids Res.* **18,** 7109–7118
22. Hou, Y.-M., Shiba, K., Mottes, C. & Schimmel, P. (1991) *Proc. Natl. Acad. Sci. U. S. A.* **88,** 976–980
23. Eriani, G., Dirheimer, G. & Gangloff, J. (1991) *Nucleic Acids Res.* **19,** 265–269
24. Hoben, P., Royal, N., Cheung, A., Yamao, F., Biemann, K. & Söll, D. (1982) *J. Biol. Chem.* **257,** 11644–11650
25. Webster, T. A., Gibson, B. W., Keng, T., Biemann, K. & Schimmel, P. (1983) *J. Biol. Chem.* **258,** 10637–10641
26. Freedman, R., Gibson, B., Donovan, D., Biemann, K., Eisenbeis, S. J., Parker, J. & Schimmel, P. (1985) *J. Biol. Chem.* **260,** 10063–10068
27. Härtlein, M. & Madern, D. (1987) *Nucleic Acids Res.* **15,** 10199–10210
28. Kawakami, K., Joensson, Y. H., Bjoerk, G. R., Ikeda, H. & Nakamura, Y. (1988) *Proc. Natl. Acad. Sci. U. S. A.* **85,** 5620–5624
29. Leveque, F., Plateau, P., Dessen, P. & Blanquet, S. (1990) *Nucleic Acids Res.* **18,** 305–312
30. Dardel, F., Fayat, G. & Blanquet, S. (1984) *J. Bacteriol.* **163,** 1115–1122
31. Mechulam, Y., Fayat, G. & Blanquet, S. (1984) *J. Bacteriol.* **163,** 787–791
32. Härtlein, M., Madern, D. & Leberman, R. (1987) *Nucleic Acids Res.* **15,** 1005–1017
33. Mayaux, J. F., Fayat, G., Fromant, M., Springer, M., Grunberg-Manago, M. & Blanquet, S. (1983) *Proc. Natl. Acad. Sci. U. S. A.* **80,** 6152–6156

[5] J. Hill and A. Tzagoloff, personal communication.

34. Hall, C. V., Van Cleemput, M., Muensch, K. H. & Yanofsky, C. (1982) *J. Biol. Chem.* **257,** 6132–6136
35. Barker, D. G., Bruton, C. J. & Winter, G. (1982) *FEBS Lett.* **150,** 419–423
36. Härtlein, M., Frank, D. & Madern, D. (1987) *Nucleic Acids Res.* **15,** 9081–9082
37. Brunie, S., Zelwer, C. & Risler, J.-L. (1990) *J. Mol. Biol.* **216,** 411–424
38. Cusack, S., Berthet-Colominas, C., Härtlein, M., Nassar, N. & Leberman, R. (1990) *Nature* **347,** 249–265
39. Rould, M. A., Perona, J. J., Söll, D. & Steitz, T. A. (1989) *Science* **246,** 1135–1142
40. Brick, P., Bhat, T. N. & Blow, D. M. (1989) *J. Mol. Biol.* **208,** 83–98
41. Ruff, M., Krishnaswamy, S., Boeglin, M., Poterszman, A., Mitschler, A., Podjarny, A., Rees, B., Thierry, J. C. & Moras, D. (1991) *Science* **252,** 1682–1688
42. Rossmann, M. G., Liljas, A., Brändén, C.-I. & Banaszàk, L. J. (1975) *The Enzymes* (Boyer, P. D., ed) 3rd Ed., Vol. 11, pp. 61–102, Academic Press, Orlando, FL
43. Blow, D. M. & Brick, P. (1985) *Nucleic Acids & Interactive Proteins 2: Biological Macromolecules & Assemblies* (Jurnak, F. A. & MacPherson, A., eds) John Wiley & Sons, New York
44. Sixma, T. K., Pronk, S. E., Kalk, K. H., Wartna, E. S., van Zanten, B. A. M., Witholt, B. & Hol, W. G. J. (1991) *Nature* **351,** 371–377
45. Jasin, M., Regan, L. & Schimmel, P. (1983) *Nature* **306,** 441–447
46. Jasin, M., Regan, L & Schimmel, P. (1984) *Cell* **36,** 1089–1095
47. Jasin, M., Regan, L. & Schimmel, P. (1985) *J. Biol. Chem.* **260,** 2226–2232
48. Nishio, K. & Kawakami, M. (1984) *J. Biochem. (Tokyo)* **96,** 1875–1881
49. Lim, W. A. & Sauer, R. T. (1989) *Nature* **339,** 31–36
50. Perona, J. J., Rould, M. A., Steitz, T. A., Risler, J.-L., Zelwer, C. & Brunie, S. (1991) *Proc. Natl. Acad. Sci. U. S. A.* **88,** 2903–2907
51. Kim, S. H., Suddath, F. L., Quigley, G. J., McPherson, A., Sussman, J. L., Wang, A. H. J. & Rich, A. (1974) *Science* **185,** 435–440
52. Robertus, J. D., Ladner, J. E., Finch, J. T., Rhodes, D., Brown, R. S., Clark, B. F. C. & Klug, A. (1974) *Nature* **250,** 546–551
53. Quigley, G. J. & Rich, A. (1976) *Science* **194,** 796–806
54. Holbrook, J. J., Liljas, A., Steindel, S. J. & Rossmann, M. G. (1975) in *The Enzymes* (Boyer, P. D., ed) 3rd Ed., Vol. 11, pp. 191–292, Academic Press, Orlando, FL
55. Lapointe, J., Duplain, L. & Proulx, M. (1986) *J. Bacteriol.* **165,** 88–93
56. Carter, C. W., Jr., Crumley, K. V., Coleman, D. E., Hage, F. & Bricogne, G. (1990) *Acta Crystallogr. A* **46,** 57–68
57. Waye, M. M. Y., Winter, G., Wilkinson, A. J. & Fersht, A. R. (1983) *EMBO J.* **2,** 1827–1829
58. Kill, K. & Schimmel, P. (1989) *Biochemistry* **28,** 2577–2586
59. Miller, W. T., Hou, Y.-M. & Schimmel, P. (1991) *Biochemistry* **30,** 2635–2641
60. Regan, L., Bowie, J. & Schimmel, P. (1987) *Science* **235,** 1651–1653
61. Park, S. J., Miller, W. T. & Schimmel, P. (1990) *Biochemistry* **29,** 9212–9218
62. Ward, W. H. J. & Fersht, A. R. (1988) *Biochemistry* **27,** 5525–5530
63. Toth, M. J. & Schimmel, P. (1986) *J. Biol. Chem.* **261,** 6643–6646
64. Burbaum, J. J. & Schimmel, P. (1991) *Biochemistry* **30,** 319–324
65. Posorske, L. H., Cohn, M., Yanagisawa, N. & Auld, D. S. (1979) *Biochim. Biophys. Acta* **576,** 128–133
66. Mayaux, J. F. & Blanquet, S. (1981) *Biochemistry* **20,** 4647–4654
67. Kisselev, L. L., Favorova, O. O., Nurbekov, M. K., Dmitrienko, S. G. & Engel'gardt, V. A. (1981) *Eur. J. Biochem.* **120,** 511–517
68. Miller, W. T., Kill, K. A. W. & Schimmel, P. (1991) *Biochemistry* **30,** 6970–6976
69. Nureki, O., Muramatsu, T., Suzuki, K., Kohda, D., Matsuzawa, H., Ohta, T., Miyazawa, T. & Yokoyama, S. (1991) *J. Biol. Chem.* **266,** 3268–3277
70. Jentoff, J. E., Smith, L. M., Fu, X., Johnson, M. & Leis, J. (1988) *Proc. Natl. Acad. Sci. U. S. A.* **85,** 7094–7098
71. Gorelick, R. J., Henderson, L. E., Hanser, J. P. & Rein, A. (1988) *Proc. Natl. Acad. Sci. U. S. A.* **85,** 8420–8424
72. Méric, C. & Goff, S. P. (1989) *J. Virol.* **63,** 1558–1568
73. Summers, M. F., South, T. L., Kim, B. & Hare, D. R. (1990) *Biochemistry* **29,** 329–340
74. Cherniack, A. D., Garriga, G., Kittle, J. D., Jr., Akins, R. A. & Lambowitz, A. M. (1990) *Cell* **62,** 745–755
75. Herbert, C. J., Labouesse, M., Dujardin, G. & Slonimski, P. P. (1988) *EMBO J.* **7,** 473–483
76. Putney, S. D. & Schimmel, P. (1981) *Nature* **291,** 632–635
77. Chow, C. M. & RajBhandary, U. L. (1989) *Mol. Cell. Biol.* **9,** 4645–4652
78. Mechulam, Y., Blanquet, S. & Fayat, G. (1987) *J. Mol. Biol.* **197,** 453–470
79. Cheung, A. Y., Watson, L. & Söll, D. (1985) *J. Bacteriol.* **161,** 212–218
80. Lambowitz, A. & Perlman, P. S. (1990) *Trends Biol. Sci.* **15,** 440–444
81. Williamson, R. M. & Oxender, D. L. (1990) *Proc. Natl. Acad. Sci. U. S. A.* **87,** 4561–4565
82. Wek, R. C., Jackson, B. M. & Hinnebusch, A. G. (1989) *Proc. Natl. Acad. Sci. U. S. A.* **86,** 4579–4583
83. Gatti, D. L. & Tzagoloff, A. (1991) *J. Mol. Biol.* **197,** 453–470
84. Winter, G., Koch, G. L. E., Hartley, B. S. & Barker, D. G. (1983) *Eur. J. Biochem.* **132,** 383–387
85. Jones, M. D., Lowe, D. M., Borgford, T. & Fersht, A. R. (1986) *Biochemistry* **25,** 1887–1891
86. Clarke, N. D., Lien, D. C. & Schimmel, P. (1988) *Science* **240,** 521–523
87. Akins, R. A. & Lambowitz, A. M. (1987) *Cell* **50,** 331–345
88. Walter, P., Gangloff, J., Bonnet, J., Boulanger, Y., Ebel, J. P. & Fasiolo, F. (1983) *Proc. Natl. Acad. Sci. U. S. A.* **80,** 2437–2441
89. Jenal, U., Rechsteiner, T., Tan, P.-Y., Bühlmann, E., Meila, L. & Leisinger, T. (1991) *J. Biol. Chem.* **266,** 10570–10577
90. Englisch, U., Englisch, S., Markmeyer, P., Schischkoff, J., Sternbach, H., Kratzin, H. & Cramer, F. (1987) *Biol. Chem. Hoppe-Seyler* **368,** 971–979
91. Fett, R. & Knippers, R. (1991) *J. Biol. Chem.* **266,** 1448–1455
92. Jacobo-Molina, A., Peterson, R. & Yang, D. C. H. (1989) *J. Biol. Chem.* **264,** 16608–16612
93. Tsui, F. W. L. & Siminovitch, L. (1987) *Nucleic Acids Res.* **15,** 3349–3367
94. Cruzan, M. E. & Arfin, S. M. (1991) *J. Biol. Chem.* **266,** 9919–9923

Minireview

THE JOURNAL OF BIOLOGICAL CHEMISTRY
Vol. 266, No. 27, Issue of September 25, pp. 17721-17724, 1991

Printed in U.S.A.

Initiation of Eukaryotic Messenger RNA Synthesis

Joan Weliky Conaway and Ronald C. Conaway

From the Program in Molecular and Cell Biology, Oklahoma Medical Research Foundation, Oklahoma City, Oklahoma 73104

The expression of a large fraction of all protein-coding genes is controlled at the transcriptional level through mechanisms involving the regulation of initiation. In eukaryotic cells, initiation of mRNA synthesis by RNA polymerase II is governed by DNA sequence elements comprising several functional classes. These include a *core promoter element*, which contains the binding site for RNA polymerase II and controls the location of the start site of transcription, and *upstream promoter elements* and *enhancers*, which regulate the rate at which RNA polymerase II initiates new rounds of transcription from the core promoter. These sequence elements direct the action of two classes of transcription factors: *initiation factors*, which are essential for initiation and which are sufficient to direct a basal level of transcription from the core promoter, and *regulatory factors*, which are not required for initiation but which mediate the action of upstream promoter elements and enhancers (1–3).

Analysis of a large number of eukaryotic genes has revealed structural diversity among core promoters. Whereas many are composed of a TATA box, located a short distance upstream of the cap site, and an initiator element, which encompasses the cap site, others, such as the terminal deoxynucleotidyl transferase, dihydrofolate reductase (DHFR),[1] and mouse ribosomal protein rpL30 promoters, lack discernible TATA boxes (4–6). To date, biochemical studies of initiation from the core region of TATA box containing promoters have progressed most rapidly. Initiation factors required for selective transcription of these promoters are being purified and characterized, and a detailed picture of the roles these factors play in assembly of an active initiation complex is emerging. This minireview will summarize recent advances in our understanding of the mechanism by which RNA polymerase II locates and binds selectively to the core region of TATA box containing promoters to form a complex capable of initiating RNA synthesis accurately at the cap site.

RNA Polymerase II

Studies of initiation by RNA polymerase II have lagged behind similar studies of bacterial polymerases. The low abundance of RNA polymerase II in most eukaryotic cells has made both its purification and its characterization difficult. Compared with bacterial RNA polymerases, which can be purified relatively easily in yields as high as 400 mg/kg of cell paste (7), RNA polymerase II is usually obtained in milligram or sub-milligram quantities from equivalent amounts of starting material (8). Until the late 1970s, the lack of DNA templates containing well defined eukaryotic promoters made biochemical studies of gene transcription by RNA polymerase II virtually impossible. Studies of the mechanism of selective transcription by bacterial RNA polymerases, on the other hand, were aided greatly by the availability of bacteriophage templates, such as T7 and λ, which contained genetically defined promoters (9). These limitations have largely been overcome by the isolation and characterization of many eukaryotic promoters and by the development of improved methods for purifying RNA polymerase II.

Over the past several years, genes encoding many of the subunits of RNA polymerase II have been cloned (for reviews, see Refs. 1, 8, 10). In addition, a picture of the overall architecture of the enzyme is emerging from electron crystallographic studies. By analyzing two-dimensional crystals of yeast RNA polymerase II, Darst *et al.* (11) have recently determined the structure of the enzyme to 16 Å (Fig. 1).

Despite these advances, however, the most severe impediment to inquiries into the molecular mechanism of initiation has arisen from the nature of the polymerase itself: purified RNA polymerase II is incapable of binding selectively to its promoter and initiating transcription without the assistance of a set of initiation factors. Unlike bacterial RNA polymerases, which, in most cases, are purified as holoenzymes that include one or more tightly associated initiation factors, RNA polymerase II is separated from its initiation factors by even the gentlest purification procedures.

Multiple Initiation Factors Direct Assembly of an Active Preinitiation Complex

The finding that RNA polymerase II lacks the ability to initiate transcription accurately from promoters prompted a search for factors that mediate this process. Fractionation of transcriptionally active extracts from cultured cells revealed that selective transcription from such TATA box-containing promoters as the AdML, conalbumin, and β-globin promoters requires the action of multiple initiation factors (12–14). Purification of these factors, however, proved to be a formidable undertaking because of their low abundance in most eukaryotic cells. In recent years, the use of more plentiful sources, such as rat liver (15) and yeast (16, 17), have combined with improved cell culture and protein purification techniques to spur rapid progress in elucidating the structures and functions of the initiation factors.

Initiation by RNA polymerase II from the core region of TATA box-containing promoters is a multistage process requiring the action of at least five initiation factors and an ATP cofactor. In the first stage, which we designate *site selection*, RNA polymerase II locates and binds selectively to the core promoter. Site selection is a compound process requiring assembly of a nucleoprotein recognition site for polymerase at the core promoter and binding of polymerase at this site. In this stage, an initiation factor, referred to as the TATA factor, first binds stably to the core promoter to form an Initial Complex. This complex serves as at least part of the recognition site for polymerase. Following site selection, additional initiation factor(s) promote formation of the *complete*, but inactive, *preinitiation complex*. Finally, in a reversible, ATP-dependent step, this intermediate is converted to an *activated complex* capable of initiating transcription rapidly and accurately from the core promoter.

Site Selection

Initial Complex Formation—The first committed step in assembly of the preinitiation complex is binding of the TATA factor to the core promoter to form the Initial Complex (Fig. 2). TATA factors have been identified in cell extracts from a wide variety of species including yeast, *Drosophila*, rat, and man. Although purification of a TATA factor from higher eukaryotes has not yet been reported, a yeast TATA factor, designated yTFIID or BTF1Y, has been purified, and its gene has been cloned (18–25). Both native and recombinant yeast TATA factors substitute for partially purified preparations of native TATA factors from higher eukaryotes in reconstituted transcription reactions *in vitro* (18, 19, 21–25); in addition, a partially purified native TATA factor from HeLa cells will replace the yeast TATA factor in a

[1] The abbreviations used are: DHFR, dihydrofolate reductase; CTD, carboxyl-terminal domain; AdML, adenovirus 2 major late; TF, transcription factor; ATPγS, adenosine 5′-*O*-(thiotriphosphate); STF, stimulatory transcription factor.

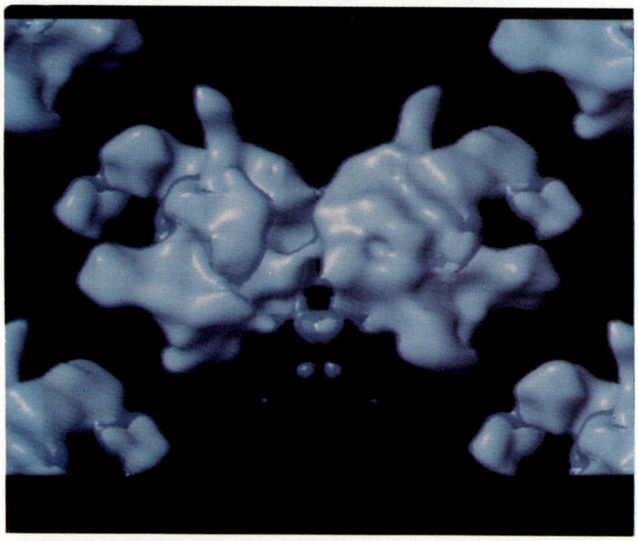

FIG. 1. **Computer-generated model of two molecules of yeast RNA polymerase II (Δ4/7 form, which lacks the fourth and seventh largest subunits), viewed along an axis perpendicular to the plane of the two-dimensional crystal.** The two molecules are related by a 2-fold vertical axis in the plane of the crystal, such that front and back views of the molecules appear side by side. The most striking feature of the structure is an armlike extension that almost completely surrounds a cylindrical channel about 25 Å in diameter. The densities in the corners of the figure are from other molecules in the unit cell. (Computer-generated image courtesy of D. S. Goodsell and A. J. Olson (72); figure courtesy of S. A. Darst and R. D. Kornberg.

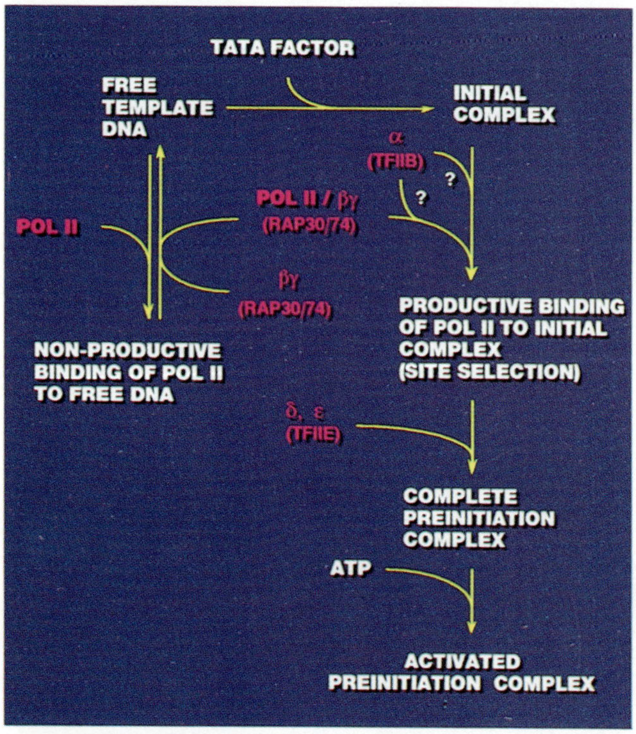

FIG. 2. **Stages in assembly of the activated preinitiation complex.** *pol II*, RNA polymerase II.

yeast transcription system (17). cDNA clones of homologous transcription factors from several sources including *Drosophila*, *Arabidopsis thaliana*, and human cells have now been isolated and shown to function *in vitro* (26–31). The cloned TATA factors from these sources range in size from 22 kDa (*Arabidopsis*) to 38 kDa (*Drosophila* and human).

While studies performed with recombinant TATA factors have provided considerable insight into the mechanism of transcription initiation by RNA polymerase II, it is doubtful that these small proteins will provide ideal models of native TATA factors from higher eukaryotes. A growing body of evidence suggests that the native and recombinant TATA factors are structurally and functionally distinct. Although there is not general agreement on the physical properties of native TATA factors from higher eukaryotes, all are significantly larger than their recombinant counterparts: the rat factor, for example, has an apparent native molecular mass of 750 kDa (32, 33); the *Drosophila* factor is more than 100 kDa (26); and the HeLa cell factor has been reported to be as small as 120 to 140 kDa (34) and as large as 17 S (13). Moreover, results of *in vitro* transcription (33, 35–37) and DNase I footprinting studies (21, 22, 26, 27, 31, 38, 39) argue that native TATA factors interact with a larger region of the core promoter than recombinant TATA factors.

Despite differences between the native and recombinant TATA factors, substantial evidence indicates that each will bind specifically to the core promoter to form the first intermediate on the pathway leading to assembly of the complete preinitiation complex. Several activities that appear to stabilize the Initial Complex have been described. These include factors in fractions of TFIIA (34, 40), [AB] (41), and STF (42) from HeLa cells, TFIIA from yeast, and ε from rat liver.

Selective Binding of RNA Polymerase II to the Initial Complex—Studies of site selection in both rat liver and HeLa cell transcription systems suggest that two transcription factors, designated α and βγ from rat liver or TFIIB and RAP30/74 (TFIIF) from HeLa cells, are required for selective binding of RNA polymerase II to the Initial Complex. α (TFIIB) activity resides in a single 32–35-kDa polypeptide (15, 44). βγ (RAP30/74 (TFIIF)) is a multisubunit protein composed of two polypeptides of approximately 30 and 70 kDa; both polypeptides are required for transcription activity (45–47). Antibodies raised against the small subunit of HeLa cell RAP30/74 (TFIIF) cross-react with the small subunit of transcription factor βγ from rat liver (48).

Early template challenge and kinetic experiments performed with partially fractionated HeLa cell transcription systems suggested (i) that free RNA polymerase II binds directly to Initial Complexes assembled at the AdML promoter and (ii) that other factors needed for initiation enter the preinitiation complex only after RNA polymerase II has bound (34, 41). Interpretation of these experiments was complicated, however, because RNA polymerase II is known to bind stably but nonselectively to free DNA (8). Thus, it was not possible to distinguish between specific interactions of polymerase with the Initial Complex and nonspecific interactions of the enzyme with free template DNA.

An electrophoretic mobility shift (gel-shift) assay was used to obtain evidence that HeLa cell TFIIB is required for binding of RNA polymerase II to the Initial Complex (40). These studies failed to detect a requirement for RAP30/74 (TFIIF) in site selection, however, perhaps because the transcription factor preparations used were relatively crude. Indeed, promoter-specific initiation was not dependent on added RAP30/74 (TFIIF), indicating that this factor had not been resolved from the other initiation factors. Using a highly purified transcription system from rat liver, we have obtained evidence that, in addition to α (TFIIB), transcription factor βγ (RAP30/74(TFIIF)) is required for selective binding of RNA polymerase II to the Initial Complex (32, 48, 49).

How do these two initiation factors direct selective binding of RNA polymerase II to the Initial Complex? Do they bind stably to the Initial Complex to form part of the recognition site for polymerase at the promoter, or do they interact with polymerase and actively promote entry of the enzyme into the preinitiation complex? Evidence from restriction site protection and competition binding experiments argues that, under buffer and ionic strength conditions optimal for transcription, α (TFIIB) from rat liver does not bind stably to the Initial Complex but, instead, actively promotes entry of RNA polymerase II into the preinitiation complex (48). An interaction of HeLa cell α (TFIIB) with the Initial Complex, however, has been observed under gel-shift assay conditions (40, 44), which have been reported to stabilize weak intermolecular interactions (50). It is quite possible, therefore, that this assay reveals weak but specific interactions between α (TFIIB) and the Initial Complex. These interactions may be significantly stronger in the presence of RNA polymerase II and

$\beta\gamma$ (RAP30/74 (TFIIF)) and are likely to play an important role in site selection.

Several lines of evidence support the model that $\beta\gamma$ (RAP30/74 (TFIIF)) promotes site selection through a direct interaction with RNA polymerase II by decreasing the affinity of the enzyme for free DNA and, in concert with α (TFIIB), increasing the affinity of the enzyme for the Initial Complex. First, binding of $\beta\gamma$ (RAP30/74 (TFIIF)) from HeLa cells to RNA polymerase II has been well characterized. This factor was initially purified by its ability to bind to an RNA polymerase II-Sepharose affinity column and was subsequently shown to co-sediment with polymerase in sucrose gradients (51). Second, rat liver $\beta\gamma$ (RAP30/74(TFIIF)) markedly inhibits nonselective binding of RNA polymerase II to free DNA, much as *Escherichia coli* σ^{70} inhibits nonspecific binding of RNA polymerase to nonpromoter sites in DNA (52). In addition, in the absence of α (TFIIB), $\beta\gamma$ (RAP30/74 (TFIIF)) inhibits nonspecific binding of RNA polymerase II to templates containing preassembled Initial Complexes (32). Third, restriction site protection experiments indicate that RNA polymerase II-dependent protection of AdML promoter sequences between the TATA box and cap site requires $\beta\gamma$ (RAP30/74 (TFIIF)), in addition to the TATA factor and α (TFIIB) (48), arguing that $\beta\gamma$ (RAP30/74 (TFIIF)) is required for selective binding of RNA polymerase II to the Initial Complex.

Other transcription factors that interact with RNA polymerase II and may play a role in site selection have been described. An activity designated Factor 5, which resembles $\beta\gamma$ (RAP30/74(TFIIF)), has been purified from *Drosophila* Kc cells and shown to bind to RNA polymerase II in solution (53). In addition, a 27-kDa transcription factor, designated BTF3, which binds to RNA polymerase II *in vitro*, was purified from HeLa cells, and its cDNA was cloned (54, 55).

Assembly and Activation of the Complete Preinitiation Complex

The Complete Preinitiation Complex—Once RNA polymerase II has bound selectively to the Initial Complex, additional initiation factors promote formation of the fully assembled, or *complete*, preinitiation complex. Two factors from rat liver, designated ϵ and δ, act at this stage. Competition binding experiments suggest that both factors ultimately become integral components of the complete preinitiation complex (49).[2] ϵ and δ appear to enter the preinitiation complex by different routes: ϵ only after RNA polymerase II has bound to the Initial Complex and δ either before or after RNA polymerase II has bound (48).[2]

Restriction site protection experiments indicate that ϵ and δ promote formation of stable protein-DNA contacts that anchor the transcription apparatus to promoter sequences near the cap site (48). These protein-DNA contacts may result from direct interactions of ϵ, δ, or both factors with promoter DNA; alternatively, ϵ or δ may stabilize interactions of RNA polymerase II and the other initiation factors with the promoter.

ϵ is a heterodimer of 34- and 58-kDa polypeptides; both polypeptides are essential for transcription activity (33). A similar factor, designated TFIIE, has been purified from HeLa cells and, like ϵ, appears to function late in assembly of the preinitiation complex. TFIIE is composed of 34- and 57-kDa polypeptides; unlike ϵ, however, the large subunit of TFIIE exhibits significant transcription activity in the absence of the small subunit (56, 57). A chromatographic fraction containing both TFIIF and TFIIE promotes formation of stable protein-DNA contacts near the cap site of the AdML promoter during the final step in preinitiation complex formation (40). TFIIE appears to interact with RNA polymerase II in solution (58), but the functional relevance of this interaction has not yet been demonstrated.

δ has been purified to near-homogeneity from rat liver nuclear extracts (59).[2] The purified factor exhibits a native molecular mass of approximately 390 kDa and has an associated DNA-dependent ATPase activity. ATPase and transcription activities co-purify with a set of eight polypeptides ranging in size from approximately 90 to 35 kDa. Reconstitution of δ from isolated polypeptides has not yet been achieved; thus, it remains to be determined whether each of these polypeptides is essential for transcription or whether some are derived from others by proteolysis. δ has no obvious counterparts in other transcription systems.

Formation of the Activated Preinitiation Complex—Transcription initiation from the core promoter has a strict requirement for an adenine nucleoside triphosphate cofactor (60–62). Several lines of evidence argue that this cofactor is required for conversion of the complete preinitiation complex to an active configuration. First, assembly of the complete preinitiation complex does not require ATP. Second, ATP is not required after synthesis of the first nine nucleotides of transcripts initiated at the AdML promoter (61). Finally, work in our laboratory (62) provided evidence that ATP functions immediately prior to RNA synthesis. We observed that inhibition of transcription by ATPγS, a potent inhibitor of the ATP-dependent step, could be prevented by brief incubation of complete preinitiation complexes with ATP or ATP analogs prior to addition of ATPγS and the ribonucleoside triphosphates needed for RNA synthesis. Further experiments indicated that the ATP-dependent step is rapidly reversible, suggesting that, in the presence of ATP, active and inactive forms of the preinitiation complex are in dynamic equilibrium.

It is not yet known how ATP activates the preinitiation complex or which components of the transcription system mediate activation. ATP may, for example, be hydrolyzed to provide energy for some crucial step in initiation or to serve as a phosphate donor in a phosphorylation reaction. On the other hand, ATP may simply bind one or more components of the transcription system to promote transition of the complete preinitiation complex to a transcriptionally active conformation.

A role for HeLa cell transcription factor RAP30/74 (TFIIF) in activation of the preinitiation complex has been proposed. Sopta and co-workers (63) reported that RAP30/74 (TFIIF) possesses an associated ATP(dATP)-dependent DNA helicase activity and suggested that this helicase might unwind the promoter to facilitate formation of an active preinitiation complex analogous to the "open" complex formed by *E. coli* RNA polymerase (86). Others (47), however, have reported that highly purified preparations of this factor lack detectable helicase activity.

The possibility that a protein kinase is involved in activation of the preinitiation complex has been investigated in light of evidence suggesting that phosphorylation of the carboxyl-terminal domain (CTD) of the largest subunit of RNA polymerase II plays a role in transcription. The CTD is rich in serine and threonine residues, which are extensively phosphorylated *in vivo* (64, 65). UV cross-linking studies indicate that, in isolated nuclei as well as in crude transcription extracts, the CTDs of polymerase molecules actively engaged in transcription are highly phosphorylated (66, 67). Recently, several groups have obtained evidence that RNA polymerase II, bound in preinitiation complexes at the AdML promoter, can be phosphorylated in the CTD by a template-associated protein kinase, immediately before or during initiation (68, 69). The template-associated protein kinase(s) has not been isolated, and a requirement for it in transcription has not yet been demonstrated directly. Moreover, interpretation of these results is complicated by the observation that RNA polymerase II lacking all or most of the CTD is capable of catalyzing accurate initiation from the AdML promoter *in vitro* (70, 71). In addition, it has been observed that, in the absence of ATP, GTP can serve as a substrate for a template-associated kinase, raising the possibility that phosphorylation of the CTD is not required for activation of the preinitiation complex but, instead, plays some role in modulating the activity of RNA polymerase II (68).

Finally, several lines of evidence suggest that, in the rat liver system, transcription factor δ could mediate activation of the preinitiation complex (59). Highly purified δ has a closely associated DNA-dependent ATPase activity, specific for ATP or dATP. The associated ATPase of δ is stimulated more strongly by DNA fragments containing the AdML or mouse interleukin-3 core promoters than by fragments containing non-promoter sequences, consistent with the notion that δ may function through a direct interaction with promoter sequences. Of the rat liver

[2] R. C. Conaway and J. W. Conaway, unpublished results.

factors required for assembly of the complete preinitiation complex, only δ is associated with measurable ATPase activity. Despite this circumstantial evidence, a direct demonstration that δ mediates activation of the preinitiation complex is lacking.

Prospects for the Future

Biochemical studies of initiation at simple, TATA box-containing promoters have provided information on the structures and activities of essential initiation factors as well as on the mechanism by which RNA polymerase II binds selectively to these promoters to form the active preinitiation complex. Future research should lead to a more detailed understanding of the roles the initiation factors play in assembly and activation of the preinitiation complex. In addition, it is likely that biochemical studies of initiation at other promoters will lead to the discovery of novel initiation factors and mechanisms. Finally, the ability to reconstitute initiation in highly purified and well characterized transcription systems should lead ultimately to an understanding of the mechanisms underlying transcriptional regulation by proteins that interact with upstream promoter elements and enhancers.

Acknowledgments—We thank Drs. I. Robert Lehman and Philip Silverman for critical comments on this manuscript. We are grateful to Drs. Roger Kornberg and Seth Darst for Fig. 1. We also thank Richard Irish and Jeff Box for artwork.

REFERENCES

1. Sawadogo, M., and Sentenac, A. (1990) *Annu. Rev. Biochem.* **59,** 711–754
2. Johnson, P. F., and McKnight, S. L. (1989) *Annu. Rev. Biochem.* **58,** 799–839
3. Struhl, K. (1989) *Annu. Rev. Biochem.* **58,** 1051–1077
4. Smale, S. T., and Baltimore, D. (1989) *Cell* **57,** 103–113
5. Means, A. L., and Farnham, P. J. (1990) *Mol. Cell. Biol.* **10,** 653–661
6. Hariharan, N., Kelley, D. E., and Perry, R. B. (1989) *Genes & Dev.* **3,** 1789–1800
7. Burgess, R. R., and Jendrisak, J. J. (1975) *Biochemistry* **14,** 4634–4638
8. Lewis, M. K., and Burgess, R. R. (1982) in *The Enzymes* (Boyer, P. D., ed) Vol. XV, pp. 109–153, Academic Press, Orlando, FL
9. Chamberlin, M. J. (1974) *Annu. Rev. Biochem.* **43,** 721–775
10. Woychik, N. A., and Young, R. A. (1990) *Trends Biochem. Sci.* **15,** 347–351
11. Darst, S. A., Edwards, A. M., Kubalek, E. W., and Kornberg, R. D. (1991) *Cell* **66,** 121–128
12. Matsui, T., Segall, J., Weill, P. A., and Roeder, R. G. (1980) *J. Biol. Chem.* **255,** 11992–11996
13. Samuels, M., Fire, A., and Sharp, P. A. (1982) *J. Biol. Chem.* **257,** 14419–14427
14. Davison, B. L., Egly, J. M., Mulvihill, E. R., and Chambon, P. (1983) *Nature* **301,** 680–686
15. Conaway, J. W., Bond, M. W., and Conaway, R. C. (1987) *J. Biol. Chem.* **262,** 8293–8297
16. Lue, N. F., and Kornberg, R. D. (1987) *Proc. Natl. Acad. Sci. U. S. A.* **84,** 8839–8843
17. Flanagan, P. M., Kelleher, R. J., Feaver, W. J., Lue, N. F., LaPointe, J. W., and Kornberg, R. D. (1990) *J. Biol. Chem.* **265,** 11105–11107
18. Buratowski, S., Hahn, S., Sharp, P. A., and Guarente, L. (1988) *Nature* **334,** 37–42
19. Cavallini, B., Huet, J., Plassat, J. L., Sentenac, A., Egly, J. M., and Chambon, P. (1988) *Nature* **334,** 77–80
20. Eisenmann, D. M., Dollard, C., and Winston, F. (1989) *Cell* **58,** 1183–1191
21. Hahn, S., Buratowski, S., Sharp, P. A., and Guarente, L. (1989) *Cell* **58,** 1173–1181
22. Horikoshi, M., Wang, C. K., Fujii, H., Cromlish, J. A., Weil, P. A., and Roeder, R. G. (1989) *Proc. Natl. Acad. Sci. U. S. A.* **86,** 4843–4847
23. Horikoshi, M., Wang, C. K., Fujii, H., Cromlish, J. A., Weil, P. A., and Roeder, R. G. (1989) *Nature* **341,** 299–303
24. Cavallini, B., Faus, I., Matthes, H., Chipoulet, J. M., Winsor, B., and Chambon, P. (1989) *Proc. Natl. Acad. Sci. U. S. A.* **89,** 9803–9807
25. Schmidt, M. C., Kao, C. C., Pei, R., and Berk, A. J. (1989) *Proc. Natl. Acad. Sci. U. S. A.* **86,** 7785–7789
26. Muhich, M. L., Iida, C. T., Horikoshi, M., Roeder, R. G., and Parker, C. S. (1990) *Proc. Natl. Acad. Sci. U. S. A.* **87,** 9148–9152
27. Hoey, T., Dynlacht, B. D., Peterson, M. G., Pugh, B. F., and Tjian, R. (1990) *Cell* **61,** 1179–1186
28. Gasch, A., Hoffmann, A., Horikoshi, M., Roeder, R. G., and Chua, N.-H. (1990) *Nature* **346,** 390–394
29. Kao, C. C., Lieberman, P. M., Schmidt, M. C., Zhou, Q., Pei, R., and Berk, A. J. (1990) *Science* **248,** 1646–1650
30. Hoffmann, A., Sinn, E., Yamamoto, T., Wang, J., Roy, A., Horikoshi, M., and Roeder, R. G. (1990) *Nature* **346,** 387–390
31. Peterson, M. G., Tanese, N., Pugh, B. F., and Tjian, R. (1990) *Science* **248,** 1625–1630
32. Conaway, J. W., Reines, D., and Conaway, R. C. (1990) *J. Biol. Chem.* **265,** 7552–7558
33. Conaway, J. W., Hanley, J. P., Garrett, K. P., and Conaway, R. C. (1991) *J. Biol. Chem.* **266,** 7804–7811
34. Reinberg, D., Horikoshi, M., and Roeder, R. G. (1987) *J. Biol. Chem.* **262,** 3322–3330
35. Conaway, J. W., Travis, E., and Conaway, R. C. (1990) *J. Biol. Chem.* **265,** 7564–7579
36. Smale, S. T., Schmidt, M. C., Berk, A. J., and Baltimore, D. (1990) *Proc. Natl. Acad. Sci. U. S. A.* **87,** 4509–4513
37. Nakatani, Y., Horikoshi, M., Brenner, M., Yamamoto, T., Besnard, F., Roeder, R. G., and Freese, E. (1990) *Nature* **348,** 86–88
38. Parker, C. S., and Topol, J. (1984) *Cell* **36,** 357–369
39. Nakajima, N., Horikoshi, M., and Roeder, R. G. (1988) *Mol. Cell Biol.* **8,** 4028–4040
40. Buratowski, S., Hahn, S., Guarente, L., and Sharp, P. A. (1989) *Cell* **56,** 549–561
41. Fire, A., Samuels, M., and Sharp, P. A. (1984) *J. Biol. Chem.* **259,** 2509–2516
42. Egly, J. M., Miyamoto, N. G., Moncollin, V., and Chambon, P. (1984) *EMBO J.* **3,** 2363–2371
43. Hahn, S., Buratowski, S., Sharp, P. A., and Guarente, L. (1989) *EMBO J.* **8,** 3379–3382
44. Maldonado, E., Ha, I., Cortes, P., Weis, L., and Reinberg, D. (1990) *Mol. Cell. Biol.* **10,** 6335–6347
45. Conaway, J. W., and Conaway, R. C. (1989) *J. Biol. Chem.* **264,** 2357–2362
46. Burton, Z. F., Ortolan, L. C., and Greenblatt, J. (1986) *EMBO J.* **5,** 2923–2930
47. Flores, O., Ha, I., and Reinberg, D. (1990) *J. Biol. Chem.* **265,** 5629–5634
48. Conaway, R. C., Garrett, K. P., Hanley, J. P., and Conaway, J. W. (1991) *Proc. Natl. Acad. Sci. U. S. A.* **88,** 6205–6209
49. Conaway, R. C., and Conaway, J. W. (1990) *J. Biol. Chem.* **265,** 7559–7563
50. Fried, M., and Crothers, D. M. (1981) *Nucleic Acids Res.* **9,** 6505–6524
51. Sopta, M., Carthew, R. W., and Greenblatt, J. (1985) *J. Biol. Chem.* **260,** 10353–10361
52. Conaway, J. W., and Conaway, R. C. (1990) *Science* **248,** 1550–1553
53. Price, D. H., Sluder, A. E., and Greenleaf, A. L. (1989) *Mol. Cell. Biol.* **9,** 1465–1475
54 Zheng, X. M., Moncollin, V., Egly, J. M., and Chambon, P. (1987) *Cell* **50,** 361–368
55. Zheng, X. M., Black, D., Chambon, P., and Egly, J. M. (1990) *Nature* **344,** 556–559
56. Ohkuma, Y., Sumimoto, H., Horikoshi, M., and Roeder, R. (1990) *Proc. Natl. Acad. Sci. U. S. A.* **87,** 9163–9167
57. Inostroza, J., Flores, O., and Reinberg, D. (1991) *J. Biol. Chem.* **266,** 9304–9308
58. Flores, O., Maldonado, E., and Reinberg, D. (1989) *J. Biol. Chem.* **264,** 8913–8921
59. Conaway, R. C., and Conaway, J. W. (1989) *Proc. Natl. Acad. Sci. U. S. A.* **86,** 7356–7360
60. Bunick, D., Zandomeni, R., Ackerman, S., and Weinmann, R. (1982) *Cell* **29,** 877–886
61. Sawadogo, M., and Roeder, R. G. (1984) *J. Biol. Chem.* **259,** 5321–5326
62. Conaway, R. C., and Conaway, J. W. (1988) *J. Biol. Chem.* **263,** 2962–2968
63. Sopta, M., Burton, Z. F., and Greenblatt, J. (1989) *Nature* **341,** 410–414
64. Corden, J. L. (1990) *Trends Biochem. Sci.* **15,** 383–387
65. Cadena, D. L., and Dahmus, M. E. (1987) *J. Biol. Chem.* **262,** 12468–12474
66. Bartholomew, B., Dahmus, M. E., and Meares, C. F. (1986) *J. Biol. Chem.* **261,** 14226–14231
67. Payne, J. M., Laybourn, P. J., and Dahmus, M. E. (1989) *J. Biol. Chem.* **264,** 19621–19629
68. Laybourn, P. J., and Dahmus, M. E. (1990) *J. Biol. Chem.* **265,** 13165–13173
69. Arias, J. A., Peterson, S. R., and Dynan, W. S. (1991) *J. Biol. Chem.* **266,** 8055–8061
70. Zehring, W. A., Lee, J. M., Weeks, J. R., Jokerst, R. S., and Greenleaf, A. L. (1988) *Proc. Natl. Acad. Sci. U. S. A.* **85,** 3698–3702
71. Kim, W. Y., and Dahmus, M. E. (1989) *J. Biol. Chem.* **264,** 3169–3176
72. Goodsell, D. S., Miam, I. S., and Olson, A. J. (1989) *J. Mol. Graphics* **7,** 41–47

THE JOURNAL OF BIOLOGICAL CHEMISTRY
Vol. 266, No. 28, Issue of October 5, pp. 18411-18414, 1991
Printed in U.S.A.

The Papillomavirus E2 Regulatory Proteins

Alison A. McBride‡, Helen Romanczuk, and Peter M. Howley

From the Laboratory of Tumor Virus Biology, National Cancer Institute, Bethesda, Maryland 20892

Papillomavirus Life Cycle

The papillomaviruses are small DNA viruses that induce benign proliferative squamous epithelial and fibroepithelial lesions (warts and papillomas) in their natural hosts. They have been isolated from a variety of animal species, and over 60 human papillomavirus types have been described so far (reviewed in Ref. 1). The papillomavirus life cycle is closely linked to keratinocyte differentiation. In the basal epithelial cells and dermal fibroblasts of a fibropapilloma the viral genome is maintained as a multicopy episome, and only "early" viral genes are expressed. Vegetative DNA replication, late gene expression, and virion production are restricted to differentiated squamous epithelial cells.

Bovine papillomavirus type 1 (BPV-1)[1] has served as the prototype for the genetic analysis of the papillomavirus functions. BPV-1 is able to transform certain rodent cell lines; within these cells the viral genome is maintained as an extrachromosomal element, and only "early" gene products are expressed. This latent, non-productive infection is thought to be analogous to the virus-host interaction characteristic of the infected basal epithelial cells or dermal fibroblasts of a fibropapilloma. Transformation of rodent cells by BPV-1 has enabled functions important for viral transformation, replication, and transcriptional regulation to be mapped. Such studies have revealed that products encoded by the BPV-1 E5 and E6 genes are required for full transformation, E1 products are necessary for viral DNA replication, and E2 polypeptides function both in replication and transcriptional regulation (reviewed in Refs. 2 and 3).

The genomic organization of the human papillomaviruses (HPVs) is very similar to BPV-1 (see Fig. 1). Extensive studies have shown that the E6 and E7 proteins from those HPV types that have been associated with cancer are able to immortalize human squamous epithelial cells. The HPV E2 genes also encode transcriptional regulatory protein(s). Little is known specifically about HPV DNA replication, but, by analogy with BPV-1, it is likely that proteins encoded by the E1 and E2 ORFs are also important for this function (reviewed in Ref. 4).

E2 Polypeptides and Transcriptional Regulation

Mutations in the BPV-1 E2 ORF are pleiotropic, disrupting transformation, replication, and transcriptional regulation (5–9). Genetic analysis of BPV-1 demonstrated that the E2 ORF encodes a transactivator which interacts directly with specific enhancer elements within the long control region (LCR) resulting in increased transcription from viral promoters P_{7940}, P_{89}, P_{2443}, and P_{3080} (10–17). The E2 ORF also encodes two transcriptional repressors that function through the same enhancer elements but inhibit E2-dependent transactivation (18–20). One of these repressors, E2-TR, is expressed from the P_{3080} promoter and translated from an internal methionine at amino acid 162 of the E2 ORF. Another repressor, E8/E2, is encoded by a spliced message that encodes 11 amino acids from the E8 ORF linked to the C-terminal 207 amino acids of E2 via the splice acceptor at nucleotide 3225. Three E2 polypeptides, with molecular masses of 48, 31, and 28 kDa, have been identified in BPV-1 transformed cells (21). These correspond to the E2, E2-TR, and E8/E2 proteins, respectively (19). The relative abundance of repressor over transactivator proteins in transformed cells (1 E2:10 E2-TR:3 E8/E2) is thought to result in a low level of viral gene expression. As described below, in certain circumstances the full-length E2 transactivator is also able to repress transcription (22, 23).

Less is known about the HPV E2 polypeptides. All of the HPVs so far sequenced encode an E2 polypeptide homologous to the transactivator. However, although cDNAs have been isolated which could potentially encode spliced or shorter E2 species, such polypeptides have not been specifically identified. The distribution of E2 DNA-binding sites in the HPV genomes differs from the pattern observed in fibropapillomaviruses such as BPV-1 (Fig. 1). In the cancer-associated human papillomaviruses HPV-16 and HPV-18, the major promoter (P_{97} and P_{105}, respectively) is negatively regulated by E2 (24–26). This repression is mediated through two E2-binding sites immediately adjacent to the TATA box of the promoters (26, 27). Occupation of these sites by E2 is thought to interfere with assembly of the transcription preinitiation complex (see Fig. 3*A* and below). E2 regulation of the E6 promoter of HPV-11 also provides evidence for both positive and negative transcriptional modulation (28, 29).

In the majority of HPV-containing cervical carcinomas the viral genome is often integrated such that the E1 and/or E2 ORFs are disrupted. It has been postulated that in the absence of these polypeptides, the E6 and E7 oncoproteins are expressed in a deregulated manner which in turn may contribute to tumor development. In support of this model, it has been shown that mutation of the promoter proximal E2-binding sites in the HPV-16 genome results in a slight but reproducible increase in the ability of the viral DNA to immortalize human keratinocytes. Moreover, mutations which disrupt the HPV-16 E1 or E2 ORF cause a further augmentation of the immortalization potential, suggesting that these ORFs play additional roles in regulating the levels or activities of the viral oncoproteins.[2]

Structural and Functional Domains of the E2 Polypeptide

A comparison of the predicted amino acid sequences of a number of papillomavirus E2 ORFs reveals that these polypeptides probably consist of three domains (30, 31). Two of the domains, consisting of about 220 amino acids at the N terminus and 90–100 amino acids at the C terminus, are relatively well conserved among papillomaviruses. However, the internal region varies both in length and amino acid sequence (Fig. 2).

The N-terminal region, which is unique to the full-length transactivator protein, encodes the activation domain (32–34). This domain is large compared with many other transactivators, and this may reflect the multifunctional role of E2 in transcription and replication. The activation domain contains two regions within the first 85 amino acids that can be predicted to form acidic amphipathic helices. These structures have been shown to be important for the activation function of a number of eukaryotic transactivators (35). Deletion analysis of BPV-1 E2 demonstrated that the region containing the potential amphipathic helices is necessary but not sufficient for transactivation (34). DNA-bound dimers of the BPV-1 transactivator protein can interact *in vitro* to form DNA loops, and this association requires the N-terminal domain of E2 (36). In addition, as described below, the transactivator can associate with the BPV-1 E1 polypeptide and is necessary for viral DNA replication (37–39). It is likely that sequences in the N-terminal domain are required because the E2

‡ To whom correspondence should be addressed. Tel.: 301-496-9486.

[1] The abbreviations used are: BPV-1, bovine papillomavirus type 1; HPV, human papillomavirus; ORF, open reading frame; LCR, long control region.

[2] H. Romanczuk and P. M. Howley, submitted for publication.

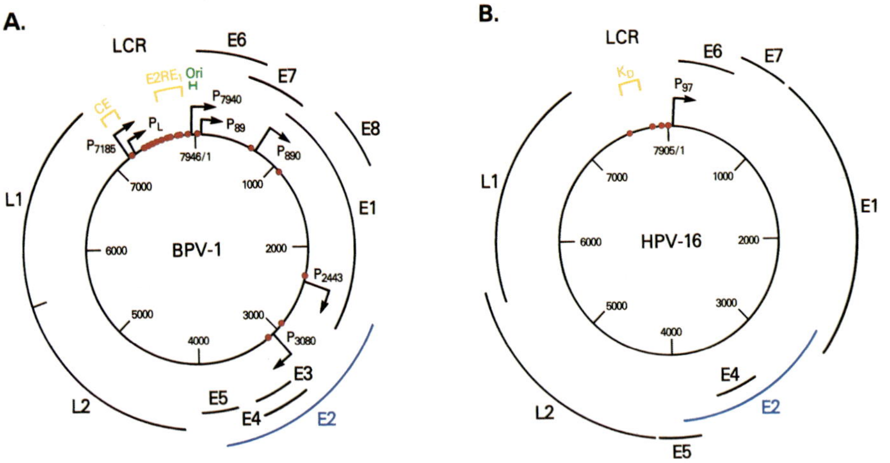

FIG. 1. **A and B, genomic organization of BPV-1 and HPV-16 DNAs.** The nucleotide numbers are shown within the circular map, and the major ORFs (designated E1-E8, L1, and L2) are shown. Promoters are indicated by *P* followed by the approximate position of the RNA start site, except for the late promoter P_L. The ACCN$_6$GGT motifs are represented by *red circles*. In BPV-1 the E2-dependent enhancer, E2RE$_1$, and a constitutive enhancer, CE, are indicated in *yellow* (13, 23). The minimal replication origin, *Ori*, is shown in *green*. In HPV-16, a cell-specific keratinocyte enhancer, K_D, is shown in *yellow* (71).

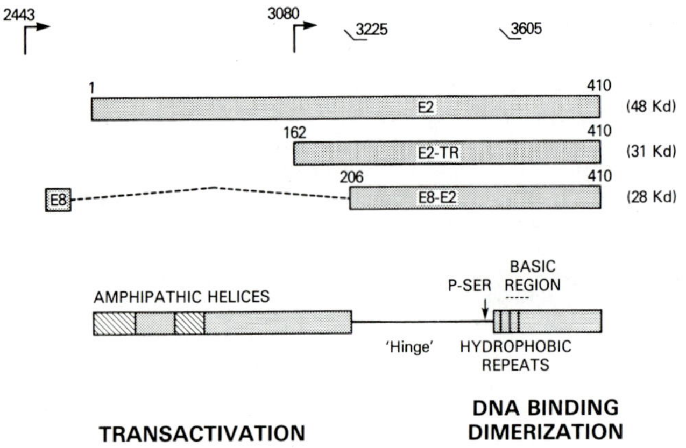

FIG. 2. **Structure of the BPV-1 E2 gene products.** A region of approximately 200 amino acids at the N terminus is relatively well conserved among the papillomaviruses and encodes the activation domain. This domain contains two regions predicted to form acidic amphipathic helices. The 85 C-terminal amino acids are also conserved and comprise the DNA-binding and dimerization domain. The basic region and hydrophobic repeats described in the text are indicated. The molecular masses of the polypeptides are shown to the *right*.

repressors cannot interact with E1 or support viral replication.

The DNA binding function is encoded by the C-terminal 85 amino acids of the BPV-1 E2 ORF, and this domain is present in all three E2 polypeptides (40–42). The E2 proteins bind to a single DNA-binding site as dimers, and dimerization is mediated through the DNA-binding domain (34, 43, 44). This C-terminal domain does not contain any obvious consensus DNA binding motifs. Some E2 proteins contain several leucine residues spaced at seven-amino acid intervals, similar to the leucine zipper dimerization structure; others contain hydrophobic residues at most of these positions. However, this region is not predicted to form an α-helix, and we have therefore designated these residues the hydrophobic heptad repeats. Overlapping the repeats is a conserved stretch of basic amino acids which likely interacts directly with the E2 DNA-binding site (Fig. 2). E2 has been demonstrated to show some cooperativity in DNA binding, and this property requires some unidentified sequences outside of the minimal DNA-binding domain (42).

The non-conserved internal region of the E2 polypeptide is designated the hinge region. It does not seem to be required for transcriptional transactivation in transient transfection assays, and may simply provide a flexible link between the DNA-binding and activation domains (33, 34). This region overlaps with the E3 and E4 ORFs, and it is possible that selective pressure on these overlapping ORFs has resulted in sequence divergence of the E2 ORF. However, as described below, this region contains two major serine phosphorylation sites of the BPV-1 E2 proteins (45).

E2 DNA-binding Sites

The E2 proteins bind specifically to a 12-base pair palindromic sequence, ACCN$_6$GGT, which is repeated several times in the papillomavirus genomes (11, 12, 46). BPV-1 has 12 sites that correspond to this consensus and an additional five closely related sequences that are also bound by E2 (47). The majority of these sites are located in the LCR, but there are several situated close to the promoters in the early region (Fig. 1). The affinity of the BPV-1 E2 protein for the various DNA-binding sites ranges over several orders of magnitude. The higher affinity sites are clustered in the LCR in regions that have been designated E2-dependent enhancer elements E2RE$_1$ and E2RE$_2$ (13).

A low level of E2-dependent activation can be obtained with one E2 binding motif in *Saccharomyces cerevisiae* and in mammalian cells. However, two E2 DNA-binding sites cooperate to constitute a strong enhancer (48–52). In enhancer elements such as E2RE$_1$, multiple binding sites cooperate to form a highly responsive element (53). Some cooperativity in DNA binding has been observed *in vitro* with the BPV-1 E2 protein, but this is not sufficient to explain the strong synergism seen *in vivo*, suggesting that additional mechanisms are probably involved (36, 42, 48).

E2 transactivates the P_{2443} and P_{3080} promoters located in the early region, and activation depends on the E2-dependent enhancers in the LCR (15–17). It was postulated that the E2-binding sites adjacent to these promoters might cooperate with those in the LCR to bring about activation. DNA-bound E2 transactivator molecules have been demonstrated to associate and form DNA loops, providing a model as to how activation could occur at a distance (36). Alternatively, binding of E2 to these sites could repress expression from the promoters in a manner analogous to the situation with the HPV-16 P_{97} and HPV-18 P_{105} promoters. However, mutation of the E2 site adjacent to the P_{2443} promoter (which has the lowest affinity of all the BPV-1 sites) has no effect on E2 regulation of this promoter (17, 54), and mutation of the two E2 sites adjacent to the P_{3080} promoter results in only a minimal decrease in E2-dependent promoter activity (17). Nevertheless, it is possible that these sites could have an important regulatory role under different circumstances. For example, the DNA binding specificity of E2 for the various motifs could change during the cell cycle, with the state of cellular differentiation, or at a particular stage of the viral life cycle.

The number and distribution of the E2-binding sites differ in the human papillomaviruses as compared with the animal papillomaviruses. In the genital papillomaviruses HPV-16 and HPV-18, only four E2-binding sites have been identified, all of which are located in the LCR (Fig. 1). As noted above, two of these sites are situated immediately upstream from the respective TATA boxes of the E6/E7 viral promoters (designated P_{97} in HPV-16 and P_{105} in HPV-18), and binding of E2 to these sites represses rather than activates transcription from these promoters (24–27). Mutation of these proximal E2-binding sites so that they no longer bind E2 results in E2-mediated activation of transcription of the P_{97} or P_{105} promoters through interaction with the remaining upstream E2-binding sites (26, 27). These results indicate that E2 modulation of HPV transcription can involve both positive and negative regulation. The final outcome may be a function of the different binding affinities of E2 for each of the sites (27, 55). Alternatively, E2 may positively regulate some other, as yet unidentified, HPV promoters.

The precise contact points of the E2 proteins on the $ACCN_6GGT$ motif have been determined (41, 55). The protein contacts the guanosine residues on either strand of the motif, which are predicted to be present on the same face of the DNA helix. The nucleotides in the non-conserved core and immediately outside the binding site determine the affinity for the E2 protein. Binding of both BPV-1 and HPV-16 E2 proteins has been shown to bend DNA, and the HPV-16 E2 preference for an A-T rich core sequence may reflect the flexibility of such a sequence (55, 56).

Mechanisms of E2 Transcriptional Regulation

As mentioned above, there are several ways in which the E2 proteins can either activate or repress transcription, depending on the position of the E2-binding sites and the nature of the E2 polypeptides. These are shown diagrammatically in Fig. 3. In one case, such as for the BPV-1 P_{7940} and P_{89} promoters, binding of the full-length E2 transactivator to a number of sites located upstream from the promoter activates transcription (13, 14). E2 can also stimulate transcription from heterologous promoters which contain several E2-binding sites located at some distance either upstream or downstream from the promoter (51). In these cases, the E2 sites behave like classical conditional enhancer elements. However, the E2 transactivator can also repress transcription when the E2 motifs are situated close to the transcriptional start site, overlapping the binding sites for essential cellular transcription factors. In the case of HPV-16 (and HPV-18), two E2 sites are located just upstream from the TATA box of the P_{97} promoter, and binding of E2 to these sites may sterically hinder the binding of the cellular factor, TFIID, and/or interfere with the assembly of the transcriptional initiation complex (26, 27, 57). There is also an essential Sp1-binding site located just upstream from the E2 motifs, but it is not known whether occupation of the E2 sites interferes with binding of Sp1 to this region of the promoter (58). In a similar situation, an E2-binding site located downstream from the BPV-1 P_{7185} promoter overlaps an Sp1-binding site. This Sp1 site is essential for P_{7185} activity, and binding of E2 polypeptides can repress expression from this promoter, probably by interfering with Sp1 binding (22, 23). Finally, as described above, the BPV-1 P_{2443} promoter has a low affinity E2-binding site located just upstream from the TATA box, but this site does not appear to play a role in the regulation of this promoter. This E2 site is adjacent to an Sp1 site which is critical for both basal and E2-dependent promoter activity, and E2 transactivation of this promoter depends on the E2 sites located 2.5 kilobases upstream in the LCR (17, 54).

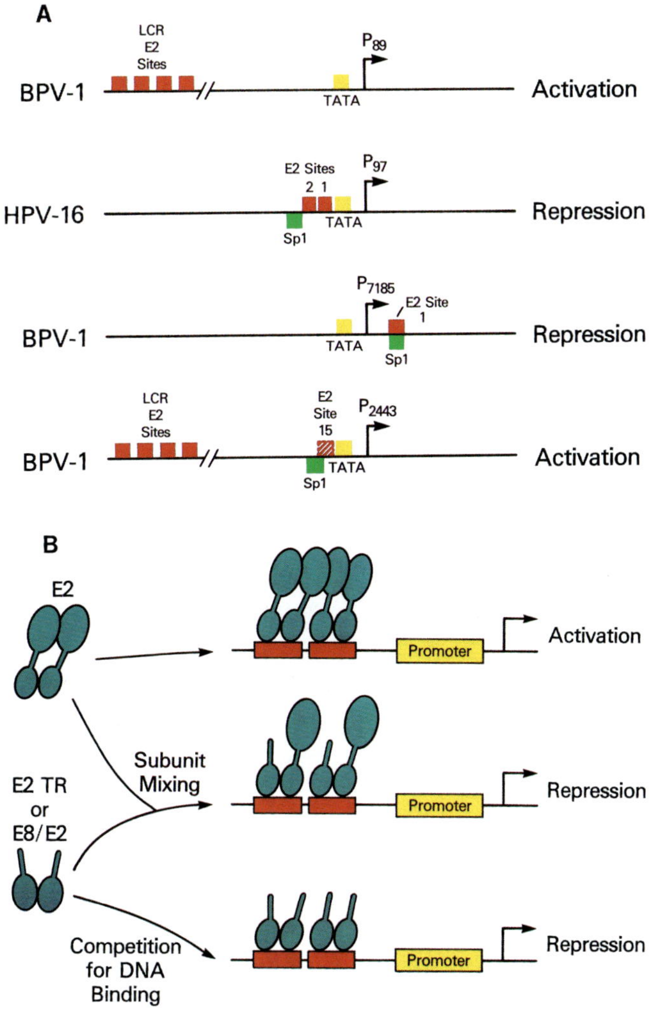

FIG. 3. **Mechanisms of E2 regulation.** *A*, regulation of transcription by position of E2 DNA-binding sites. Four examples of papillomavirus promoters are shown. Transcriptional activation or repression depends on the position and proximity of the E2 motifs with respect to the binding sites of cellular factors. *B*, regulation of transcription by different E2 polypeptide species. Two mechanisms by which the E2 repressors could inhibit transactivation by the full-length E2 protein are shown.

The mechanism by which E2 stimulates transcription has not yet been elucidated, but it is likely that activation results from a direct or indirect interaction of E2 with some component of the basic transcriptional machinery. E2 is unable to activate minimal promoters containing only a TATA box but requires additional promoter proximal elements which bind factors such as Sp1 (59). There are Sp1 sites located close to many papillomavirus promoters, and several of these sites have been shown to be essential for promoter activity (17, 23, 54, 58). Many of these sites are adjacent to or overlap E2 binding motifs and may play a role in E2-mediated regulation. Sp1 can interact directly with the full-length E2 polypeptide and can enhance E2 binding to weak affinity DNA sites (17). Binding of Sp1 and E2 to sites spaced widely apart on a DNA fragment results in the formation of DNA loops *in vitro* and provides a conceivable mechanism by which the LCR E2-binding sites can activate transcription from promoters located far downstream in the early region, such as P_{2443} and P_{3080}. However, it seems unlikely that the interaction between E2 and Sp1 is exclusive because E2 is able to synergize with other promoter proximal elements (59).

Modulation of papillomaviral transcription also depends on the nature and type of interaction of the E2 polypeptides (Fig. 3*B*). There are several mechanisms by which the E2 repressor proteins could inhibit transcriptional activation mediated by the full-length transactivator. The repressors could bind and titrate out essential cellular factors required for activation. Alternatively, the repressors could block E2 transactivation by binding competitively to the $ACCN_6GGT$ DNA-binding site or by forming potentially inactive heterodimers with the transactivator species. Since all three E2 regulatory factors share the C-terminal domain required for DNA binding and dimerization, the latter two models are likely mechanisms for E2 repression (Fig. 3*B*).

E2 Polypeptides and Viral DNA Replication

It has recently been shown conclusively that the full-length products of the E1 and E2 ORFs are both necessary and sufficient for BPV-1 transient DNA replication (39). The E1 polypeptide has some homology with SV40 and polyomavirus T antigens and binds ATP and binds DNA nonspecifically (38, 60, 62). In addition, the full-length E1 and E2 polypeptides form a complex capable of binding to E2 DNA sites, suggesting that E2 might direct the E1 replication protein to the viral origin of replication (37, 38). In support of this hypothesis, the minimal origin required for transient replication has been mapped to a 105-base pair fragment spanning the *Hpa*I restriction site at nucleotide 1 of the

BPV-1 genome.[3] This fragment also contains E2-binding sites 11 and 12. However, subsequent studies have shown that the E1 polypeptide does in fact bind specifically to a region spanning the *Hpa*I site (63) and that the adjacent E2 sites are not required for transient DNA replication.[3] The role of E2 in replication is therefore less clear. It may function by stabilizing the interaction of E1 with DNA or by altering the chromatin structure around the origin of replication thus allowing greater accessibility of cellular replication factors. In support of the former hypothesis, it has recently been shown that cellular extracts supplemented with purified BPV-1 E1 and E2 proteins can support replication of exogenously added papillomavirus DNA. The E2 transactivator stimulates binding of the E1 replication protein to the origin of replication and thereby activates DNA replication (64).

Only the full-length E2 protein is able to complex with E1, but it is possible that the E2 repressors could regulate replication by competitively binding to E2 sites adjacent to the replication origin or by complexing with the E2 transactivator. In support of this model, a BPV-1 mutant that is unable to express the E2-TR repressor replicates at a high copy number (65, 66).

Regulation of E2 Function

The abundance and activities of the E2 polypeptides can be regulated at several levels. As described above, the promoters from which the E2 gene products are thought to be expressed are themselves E2-responsive and therefore could potentially be autoregulated by the E2 proteins (15–17). In addition, the functions of the E2 transactivator can be modulated by association or competition with the shorter repressor species.

The E2 proteins may also be regulated by post-translational modifications. The E2 polypeptides are phosphorylated primarily on two serine residues at amino acids 298 and 301 in a region of the hinge immediately adjacent to the DNA-binding domain and present in all three E2 polypeptide species (45). Substitution of serine 301 with an alanine residue results in a virus that replicates to a much higher copy number than that of the wild-type virus (67). This suggests that E2 phosphorylation may play a direct role in viral DNA replication. In support of this hypothesis, E1 binds preferentially to the underphosphorylated form of E2, suggesting that phosphorylation may modulate the E1-E2 protein interaction (68). Alternatively, E2 phosphorylation may modulate replication indirectly by regulating viral gene expression.

While it seems likely that an E1-E2 protein complex is required for viral DNA replication, there is little data on the effect of this complex on the transcriptional properties of the E2 transactivator. It has been postulated that the E1 ORF encodes a transcriptional repressor because BPV-1 E1 mutants exhibit increased viral transcription (69, 70); however, the mechanism by which this phenotype arises has not yet been elucidated.

REFERENCES

1. De Villiers, E.-M. (1989) *J. Virol.* **63,** 4898–4903
2. DiMaio, D., and Neary, K. (1990) in *Papillomaviruses and Human Cancer* (Pfister, H., ed) pp. 113–144, CRC Press, Inc., Boca Raton, FL
3. Lambert, P. F. (1991) *J. Virol.* **65,** 3417–3420
4. Münger, K., Phelps, W. C., and Howley, P. M. (1990) in *The Cellular and Molecular Biology of Human Carcinogenesis* (Boutwell, R. K., and Riegel, I. L., eds) pp. 223–254, Academic Press, San Diego
5. Rabson, M. S., Yee, C., Yang, Y. C., and Howley, P. M. (1986) *J. Virol.* **60,** 626–634
6. DiMaio, D. (1986) *J. Virol.* **57,** 475–480
7. Groff, D. E., and Lancaster, W. D. (1986) *Virology* **150,** 221–230
8. Hermonat, P. L., and Howley, P. M. (1987) *J. Virol.* **61,** 3889–3895
9. DiMaio, D., and Settleman, J. (1988) *EMBO J.* **7,** 1197–1204
10. Spalholz, B. A., Yang, Y. C., and Howley, P. M. (1985) *Cell* **42,** 183–191
11. Moskaluk, C., and Bastia, D. (1987) *Proc. Natl. Acad. Sci. U. S. A.* **84,** 1215–1218
12. Androphy, E. J., Lowy, D. R., and Schiller, J. T. (1987) *Nature* **325,** 70
13. Spalholz, B. A., Lambert, P. F., Yee, C. L., and Howley, P. M. (1987) *J. Virol.* **61,** 2128–2137
14. Haugen, T. H., Cripe, T. P., Ginder, G. D., Karin, M., and Turek, L. P. (1987) *EMBO J.* **6,** 145–152
15. Hermonat, P. L., Spalholz, B. A., and Howley, P. M. (1988) *EMBO J.* **7,** 2815–2822
16. Prakash, S. S., Horwitz, B. H., Zibello, T., Settleman, J., and DiMaio, D. (1988) *J. Virol.* **62,** 3608–3613
17. Li, R., Knight, J. D., Jackson, S. P., Tjian, R., and Botchan, M. R. (1991) *Cell* **65,** 493–505
18. Lambert, P. F., Spalholz, B. A., and Howley, P. M. (1987) *Cell* **50,** 69–78
19. Lambert, P. F., Hubbert, N. L., Howley, P. M., and Schiller, J. T. (1989) *J. Virol.* **63,** 3151–3154
20. Choe, J., Vaillancourt, P., Stenlund, A., and Botchan, M. (1989) *J. Virol.* **63,** 1743–1755
21. Hubbert, N. L., Schiller, J. T., Lowy, D. R., and Androphy, E. J. (1988) *Proc. Natl. Acad. Sci. U. S. A.* **85,** 5864–5868
22. Stenlund, A., and Botchan, M. R. (1990) *Genes & Dev.* **4,** 123–136
23. Vande Pol, S. B., and Howley, P. M. (1990) *J. Virol.* **64,** 5420–5429
24. Thierry, F., and Yaniv, M. (1987) *EMBO J.* **6,** 3391–3397
25. Bernard, B. A., Bailly, C., Lenoir, M.-C., Darmon, M., Thierry, F., and Yaniv, M. (1989) *J. Virol.* **63,** 4317–4324
26. Romanczuk, H., Thierry, F., and Howley, P. M. (1990) *J. Virol.* **64,** 2849
27. Thierry, F., and Howley, P. M. (1991) *New Biol.* **3,** 1–11
28. Chin, M. T., Broker, T. R., and Chow, L. T. (1989) *J. Virol.* **63,** 2967–2976
29. Chin, M. T., Hirochika, R., Hirochika, H., Broker, T. R., and Chow, L. T. (1988) *J. Virol.* **62,** 2994–3002
30. Baker, C. C. (1987) in *The Papovaviridae 2: The Papillomaviruses* (Salzman, N. P., and Howley, P. M., eds) pp. 321–385, Plenum Press, New York
31. Giri, I., Danos, O., and Yaniv, M. (1985) *Proc. Natl. Acad. Sci. U. S. A.* **82,** 1580–1584
32. Haugen, T. H., Turek, L. P., Mercurio, F. M., Cripe, T. P., Olson, B. J., Anderson, R. D., Seidl, D., Karin, M., and Schiller, J. (1988) *EMBO J.* **7,** 4245–4253
33. Giri, I., and Yaniv, M. (1988) *EMBO J.* **7,** 2823–2829
34. McBride, A. A., Byrne, J. C., and Howley, P. M. (1989) *Proc. Natl. Acad. Sci. U. S. A.* **86,** 510–514
35. Giniger, E., and Ptashne, M. (1987) *Nature* **330,** 670–672
36. Knight, J. D., Li, R., and Botchan, M. (1991) *Proc. Natl. Acad. Sci. U. S. A.* **88,** 3204–3208
37. Mohr, I. J., Clark, R., Sun, S., Androphy, E. J., MacPherson, P., and Botchan, M. R. (1990) *Science* **250,** 1694–1699
38. Blitz, I. L., and Laimins, L. A. (1991) *J. Virol.* **65,** 649–656
39. Ustav, M., and Stenlund, A. (1991) *EMBO J.* **10,** 449–457
40. McBride, A. A., Schlegel, R., and Howley, P. M. (1988) *EMBO J.* **7,** 533
41. Moskaluk, C. A., and Bastia, D. (1988) *J. Virol.* **62,** 1925–1931
42. Monini, P., Grossman, S. R., Pepinsky, B., Androphy, E. J., and Laimins, L. A. (1991) *J. Virol.* **65,** 2124–2130
43. Dostatni, N., Thierry, F., and Yaniv, M. (1988) *EMBO J.* **7,** 3807–3816
44. Moskaluk, C. A., and Bastia, D. (1989) *Virology* **169,** 236–238
45. McBride, A. A., Bolen, J. B., and Howley, P. M. (1989) *J. Virol.* **63,** 5076
46. Hawley-Nelson, P., Androphy, E. J., Lowy, D. R., and Schiller, J. T. (1988) *EMBO J.* **7,** 525–531
47. Li, R., Knight, J., Bream, G., Stenlund, A., and Botchan, M. (1989) *Genes & Dev.* **3,** 510–526
48. Lambert, P. F., Dostatni, N., McBride, A. A., Yaniv, M., Howley, P. M., and Arcangioli, B. (1989) *Genes & Dev.* **3,** 38–48
49. Morrissey, L. C., Barsoum, J., and Androphy, E. J. (1989) *J. Virol.* **63,** 4422–4425
50. Stanway, C. A., Sowden, M. P., Wilson, L. E., Kingsman, A. J., and Kingsman, S. M. (1989) *Nucleic Acids Res.* **17,** 2187–2196
51. Thierry, F., Dostatni, N., Arnos, F., and Yaniv, M. (1990) *Mol. Cell. Biol.* **10,** 4431–4437
52. Gauthier, J-M., Dostatni, N., Lusky, M., and Yaniv, M. (1991) *New Biol.* **3,** 498–509
53. Spalholz, B. A., Byrne, J. C., and Howley, P. M. (1988) *J. Virol.* **62,** 3143
54. Spalholz, B. A., Vande Pol, S. B., and Howley, P. M. (1991) *J. Virol.* **65,** 743–753
55. Bedrosian, C. L., and Bastia, D. (1990) *Virology* **174,** 557–575
56. Moskaluk, C., and Bastia, D. (1988) *Proc. Natl. Acad. Sci. U. S. A.* **85,** 1826–1830
57. Dostatni, N., Lambert, P., Sousa, R., Ham, J., Howley, P., and Yaniv, M. (1991) *Genes & Dev.*, in press
58. Gloss, B., and Bernard, H.-U. (1990) *J. Virol.* **64,** 5577–5584
59. Ham, J., Dostatni, N., Arnos, F., and Yaniv, M. (1991) *EMBO J.* **10,** 2931–2940
60. Clertant, P., and Seif, I. (1984) *Nature* **311,** 276–279
61. Sun, S., Thorner, L., Lentz, M., MacPherson, P., and Botchan, M. (1990) *J. Virol.* **64,** 5093–5105
62. Santucci, S., Androphy, E. J., Bonne-Andra, C., and Clertant, P. (1990) *J. Virol.* **64,** 6027–6039
63. Wilson, V. G., and Ludes-meyer, J. (1991) *J. Virol.*, in press
64. Yang, L., Li, R., Mohr, I. J., Clark, R., and Botchan, M. R. (1991) *Nature*, in press
65. Lambert, P. F., Monk, B. C., and Howley, P. M. (1990) *J. Virol.* **64,** 950
66. Riese, D. J., II, Settleman, J., Neary, K., and DiMaio, D. (1990) *J. Virol.* **64,** 944–949
67. McBride, A. A., and Howley, P. M. (1991) *J. Virol.*, in press
68. Lusky, M., and Fontane, E. (1991) *Proc. Natl. Acad. Sci. U. S. A.* **88,** 6363
69. Lambert, P. F., and Howley, P. M. (1988) *J. Virol.* **62,** 4009–4015
70. Schiller, J. T., Kleiner, E., Androphy, E. J., Lowy, D. R., and Pfister, H. (1989) *J. Virol.* **63,** 1775–1782
71. Cripe, T. P., Haugen, T. H., Turk, J. P., Tabatabai, F., Schmid, P. G., III, Dürst, M., Gissmann, L., Roman, A., and Turek, L. P. (1987) *EMBO J.* **6,** 3745–3753

[3] M. Ustav, E. Ustav, P. Szymanski, and A. Stenlund, submitted for publication.

Minireview

THE JOURNAL OF BIOLOGICAL CHEMISTRY
Vol. 266, No. 29, Issue of October 15, pp. 19127–19130, 1991

Printed in U.S.A.

DNA Polymerase III Holoenzyme

COMPONENTS, STRUCTURE, AND MECHANISM OF A TRUE REPLICATIVE COMPLEX*

Charles S. McHenry

From the Department of Biochemistry, Biophysics, and Genetics, University of Colorado Health Sciences Center, Denver, Colorado 80262

The DNA polymerase III holoenzyme is the replicative polymerase of *Escherichia coli*, responsible for synthesis of the majority of the chromosome (for a more extensive review, see Ref. 1). The replicative role of the enzyme has been established through both biochemical and genetic criteria. Holoenzyme[1] has been biochemically defined and purified using natural chromosomal assays. Only the holoenzyme form of pol III will efficiently replicate the DNA of single-stranded bacteriophages *in vitro* in the presence of other known *E. coli* replicative proteins (2–4). Similarly, only the holoenzyme will function in the replication of bacteriophage λ, plasmids, and DNA containing the *E. coli* replicative origin (5–8). Genetic studies also support assignment of the major replicative role to the holoenzyme. The holoenzyme appears to contain up to 10 different subunits (Table I). Temperature-sensitive, conditionally lethal mutations in four replication genes can be correlated with defects in five of the subunits of holoenzyme; the genes for the remaining five subunits remain unknown (Table I).

There are at least three distinct polymerases in *E. coli*, yet only holoenzyme appears to play a major replicative role. What are the special features of the holoenzyme that confer its unique role in replication? Work to date suggests that an ability to rapidly elongate in a highly processive mode at physiological levels of salt, to utilize a long single-stranded template coated with the single-stranded DNA-binding protein, to jump over obstacles created by annealed oligonucleotides, to interact with other proteins of the replicative apparatus, and to coordinate replication through an asymmetric dimeric structure are critical to its unique functions. Many of these features are conserved between bacterial and mammalian systems, suggesting that insight gained through studies with the holoenzyme may be transferable to a variety of life forms.

Multiple DNA Polymerase III Forms

The holoenzyme can be biochemically resolved into a series of successively simpler forms (Table II). Having these multiple forms available has facilitated assignment of functions to individual holoenzyme subunits and assemblies. While it is possible that some polymerase subassemblies may exist in the cell free by themselves, it is unlikely that they make significant synthetic contributions *in vivo* given their extreme sensitivity to physiological ionic strengths (42, 43).

Processivity

Studies of the processivities of the multiple pol III forms have revealed contributions of individual subunits (Table II) (38, 40, 41). The multiple forms of pol III exhibit strikingly different processivities. The processivity of the core pol III is nearly distributive under physiological ionic strength. Processivity is enhanced by addition of the τ subunit to form pol III′. pol III′ achieves maximum processivity in the presence of physiological concentrations of spermidine, an agent that inhibits the core pol III. Addition of the $\gamma\delta$ complex[2] (γ, δ, δ', ψ, and χ) to pol III′ to form pol III* further increases processivity in the presence of single-stranded DNA-binding protein. The holoenzyme is exceedingly processive, having the capability of remaining stably bound to a template for 30–40 min, the time required for replication of the entire *E. coli* chromosome (1, 38, 39, 44). The processivity of the proofreading exonuclease responds like the elongation activity to the addition of holoenzyme auxiliary subunits (45). Thus, a progression in processivities is observed that parallels the structural complexity of the corresponding enzyme form.

Initiation Complex Formation and a Kinetic Barrier to Polymerase Cycling

To achieve high processivity, the holoenzyme requires ATP and primed DNA to form a stable initiation complex (2, 38). Initiation complexes can be isolated by gel filtration and upon addition of dNTPs form a complete RFII in 10–15 s (2, 27). β (and presumably the remaining holoenzyme components) remains stably associated with both initiation and termination complexes (RFII) for at least 30 min (44). This observation revealed a problem in the *in vitro* system. Okazaki fragments are made on the lagging strand of the replication fork each second at 37 °C. This requires the lagging strand polymerase to bind to a primer, synthesize a 1000-nucleotide Okazaki fragment, dissociate, and bind to the next primer each second. The 30 min we observe for release of holoenzyme *in vitro* is considerably slower than the fraction of a second required *in vivo*.

At the time that we were struggling with the issue created by the kinetic barrier to recycling, we discovered that pol III′ was dimeric and inferred that the holoenzyme was as well (18). This had important implications regarding the cycling problem when considered in the context of Bruce Alberts' model for cycling during bacteriophage T4 replication. Alberts (46, 47) proposed that at the replication fork, T4 polymerase formed a dimer of leading and lagging strand polymerases.[3] This would provide a mechanism for retargeting the lagging strand polymerase to the next primer for Okazaki fragment synthesis rather than permitting the polymerase to diffuse to another fork. It occurred to us that the Alberts model might

* Work from the author's laboratory was supported by grants from the American Cancer Society, the National Institute of General Medical Sciences, and the Lucille P. Markey Charitable Trust.

[1] The abbreviations used are: holoenzyme, DNA polymerase III holoenzyme; pol III, DNA polymerase III; ATPγS, adenosine 5′-*O*-(thiotriphosphate).

[2] The $\gamma\delta$ complex was initially purified as an activity that permitted reconstitution of holoenzyme using core DNA polymerase III and the β subunit (4). With improvements in gel systems, added subunits became visible in holoenzyme (13) and the $\gamma\delta$ complex (29). The five-subunit structure of the $\gamma\delta$ complex originally purified in 1976 (4, 29) was identical to the five-subunit γ complex described in 1988 (30). In this minireview, I use the original nomenclature for the $\gamma\delta$ complex to refer to the five-subunit complex.

[3] Subsequent to the proposal of the Alberts dimeric T4 model, the Kornberg laboratory (48, 49) also proposed that the proteins at the *E. coli* replication fork might coordinately replicate the leading and lagging strand through a dimeric replicative complex. This model, however, did not entail an asymmetric polymerase with functionally or structurally distinguishable leading and lagging strand halves.

provide a solution to the *E. coli* cycling problem and an explanation for our observations. If the replicative complex were dimeric, then an opportunity for communication between polymerase halves would be provided. To guarantee a properly assembled complex continuously associated with the replication fork, the lagging strand polymerase might only be able to dissociate when the leading strand polymerase is in a productive elongation conformation. If this were true, then dissociation of the lagging strand polymerase might be slow when a uniform population of single-stranded templates were used since they would all be completed at nearly the same time. A mechanism permitting communication between the leading and lagging strand polymerases might be useful in coordinating leading with lagging strand replication, as I have discussed elsewhere (50).

The Asymmetric Dimer Hypothesis

To further investigate initiation complexes, we explored the use of ATP analogs. ATPγS was found to substitute for ATP in initiation complex formation, but only partially. In spite of a greater efficacy at low nucleotide concentrations, ATPγS only supported formation of one-half as much complex as ATP. The same effect was observed upon reversal of the reaction. ATPγS caused one-half of the initiation complexes to dissociate; ATP exhibited no effect. Based on this observation we first proposed the asymmetric dimer hypothesis, a modification of previous notions about dimeric polymerases (51, 52). We suggested that ATPγS might be revealing a functional asymmetry in the holoenzyme consistent with distinct leading and lagging strand polymerase, with ATPγS having differential effects on the two halves. Such an asymmetry would be consistent with the asymmetric functional requirement of polymerases at replication forks. The leading strand polymerase, once associated with the replication fork, need not dissociate until the entire chromosome is replicated, a process that takes approximately 40 min in *E. coli.* In contrast, the lagging strand polymerase needs to dissociate from the completed product, reassociate with the next primer, and synthesize an entire Okazaki fragment each second in the cycle depicted in Fig. 1. It would appear to be advantageous for a cell to contain an asymmetric dimeric polymerase with an extremely processive leading strand polymerase and a lagging strand polymerase with modified properties that permit rapid cycling.

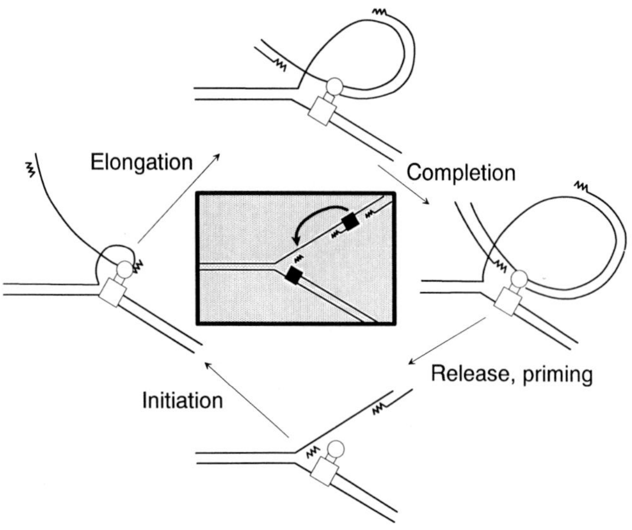

FIG. 1. **Cycling of an asymmetric dimeric polymerase at a replication fork.** The leading strand polymerase (□) remains continuously clamped to the template, while the lagging strand polymerase (○) can rapidly recycle mediated, in part, by its looser template interactions (*inset*, standard depiction of a replication fork with equivalent leading and lagging strand polymerase). In both models, the dissociated lagging strand polymerase must recycle to the next primer synthesized at the fork.

TABLE I
Subunits of the DNA polymerase III holoenzyme

Subunit	M_r	Structural gene	Ref.
α	129,900	*dna*E	9–14
τ	71,100	*dna*X	15–20
γ	47,500	*dna*X[a]	4, 16, 17, 19–24
β	40,600	*dna*N	4, 25–28
δ	34,000	?	4
δ'	32,000	?	29–31
ϵ	27,500	*dna*Q (*mut*D)	32–37
ψ	14,000–16,000	?	1, 29–31
χ	12,000–14,000	?	1, 29–31
θ	10,000	?	13

[a] Formerly *dna*Z.

TABLE II
DNA polymerase III forms

Form	Subunit composition	Processivity	Ref.
Holoenzyme	$\alpha, \epsilon, \theta, \tau, \gamma, \delta, \delta', \chi, \psi, \beta$	>5000, 150,000	38, 39
Reconstituted holoenzyme minus ϵ, θ, τ	$\alpha, \gamma, \delta, \delta', \chi, \psi, \beta$	1–3,000	40
Pol III*	$\alpha, \epsilon, \theta, \tau, \gamma, \delta, \delta', \chi, \psi$	200	41
Pol III′	$\alpha, \epsilon, \theta, \tau$	60	41
Pol III (core)	α, ϵ, θ	10	38

Although it is attractive and has won widespread acceptance, the asymmetric dimer hypothesis remains unproven. It has not yet been rigorously demonstrated that the holoenzyme has two polymerase entities that can function in concert at a replication fork. Even physical characterizations of holoenzyme as a dimer (53) are subject to question, since they did not contain corrections for the high frictional coefficients of elongated polymerases that have been observed in simpler forms (18). However, considerable supportive data for the hypothesis have been obtained. One prediction of the model is that the enzyme is not only dimeric, but interactive, so that the leading and lagging strand polymerases can communicate during coordinated replication. Strong positive cooperativity has been demonstrated between the ATP binding sites for initiation complex formation (29).

Additional support for the asymmetric dimer model was derived from studies of the γ and τ subunits of the pol III holoenzyme. τ and γ are both products of the *dna*X gene (16, 17). We first proposed that translational frameshifting accounted for generation of these two subunits from one gene and, based on conjecture and sequence data, proposed a potential site (1). The experimental results from our laboratory and others verified a frameshifting mechanism, but at the sequence AAAAAAG followed by a strong hairpin, similar to the frameshifting site in retroviruses (24, 54–56). The AAAAAAG sequence had been identified earlier as a "shifty" sequence in model constructs (57). The sequence of *dna*X suggested a tight DNA binding domain in the carboxyl terminus of the gene within sequences contained in τ but not γ (58). This finding, coupled with the knowledge that τ, but not γ, binds to the pol III core and forms a polymerase of increased processivity, led to the suggestion that τ may be in the leading strand half conferring high processivity, and γ in the lagging strand half permitting more rapid cycling, mediated by looser γ-template interactions (58, 59). In the future, it will be interesting to learn whether frameshifting is regulated to guarantee an appropriate ratio of leading and lagging strand components.

Recent work has provided further support for the asymmetric dimer hypothesis. τ and γ appear to compete for the same binding sites within holoenzyme, suggesting a similar role for the two proteins (53). Both τ and γ are present in every active holoenzyme molecule (59). This eliminates the possibility that τ and γ randomly assort during holoenzyme assembly *in vivo* as they apparently do *in vitro*. The mechanism that operates to ensure an asymmetric structure *in vivo* is unclear. Perhaps the holoenzyme assembles in the cell at the origin of replication where asymmetric protein-protein interactions at the initiating replication fork direct asymmetric assembly. Consistent with such a mechanism, genetic evidence has been obtained for an interaction between the origin-specific *dna*A gene product and a *dna*X product, either γ and/or τ (60).

The τ and γ subunits appear to contain the site that sets the ATP-dependent clamp on primed DNA. A consensus ATP binding sequence is located within the amino-terminal sequences common to both proteins (19). Both τ and γ bind ATP with a dissociation constant of ~2 μM (61). τ, but not γ, is a DNA-dependent ATPase by itself (61, 62).

τ and γ have differential interactions with replication proteins. γ is isolated in a complex with δ, δ′, ψ, and χ (29, 30). τ has only been isolated as a stable complex with pol III or by itself. In a technically difficult and elegant study, O'Donnell and Studwell (31) resolved the γδ complex to yield δ and δ′. They demonstrated that δ stimulates γ and τ, while δ′ only stimulates τ in formation of preinitiation complexes in the presence of β.

The above information suggests related but discrete roles for the γ and τ subunits. This is consistent with the original hypothesis of an asymmetric dimer and supports the notion that an asymmetric placement of the τ and γ subunits within the replicative complex determines both structural and functional asymmetry. Among the critical experiments required to test the unity of these two hypotheses are (i) a determination of whether *either* γ *or* τ complexes *uniquely* use ATPγS to form initiation complexes and (ii) a direct determination of the subunit arrangements within the pol III holoenzyme.

If the holoenzyme does exist as an asymmetric dimer, how does it function to rapidly recycle on the lagging strand of the replication fork? Without excess auxiliary subunits, holoenzyme cycles very slowly. Using different preparations of holoenzyme and differing reaction conditions, times from 30 min to 1–2 min have been estimated (42, 44, 63). Although quantitatively different and largely dependent upon the ratio of excess auxiliary subunits present, they are drastically longer than the value of under 1 s required *in vivo*. O'Donnell and colleagues (31, 63) have demonstrated that formation of a preinitiation complex can markedly facilitate the cycling of polymerase core from a completed initiation complex to the next primer. In a corollary of the asymmetric dimer hypothesis, they have proposed that this mechanism might be exploited *in vitro* to facilitate rapid cycling by a polymerase that contains two γδ complexes in the lagging strand half and a τδ′ complex in the leading strand half (31, 63, 64). The two γδ complexes within the holoenzyme could alternate between completed Okazaki fragments and the next primer. Whether the stoichiometry of these components within the holoenzyme supports this possibility is not yet certain. Although formation of a preinitiation complex bypasses a rate-limiting step in initiation complex formation, questions regarding the availability of γδ complex to enable such a reaction remain. For the first γδ complex used in the synthesis of two Okazaki fragments to be reused for a third fragment in a series along the lagging strand of the replication fork requires for it to be released from the terminus of the first fragment. The release of γδ complex from products appears to be very slow relative to the required rates of polymerase cycling (31, 63). Thus, the cycling problem is just passed back one step and becomes related to the question originally stated at the holoenzyme level. How is γδ complex released and recycled at a rate of 1/s so that it is available for the next Okazaki fragment synthesized? The answer may be found in special interactions found within the holoenzyme.

Another special feature of the holoenzyme, initially thought to be involved in polymerase cycling, is its ability to jump over primers (65, 66). Holoenzyme can very rapidly transfer intramolecularly between primers, provided they are ahead of the advancing polymerase (66). Since this direction is the opposite of that required for lagging strand cycling, it is probably not involved in cycling, but instead in jumping over annealed oligonucleotides at the replication fork that would otherwise stall polymerase movement. Consistent with this notion, τ, the subunit proposed to be associated with the leading strand half of holoenzyme, is apparently required for primer jumping (67).

Structure of the DNA Polymerase III Holoenzyme

Central to the understanding of cycling, asymmetric dimers, and related issues is the determination of the structure of the pol III holoenzyme at the level of subunit arrangements and subunit-primer-template contact. One could expect protein-protein cross-linking, site-specific protein-DNA cross-linking, fluorescence energy transfer between labeled subunits and sites, electron microscopy, and eventually, x-ray crystallography all to make contributions to this important endeavor.

Presently, our understanding of the structure of the holoenzyme is limited to information gained through functional interactions detected and suppressor mutations. A preliminary model for holoenzyme subunit-subunit interactions is shown in Fig. 2. α and ε form an isolable complex upon mixing (68). Mutations in the structural gene for ε have also been found that suppress *dna*E (α) mutations (69). The assumption usually made is that suppressor mutations arise through modification of a subunit that interacts directly with the suppressed mutant gene product. pol III core (αεθ) is isolable, but we don't know with which subunit(s) θ interacts (13). τ can be isolated in a complex with pol III core; the contacted subunit is unknown (18). Suppressor data suggest an interaction between γ and/or τ and β (70). Either γδ or τ in the presence of either δ or δ′ can transfer β to primed DNA to form a preinitiation complex (31). Suppressor data indicate an interaction between β and α (71). Reinforcing this conclusion, we also know that β can interact with and increase the processivity of core pol III (72). Genetic evidence for α-α

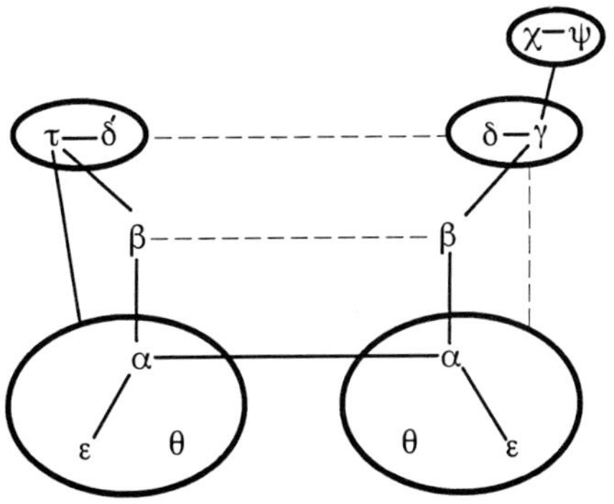

FIG. 2. **Minimal subunit interactions within the DNA polymerase III holoenzyme.** Subunits are designated by *Greek letters*. *Lines* contacting a subunit indicate a direct interaction with that subunit. *Lines* contacting an *ellipse* indicate an interaction with the complex; the contacted subunit within the complex is uncertain. *Dotted lines* represent less certain interactions.

interaction and pol III holoenzyme being a dimer was obtained through *dna*E interallelic complementation (73). From biochemical studies, it is known that δ can interact with γ and either δ or δ' with τ (31). χ and ψ have been isolated in a complex with γ (31).

My laboratory has taken a cross-linking approach to identify specific subunit-subunit and subunit-primer-template contacts and a parallel approach using fluorescence energy transfer to identify distances between sites within replication complexes. In initial studies using fluorescence energy transfer, we have focused on the β subunit. β exists as a dimer free in solution by itself. Addition of Mg^{2+} at physiological concentrations triggers dissociation and an accompanying conformational change in the vicinity of Cys_{333} on the distal ends of the β dimer (74, 75). Within the initiation complex, β exists in a ratio of 1:1 with primer, with Cys_{333} located ~65 Å from the antepenultimate nucleotide of the primer.[4] Continuation of these studies should permit us to assign the subunit arrangement within holoenzyme. This knowledge coupled with that gained through sophisticated functional studies being conducted in several laboratories should lead to an understanding of the individual roles of subunits within the complex replicative apparatus and the cooperation that occurs between components enabling a coordinated synthetic reaction to occur at the replication fork.

Note Added in Proof—Recent work directed toward assigning the structural genes for the last five holoenzyme components has revealed two genes that have already been identified as open reading frames.[5] The gene for δ is located immediately downstream of *rlpB* at 16 min (76), and the structural gene for ψ is located between *xerB* and *valS* at 97 min (77). This work has also revealed that the structural gene for δ' is distinct from δ.

REFERENCES

1. McHenry, C. (1988) *Annu. Rev. Biochem.* **57,** 519–550
2. Wickner, W., and Kornberg, A. (1973) *Proc. Natl. Acad. Sci. U. S. A.* **70,** 3679–3683
3. Hurwitz, J., and Wickner, S. (1974) *Proc. Natl. Acad. Sci. U. S. A.* **71,** 6–10
4. McHenry, C., and Kornberg, A. (1977) *J. Biol. Chem.* **252,** 6478–6484
5. Mensa-Wilmot, K., Seaby, R., Alfano, C., Wold, M. S., Gomes, B., and McMacken, R. (1989) *J. Biol. Chem.* **264,** 2853–2861
6. Minden, J. S., and Marians, K. J. (1985) *J. Biol. Chem.* **260,** 9316–9325
7. Lanka, E., Scherzinger, E., Guenther, E., and Schuster, H. (1979) *Proc. Natl. Acad. Sci. U. S. A.* **76,** 3632–3636
8. Kaguni, J. M., and Kornberg, A. (1984) *Cell* **38,** 183–190
9. Gefter, M. L., Hirota, Y., Kornberg, T., Wechsler, J. A., and Barnoux, C. (1971) *Proc. Natl. Acad. Sci. U. S. A.* **68,** 3150–3153
10. Welch, M. M., and McHenry, C. S. (1982) *J. Bacteriol.* **152,** 351–356
11. Livingston, D. M., Hinkle, D. C., and Richardson, C. C. (1975) *J. Biol. Chem.* **250,** 461–469
12. Otto, B., Bonhoeffer, F., and Schaller, H. (1973) *Eur. J. Biochem.* **34,** 440–447
13. McHenry, C. S., and Crow, W. (1979) *J. Biol. Chem.* **254,** 1748–1753
14. Tomasiewicz, H., and McHenry, C. (1987) *J. Bacteriol.* **169,** 5735–5744
15. Henson, J. M., Chu, H., Irwin, C. A., and Walker, J. R. (1979) *Genetics* **92,** 1041–1059
16. Kodaira, M., Biswas, S. B., and Kornberg, A. (1983) *Mol. Gen. Genet.* **192,** 80–86
17. Mullin, D. A., Woldringh, C. L., Henson, J. M., and Walker, J. R. (1983) *Mol. Gen. Genet.* **192,** 73–79
18. McHenry, C. S. (1982) *J. Biol. Chem.* **257,** 2657–2663
19. Yin, K. C., Blinkowa, A., and Walker, J. R. (1986) *Nucleic Acids Res.* **14,** 6541–6549
20. Flower, A. M., and McHenry, C. S. (1986) *Nucleic Acids Res.* **14,** 8091–8101
21. Truitt, C. L., and Walker, J. R. (1974) *Biochem. Biophys. Res. Commun.* **61,** 1036–1042
22. Filip, C. C., Allen, J. S., and Gustafson, R. A. (1974) *J. Bacteriol.* **119,** 443–449
23. Hübscher, U., and Kornberg, A. (1980) *J. Biol. Chem.* **255,** 11698–11703
24. Flower, A. M., and McHenry, C. S. (1990) *Proc. Natl. Acad. Sci. U. S. A.* **87,** 3713–3717
25. Sakakibara, Y., and Mizukami, T. (1980) *Mol. Gen. Genet.* **178,** 541–553

[4] M. A. Griep and C. S. McHenry, submitted for publication.
[5] J. Carter, M. Franden, R. Aebersold, and C. McHenry, unpublished data.

26. Burgers, P. M. J., Kornberg, A., and Sakakibara, Y. (1981) *Proc. Natl. Acad. Sci. U. S. A.* **78,** 5391–5395
27. Johanson, K. O., and McHenry, C. S. (1980) *J. Biol. Chem.* **255,** 10984–10990
28. Ohmori, H., Kimura, M., Nagata, T., and Sakakibara, Y. (1984) *Gene (Amst.)* **28,** 159–170
29. McHenry, C. S., Oberfelder, R., Johanson, K., Tomasiewicz, H., and Franden, A. (1986) in *DNA Replication and Recombination* (Kelly, T., and McMacken, R., eds) pp. 47–62, Alan R. Liss, Inc., New York
30. Maki, S., and Kornberg, A. (1988) *J. Biol. Chem.* **263,** 6555–6560
31. O'Donnell, M., and Studwell, P. S. (1990) *J. Biol. Chem.* **265,** 1179–1187
32. Horiuchi, T., Maki, H., and Sekiguchi, M. (1978) *Mol. Gen. Genet.* **163,** 277–283
33. Cox, E. C., and Horner, D. L. (1983) *Proc. Natl. Acad. Sci. U. S. A.* **80,** 2295–2299
34. Echols, H., Lu, C., and Burgers, P. M. J. (1983) *Proc. Natl. Acad. Sci. U. S. A.* **80,** 2189–2192
35. Scheuermann, R., Tam, S., Burgers, P. M. J., Lu, C., and Echols, H. (1983) *Proc. Natl. Acad. Sci. U. S. A.* **80,** 7085–7089
36. DeFrancesco, R., Bhatnagar, S. K., Brown, A., and Bessman, M. J. (1984) *J. Biol. Chem.* **259,** 5567–5573
37. Maki, H., Horiuchi, T., and Sekiguchi, M. (1983) *Proc. Natl. Acad. Sci. U. S. A.* **80,** 7137–7141
38. Fay, P. J., Johanson, K. O., McHenry, C. S., and Bambara, R. A. (1981) *J. Biol. Chem.* **256,** 976–983
39. Mok, M., and Marians, K. J. (1987) *J. Biol. Chem.* **262,** 16644–16654
40. Studwell, P. S., and O'Donnell, M. (1990) *J. Biol. Chem.* **265,** 1171–1178
41. Fay, P. J., Johanson, K. O., McHenry, C. S., and Bambara, R. (1982) *J. Biol. Chem.* **257,** 5692–5699
42. Burgers, P. M. J., and Kornberg, A. (1982) *J. Biol. Chem.* **257,** 11474–11478
43. Griep, M. A., and McHenry, C. S. (1989) *J. Biol. Chem.* **264,** 11294–11301
44. Johanson, K. O., and McHenry, C. S. (1982) *J. Biol. Chem.* **257,** 12310–12315
45. Reems, J., Griep, M., and McHenry, C. (1991) *J. Biol. Chem.* **266,** 4878–4882
46. Alberts, B., Morris, C., Mace, D., Sinha, N., Bittner, M., and Moran, L. (1975) in *DNA Synthesis and Its Regulation* (Goulian, M., and Hanawalt, P., eds) pp. 241–269, W. A. Benjamin, Menlo Park, CA
47. Sinha, N. K., Morris, C. F., and Alberts, B. M. (1980) *J. Biol. Chem.* **255,** 4290–4303
48. Arai, N., and Kornberg, A. (1981) *J. Biol. Chem.* **256,** 5294–5298
49. Kornberg, A. (1982) *1982 Supplement to DNA Replication*, W. H. Freeman & Co., San Francisco
50. McHenry, C. S., Tomasiewicz, H., Griep, M., Fürste, J., and Flower, A. (1988) in *DNA Replication and Mutagenesis* (Moses, R. E., and Summers, W. C., eds) pp. 14–26, American Society for Microbiology, Wash., D. C.
51. McHenry, C. S., and Johanson, K. O. (1984) in *Proteins Involved in DNA Replication* (Hübscher, U., and Spadari, S., eds) pp. 315–319, Plenum Publishing Corp., New York
52. Johanson, K. O., and McHenry, C. S. (1984) *J. Biol. Chem.* **259,** 4589–4595
53. Maki, H., Maki, S., and Kornberg, A. (1988) *J. Biol. Chem.* **263,** 6570–6578
54. McHenry, C., Griep, M., Tomasiewicz, H., and Bradley, M. (1989) in *Molecular Mechanisms in DNA Replication and Recombination* (Richardson, C., and Lehman, I. R., eds) pp. 115–126, Alan R. Liss, Inc., New York
55. Tsuchihashi, Z., and Kornberg, A. (1990) *Proc. Natl. Acad. Sci. U. S. A.* **87,** 2516–2520
56. Blinkowa, A. L., and Walker, J. R. (1990) *Nucleic Acids Res.* **18,** 1725–1729
57. Weiss, R., Dunn, D., Shuh, M., Atkins, J., and Gesteland, R. (1989) *New Biologist* **1,** 159–169
58. McHenry, C. S., Flower, A. M., and Hawker, J. R. (1987) *Cancer Cells (Cold Spring Harbor)* **6,** 35–41
59. Hawker, J. R., and McHenry, C. S. (1987) *J. Biol. Chem.* **262,** 12722–12727
60. Blinkowa, A., and Walker, J. (1983) *J. Bacteriol.* **153,** 535–538
61. Tsuchihashi, Z., and Kornberg, A. (1989) *J. Biol. Chem.* **264,** 17790–17795
62. Lee, S. H., and Walker, J. R. (1987) *Proc. Natl. Acad. Sci. U. S. A.* **84,** 2713–2717
63. O'Donnell, M. E. (1987) *J. Biol. Chem.* **262,** 16558–16565
64. Studwell, P. S., Stukenberg, P. T., Onrust, R., Skangalis, M., and O'Donnell, M. (1990) in *Molecular Mechanisms in DNA Replication and Recombination* (Richardson, C. C., and Lehman, I. R., eds) pp. 153–164, Alan R. Liss, Inc., New York
65. Burgers, P. M. J., and Kornberg, A. (1983) *J. Biol. Chem.* **258,** 7669–7675
66. O'Donnell, M. E., and Kornberg, A. (1985) *J. Biol. Chem.* **260,** 12875–12883
67. Maki, S., and Kornberg, A. (1988) *J. Biol. Chem.* **263,** 6560–6569
68. Maki, H., and Kornberg, A. (1987) *Proc. Natl. Acad. Sci. U. S. A.* **84,** 4389–4392
69. Maurer, R., Osmond, B. C., and Botstein, D. (1984) *Genetics* **108,** 25–38
70. Engstrom, J., Wong, A., and Maurer, R. (1986) *Genetics* **113,** 499–516
71. Kuwabara, N., and Uchida, H. (1981) *Proc. Natl. Acad. Sci. U. S. A.* **78,** 5764–5767
72. LaDuca, R. J., Crute, J. J., McHenry, C. S., and Bambara, R. A. (1986) *J. Biol. Chem.* **261,** 7550–7557
73. Bryan, S., Hagensee, M. E., and Moses, R. E. (1988) in *DNA Replication and Mutagenesis* (Moses, R. E., and Summers, W. C., eds) pp. 305–313, American Society for Microbiology, Wash., D. C.
74. Griep, M. A., and McHenry, C. S. (1988) *Biochemistry* **27,** 5210–5215
75. Griep, M. A., and McHenry, C. S. (1990) *J. Biol. Chem.* **265,** 20356–20363
76. Takase, I., Ishino, F., Wachi, M., Kamata, H., Doi, M., Asoh, S., Matsuzawa, H., Ohta, T., and Matsuhashi, M. (1987) *J. Bacteriol.* **169,** 5692–5699
77. Stirling, C., Colloms, S., Collins, J., Szatmari, G., and Sherratt, D. (1989) *EMBO J.* **8,** 1623–1627

Minireview

THE JOURNAL OF BIOLOGICAL CHEMISTRY
Vol. 266, No. 30, Issue of October 25, pp. 19867-19870, 1991

Printed in U.S.A.

Structural Features in Eukaryotic mRNAs That Modulate the Initiation of Translation*

Marilyn Kozak

From the Department of Biochemistry, University of Medicine and Dentistry of New Jersey, Piscataway, New Jersey 08854

In higher eukaryotes, translation is modulated at the level of initiation by five aspects of mRNA structure: (i) the m7G cap; (ii) the primary sequence or context surrounding the AUG codon; (iii) the position of the AUG codon, *i.e.* whether or not it is "first"; (iv) secondary structure both upstream and downstream from the AUG codon; and (v) leader length. Here I briefly discuss how experimental manipulation of these features affects the fidelity and/or efficiency of initiation. Elsewhere (1) I discuss the extent to which natural mRNA leader sequences conform to these experimentally determined requirements for initiation.

Although my primary concern is to document the occurrence and consequences of the five structural elements in mRNAs, I will allude now and then to why each feature has the effect it does. The explanations will invariably hark back to the scanning process by which ribosomes are thought to initiate translation. In its simplest form, the scanning model (2) postulates that a 40 S ribosomal subunit, carrying Met-$tRNA_i^{Met}$ and an imperfectly defined set of initiation factors (3), enters at the 5′-end of the mRNA and migrates linearly until it reaches the first AUG codon, whereupon a 60 S ribosomal subunit joins and the first peptide bond is formed (Fig. 1). Evidence in support of the scanning model has been adduced previously (2, 4, 5). The most recent evidence includes the apparent queuing of 40 S ribosomal subunits on a long, unstructured leader sequence (6) and the demonstration that 40 S subunits stall on the 5′-side of a stable hairpin structure introduced between the cap and the AUG codon (7).

m7G Cap

That the ubiquitous m7G cap increases the efficiency of translation *in vitro* was first shown by Shatkin (8) and has been confirmed many times since. While the dependence of *in vitro* translation on the m7G cap may vary with the choice of reaction conditions, *in vivo* translation of most mRNAs is stringently cap-dependent, as shown in studies with vesicular stomatitis virus mutants that are defective in methylation (9) and with various other test systems (10, 11). Parenthetically, the experiments with vesicular stomatitis virus distinguish nicely between the ability of the cap to stabilize transcripts and its ability to stimulate translation; a guanylylated, unmethylated cap is sufficient to protect transcripts from 5′-exonucleases, but N-7 methylation is essential for efficient translation. A methylated cap and the associated cap-binding protein may be less important for the translation of mRNAs that have a rather long, unstructured leader sequence (6, 12), but such mRNAs are rare. Among animal cells and viruses, the only mRNAs that clearly are translated without benefit of a cap derive from picornaviruses.

* Research in my laboratory is supported by National Institutes of Health Grant GM33915.

Context

Systematic mutagenesis of nucleotides in the vicinity of the AUG codon revealed that GCC$^{A}_{G}$CC<u>AUG</u>G is the optimal context for initiation of translation in cultured monkey cells (13, 14). (The A of the AUG codon is designated +1, with positive and negative integers proceeding 3′ and 5′, respectively.) A purine, preferably A, in position −3 and a G in position +4 have the strongest effects, modulating translation at least 10-fold; the smaller effects of other nucleotides near the AUG codon are seen most easily in the absence of A^{-3} and G^{+4}. To be effective, the GCCACC motif must abut the AUG codon. Shifting the motif by just one nucleotide to the left or right abolishes its facilitating effect (13, 14). The strong contributions of A or G in position −3 and G in position +4, deduced initially in transfection assays with COS cells, have been confirmed in experiments with transformed plants (15, 16) and with standard *in vitro* translation systems from plants (17) and animals (17–21). One set of constructs used for the *in vitro* translation experiments was designed with two in-frame AUG codons, positioned to produce "long" and "short" versions of chloramphenicol acetyltransferase (17). The experiments carried out with those constructs revealed that a suboptimal context around the first AUG codon causes some 40 S ribosomal subunits to bypass the first AUG and initiate instead at the second AUG codon. Thus, context affects the fidelity as well as the efficiency of initiation. When the first AUG codon lies in a weak context, it is recognized inefficiently irrespective of the mRNA concentration and irrespective of the presence or absence of competing mRNAs (17), consistent with the scanning model which postulates that recognition of the AUG codon occurs *after* the competition-sensitive binding of the 40 S ribosome to the 5′-end of the mRNA. The leaky scanning that results from a suboptimal context around the first AUG codon enables some viral mRNAs to produce two proteins by initiating at the first and second AUG codons, as described elsewhere (1, 22).

The experimentally determined optimal context for initiation (13, 14) matches the consensus sequence derived from inspection of 699 vertebrate mRNA sequences (23). Except for yeasts (24), other eukaryotic organisms that have been examined show context effects similar to those described for vertebrates. Thus, ACC in positions −3 to −1 promotes translation in *Drosophila* (25), although the actual consensus sequence in flies differs slightly from mammals. Plant mRNAs have the expected purine in position −3 (in 93% of 252 mRNA sequences examined)[1] and the expected G in position +4 (in 74% of the mRNAs examined); and those conserved nucleotides augment translation in plants (15–17). A recent compilation of translational start sites in protozoa also shows a strong preference for A in position −3 (26).

Position

The scanning model predicts that ribosomes should initiate at the *first* AUG codon in a good context, a prediction that is upheld by most (perhaps 90% of) vertebrate mRNAs (23). (The number cannot be fixed more precisely because reports of cDNA sequences with upstream AUG codons often turn out to reflect errors or misinterpretations, as documented elsewhere (1, 2, 23).) The importance of position in determin-

[1] M. Kozak, unpublished compilation.

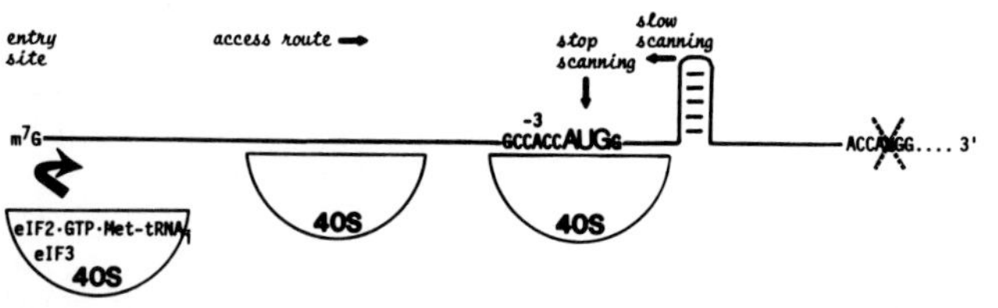

FIG. 1. **The scanning model for initiation of translation in eukaryotes.** The 40 S ribosomal subunit, carrying Met-$tRNA_i^{Met}$ and initiation factors, binds initially near the capped 5′-end of the mRNA and then migrates linearly to the first AUG codon. The contributions of context and downstream secondary structure are discussed in the text.

ing the site of initiation has been shown experimentally by introducing AUG codons upstream from the normal start site: insertion of a strong, upstream, out-of-frame AUG codon dramatically inhibits translation (2) while a strong, upstream, in-frame AUG codon supplants the original site of initiation (2, 27). A common mistake in trying to deduce translational start sites is to screen an entire cDNA sequence for the AUG codon that best matches the consensus sequence. That approach misses the point that the scanning 40 S ribosome evaluates AUG codons sequentially; the position of an AUG codon, relative to the 5′-end of the mRNA, is as important as its context. Indeed, position is more important since *some* ribosomes may initiate at the first AUG codon even when it occurs in the weakest context; and an A in position −3, irrespective of the rest of the context,[2] is usually sufficient for *most* ribosomes to select the first AUG codon.

While efficient translation thus requires that spurious upstream AUG codons be avoided, under some circumstances an upstream AUG codon in a favorable context will reduce but not abolish initiation from downstream. This happens when the first AUG codon is followed shortly by a terminator codon, creating a small open reading frame (ORF)[3] at the 5′-end of the mRNA. The simplest explanation is that, after an 80 S ribosome translates the 5′-"mini-cistron," the 40 S ribosomal subunit remains bound to the mRNA, resumes scanning, and may *reinitiate* at another AUG codon downstream. Some rudimentary rules for reinitiation have been deduced for mammals (28) as well as yeasts (29). In eukaryotic systems, expanding the distance between the 5′- and 3′-cistrons increases the efficiency of reinitiation, in contrast with bacterial systems where overlapping of the affected cistrons often enhances reinitiation. An important caveat is that reinitiation has been shown to occur fairly efficiently in eukaryotes only when the 5′-ORF is short (28–31). Thus, there are no naturally occurring bicistronic mRNAs from yeasts or mammals that express two *full-length* proteins from nonoverlapping cistrons,[4] and artificially constructed bicistronic transcripts of that sort allow only very inefficient translation of the second cistron (32–34). There are many plant and animal virus mRNAs that are structurally bicistronic, encoding two full-length proteins in nonoverlapping ORFs, but they are functionally monocistronic, *i.e.* only the 5′-proximal ORF is translated (35). The reason why eukaryotic ribosomes can reinitiate efficiently only near the 5′-end of the mRNA, *i.e.* after translating only a short 5′-ORF, might have to do with the kinetics of release of initiation factors (28).

Secondary Structure: Positive Effects

While there have long been hints that secondary structure[5] in mRNAs can reduce the efficiency of translation, it was surprising to find that a small amount of secondary structure near the start of the coding sequence can actually enhance recognition of the preceding AUG codon. Downstream secondary structure apparently contributes to the fidelity of initiation by preventing the 40 S ribosome from scanning too fast or too far. The effect is striking in that the introduction of downstream secondary structure (ΔG, −19 kcal/mol) can completely suppress the leaky scanning that otherwise occurs when the first AUG codon lies in an unfavorable context (36). The simplest rationalization is that downstream secondary structure slows scanning, thereby providing more time for recognition of the preceding AUG codon. The maximal effect was seen when 14 nucleotides intervened between the base of the hairpin and the preceding AUG codon, and that number fits nicely with RNase protection experiments which mapped the leading edge of an initiating ribosome 12–15 nucleotides 3′ of the AUG codon. Thus, our working hypothesis is that a hairpin located 12–15 nucleotides downstream from the AUG codon causes the 40 S ribosomal subunit to stall momentarily with its AUG-recognition center right over the AUG codon, thereby facilitating initiation.

There are several situations in which this positive effect of secondary structure might be important. Although 97% of vertebrate mRNAs have the required purine in position −3, very few possess the full consensus sequence. Thus, some feature in addition to primary structure would seem to be required to explain the usual absence of leakiness, and the possibility that secondary structure near the start of the coding sequence compensates for the less-than-perfect context around the AUG codon seems an attractive solution. The contribution of downstream secondary structure might be especially important for the handful of vertebrate mRNAs that initiate translation at an AUG codon in a very weak context (*i.e.* lacking both a purine in position −3 and G in position +4 (23)) or at a non-AUG codon. The experimental imposition of appropriately positioned secondary structure indeed increases initiation from cryptic non-AUG codons in test cases (36), and nearly all of the mRNAs that naturally support initiation from upstream non-AUG codons have extraordinarily GC-rich (hence highly structured) leader sequences (Fig. 2).

Secondary Structure: Negative Effects

In contrast with the positive effects of secondary structure introduced downstream from the AUG codon, stem-and-loop structures introduced between the cap and the AUG codon never facilitate initiation. Whether or not secondary structure upstream from the AUG codon impairs translation depends on the strength and position of the hairpin. A summary of the rules follows. (i) A modest amount of secondary structure near the cap (*i.e.* within the first 12 nucleotides) can drastically inhibit translation (7). Secondary structure in this position has been shown to prevent mRNA from binding to 40 S ribosomes (7), as expected if the 5′-end of the mRNA is the entry site for ribosomes (Fig. 1). The much discussed inhibition of ferretin mRNA translation by the IRE-binding protein (48, 49) might be attributable to stabilization by the repressor protein of 5′ secondary structure and consequent inhibition of ribosome entry. (ii) When secondary structure occurs sufficiently far from the cap that the initial binding of 40 S ribosomal subunits is not impaired, the stability of the hairpin determines whether or not it inhibits scanning. Stem-and-loop structures with a free energy of −30 kcal/mol positioned 50 or 60 nucleotides from the cap did not impair translation in COS cells (50) or in cell-free extracts (7). This seems remarkable inasmuch as base-paired structures as slight as −12 kcal/mol can drastically impair initiation in prokaryotic systems (51). It is clear that 40 S ribosomal subunits get past a −30 kcal/mol hairpin by migrating through it, rather than jumping over it, inasmuch as AUG codons that were buried

[2] This lack of dependence on the complete consensus sequence may be explained by the compensatory effect of downstream secondary structure, as discussed in the next section.

[3] The abbreviation used is: ORF, open reading frame.

[4] A few reported exceptions to this rule, discussed in Ref. 1, are not compelling.

[5] The discussion of secondary structure here and in the next section is limited to cases in which stem-and-loop structures were proven to exist by biochemical or genetic means.

Start upstream translation

L-myc, human	TGCAAGCTGGTGGGGTTGGGGAGGAACGAGAGCCCGGCA(16)GACCCGGGGACACCTCCTTCGCCCGGCCGG
c-myc, mouse	TAGACGCTGGATTTTTTTCGGGTAGTGGAAAACCAGCCTCCCGCGACGATGCCCCTCAACGTTAGCTTCACC
ltk tyrosine kinase	GAGAGGCTGGAGACCCGCGCGGCGGCGCCGGGCAGCGGCGGGAGAGGAGGCGCGGCAGGCGGTGATACTTCTG
int-2 gene, mouse	GCCGGCCTGGCGCGCGGGCGTGTGCTCCCAGCGCCGCGCCTTCGTGAGACCCGCGCTGGCGCAGCAGCC
p88[krox-24], mouse	TCCACCACGGGCCGCGGCTACCGCCAGCCTGGGGGCCCA(47)AACCCCCGGCGAG(18)GGCCCCGGGCTG
basic fibroblast GF	GGGAGGCTGGGGGGCCGGGGCCGGGGCCGTCCCCCGGAG(25)GGGGGACGGCGGCTCCCCGCG
pim-1, mouse	GCAGCCCTGGGTCCCGCAGCGCCTCTCGCCCTGCCGCCTCCCGCACTGCCTGACCCAGCCGGCGAACCCGCCCG
*AAV capsid protein B	GTTAAGACGGCTCCGGGAAAAAAGAGGCCGGTAGAGCACTCTCCTGTGGAGCCAGACTCCTCCTCGGGAACC
*MuLV gp85[gag]	GCAACCCTGGGAGACGTCCCAGGGACTTCGGGGGCCGTTTTTGTGGCCCGACCTGAG
*Equine IAV tat protein	TTGAACCTGGCTGATCGTAGGATCCCCGGGACAGCAGAGGAGAACTTACAGAAGTCTTCTGGAGGTGTTCCTGG

-3 +4

FIG. 2. **Sequences of eukaryotic mRNAs that initiate translation at upstream non-AUG codons, usually in addition to initiating at the first AUG codon.** The resulting N-terminally extended polypeptides produced by the three viral mRNAs (marked with *asterisks*) serve unique functions. In contrast, most cellular proteins initiated from upstream non-AUG codons have not been shown to mediate unique, essential functions; and their synthesis might be the inadvertent result of the way eukaryotic ribosomes reach the AUG codon (see text). *Dots* placed above each sequence (and below, when there is an alternative base-pairing scheme) indicate nucleotides that might anneal to form stem-and-loop structures; the overall GC-richness of most of these leader sequences makes many alternative pairings possible, however. In addition to the postulated contribution of downstream secondary structure, initiation at non-AUG codons requires a favorable primary sequence, especially in positions −3 and +4, as noted in the figure. Abbreviations: *GF,* growth factor; *AAV,* adeno-associated virus; *MuLV,* murine leukemia virus; *IAV,* infectious anemia (lenti)virus. The sequences are from Refs. 37–46. Not shown is the transcription enhancer factor TEF-1, which appears to initiate translation exclusively at an AUU codon by a mechanism that may be unique (47). For *ltk*, the upstream CUG initiator codon resides in an intron; AUG is the unique initiator codon in the mature mRNA (71).

in some of the hairpin structures became accessible to ribosomes (50). This raises the question of how such structures get melted. The possibility that certain initiation factor(s) may have helicase activity has been raised (52) and is evaluated elsewhere (4). The short answer is that it isn't known whether individual initiation factors or the 40 S ribosome-plus-factor complex actively unwinds secondary structures or whether the 40 S complex simply waits for such structures to breathe and then advances. Regardless of the mechanism, the ability of 40 S initiation complexes to penetrate base-paired structures has limits, as discussed next. (iii) Translation is profoundly inhibited *in vivo* (50) and *in vitro* (7) upon inserting into the 5′-noncoding domain a stem-and-loop structure with a free energy of −50 or −60 kcal/mol. Translation is inhibited even when the hairpin impinges on neither the cap nor the AUG codon. A very stable base-paired structure apparently inhibits translation by blocking the migration of 40 S ribosomes, as evidenced by RNase protection experiments which showed a 40 S ribosome trapped on the 5′-side of the hairpin (7). In contrast with the inability of 40 S ribosomal subunits to unwind a stem-and-loop structure of −60 kcal/mol, 80 S elongating ribosomes can, to some extent, penetrate such structures (7).

(iv) Some effects of secondary structure on the initiation of translation might be regulable, but this idea is much more speculative than the preceding points. It may be pertinent that a −30 kcal/mol hairpin (which normally does not inhibit translation, as explained above) becomes inhibitory when cells in culture are subjected to hypertonic stress (53). A somewhat related issue is whether the inhibitory effects of secondary structure are more pronounced in some cell types than in others. That question was raised by Muller and Witte (54) but not really answered, inasmuch as each time they switched cell types they also switched vectors. (If the vector-derived portion of the 5′-noncoding sequence is unstructured to begin with, as in baculovirus and riboprobe vectors, the introduction of a structure-prone leader sequence might be expected to inhibit translation more profoundly than if a bit more secondary structure is added to an already structured leader!) The recent report (55) that structure-prone mRNAs are translated more readily after fertilization of *Xenopus* eggs could indicate some novel, developmentally regulated helicase activity, although it is possible that covalent modification of hairpin-containing mRNAs by a previously recognized unwinding activity (56) is responsible for the enhanced translation.

Leader Length

Recognition of the first AUG codon may be impaired when it is positioned too close to the cap (57–59). When this issue was explored systematically, using synthetic transcripts in which the first AUG codon was in a favorable context, about half of the ribosomes bypassed the first AUG codon and initiated instead at a downstream site when the first AUG occurred within 12 nucleotides of the cap (60). The leakiness was suppressed when the leader sequence was lengthened to 20 nucleotides or when a modest amount of secondary structure was introduced downstream from the first AUG codon (60). In natural mRNAs the amount of secondary structure near the beginning of the coding region varies, making it hard to say *a priori* whether a particular short leader sequence will pose a problem.

Further lengthening of the 5′-noncoding sequence beyond the 20 or so nucleotides required for the fidelity of initiation can dramatically increase the efficiency of translation *in vitro* (6). The increased efficiency was clearly attributable to leader length, rather than to any particular sequence, inasmuch as insertion of three different synthetic oligonucleotides, each 60 nucleotides long, stimulated translation identically (6). The only feature common to all three sequences was a paucity of G residues, which ensured against the formation of secondary structure. The efficiency of translation *in vitro* was proportional to leader length in the range of 17 to about 80 nucleotides (6). Augmentation of translation by long, synthetic leader sequences mimics the effects of certain natural leader sequences (61–65). The fact that the precise sequence of these translational "enhancers" is not critical (6, 61–64) makes it unlikely that their facilitating effect on translation is mediated by proteins that recognize particular sequence motifs. Rather, the observed loading of extra 40 S ribosomal subunits on long leader sequences (6) seems likely to underlie the improvement in translation. In keeping with that interpretation, a long unstructured leader sequence augments translation only when it is at the exact 5′-end of the transcript (6, 53, 63). The introduction of a moderately long, unstructured, synthetic leader sequence turns out to be an easy way to increase the efficiency of *in vitro* expression vectors (6).

Closing Notes

The trick to identifying elements within 5′-noncoding sequences that modulate translation is to isolate each feature. There are many ways that the effects of context, upstream AUG codons, etc., might not be seen. Mutations that change the primary sequence around the initiator codon may have little effect if the sequence 3′ of the AUG codon is structured, inasmuch as downstream secondary structure compensates for absence of the preferred context (36). Since features in addition to context modulate initiation, as described herein, it follows that virtually nothing can be learned from comparing two completely different leader sequences in which context is only one of many variables (66). When matched mRNAs are compared,[6] converting a good context to a poorer one usually reduces the translational yield, but converting a poor context to a better one may not increase the yield of protein if some other step (such as elongation (67) or protein processing (68)) is limiting. Biological assays, albeit very sensitive, are so many steps removed from translation that failure to see the expected effects of context are difficult to interpret (69). In cases where less mRNA accumulates in cells under conditions (such as the imposition of a poor context) that impair translation, the gesture of "correcting" protein yields for differences in mRNA levels may ablate the perceived effects of context on translation (25).[7] Thus, the best systems for testing effects of context are those in which differences in translatability do not affect mRNA stability. Context effects and inhibition by upstream AUG codons may also be missed if inappropriate reaction conditions are used for *in vitro* translation. While the conditions recommended by some commercial suppliers of reticulocyte lysates give excellent overall incorporation of amino acids, in some cases they do not support the proper selection of initiator codons (70). With mRNAs that are translated poorly due to extensive secondary structure near the 5′-end, manipulating upstream AUG codons may be expected to have little impact because the effects of secondary structure dominate. Finally, the possibility of reinitiation complicates predictions about the effects of removing upstream AUG codons, since certain ORFs (those that favor reinitiation) can actually facilitate translation by blocking access to other upstream ORFs that are less conducive to reinitiation (28, 29).

The extent to which natural mRNA leader sequences conform to these experimentally determined requirements for initiation is examined elsewhere (1). A surprising realization is that, although nearly all vertebrate mRNAs have features that ensure the fidelity of initiation, many mRNAs that encode critical regulatory proteins do not appear to be designed for efficient translation. Thus, throttling at the level of translation may be an important component of gene regulation in vertebrates.

REFERENCES

1. Kozak, M. (1991) *J. Cell Biol.*, in press
2. Kozak, M. (1989) *J. Cell Biol.* **108,** 229–241
3. Pain, V. M. (1986) *Biochem. J.* **235,** 625–637
4. Kozak, M. (1990) in *Post-transcriptional Control of Gene Expression* (McCarthy, J. E. G., and Tuite, M. F., eds) pp. 227–236, Springer-Verlag, Berlin
5. Cigan, A. M., Feng, L., and Donahue, T. F. (1988) *Science* **242,** 93–97
6. Kozak, M. (1991) *Gene Expression* **1,** 117–125
7. Kozak, M. (1989) *Mol. Cell. Biol.* **9,** 5134–5142

[6] Indeed, interpretation can be complicated even when context is the only variable. Because the consensus sequence for initiation is GC-rich, mutations that improve the primary structure may inadvertently increase secondary structure to a point that becomes inhibitory. Some evidence of this was seen in Ref. 17, but the facilitating effect of context could nevertheless be discerned in those experiments by monitoring initiation from the first *versus* the second AUG codon.

[7] The problem is that a defect in translation sometimes accelerates mRNA degradation, in which case expressing the yield of protein as a function of the (lowered) steady-state level of mRNA makes the efficiency of translation appear greater than it really is.

8. Shatkin, A. J. (1976) *Cell* **9,** 645–653
9. Horikami, S. M., De Ferra, F., and Moyer, S. A. (1984) *Virology* **138,** 1–15
10. Fuerst, T. R., and Moss, B. (1989) *J. Mol. Biol.* **206,** 333–348
11. Malone, R. W., Felgner, P. L., and Verma, I. M. (1989) *Proc. Natl. Acad. Sci. U. S. A.* **86,** 6077–6081
12. Gallie, D. R., Lucas, W. J., and Walbot, V. (1989) *Plant Cell* **1,** 301–311
13. Kozak, M. (1986) *Cell* **44,** 283–292
14. Kozak, M. (1987) *J. Mol. Biol.* **196,** 947–950
15. Jones, J. D. G., Dean, C., Gidoni, D., Gilbert, D., Bond-Nutter, D., Lee, R., Bedbrook, J., and Dunsmuir, P. (1988) *Mol. Gen. Genet.* **212,** 536–542
16. Taylor, J. L., Jones, J. D. G., Sandler, S., Mueller, G. M., Bedbrook, J., and Dunsmuir, P. (1987) *Mol. Gen. Genet.* **210,** 572–577
17. Kozak, M. (1989) *Mol. Cell. Biol.* **9,** 5073–5080
18. Himmler, A., Drechsel, D., Kirschner, M. W., and Martin, D. W., Jr. (1989) *Mol. Cell. Biol.* **9,** 1381–1388
19. Query, C. C., Bentley, R. C., and Keene, J. D. (1989) *Cell* **57,** 89–101
20. Stirzaker, S. C., Whitfeld, P. L., Christie, D. L., Bellamy, A. R., and Both, G. W. (1987) *J. Cell Biol.* **105,** 2897–2903
21. Sundan, A., Evensen, G., Hornes, E., and Mathiesen, A. (1989) *Nucleic Acids Res.* **17,** 1717–1732
22. Kozak, M. (1986) *Cell* **47,** 481–483
23. Kozak, M. (1987) *Nucleic Acids Res.* **15,** 8125–8148
24. Cigan, A. M., Pabich, E. K., and Donahue, T. F. (1988) *Mol. Cell. Biol.* **8,** 2964–2975
25. Feng, Y., Gunter, L. E., Organ, E. L., and Cavener, D. R. (1991) *Mol. Cell. Biol.* **11,** 2149–2153
26. Yamauchi, K. (1991) *Nucleic Acids Res.* **19,** 2715–2720
27. Kozak, M. (1983) *Cell* **34,** 971–978
28. Kozak, M. (1987) *Mol. Cell. Biol.* **7,** 3438–3445
29. Abastado, J-P., Miller, P. F., Jackson, B. M., and Hinnebusch, A. G. (1991) *Mol. Cell. Biol.* **11,** 486–496
30. Werner, M., Feller, A., Messenguy, F., and Pierard, A. (1987) *Cell* **49,** 805–813
31. Sedman, S. A., and Mertz, J. E. (1988) *J. Virol.* **62,** 954–961
32. Angenon, G., Uotila, J., Kurkela, S. A., Teeri, T. H., Botterman, J., Van Montagu, M., and Depicker, A. (1989) *Mol. Cell. Biol.* **9,** 5676–5684
33. Hasemann, C. A., and Capra, J. D. (1990) *Proc. Natl. Acad. Sci. U. S. A.* **87,** 3942–3946
34. Kaufman, R. J., Murtha, P., and Davies, M. V. (1987) *EMBO J.* **6,** 187–193
35. Kozak, M. (1986) *Adv. Virus Res.* **31,** 229–292
36. Kozak, M. (1990) *Proc. Natl. Acad. Sci. U. S. A.* **87,** 8301–8305
37. Dosaka-Akita, H., Rosenberg, R. K., Minna, J. D., and Birrer, M. J. (1991) *Oncogene* **6,** 371–378
38. Hann, S. R., King, M. W., Bentley, D. L., Anderson, C. W., and Eisenman, R. N. (1988) *Cell* **52,** 185–195
39. Bernards, A., and de la Monte, S. M. (1990) *EMBO J.* **9,** 2279–2287
40. Acland, P., Dixon, M., Peters, G., and Dickson, C. (1990) *Nature* **343,** 662–665
41. Lemaire, P., Vesque, C., Schmitt, J., Stunnenberg, H., Frank, R., and Charnay, P. (1990) *Mol. Cell. Biol.* **10,** 3456–3467
42. Florkiewicz, R. Z., and Sommer, A. (1989) *Proc. Natl. Acad. Sci. U. S. A.* **86,** 3978–3981
43. Saris, C. J. M., Domen, J., and Berns, A. (1991) *EMBO J.* **10,** 655–664
44. Becerra, S. P., Rose, J. A., Hardy, M., Baroudy, B. M., and Anderson, C. W. (1985) *Proc. Natl. Acad. Sci. U. S. A.* **82,** 7919–7923
45. Prats, A.-C., De Billy, G., Wang, P., and Darlix, J.-L. (1989) *J. Mol. Biol.* **205,** 363–372
46. Stephens, R. M., Derse, D., and Rice, N. R. (1990) *J. Virol.* **64,** 3716–3725
47. Xiao, J. H., Davidson, I., Matthes, H., Garnier, J.-M., and Chambon, P. (1991) *Cell* **65,** 551–568
48. Goossen, B., Caughman, S. W., Harford, J. B., Klausner, R. D., and Hentze, M. W. (1990) *EMBO J.* **9,** 4127–4133
49. Harrell, C. M., McKenzie, A. R., Patino, M. M., Walden, W. E., and Theil, E. C. (1991) *Proc. Natl. Acad. Sci. U. S. A.* **88,** 4166–4170
50. Kozak, M. (1986) *Proc. Natl. Acad. Sci. U. S. A.* **83,** 2850–2854
51. Hall, M. N., Gabay, J., Débarbouillé, M., and Schwartz, M. (1982) *Nature* **295,** 616–618
52. Rozen, F., Edery, I., Meerovitch, K., Dever, T. E., Merrick, W. C., and Sonenberg, N. (1990) *Mol. Cell. Biol.* **10,** 1134–1144
53. Kozak, M. (1988) *Mol. Cell. Biol.* **8,** 2737–2744
54. Muller, A. J., and Witte, O. N. (1989) *Mol. Cell. Biol.* **9,** 5234–5238
55. Fu, L., Ye, R., Browder, L. W., and Johnston, R. N. (1991) *Science* **251,** 807–810
56. Bass, B. L., and Weintraub, H. (1988) *Cell* **55,** 1089–1098
57. Dabrowski, C., and Alwine, J. C. (1988) *J. Virol.* **62,** 3182–3192
58. Sedman, S. A., Gelembiuk, G. W., and Mertz, J. E. (1990) *J. Virol.* **64,** 453–457
59. Gillman, E. C., Slusher, L. B., Martin, N. C., and Hopper, A. K. (1991) *Mol. Cell. Biol.* **11,** 2382–2390
60. Kozak, M. (1991) *Gene Expression* **1,** 111–115
61. Sleat, D. E., Hull, R., Turner, P. C., and Wilson, T. M. A. (1988) *Eur. J. Biochem.* **175,** 75–86
62. Dolph, P. J., Huang, J., and Schneider, R. J. (1990) *J. Virol.* **64,** 2669–2677
63. McGarry, T. J., and Lindquist, S. (1985) *Cell* **42,** 903–911
64. Schöffl, F., Rieping, M., Baumann, G., Bevan, M., and Angermüller, S. (1989) *Mol. Gen. Genet.* **217,** 246–253
65. Hooft van Iddekinge, B. J. L., Smith, G. E., and Summers, M. D. (1983) *Virology* **131,** 561–565
66. Falcone, D., and Andrews, D. W. (1991) *Mol. Cell. Biol.* **11,** 2656–2664
67. Fajardo, J. E., and Shatkin, A. J. (1990) *Proc. Natl. Acad. Sci. U. S. A.* **87,** 328–332
68. Sarver, N., Ricca, G. A., Link, J., Nathan, M. H., Newman, J., and Drohan, W. N. (1987) *DNA* **6,** 553–564
69. Li, M.-G., and Starlinger, P. (1990) *Proc. Natl. Acad. Sci. U. S. A.* **87,** 6044–6048
70. Kozak, M. (1990) *Nucleic Acids Res.* **18,** 2828
71. Krolewski, J. J., and Dalla-Favera, R. (1991) *EMBO J.* **10,** 2911

Minireview

THE JOURNAL OF BIOLOGICAL CHEMISTRY
Vol. 266, No. 31, Issue of November 5, pp. 20579-20582, 1991

Printed in U.S.A.

Molecular Genetics of Alzheimer Disease Amyloid*

Rudolph E. Tanzi‡, Peter St. George-Hyslop§, and James F. Gusella

From the Molecular Neurogenetics Laboratory, Massachusetts General Hospital, Department of Genetics, Harvard Medical School, Charlestown, Massachusetts 02129

Alzheimer disease is a late onset neurodegenerative disorder characterized by global cognitive decline. Definitive diagnosis of Alzheimer disease requires either the autopsy or biopsy confirmation of specific lesions formed by insoluble proteinaceous fibers in the brain: extracellular deposits of amyloid in senile plaques and cerebral blood vessels, and intracellular neurofibrillary tangles in the cytoplasm of neurons (1, 2). Although intracellular amyloid has been observed in intimate association with neurofibrillary tangles, the relationship between tangles and senile plaques remains unclear (3).

The etiology of Alzheimer disease is unknown. Although most cases appear sporadic, epidemiological surveys have demonstrated an increased risk in first- and second-degree relatives of Alzheimer disease patients (4–10). A small proportion of Alzheimer disease cases occur in families displaying autosomal dominant inheritance of the disorder. In these instances, termed familial Alzheimer disease, the disease seems to be triggered by a genetic defect (Fig. 1). Considerable controversy surrounds the question of what proportion of Alzheimer disease is familial (5–100% (7, 10)), in part because it is often difficult to assess whether familial aggregation of Alzheimer disease in late onset pedigrees reflects genetic inheritance, predisposition, or clustering of a sporadic form of Alzheimer disease resulting from an environmental origin or ascertainment bias. Most cases of Alzheimer disease display onset beyond the age of 65, but in the largest familial Alzheimer disease kindreds the disorder appears earlier in life (age 40–50). Familial Alzheimer disease is otherwise clinically and neuropathologically indistinguishable from cases with later onset.

Genetic Linkage Studies of Familial Alzheimer Disease

In addition to Alzheimer disease, senile plaques and neurofibrillary tangles are also found in older (over 35 years) patients with Down syndrome (trisomy 21 (2, 11)). Consequently, chromosome 21 became one of the first targets of genetic linkage studies aimed at localizing the familial Alzheimer disease gene defect. This approach requires tracing the inheritance of polymorphic genetic markers in families with the disorder. The advent of recombinant DNA technology has allowed the direct detection of DNA sequence differences. These may often be assayed as restriction fragment length polymorphisms and provide an abundant source of DNA markers. If a genetic marker is located close to a disease gene on a given chromosome, the two will show a correlated pattern of inheritance, with a particular marker allele segregating (or showing "linkage") with the disorder in an individual family. The significance of an observed pattern of correlated inheritance is assessed by "maximum likelihood analysis" to calculate a "log of the odds" (LOD)[1] score. The LOD score represents $\log_{10}$ of the likelihood of linkage relative to the likelihood of random segregation for a given data set. An LOD score in excess of +3 (1,000:1 odds in favor of linkage) is considered as proof that the genetic marker and the gene defect reside on the same chromosome and are linked.

In 1987, two DNA markers on the proximal long arm (q arm) of chromosome 21, *D21S16* and *D21S1/D21S11*, were found to display genetic linkage to the disorder in a study of four large familial Alzheimer disease pedigrees exhibiting an early age of onset (<65 years (12)). An example of one of these pedigrees is shown in Fig. 1. Concurrently, the principal component of Alzheimer's associated amyloid, βA4, a 42-amino acid peptide, was found to derive from a larger protein precursor (the amyloid β protein precursor, APP) encoded by a gene (*APP*) on chromosome 21q (13–16). However, when the *APP* gene was directly tested for genetic linkage with familial Alzheimer disease, several recombinants between APP and familial Alzheimer disease were observed (17, 18), suggesting that it was not the site of the primary genetic defect in familial Alzheimer disease.

Following these reports, several groups endeavored to test the same chromosome 21 DNA markers in additional familial Alzheimer disease kindreds. Goate *et al.* (19) obtained evidence supporting the presence of a familial Alzheimer disease gene defect centromeric to the markers, *D21S1/S11* and *D21S16*. Another group was able to detect probable linkage between familial Alzheimer disease and another centromeric marker, *D21S13* (20), closely linked to *D21S16*. These results provided further support for the initial finding of a familial Alzheimer disease gene defect on 21q. On the other hand, Schellenberg *et al.* (21) reported the possibility of non-allelic heterogeneity based on the exclusion of familial Alzheimer disease from the *D21S1/D21S11* region in a set of pedigrees of predominantly Volga German descent. Pericak-Vance *et al.* (22) suggested the possibility of a second familial Alzheimer disease locus based on the lack of linkage between *D21S1/D21S11* and familial Alzheimer disease in pedigrees manifesting a relatively late age-of-onset, although an early onset familial Alzheimer disease pedigree displayed positive linkage to *D21S1/D21S11* and *D21S16* in the same study.

An impressive international collaborative effort was mounted to type 48 familial Alzheimer disease pedigrees for five DNA polymorphisms at three loci from the proximal half of chromosome 21q (23). *D21S1/D21S11* and *D21S16/D21S13* both gave LOD score values for linkage to familial Alzheimer disease that were greater than 3.0 supporting the original finding of linkage to chromosome 21. In addition, the

* Alzheimer disease and chromosome 21-related investigations in our laboratories are supported by grants from the American Health Assistance Foundation, the Metropolitan Life Foundation, the Alzheimer's Association (Mrs. Frank L. Harrington pilot research grant), and National Institutes of Health Grant HG00317.

‡ Recipient of a French Foundation fellowship.

§ Present address: Tanz Neuroscience Bldg., University of Toronto, 6 Queen's Park Crescent, Toronto, Ontario M5S 1A8, Canada.

[1] The abbreviations used are: LOD, log of the odds; APP, amyloid β protein precursor; KPI, Kunitz protease inhibitor.

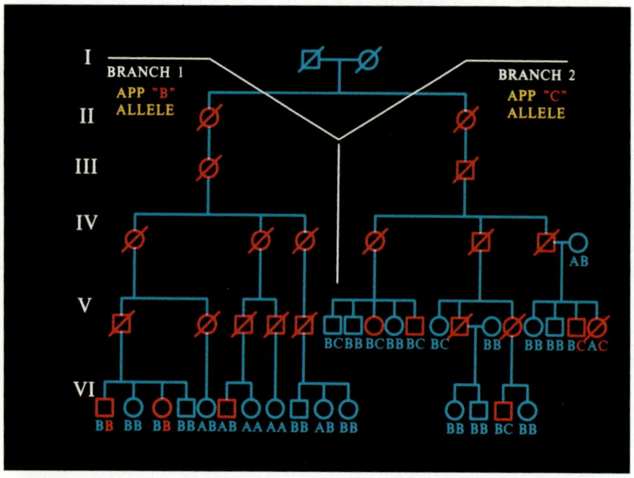

FIG. 1. **The Italian familial Alzheimer disease kindred (FAD4 (12)) and APP genotypes indicating an apparent recombination event between APP and the familial Alzheimer disease gene defect.** Affected individuals are shown in *red* while those without the Alzheimer disease phenotype are shown in *blue. Slashed symbols* indicate individuals who are deceased. The APP genotypes are shown under each individual for whom DNA is available for typing. The APP alleles represent a combination of two *Eco*RI RFLPs detected by the APP cDNAs, FB68L and HL124 (17). Where the phase of the disease gene with respect to the APP RFLP can be determined unequivocally, the allele of the marker segregating with the defect is shown in *red.* A single apparent recombination event is revealed by comparing the four-member sibship of Branch I, generation VI, with the four-member sibship of Branch II, generation V. In the former, the "B" allele of APP is present on the familial Alzheimer disease chromosome while in the latter the "C" allele of APP is detected. Although the relative phase of the defect and the APP alleles cannot be assigned with certainty in the other affected members of each branch, the crossover event +probably occurred in the multiple generations separating the two branches of the pedigree.

Affected Pedigree Member method (24), which examines the frequency with which affected family members "share" alleles of a specific polymorphisms, confirmed the linkage results by demonstrating a significant association between familial Alzheimer disease and the three chromosome 21 loci.

This study also revealed that only a subset of the 48 pedigrees contributed positively to the overall data set and demonstrated that familial Alzheimer disease is genetically heterogeneous. While some pedigrees with early onset (<65 years) displayed positive linkage with chromosome 21, late onset pedigrees were unlinked. Multipoint analysis aimed at localizing the gene defect relative to the two linked loci yielded peak LOD scores of approximately 5.0 for the early onset pedigrees in two distinct regions of chromosome 21, one proximal to *D21S16/D21S13* (the centromeric region) and the other distal to *D21S1/D21S11* (the *APP* region). The presence of two multipoint peaks suggests the possibility that the genetic defects in different chromosome 21-linked familial Alzheimer disease kindreds could be at distinct locations on chromosome 21. However, it could also be due to the confounding effect that would be caused if a portion of the early onset pedigrees are, in fact, not linked to chromosome 21. The confirmed existence of non-allelic genetic heterogeneity in familial Alzheimer disease has prompted further scanning of the genome for additional familial Alzheimer disease loci. To date, a locus on chromosome 19 has been suggested as a possible predisposing or causative factor in late onset Alzheimer disease, but this remains to be confirmed (25).

The Amyloid β Peptide Precursor

An examination of the pathological profile of a genetic disease can potentially lead to candidate genes for the site of the defect. Accordingly, a major focus in studies of familial Alzheimer disease has been the locus encoding the protein precursor (APP) of the 4.2-kDa peptide, βA4, which is the chief component of amyloid in senile plaques and cerebrovascular deposits (2). APP mRNA can be alternatively spliced into three principal transcripts named according to the number of amino acids they encode. The original full-length APP gene isolated, APP695, is expressed exclusively in the brain (14, 16) while two larger forms, APP751 and APP770, are also expressed in peripheral tissues (26, 27). APP751 and APP770 contain a 56-amino acid domain that interrupts residue 289 of APP695 (26–28). This domain represents a Kunitz-type serine protease inhibitor.

APP is an integral membrane-bound protein (14) with the Kunitz protease inhibitor (KPI) domain located in the large extracellular portion of the molecule (Fig. 2). While one-third of the βA4 domain resides in the transmembrane region of the molecule, the remaining two-thirds is situated in the extracellular space. The extracellular portion of APP is both *O*- and *N*-glycosylated and can be cleaved for release into the extracellular space as a large soluble protein (110 kDa) (29, 30). The secreted form of APP containing the KPI domain is identical to the serine protease inhibitor, protease nexin II (31, 32). Protease nexin II inhibits Factor XIa of the coagulation pathway, chymotrypsin, and the kallikreins, 7 S nerve growth factor γ-subunit and epidermal growth factor-binding protein. In peripheral cell cultures, the secreted soluble derivatives of APP contain only a portion of the βA4 peptide while the remainder is contained in a membrane-bound C-terminal fragment (33, 34). Thus, normal processing of APP destroys the integrity of the βA4 domain, presumably precluding amyloid formation and raising the possibility that the generation of βA4 and amyloidosis may be the consequence of abnormal processing of APP.

The *APP* gene maps to chromosome 21 at the border of cytogenetic bands 21q21.3 and 21q22 (35). Consequently, the presence of βA4 amyloid deposits in the brains of patients

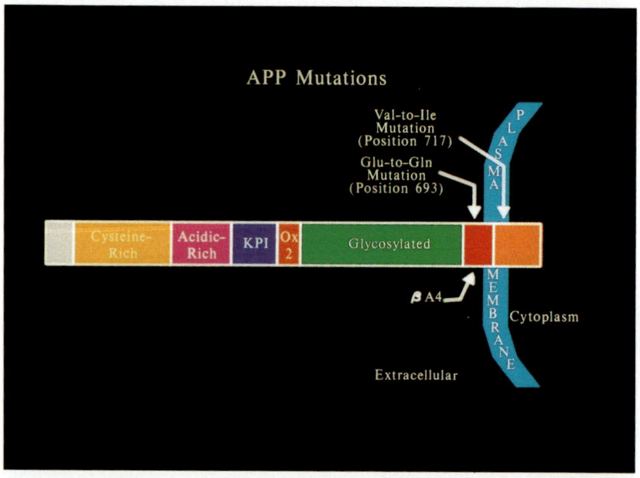

FIG. 2. **Schematic representation of APP and the mutations found in the portion of the molecule encoded by exon 17.** The extracellular portion of the molecule contains a signal peptide (shown in *white,* but actually removed during insertion into the membrane), a cysteine-rich region, an acidic region, and a large glycosylated region. Between the two latter segments are two domains: encoded by alternatively spliced exons, the KPI domain and OX-2. The extracellular portion of APP contains roughly two-thirds of the βA4 region. The remainder of βA4 resides within the membrane-spanning domain; the C-terminal portion of APP is cytoplasmic. The mutations at positions 693 and 717 (of APP770) represent putative sites for the defects in HCHWA-D and in an inherited form of Alzheimer disease, respectively.

with Down syndrome (trisomy 21) might be most easily explained by gene dosage. With the exception of one preliminary and apparently premature report (36), increased APP gene dosage in the germ line does not appear to account for the presence of amyloid in Alzheimer disease or familial Alzheimer disease (37–39).

Mutations in APP

If the *APP* gene actually harbored the defect leading to familial Alzheimer disease, one would expect to observe no crossovers between the two loci in a direct two-point genetic linkage test. When tested in the four original familial Alzheimer disease families used to demonstrate linkage to chromosome 21 and other pedigrees, multiple recombination events were observed between APP and familial Alzheimer disease in affected individuals (12, 13) (see example in Fig. 1). The presence of at least one crossover in each of the four families indicated that the APP gene is most likely not the site of the defect in any of these pedigrees. Maximum likelihood analysis indicated that the *APP* gene was at least 8 cm away from the familial Alzheimer disease defect (12).

The recent demonstration that familial Alzheimer disease is genetically heterogeneous renewed the possibility that the *APP* gene might still be the site of the defect in some familial Alzheimer disease pedigrees, especially those displaying linkage to chromosome 21 and no crossovers with *APP*. Goate *et al.* (40) re-examined *APP* genetic linkage in six familial Alzheimer disease families and found two pedigrees with no crossovers between *APP* and the disorder. When a portion of exon 17 of the APP gene from affected individuals in these two pedigrees was sequenced, a conservative amino acid change was discovered. Exon 17 was chosen on the basis of the recent report that a mutation in this exon, which encodes the βA4 region, represents the gene defect in the rare Dutch disorder, hereditary cerebral hemorrhage with amyloidosis (HCHWA-D, Ref. 41) (see Fig. 2). In HCHWA-D, βA4-type amyloid accumulates in large amounts in cerebral vessels and ultimately leads to stroke and death by the fifth or sixth decade of life (42).

The apparent missense mutation in exon 17 discovered in familial Alzheimer disease caused the substitution of an isoleucine for a valine at residue 717 (according to APP770 sequence) within the transmembrane region immediately C-terminal to the βA4 domain. This base substitution co-segregated with familial Alzheimer disease in two separate pedigrees but was absent from eight other familial Alzheimer disease families and from 200 normal chromosomes analyzed (40). Although it has now been found in four additional familial Alzheimer disease kindreds (43), the vast majority of familial Alzheimer disease pedigrees of both early and late onset do not exhibit this mutation (40, 44, 45). While it is conceivable that this alteration could represent a very rare, neutral polymorphism, it seems far more likely that, at least in some cases, mutation at the APP locus does cause familial Alzheimer disease.

In all but one of the six families reported with this mutation, a distinct neuropathological profile has been seen (40, 43, 46) including cortical diffuse Lewy bodies and abundant congophilic amyloid angiopathy leading in some cases to stroke. This latter feature is reminiscent of the phenotype of patients with Dutch HCHWA-D in which dementia sometimes follows cerebral strokes. Thus, the two known APP mutations might be viewed as allelic forms of the same disease characterized by prominent vascular involvement.

It has been suggested that the slightly more hydrophobic isoleucine residue replacing the valine at position 717 may alter the anchoring of APP in the membrane (40). Along these lines, an interesting comparison may be drawn with the dominant mutation, *deg-1(u38)*, which results in neuronal degeneration in *Caenorhabditis elegans* (47). This mutation ultimately leads to death with late onset long after neuronal synapses have been formed. The *deg-1* gene appears to encode a membrane-associated protein; it has been suggested that a mutation in this gene might lead to compromised membrane integrity and cell lysis. A similar phenotype is observed with mutations in the *mec-4* gene which is a close homologue of the *deg-1* gene (48). Mutations in both the *mec-4* and *deg-1(u38)* genes involve the same residue, Ala-442, in the hydrophobic domain of the protein (48). Substitutions of Thr or Val for Ala are sufficient to induce degeneration apparently via steric hindrance. Driscoll and colleagues (48) have found that cell viability is maintained if the amino acid in position 442 contains a side chain no larger than a sulfhydryl on the β carbon. Interestingly, Ala-442 is predicted to be contained in a β turn structure close to or within the hydrophobic membrane-spanning domain. Likewise, both mutations identified in APP occur in the hydrophobic βA4 domain either close to or within the trans-membrane domain. The APP717 mutation associated with familial Alzheimer disease replaces a valine with an isoleucine. The latter residue possesses a larger nonpolar R group. By analogy to the proposed action of Ala-442 mutations in *C. elegans*, the APP717 substitution could be envisioned to disrupt membrane integrity resulting in the release of unprocessed APP. Consequently, the βA4 domain would not undergo "normal" cleavage and become potentially amenable to amyloid formation.

An alternative explanation for the role of the APP717 mutation in familial Alzheimer disease pathogenesis involves the disruption of a putative regulatory stem-loop structure in APP mRNA (49). The mutation site is flanked by 26 nucleotides capable of forming a stem-loop structure which contains the consensus sequence CAGUGA characteristic of the iron-responsive elements that regulate translation of ferritin and transferrin receptor in response to iron concentration (50). The putative iron-responsive element in APP is situated precisely at residues 40–42 of the βA4 domain. The destabilizing effect imparted on the stem-loop structure by the APP717 substitution could, by analogy to ferritin and transferrin receptor, affect translational regulation of APP resulting in either altered levels of APP protein or abnormally truncated products. Either the overproduction of APP or the production of truncated APP molecules containing intact βA4 domains and lacking cytoplasmic portions could conceivably lead to amyloid formation.

Summary

It is not yet clear whether linkage of familial Alzheimer disease to chromosome 21 could be explained entirely by mutations in APP. Goate *et al.* (40) have raised the possibility that reports of recombination between APP and familial Alzheimer disease in chromosome 21-linked pedigrees may have been in error because of factors such as mistyping, misdiagnosis, non-paternity, and phenocopy. The largest family used in the original report of linkage of familial Alzheimer disease to chromosome 21 (12) is shown in Fig. 1. It yields a peak LOD score of 2.94 at 15 cM with the proximal chromosome 21 marker *D21S52* located approximately 15 cm from APP. This early onset (40–50 years) familial Alzheimer disease pedigree contains two branches in which different APP alleles appear to segregate with Alzheimer disease. Since it is highly unlikely that all affected individuals in each branch were mistyped and/or misdiagnosed, this obligate crossover event implies several possibilities. 1) APP is not the site of the defect in this familial Alzheimer disease pedigree for which there is suggestive but not significant evidence for linkage to chromosome 21; or 2) APP is the site of the defect, and the apparent crossover event is due either to intragenic recombination within the relatively large APP gene (approximately 180 kilobases) or to the introduction into this pedigree of a second familial Alzheimer disease gene. The complete sequencing of all exons and the promotor of APP in affected individuals of this pedigree should resolve whether APP is the site of the gene defect in this family.

Continued linkage analysis of familial Alzheimer disease pedigrees with highly informative polymorphisms (*e.g.* simple sequence repeats) for APP and other chromosome 21 loci should not only allow the identification of familial Alzheimer disease kindreds potentially harboring APP mutations but also resolve the issue of whether a familial Alzheimer disease gene defect distinct from APP is also located on chromosome 21. In view of the complex genetic nature of inherited Alzheimer disease, an extensive collaborative effort of multiple groups to methodically screen familial Alzheimer disease families with loci spanning the human genome will probably be necessary to accelerate the localization and identification of both causative and predisposing familial Alzheimer disease gene defects. However, given the important role played by *APP* and the presence of a potentially causative mutation in this gene in a few familial Alzheimer disease families, attempts to define genetic, biochemical, and environmental factors affecting expression of the *APP* gene and the processing of its product, including the possibility of additional mutations in the *APP* gene, offer a alternative approach to unraveling the enigmatic etiology of Alzheimer disease.

REFERENCES

1. Terry, R. D., and Katzman, R. (1983) *Ann. Neurol.* **14,** 497
2. Glenner, G. G., and Wong, C. W. (1984) *Biochem. Biophys. Res. Commun.* **120,** 885–890
3. Masters, C. L., Multhaup, G., Simms, G., Pottgiesser, J., Martins, R. N., and Beyreuther, K. (1985) *EMBO J.* **4,** 2757–2760
4. Appel, S. (1981) *Ann. Neurol.* **20,** 499–505
5. Amaducci, L. A., Fratiglioni, L., and Rocca, W. A. (1986) *Neurology* **36,** 922–931
6. Mohs, R. C., Breitner, J. C., Silverman, J. M., and Davis, K. L. (1987) *Arch. Gen. Psychiatry* **44,** 405–408
7. Breitner, J. C., Silverman, J. S., Mohs, R. C., and David, K. L. (1988) *Neurology* **38,** 207–212
8. Farrer, L. A., O'Sullivan, D. M., Cupples, L. A., Growdon, J. H., and Myers, R. H. (1989) *Ann. Neurol.* **25,** 485–493
9. Farrer, L. A., Myers, R. H., Connor, L., Cupples, A., and Growdon, J. H. (1991) *Am. J. Hum. Genet.* **48,** 1026–1033
10. St. George-Hyslop, P. H., Myers, R. D., Haines, J. L., Farrer, L. A., Tanzi, R. E., Abe, K., James, M. F., Conneally, P. M., Polinsky, R. J., and Gusella, J. F. (1989) *Neurobiol. Aging* **10,** 417–425
11. Heston, L. L., Mastri, A. R., Anderson, E., and White, J. (1981) *Arch. Gen. Psychiatry* **31,** 1085–1090
12. St. George-Hyslop, P. H., Tanzi, R. E., Polinsky, R. J., Haines, J. L., Nee, L., Watkins, P. C., Myers, R. H., Feldman, R. G., Pollen, D., Drachman, D., Growdon, J., Bruni, A., Foncin, J-F., Salmon, D., Frommelt, P., Amaducci, L., Sorbi, S., Piacentini, S., Stewart, G. D., Hobbs, W. J., Conneally, P. M., and Gusella, J. F. (1987) *Science* **235,** 885–889
13. Goldgaber, D., Lerman, J. I., McBride, O. W., Saffiotti, U., and Gajdusek, D. C. (1987) *Science* (1987) **235,** 877–880
14. Kang, J., Lemaire, H. G., Unterbeck, A., Salbaum, J., Masters, L., Grzeschik, K. H., Multhaup, G., Beyreuther, K., and Muller-Hill, B. (1987) *Nature* **325,** 733–736
15. Robakis, N. K., Ramakrishna, N., Wolfe, G., and Wisniewski, H. M. (1987) *Proc. Natl. Acad. Sci. U. S. A.* **84,** 4190–4194
16. Tanzi, R. E., Gusella, J. F., Watkins, P. C., Bruns, G. A., St. George-Hyslop, P., VanKeuren, M. L., Patterson, S. P., Kurnit, D. M., and Neve, R. L. (1987) *Science* **235,** 880–884
17. Tanzi, R. E., St. George-Hyslop, P. H., Haines, J. L., Polinsky, R. J., Nee, L., Foncin, J-F., Neve, R. L., McClatchey, A. I., Conneally, P. M., and Gusella, J. F. (1987) *Nature* **329,** 156–157
18. Van Broeckhoven, C., Genthe, C. A., Vandenberghe, B., Horstemke, B., Backhovens, P., Raeymaekers, P., Van Hul, W., Wehnert, J., Gheuens, P., Cras, P., Bruyland, M., Martin, J. J., Salbaum, M., Multhaup, G., Masters, C. L., Beyreuther, K., Gurling, H., Mullan, M. J., Holland, A., Barton, N., Irving, N., Williamson, R., Richards, S. J., and Hardy, J. A. (1987) *Nature* **329,** 153–155
19. Goate, A. M., Haynes, A. R., Owen, M. J., Farrall, M., James, L. A., Lai, L. Y. C., Mullan, M. J., Roques, P., Rossor, M. N., Williamson, R., and Hardy, J. A. (1989) *Lancet* **i,** 352–355
20. Van Broeckhoven, C., van Hul, W., Backhoeven, H., Van Camp, G., Stinissen, P., Wehnert, A., Raeymaekers, P., De Winter, G., Bruyland, M., Gheuens, J., Martin, J. J., and Vandenberghe, B. (1988) *Am. J. Hum. Genet.* **43,** 8205
21. Schellenberg, G. D., Bird, T. D., Wijsman, E. M., Moore, D. K., Boehnke, M., Bryant, E. M., Lampe, T. H., Nochlin, D., Sumi, S. M., Deeb, S. S., Beyreuther, K., and Martin, G. M. (1988) *Science* **241,** 1507–1510
22. Pericak-Vance, M. A., Yamakoa, L. H., Haynes, C. S., Gaskell, P. C., Hung, W-Y., Clark, C. M., Heyman, A. L., and Roses, A. D. (1988) *Exp. Neurol.* **102,** 271–279
23. St. George-Hyslop, P., and the familial Alzheimer disease Collaborative Group (1990) *Nature* **347,** 194–197
24. Weeks, D., and Lange, K. (1988) *Am. J. Hum. Genet.* **42,** 315–326
25. Pericak-Vance, M. A., Bebout, J. L., Gaskell, P. C., Yamaoka, L. H., Hung, W-Y., Alberts, M. J., Walker, A. P., Bartlett, R. J., Haynes, C. A., Welsh, K. A., Earl, N. L., Heyman, A., Clark, C. M., and Roses, A. D. (1991) *Am. J. Hum. Genet.* **48,** 1034–1050
26. Tanzi, R. E., McClatchey, A. I., Lamperti, E. D., Villa-Komaroff, L., Gusella, J. F., and Neve, R. L. (1988) *Nature* **331,** 528–530
27. Ponte, P., Gonzalez-DeWhitt, P., Schilling, J., Miller, J., Hsu, D., Greenberg, B., Davis, K., Wallace, W., Lieberburg, I., Fuller, F., and Cordell, B. (1988) *Nature* **331,** 525–527
28. Kitaguchi, N., Takahashi, Y., Tokushima, Y., Shiojiri, S., and Ito, H. (1988) *Nature* **331,** 530–532
29. Weidemann, A., Konig, G., Bunke, D., Fischer, P., Salbaum, M. J., Masters, C., and Beyreuther, K. (1989) *Cell* **57,** 115–126
30. Esch, F. S., Keim, P. S., Beattie, E. C., Blacher, R. W., Culwell, A. R., Oltersdorf, T., McClure, D., and Ward, P. J. (1990) *Science* **248,** 1122–1124
31. Oltersdorf, T., Fritz, F. C., Schenk, D. B., Lieberberg, I., Johnson-Wood, K. L., Beattie, E. C., Ward, P. J., Blacher, R. W., Dovey, H. F., and Sinha, S. (1989) *Nature* **341,** 144–147
32. Van Nostrand, W. E., Wagner, S. L., Suzuki, M., Choi, B. H., Farrow, J. S., Geddes, J. W., Cotman, C. W., and Cunningham, D. D. (1989) *Nature* **341,** 546–549
33. Sisodia, S. S., Koo, E. H., Beyreuther, K., Unterbeck, A., and Price, D. L. (1990) *Science* **248,** 492–495
34. Palmert, M. R., Podlisney, M. B., Witker, D. S., Oltersdorf, T., Younkin, L. H., Selkoe, D. J., and Younkin, S. G. (1989) *Proc. Natl. Acad. Sci. U. S. A.* **86,** 6338–6342
35. Patterson, D., Gardiner, K., Kao, F-T., Tanzi, R., Watkins, P., and Gusella, J. (1988) *Proc. Natl. Acad. Sci. U. S. A.* **85,** 8266–8270
36. Delabar, J-M., Goldgaber, D., Lamour, Y., Nicole, A., Huret, J-L., de Grouchy, J., Brown, P., Gajdusek, D. C., and Sinet, P-M. (1987) *Science* **235,** 1390–1392
37. Tanzi, R. E., Bird, E. D., Latt, S. A., and Neve, R. L. (1987) *Science* **238,** 666–669
38. St. George-Hyslop, P. H., Tanzi, R. E., Polinsky, R. J., Neve, R. L., Pollen, D., Drachman, D., Growdon, J., Cupples, L. A., Nee, L., Myers, R. H., O'Sullivan, D., Watkins, P. C., Amos, J. A., Deutsch, C. K., Bodfish, J. W., Kinsbourne, M., Feldman, R. G., Bruni, A., Amaducci, L., Foncin, J-F., and Gusella, J. F. (1987) *Science* **238,** 664–666
39. Podlisney, M., Lee, G., and Selkoe, D. (1987) *Science* **238,** 669–671
40. Goate, A. M., Chartier-Harlin, M. C., Mullan, M. C., Brown, J., Crawford, F., Fidani, L., Guiffra, A., Haynes, A., Irving, N., James, L., Mant, R., Newton, P., Rooke, K., Roques, P., Talbot, C., Williamson, R., Rossor, M., Owen, M., and Hardy, J. (1991) *Nature* **349,** 704–706
41. Levy, E., Carman, M. D., Fernandez-Madrid, I. J., Power, M. D., Lieberburg, I., Sjoerd, G., van Duinen, S. G., Bots, G., Luyendijk, W., and Frangione, B. (1990) *Science* **248,** 1124–1126
42. van Duinen, S. G., Castano, E. M., Prelli, F., Bots, G. T., Luyendijk, W., and Frangione, B. (1987) *Proc. Natl. Acad. Sci. U. S. A.* **84,** 5991–5994
43. Naruse, S., Igarashi, S., Kobayashi, H., Aoki, K., Inuzuki, I., Kaneko, K., Shimizu, T., Iihara, K., Kojima, T., Miyatake, T., and Tsuji, T. (1991) *Lancet* **337,** 978–979
44. Van Duijn, C. M., Hendriks, L., Cruts, M., Hardy, J. A., Hofman, A., and Van Broeckhoven, C. (1991) *Lancet* **337,** 978
45. Schellenberg, G. D., Anderson, L., O'Dahl, S., Wisjman, E. M., Sadovnick, A. D., Ball, M. J., Larson, E. B., Kukull, W. A., Martin, G. M., Roses, A. D., and Bird, T. D. (1991) *Am. J. Hum. Genet.* **49,** 511–517
46. Hardy, J. A., Mullan, M., Chartier-Harlin, M-C., Brown, J., Goate, A., Rossor, M., Collinge, J., Roberts, G., Luthert, P., Lantos, P., Naruse, S., Kaneko, K., Tsuji, T., Miyatake, T., Shimizu, T., Kojima, T., Nakano, I., Yoshioka, K., Sakaki, Y., Miki, T., Katsuya, T., Ogihara, T., Roses, A., Pericak-Vance, M., Haan, J., Roos, R., Lucotte, G., and David, F. (1991) *Lancet* **337,** 1342–1343
47. Chalfie, M., and Wolinsky, E. (1990) *Nature* **345,** 410–416
48. Driscoll, M., and Chalfie, M. (1991) *Nature* **349,** 588–593
49. Tanzi, R. E., and Hyman, B. T. (1991) *Nature* **350,** 564
50. Leibold, E. A., and Munro, H. N. (1987) *J. Biol. Chem.* **262,** 7335–7341

Minireview

THE JOURNAL OF BIOLOGICAL CHEMISTRY
Vol. 266, No. 32, Issue of November 15, pp. 21327–21330, 1991

Printed in U.S.A.

Lysosomal Membrane Glycoproteins

STRUCTURE, BIOSYNTHESIS, AND INTRACELLULAR TRAFFICKING*

Minoru Fukuda

From the La Jolla Cancer Research Foundation, Cancer Research Center, La Jolla, California 92037

Lysosomes serve as the major digestive compartment of mammalian cells. They are responsible for the degradation of foreign materials internalized by endocytosis and intracellular material delivered to lysosomes during autophagocytosis (1, 2). In the past several years, significant progress has been made in understanding the biosynthesis and targeting of lysosomal enzymes, and the findings can be summarized as follows (for review see Refs. 3 and 4). Asparagine-linked high mannose oligosaccharides on newly synthesized lysosomal acid hydrolases acquire phosphate groups in the Golgi apparatus. The resulting mannose 6-phosphate groups then serve as a specific recognition marker for the binding of lysosomal enzymes to mannose 6-phosphate receptors located in the Golgi apparatus. The receptor-lysosomal enzyme complex is translocated to a prelysosomal compartment where the complex is dissociated by the low pH (pH ~ 5.5). The lysosomal enzymes are continuously packed into lysosomes, whereas the mannose 6-phosphate receptors recycle back to the Golgi apparatus. Some of the mannose 6-phosphate receptors reach the cell surface and can then deliver lysosomal enzymes from the cell surface to the lysosomes through the endocytic pathway.

In contrast to the extensive knowledge of lysosomal acid hydrolases, much less is known about the components of lysosomal membranes. The lysosomal membrane plays a vital role in the proper function of lysosomes by sequestering numerous acid hydrolases from the rest of the cytoplasmic components. The lysosomal membrane is presumably involved in various important functions of the lysosomes, such as its resistance to degradation by lysosomal hydrolases and its ability to interact and fuse specifically with other membrane organelles, including endosomes, phagosomes, and plasma membranes. The lysosomal membrane also maintains an acidic intralysosomal environment and transports amino acids and mono- and oligosaccharides produced by lysosomal hydrolases (for review, see Refs. 1–3).

In order to understand the components of the lysosomal membrane, initial attempts were made to identify lysosomal membrane proteins which are unique to lysosomes and not present in the plasma membrane. Burnside and Schneider (5) and Ohsumi *et al.* (6) identified two of such glycoproteins: those with M_r ~60,000 and those with M_r ~90,000–110,000. Other studies, however, were directed to identify lysosomal membrane glycoproteins by producing monoclonal antibodies against purified lysosomal membrane. By this approach, Barriocanal *et al.* (7) and Lewis *et al.* (8) identified three different groups of lysosomal membrane glycoproteins: those with M_r ~90,000–120,000; those with M_r ~72,000; and those with M_r ~27,000. Among these, the glycoproteins with M_r ~90,000–120,000 were found independently by different workers as the major components of the lysosomal membranes. First, Chen *et al.* (9) found that there are two different glycoproteins with these molecular weights, termed lamp-1 and lamp-2, with the latter shown to be identical to Mac-3 (10). Lippincott-Schwartz and Fambrough (11) identified chicken lamp-1 (LEP-100) as a membrane glycoprotein, which traverses from plasma membrane to lysosomes. On the other hand, lamp-1 and lamp-2 were isolated also as the major carriers for poly-*N*-acetyllactosamines (12–14) that are very complex carbohydrates with various ligand and antigenic structures (15). Until now, much more extensive studies have therefore been made on lamp-1 and lamp-2 molecules than on other lysosomal membrane glycoproteins. Hereafter, this review mainly summarizes the current knowledge on lamp-1 and lamp-2, and when necessary, the studies on other lysosomal membrane glycoproteins will be mentioned briefly.

Structure of Lysosomal Membrane Glycoproteins

lamp-1 and lamp-2 have been isolated from various species, and the cDNAs encoding their polypeptide chains have been isolated from human (12, 14), mouse (16, 17), rat (18–20), and chicken (21) cells. The deduced amino acid sequences revealed the following characteristics for these molecules (see Fig. 1 and Table I). (*a*) The glycoproteins consist of a polypeptide core of ~40 kDa. A large part of the molecule is located in the lumenal side of lysosomes, and this domain is connected to the transmembrane domain, which is extended to a short cytoplasmic tail. lamp-1 and lamp-2 are homologous to each other, and they share a similarity in these domain structures. (*b*) The intralumenal domain contains 347 (mouse) to 361 (chicken) amino acid residues. The intralumenal portion can be divided into two internally homologous domains separated by a region rich in proline residues. (*c*) Each homologous domain contains 4 half-cystine residues. The neighboring half-cystine residues are connected to each other, forming four disulfide loops in the intralumenal portion (22, 23). The positions of the cysteine residues are conserved well between lamp-1 and lamp-2 molecules and among those from different species. These disulfide bonds in lamp molecules are different from those in members of the immunoglobulin superfamily. (*d*) The proline-rich region is apparently homologous in both amino acid sequence and structure to the immunoglobulin hinge region (12). In fact, the examination of the supermolecular structure by rotary shadowing electron microscopy revealed that the hinge is free to move (22). The hinge region of lamp-1 molecules is enriched with proline and serine residues, whereas that of lamp-2 molecules is enriched with proline and threonine residues. (*e*) The intralumenal domain is extensively decorated by a significant number (16–20) of *N*-glycans, some of which are very complex poly-*N*-acetyllactosamines (see below). Lamps are apparently among the most densely *N*-glycosylated glycoproteins so far reported. Recent studies indicate that both lamp-1 and lamp-2 also contain *O*-glycans, some of which are poly-*N*-acetyllactosamines.[1] The carbohydrate constitutes 55–65% of the total mass in the lamp molecules. (*f*) Comparison of the amino acid sequences of lamp-1 and lamp-2 revealed the following interesting features. The human lamp-1 is more homologous to chicken lamp-1 (51.5% identity) than to human lamp-2 (36.7%). Similarly, lamp-2 from one species is more homologous to lamp-2 from other species than to lamp-1 of the same species. These results suggest that lamp-1 and lamp-2 diverged relatively early in the evolution but that lamp-1 and lamp-2 structures have been conserved during evolution. In fact, it has been shown that human lamp-1 and lamp-2 are encoded by separate genes located on chromosomes 13q34 and Xq24–25, respectively (24). The results also suggest that lamp-1 and lamp-2 have distinctly separate functions. (*g*) The recently elucidated structure of genomic DNA that encodes chicken lamp-1 has the following characteristics (25). First, the exons of lamp-1 gene can be divided into two clusters. Each cluster, encoding the NH_2-terminal half or the COOH-

* The work from our laboratory was supported by Grant CA48737 and in part by Grant CA33895 awarded from the National Cancer Institute.

[1] K. Maemura and M. Fukuda, manuscript in preparation.

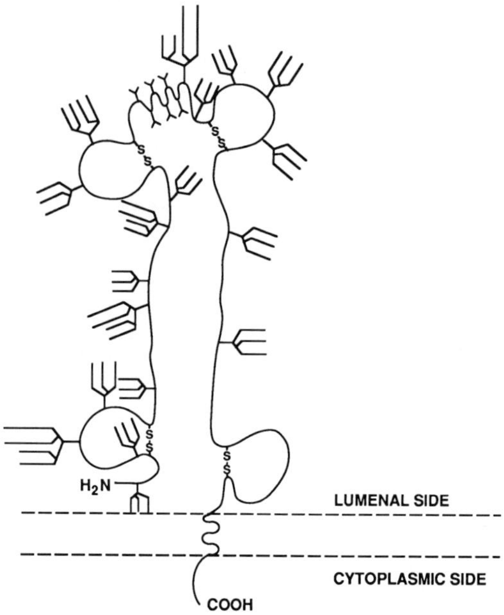

FIG. 1. **Depicted structure of lamp-1.** *N*-Glycans and *O*-glycans are indicated by ⋲ and *Y*, respectively. Some of the *N*-glycosylation sites are modified by bulky poly-*N*-acetyllactosamines, while *O*-glycans are almost exclusively attached at the hinge region (S. R. Carlsson and M. Fukuda, manuscript in preparation). Each loop is made by a disulfide bond. The majority of the molecule resides in the lumenal side of lysosomes. Since the hinge region can freely move, lamp-1 probably has more than one configuration. lamp-2 has an almost identical structure. The structure depicted is adapted from Ref. 23 and modified by incorporating more recent structural information (for details, see the text).

TABLE I
Nomenclature and properties of the major lysosomal membrane glycoproteins

Properties	lamp-1	lamp-2	lamp-3[a]	Limp II
Synonyms				
Human	hlamp-1	hlamp-2	CD63 ME491	
Rat	LIMP III lgp120	LIMP IV lgp110	LIMP I	lgp85
Mouse	mLAMP-1	mLAMP-2		
Chicken	LEP100			
Apparent molecular mass (kDa)	90–120	95–120	30–55	60–85
Copies/cell ($\times 10^{-4}$)[b]	30–60	20–60	<20	<20
Polypeptide chain (residues)	382–396	380–389	238	478
Number of *N*-glycans attached	17–20	16–17	3	Up to 11
O-Glycans	+	+	?	?
Carbohydrate content (%)	55–65	55–65	25–55	20–45
Gly-Tyr motif	Yes	Yes	Yes	No

[a] Limp I, CD63, or ME491 is now called lamp-3, since it shares the same cytoplasmic Gly-Tyr motif as lamp-1 and lamp-2.
[b] See also Granger *et al.* (20).

terminal half of the intralumenal domain of the protein, was shown to be homologous to each other in the amino acid sequences. These two clusters of exons are connected by an exon encoding the hinge region. At the 3′-end of the genome, one exon encodes for the transmembrane and cytoplasmic domain of the protein. These results favor the hypothesis that a primordial gene encoding the NH_2- or COOH-terminal half of the intralumenal portion was duplicated, and then a gene segment encoding the hinge region was inserted during or after such duplication (14, 25).

More limited information is available for other lysosomal membrane glycoproteins. The deduced amino acid sequences derived from cDNA sequences revealed the following characteristics for those glycoproteins (see Table I).

One class of glycoproteins in the molecular weight range of M_r ~30,000–54,000, previously called Limp III, contains a core peptide of ~25 kDa and three *N*-glycosylation sites (26). This glycoprotein, found to be identical to CD63, likely traverses the lipid bilayer four times, and the signal peptide is apparently not cleaved. All three *N*-glycosylation sites are close to the lipid bilayer and contain poly-*N*-acetyllactosaminyl structures, as evidenced by the heterogenous molecular weights. Although there is no overall similarity with a member of the lamp-1 or lamp-2 family, it contains the Gly-Tyr motif at its cytoplasmic tail (see below).

Another class of glycoproteins with M_r ~74,000, previously called Limp II, has 477 amino acids with two transmembrane portions; one is uncleavable signal peptide and the other is the transmembrane portion near the COOH terminus (27, 28). The presumed intralumenal portion contains 11 potential *N*-glycosylation sites, many of which must be utilized since the mature glycoprotein has an apparent molecular weight of 74,000–85,000. It has 5 cysteine residues, some of which participate in disulfide loop formation. This glycoprotein is distinctly different from the above two different classes of lysosomal membrane glycoproteins, since the cytoplasmic segment has an entirely different amino acid sequence (27, 28). Accordingly, we would like to rename Limp III as lamp-3, whereas we will wait to rename Limp II until more lysosomal membrane glycoproteins are isolated (Table I).

As a whole, however, those lysosomal membrane glycoproteins so far characterized are all extensively glycosylated in the lumenal portion of the molecule. It may also be of significance that both lamp-3 and Limp II have a molecular architecture as type III membrane glycoproteins (29), similar to erythrocyte band 3, which was first shown to contain poly-*N*-acetyllactosamine (15).

Intracellular Traffic of lamp-1 and lamp-2; the Role of the Cytoplasmic Domain

Several observations suggest that the traffic of lysosomal membrane glycoproteins to lysosomes is mediated by mechanisms different from those for lysosomal enzymes. First, it has been shown that the lysosomal membrane glycoproteins reach the lysosome independently of the attached *N*-glycans (7). Second, subcellular localization examined by immunoelectron microscopy indicates that lysosomal membrane glycoproteins reside primarily in lysosomes, whereas mannose 6-phosphate receptors reside mainly in the Golgi complex and prelysosomal compartments (30). These results indicate that the molecular signal for targeting of lysosomal membrane glycoproteins probably resides in the peptide moiety and differs from those for lysosomal enzymes. It was assumed that the peptide signal essential for lysosomal trafficking is the portion where a significant homology exists between lamp-1 and lamp-2 molecules. In this respect, the cytoplasmic segment is the prime candidate for a peptide signal, since the His-Ala-Gly-Tyr sequence is identical in human lamp-1 and lamp-2 molecules (Fig. 2). By expressing human lamp-1 cDNA in monkey COS-1 cells, it was first elucidated that the tyrosine residue in the cytoplasmic tail is essential for the delivery of lamp molecules to lysosomes. Further, it was demonstrated that the cytoplasmic tail enables a reporter molecule to be delivered to the lysosomes. These results demonstrated that the tyrosine residue in conjunction with the cytoplasmic tail gives a sufficient signal for trafficking of lamp molecules to lysosomes (31).

Other studies revealed that the cytoplasmic tyrosine is essential for trafficking during receptor-mediated endocytosis, involving receptors such as low density lipoprotein (32), transferrin (33), and mannose 6-phosphate (34) receptors. In fact, a small portion of the expressed lamp molecule is present on the cell surface, and its transport to lysosomes was shown to be also dependent on the cytoplasmic tyrosine at a particular position (31). Similarly, influenza virus hemagglutinin can be endocytosed when a tyrosine is introduced at its particular cytoplasmic position (35). Peters *et al.* (36) obtained almost identical results on lysosomal acid phosphatase, LAP. It is possible that this functional tyrosine residue is in an exposed tight turn of the amino acid sequence, as demonstrated recently for transferrin receptor (37). These results indicate that there are two possible routes for lamp molecules to reach the lysosomes. First, lamp is sorted from other membrane and soluble proteins at the *trans*-Golgi cisternae. In this pathway, newly synthesized lamp molecules will be transported to late endosomes and then to lysosomes together with newly synthesized lysosomal enzymes, which are initially complexed with mannose 6-phosphate receptor. In the alternative pathway, lamp

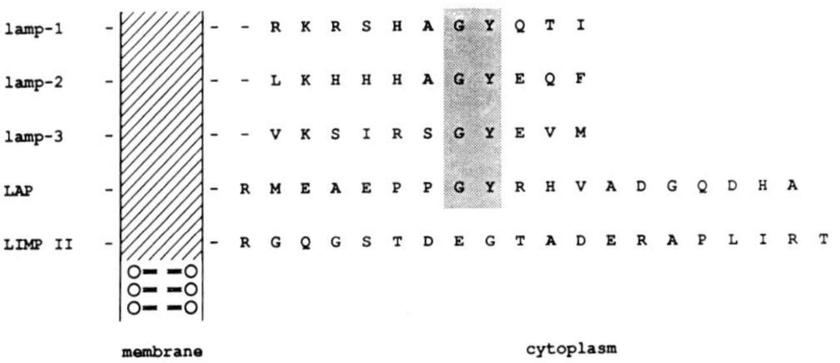

FIG. 2. **Comparison of cytoplasmic tail sequences of membrane proteins transported to lysosomes.** The sequences shown are for human lamp-1, lamp-2, lamp-3, and LAP. Although LAP is eventually processed to a soluble enzyme in lysosomes, it is initially delivered to lysosomes as a membrane protein. lamp-1, lamp-2, lamp-3 (also called CD63, ME491, or LIMP I), and LAP share the Gly-Tyr motif, while LIMP II has a distinctly different cytoplasmic tail.

molecules are first transported to the cell surface along with the secretory protein pathway and then sorted from plasma membrane proteins by selective internalization, followed by transport to dense lysosomes.

The studies by Braun *et al.* (38) on LAP[2] support the latter possibility. LAP could be detected on the cell surface of baby hamster kidney cells where human LAP was overly expressed. Those LAP molecules on the cell surface were eventually transported to dense lysosomes. In contrast, Tanaka *et al.* (39) demonstrated that the majority of LAP synthesized *in situ* is transported to lysosomes without going through the expression on the cell surface. Similarly, the kinetic studies on the biosynthesis of lamp molecules in mouse macrophages and rat fibroblasts support the first model (40, 41). Those studies, however, did not examine the fate of lamp molecules on the cell surface, which were probably present in minute quantity. Recent studies included the analysis of lamp molecules on the cell surface and provided evidence that the majority of newly synthesized lamp molecules is directly transported to the lysosomes and that lamp molecules expressed on the cell surface are transported to the lysosomes much later.[3] It is then reasonable to conclude that the majority of lysosomal membrane glycoproteins is directly transported to the lysosomes. When a small portion of lamp molecules is transported to the cell surface, these molecules are eventually retrieved through the endocytic pathway. In both the direct pathway to the lysosomes and the retrieval pathway, the cytoplasmic tail appears to provide critical signals for the transport of lysosomal membrane glycoproteins. It is noteworthy that Limp II has a distinctly different cytoplasmic tail from those of lamp-1, lamp-2, and lamp-3 (Fig. 2), suggesting that this molecule on the cell surface may not be retrieved by the endocytic pathway (see also Ref. 69).

Surface Expression of Lamp Molecules

There has been some controversy whether or not lamp molecules are expressed on the cell surface. Examination by immunoelectron microscopy generally failed to detect lamp molecules on the cell surface, probably because of its relatively low sensitivity (41). In contrast, light microscopic examination of immunofluorescence labeling detected lamp molecules on the cell surface, although its quantity varies depending on the cell type (11, 42, 43). Thus it is appropriate to conclude from these studies that a portion of lamp molecules is expressed on the cell surface. It has been shown recently that cytoplasmic tyrosine is specifically recognized by HA-II adaptor proteins present in coated pits on the plasma membrane (44). The complex of HA-II adaptor proteins and receptors for low density lipoprotein, poly-Ig, or mannose 6-phosphate is then transported through the endocytic pathway to late endosomes. It is then reasonable to assume that a similar adaptor protein or HA-I adaptor recognizes the cytoplasmic tail of lysosomal membrane glycoproteins, and their complex, formed most likely at the *trans*-Golgi cisternae, is transported to the lysosomes (or prelysosomes). The assumption of the presence of such adaptor(s) can very well explain the variable expression of lamp molecules on the cell surface. If lysosomal membrane glycoproteins are expressed in an excessive amount, a portion of lysosomal membrane glycoproteins escapes from the complex formation with adaptors, and these molecules are transported through a default pathway to the plasma membrane. This hypothesis also explains the reason why the majority of LAP, that were overexpressed on baby hamster kidney cells, did not directly go to the lysosomes.

In certain cases, however, the surface expression of lamp molecules could have more significance. It has been shown recently that lamp-1 and CD63 (or lamp-3) can be detected on the surface of platelets after platelets are activated (45, 46). Expression of lamp-1 and lamp-3 on the cell surface is likely a by-product of lysosome-plasma membrane fusion but probably also protects the plasma membrane from degradation by lysosomal hydrolases. Similar results can be observed in the secretory granules present in cytotoxic T-lymphocytes. Cytotoxic T-lymphocytes exocytose granules during specific interaction with target cells. Such granules contain a family of serine esterases and the lethal protein, perforin. Recent immunocytochemical examinations revealed that lamp-1, lamp-2, and lamp-3 are present on granule-delimiting outer membrane in mature granules, while mannose 6-phosphate receptor can also be detected in immature granules (47). After granules are fused with the plasma membrane, lysosomal membrane glycoproteins apparently protect the plasma membrane from the content of granules. These results indicate that lysosomal membrane glycoproteins are utilized when cytoplasm and cell membrane need to be protected from hydrolases and other lethal proteins by sequestering those proteins in the separate vesicles. In relation to these findings, it is worth mentioning the recent results obtained on MHC class II molecules. MHC class II molecules preferentially bind peptides derived from exogenous antigens degraded in the endocytic pathway. Immunoelectron microscopic examination revealed that MHC class II molecules are transported to structures with lysosomal characteristics, which are characterized by the presence of lamp-1, lamp-2, and lamp-3, and are segregated from other molecules, including the MHC I molecule, probably at the Golgi complex (48). These results also reveal that lysosomal membrane glycoproteins may play critical roles in membrane dynamics when hydrolases need to be delivered in a specific way. Further studies will be expected to continuously unveil new compartments of lysosomal nature, by using antibodies specific to these lamp molecules.

Role of Carbohydrates in Lamp Molecules

As mentioned above, lamp molecules are unique by having a significant amount of poly-*N*-acetyllactosamines, which are composed of (Galβ1→4GlcNAcβ1→3)$_n$ (*N*-acetyllactosamine) repeats in side chains. These long side chains appear to be better acceptors for various glycosyltransferases, including fucosyltransferases. This is the reason why the termini of poly-*N*-acetyllactosamines often contain unique structures such as sialyl Le^X, NeuNAcα2→3Galβ1→4(Fucα1→3)GlcNAc (for review, see Ref. 15).

Lamp molecules have been identified as the major carriers for poly-*N*-acetyllactosamines in many cells. The half-life of lamp-1 molecules was appreciably decreased when they were synthesized in the presence of tunicamycin, which inhibits *N*-glycosylation (7). The half-lives of lamp-1 and lamp-2 were dramatically increased when HL-60 cells were induced to differentiate into granulocytic cells (49). During this differentiation, more of *N*-glycosylation sites acquire poly-*N*-acetyllactosamines, although there are sites that are preferentially modified by poly-*N*-acetyllactosamines (50). Lamp molecules are presumably resistant to various hydrolytic enzymes in the lysosomes by having very complex carbohydrates, such as poly-*N*-acetyllactosamines. In fact, the abundance of lamp-1 and lamp-2 molecules is so high that lamp molecules may form a nearly continuous coat on the inner surface of the lysosomal membrane and serve as a barrier to soluble hydrolases (20). These results indicate that poly-*N*-acetyllactosamines attached to lysosomal membrane glycoproteins play a critical role in maintaining the stability in the

[2] The abbreviations used are: LAP, lysosomal acid phosphatase; MHC, major histocompatibility complex; HA-I and -II, hydroxyapatite group I and II.
[3] S. R. Carlsson and M. Fukuda, submitted for publication.

lysosomes. It is reasonable that differentiated HL-60 cells have more stable lamp molecules, since these cells are more phagocytic and require more functional lysosomes. On the other hand, the poly-*N*-acetyllactosamine content of lamp molecules is decreased when CaCo-2 colonic tumor cells are differentiated into cells which are presumably less phagocytic (51).

These observations should also be evaluated in conjunction with other reports that tumor cells, including highly metastatic tumor cells, express more poly-*N*-acetyllactosamines than normal counterparts or poorly metastatic counterparts (52–55). Conversely, several glycosylation inhibitors, which inhibit poly-*N*-acetyllactosamine formation, dramatically reduce the tumor formation when they are administered into animals (56–58). In parallel to those studies, it has been observed that the amount of sialyl Le^X and sialyl Le^a structure, NeuNAcα2→3Galβ1→3(Fucα1→4)GlcNAc, is drastically increased on the cell surface of various tumors *in situ* (59, 60). Recent studies indicate that adhesive molecules present on endothelial cells (ELAM-1) and platelets (GMP-140) actually bind to sialyl Le^X on granulocytes and monocytes when those cells are recruited to inflammatory sites (61–65). It was also reported that some tumor cells apparently bind to endothelial cells or platelets through ELAM-1 or GMP-140-mediated interaction (66, 67). These results lead to the possibility that highly metastatic tumor cells bind more efficiently to endothelial cells at metastatic sites, because those cells are enriched with sialyl Le^X and sialyl Le^a structures, which serve as ligands for ELAM-1 and GMP-140 molecules on endothelial cells and platelets (61–65, 68).

As mentioned above, a small amount of the total number of lamp molecules can be detected on the cell surface. Considering that lamp molecules are the major sialoglycoproteins in cells, such expression still constitutes more than 5×10^4 lamp molecules on the cell surface. Further, it was found that highly metastatic tumor cells express more lamp molecules on the cell surface than poorly metastatic cells.[4] These results suggest that lamp molecules are the major carriers for poly-*N*-acetyllactosamines also on the cell surface, which carry various ligand structures for adhesive molecules.

One of the unique characteristics of lamp molecules is having poly-*N*-acetyllactosamines. Such unique carbohydrates apparently play roles in the cytoplasm to protect lysosomal membranes and on the cell surface to provide ligand structures for cell-adhesive molecules. It is expected that understanding the roles of lysosomal membrane glycoproteins as having poly-*N*-acetyllactosamines will continuously lead us into new avenues in molecular cell biology.

Acknowledgments—I thank all colleagues and collaborators who made our work possible.

REFERENCES

1. de Duve, C. (1983) *Eur. J. Biochem.* **137,** 391–397
2. Holtzman, E. (1989) *Lysosomes*, Plenum Press, New York
3. Kornfeld, S., and Mellman, I. (1989) *Annu. Rev. Cell Biol.* **5,** 483–525
4. Damus, N. M., Lobel, P., and Kornfeld, S. (1989) *J. Biol. Chem.* **264,** 12115–12118
5. Burnside, J., and Schneider, D. L. (1982) *Biochem. J.* **204,** 525–534
6. Ohsumi, Y., Ishikawa, T., and Kato, K. (1983) *J. Biochem.* (*Tokyo*) **93,** 547–556
7. Barriocanal, J. G., Bonifacino, J. S., Yuan, L., and Sandoval, I. V. (1986) *J. Biol. Chem.* **261,** 16755–16763
8. Lewis, V., Green, S. A., Marsh, M., Vihko, P., Helenius, A., and Mellman, I. (1985) *J. Cell Biol.* **100,** 1839–1847
9. Chen, J. W., Murphy, T. L., Willingham, M. C., Pastan, I., and August, J. T. (1985) *J. Cell Biol.* **101,** 85–95
10. Ho, M.-K., and Springer, T. A. (1983) *J. Biol. Chem.* **258,** 636–642
11. Lippincott-Schwartz, J., and Fambrough, D. M. (1987) *Cell* **49,** 669–677
12. Viitala, J., Carlsson, S. R., Siebert, P. D., and Fukuda, M. (1988) *Proc. Natl. Acad. Sci. U. S. A.* **85,** 3743–3747
13. Carlsson, S. R., Roth, J., Piller, F., and Fukuda, M. (1988) *J. Biol. Chem.* **263,** 18911–18919
14. Fukuda, M., Viitala, J., Matteson, J., and Carlsson, S. R. (1988) *J. Biol. Chem.* **263,** 18920–18928
15. Fukuda, M. (1985) *Biochim. Biophys. Acta* **780,** 119–150
16. Chen, J. W., Cha, Y., Yuksel, K. U., Gracy, R. W., and August, J. T. (1988) *J. Biol. Chem.* **263,** 8754–8758

[4] O. Saitoh, W.-C. Wang, R. Lotan, and M. Fukuda, submitted for publication.

17. Cha, Y., Holland, S. M., and August, J. T. (1990) *J. Biol. Chem.* **265,** 5008–5013
18. Howe, C. L., Granger, B. L., Hull, M., Green, S. A., Gabel, C. A., Helenius, A., and Mellman, I. (1988) *Proc. Natl. Acad. Sci. U. S. A.* **85,** 7577–7581
19. Himeno, M., Noguchi, Y., Sasaki, H., Tanaka, Y., Furuno, K., Kono, A., Sasaki, Y., and Kato, K. (1989) *FEBS Lett.* **244,** 351–356
20. Granger, B. L., Green, S. A., Gabel, C. A., Howe, C. L., Mellman, I., and Helenius, A. (1990) *J. Biol. Chem.* **265,** 12036–12043
21. Fambrough, D. M., Takeyasu, K., Lippincott-Schwarz, J., Spiegel, N. R., and Somerville, D. (1988) *J. Cell Biol.* **106,** 61–67
22. Carlsson, S. R., and Fukuda, M. (1989) *J. Biol. Chem.* **264,** 20526–20531
23. Arterburn, L. M., Earles, B. J., and August, J. T. (1990) *J. Biol. Chem.* **265,** 7419–7423
24. Mattei, M.-G,. Matteson, J., Chen, J. W., Williams, M. A., and Fukuda, M. (1990) *J. Biol. Chem.* **265,** 7548–7551
25. Zot, A. S., and Fambrough, D. M. (1990) *J. Biol. Chem.* **265,** 20988–20995
26. Metzelaar, M. J., Wijngaard, P. C. J., Peters, P. J., Sixma, J. J., Nienhuis, H. K., and Clevers, H. C. (1991) *J. Biol. Chem.* **266,** 3239–3245
27. Fujita, H., Ezaki, J., Noguchi, Y., Kono, A., Himeno, M., and Kato, K. (1991) *Biochem. Biophys. Res. Commun.* **178,** 444–452
28. Vega, M. A., Segui-Real, B., Garcia, A. J., Calés, C., Rodriguez, F., Vanderkerckhove, J., and Sandoval, I. V. (1991) *J. Biol. Chem.* **266,** 16818–16824
29. Wickner, W. T., and Lodish, H. F. (1985) *Science* **230,** 400–407
30. Griffiths, G., Hoflack, B., Simons, K., Mellman, I., and Kornfeld, S. (1988) *Cell* **52,** 329–341
31. Williams, M. A., and Fukuda, M. (1990) *J. Cell Biol.* **111,** 955–966
32. Chen, W.-J., Goldstein, J. L., and Brown, M. S. (1990) *J. Biol. Chem.* **265,** 3116–3123
33. Jing, S., Spencer, T., Miller, K., Hopkins, C., and Trowbridge, I. S. (1990) *J. Cell Biol.* **110,** 283–294
34. Lobel, P., Fujimoto, K., Ye, R. D., Griffiths, G., and Kornfeld, S. (1989) *Cell* **57,** 787–796
35. Lazarovitz, J., and Roth, M. (1988) *Cell* **53,** 743–752
36. Peters, C., Braun, M., Weber, B., Wendland, M., Schmidt, B., Pohlmann, R., Waheed, A., and Von Figura, K. (1990) *EMBO J.* **9,** 3497–3506
37. Collawn, J. F., Stangel, M., Kuhn, L. A., Esekogwu, V., Jing, S., Trowbridge, I. A., and Tainer, J. A. (1990) *Cell* **63,** 1061–1072
38. Braun, M., Waheed, A., and Von Figura, K. (1989) *EMBO J.* **8,** 3633–3640
39. Tanaka, I., Yano, S., Furuno, K., Ishikawa, T., Himeno, M., and Kato, K. (1990) *Biochem. Biophys. Res. Commun.* **170,** 1067–1073
40. D'Souza, M. P., and August, J. T. (1986) *Arch. Biochem. Biophys.* **249,** 522–532
41. Green, S. A., Zimmer, K-P., Griffiths, G., and Mellman, I. (1987) *J. Cell Biol.* **105,** 1227–1240
42. Mane, S. M., Marzella, L., Bainton, D. F., Holt, V. K., Cha, Y., Hildreth, J. E. K., and August, J. T. (1989) *Arch. Biochem. Biophys.* **268,** 360–378
43. Amons, B., and Lotan, R. (1990) *J. Biol. Chem.* **265,** 19192–19198
44. Glickman, J. N., Conibear, E., and Pearse, B. M. F. (1989) *EMBO J.* **8,** 1041–1047
45. Nieuwenhuis, H. L., Van Oosterhout, J. J. G., Rozemuller, E., Van Iwaarden, F., and Sixma, J. J. (1987) *Blood* **70,** 838–845
46. Febbraio, M., and Silverstein, R. L. (1990) *J. Biol. Chem.* **265,** 18531–18537
47. Peters, P. J., Borst, J., Oorschot, V., Fukuda, M., Kräbenbül, O., Tschopp, J., Slot, J. W., and Geuze, H. J. (1991) *J. Exp. Med.* **173,** 1099–1109
48. Peters, P. J., Neefjes, J. J., Oorschot, V., Ploegh, H. L., and Geuze, H. J. (1991) *Nature* **349,** 669–676
49. Lee, N., Wang, W.-C., and Fukuda, M. (1990) *J. Biol. Chem.* **265,** 20476–20487
50. Carlsson, S. R., and Fukuda, M. (1990) *J. Biol. Chem.* **260,** 20488–20495
51. Youakim, A., Romero, P. A., Yee, K., Carlsson, S. R., Fukuda, M., and Herscovics, A. (1989) *Cancer Res.* **49,** 6889–6895
52. Yamashita, K., Ohkura, T., Tachibana, Y., Takasaki, S., and Kobata, A. (1984) *J. Biol. Chem.* **259,** 10834–10840
53. Pierce, M., and Arango, J. (1986) *J. Biol. Chem.* **261,** 10772–10777
54. Hubbard, S. C. (1987) *J. Biol. Chem.* **262,** 16403–16411
55. Yousefi, S., Higgins, E., Daoling, Z., Pollex-Krüger, A., Hindsgaul, O., and Dennis, J. W. (1991) *J. Biol. Chem.* **266,** 1772–1782
56. Irimura, T., Gonzalez, R., and Nicolson, G. L. (1981) *Cancer Res.* **41,** 3411–3418
57. Humphries, M. J., Matsumoto, K., White, S. L., and Olden, K. (1986) *Proc. Natl. Acad. Sci. U. S. A.* **83,** 1752–1756
58. Dennis, J. W. (1986) *Cancer Res.* **46,** 5131–5136
59. Magnani, J. L., Nilsson, B., Brockhaus, M., Zopf, D., Steplewski, Z., Koprowski, H., and Ginsburg, V. (1982) *J. Biol. Chem.* **257,** 14365–14369
60. Fukushima, K., Hirota, M., Terasaki, P. I., Wakisaka, A., Togashi, H., Chia, D., Suyama, N., Fukushi, Y., Nudelman, E., and Hakomori, S. (1984) *Cancer Res.* **44,** 5279–5285
61. Lowe, J. B., Stoolman, L. M., Nair, R. P., Larsen, R. D., Berhend, T. L., and Marks, R. M. (1990) *Cell* **63,** 475–484
62. Phillips, M. L., Nudelman, E., Gaeta, F. C. A., Perez, M., Singhal, A. K., Hakomori, S-I., and Paulson, J. C. (1990) *Science* **250,** 1130–1132
63. Walz, G., Aruffo, A., Kolanus, W., Bevilacqua, M., and Seed, B. (1990) *Science* **250,** 1132–1135
64. Larsen, E., Palabrica, T., Sajer, S., Gilbert, G. E., Wagner, D. D., Furie, B. C., and Furie, B. (1990) *Cell* **63,** 467–474
65. Polley, M. J., Phillips, M. L., Wayner, E., Nudelman, E., Singhal, A. K., Hakomori, S.-I., and Paulson, J. C. (1991) *Proc. Natl. Acad. Sci. U. S. A.* **88,** 6224–6228
66. Rice, G. E., and Bevilacqua, M. P. (1989) *Science* **246,** 1303–1306
67. Hession, C., Osborn, L., Goff, D., Chi-Rosso, G., Vassallo, L., Pasek, M., Pittack, L., Tizard, R., Goelz, G., McCarthy, K., Hopple, S., and Lobb, R. (1989) *Proc. Natl. Acad. Sci. U. S. A.* **87,** 1673–1677
68. Berg, E. L., Robinson, M. K., Mansson, O., Butcher, E. C., and Magnani, J. L. (1991) *J. Biol. Chem.* **266,** 14869–14872
69. Vega, M. A., Rodriguez, F., Segui, B., Calés, C., Alcalde, J., and Sandoval, I. V. (1991) *J. Biol. Chem.* **266,** 16269–16272

THE JOURNAL OF BIOLOGICAL CHEMISTRY
Vol. 266, No. 33, Issue of November 25, pp. 22067-22070, 1991

Printed in U.S.A.

Minireview

Conotoxins*

Baldomero M. Olivera‡, Jean Rivier§, Jamie K. Scott¶, David R. Hillyard‡, and Lourdes J. Cruz‡||

From the ‡Departments of Biology and Pathology, University of Utah, Salt Lake City, Utah 84112, the §Clayton Foundation Laboratories for Peptide Biology, Salk Institute, LaJolla, California 92057, the ¶Department of Biology, University of Missouri, Columbia, Missouri 65211, and ||The Marine Science Institute, University of the Philippines, Quezon City, Philippines 1101

Many successful animal and plant families have developed distinctive biochemical strategies; one of the more unusual examples is found in a group of marine gastropods, the cone snails (*Conus*) (1). These animals have evolved a specialized biochemistry of small constrained peptides, the conotoxins. These peptides are the direct translation products of genes (2). However, because they are small enough for direct chemical synthesis and sufficiently constrained for three-dimensional conformation determination, conotoxins bridge protein chemistry and molecular genetics. Furthermore, the strategy that the cone snails have evolved over millions of years for the generation and design of an enormous array of small peptide ligands, each with high affinity and specificity for a particular receptor protein target, may be adaptable for use *in vitro*.

Natural History of Cone Snails

The focus of this minireview is the small peptides made in the venoms of the cone snails (*Conus*). On a geological time scale, the true cones are a recently evolved group. The oldest verifiable *Conus* fossils occur well after the Cretaceous extinction (3), an event resulting in the disappearance of dinosaurs on land and the ammonites in marine environments. Just as the extinction of dinosaurs provided an opportunity for the rise of the mammals, the extinction of the ammonites was probably a key factor for the success of *Conus*. Ammonites were believed to be among dominant predators in rich, shallow water marine communities, an ecological niche occupied by the cone snails today. The genus *Conus* has been expanding at an impressive rate; the ~500 living species make it perhaps the largest single molluscan genus (see Fig. 1).

Although individual *Conus* species can be highly specialized, as a whole the genus shows a remarkably broad phylogenetic range of prey. At least five different phyla of animals are envenomated by cone snails; there are large numbers of *Conus* species which feed only on polychaete worms, other snails, or fish (4). Slow moving snails might not be expected to capture fish successfully, but dozens of *Conus* species eat nothing else. Observing a fish-hunting cone such as *Conus striatus* capture prey is a memorable sight. In the presence of fish, the snail extends its long threadlike proboscis which serves as a fishing line. A hollow, arrow-shaped tooth is ejected at the tip of the proboscis and is used to harpoon the fish (see Fig. 2) and inject the venom. The fish typically jerks suddenly after being struck but remains tethered through the proboscis. A good strike causes the fish to be immobilized within 1 or 2 s, unable to use its major fins. Total paralysis is effected a few seconds later, but often the fish has been engulfed by the snail into its distensible stomach even before this has occurred. The potent venom is made in a long duct and expelled using a muscular bulb. Although the ~500 *Conus* species hunt different prey and have different foraging strategies, all inject venom through a harpoon-like tooth to immobilize prey. One species, *Conus geographus*, is so venomous that two-thirds of human stinging cases are fatal.

Overview of Conus Peptides

The biologically active agents in *Conus* venoms are unusually small peptides, 10–30 amino acids in length. Most peptides are multiply disulfide-bonded; small loops of 1–6 amino acids are interspersed between the disulfide-bonded Cys residues. There is a large array of different peptides in every venom, and each appears to be specifically targeted to a particular receptor. The profile shown in Fig. 3 is typically obtained upon analysis of a *Conus* venom fraction; a wide range of biological activities is observed. Physiological targets have been identified for several peptides found in *Conus* venoms (see Table I and Refs. 1 and 5). However, for most peptides in *Conus* venoms already biochemically characterized (over 70 peptides so far, from 10 venoms), the receptor targets remain unknown. The full complexity of any single *Conus* venom has not yet been determined; there may well be over 100 different peptides in the more complex venoms.

Each *Conus* species has a venom with a distinct pharmacological profile. For example, a major component of *C. geographus* venom is conantokin-G, which causes sleep in young mice and hyperactivity in older mice and targets to the NMDA[1] receptor (6, 7). The venom of another fish hunter, *C. striatus* does not exhibit this activity. Conversely, *C. striatus* venom has a major excitotoxin not present in *C. geographus* venom. Although both species make peptides targeted to the acetylcholine receptor and to voltage-sensitive calcium channels, each venom has a large subset of pharmacologically distinct entities. Additional pharmacologically active factors from different *Conus* venoms have been described including agents with α-adrenergic (8) and cholinomimetic (9) effects, as well as purified components with potent effects on smooth and cardiac muscle systems (10–13). No detailed sequence information has yet been published for these, but they appear to be distinct from the conotoxin classes in Table I.

Despite the great diversity of peptides in *Conus* venoms, one striking structural feature is the pattern of Cys residues. A large fraction of *Conus* peptides exhibits one of three characteristic arrangements of cysteine residues: the "standard" 2-loop, 3-loop, and 4-loop conotoxin frameworks (see Table II). The major 4-loop framework (C- - -C- - -CC- - -C- - -C) has been identified in over 20 *Conus* peptides with a wide range of pharmacological effects. Alternative arrangements, some characteristic of Cys-rich peptides in other systems, are not found. For example, an alternative 4-loop framework found in mammalian defensins (14) (C- - -C- - -C- - -C- - -CC) is not present in any *Conus* peptide.

There are a number of *Conus* peptides that lack disulfide bonding altogether. These may assume specific conformations through mechanisms other than multiple disulfide linkages. An example are the conantokins, which target to NMDA receptors. In these peptides γ-carboxyglutamate residues are believed to induce an α-helical conformation in the presence of calcium ions (15). Thus, although the great majority of venom peptides have multiple disulfide bonds, there may be alternative strategies for stabilizing high affinity binding conformations in a minor fraction of *Conus* peptides.

Hypervariability of Conotoxin Homologs

Analysis of cDNA clones of conotoxins has led to the conclusion that a specialized genetic mechanism has evolved in *Conus*

* The research on conotoxins described in this review was primarily supported by Grant GM22737 from the National Institutes of Health, and in part by Contract N0014-88-K0178 from the Office of Naval Research (to B. M. O.) and the International Foundation for Science, Stockholm, Sweden (to L. J. C.).

[1] The abbreviations used are: NMDA, *N*-methyl-D-aspartate; HPLC, high performance liquid chromatography.

FIG. 1. **Peptide specialists: some of the ~500 different cone snail species.** Each *Conus* species produces a venom with its own characteristic set of diverse small constrained peptides. Although most venoms have not yet been biochemically characterized, each should yield distinctive peptide ligands which specifically bind cell-surface receptors or ion channels. There is remarkable hypervariability between peptide sequences from venoms of different cone species. Possibly, a similar hypervariability generating mechanism serves to produce the strikingly diverse shell patterns as well. Photograph by Kerry Matz.

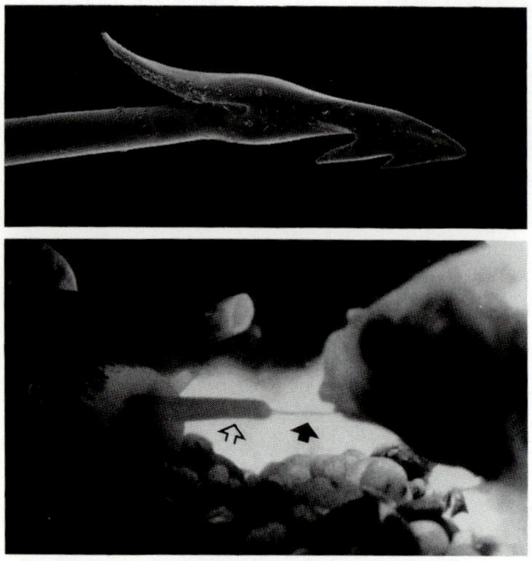

FIG. 2. *Top panel*, the tip of the harpoon-like tooth of *Conus obscurus*. The barbed, hollow tooth is used for injecting venom into the fish prey. Scanning electron micrograph by Dr. Ed King and Chris Hopkins. *Lower panel*, a specimen of *C. striatus* has harpooned a fish which is immobilized and is being drawn toward the mouth of the snail. The *filled arrow* indicates the harpoon tooth through which venom was injected; the *empty arrow* shows the proboscis which has been largely pulled back into the mouth of the snail. The structure at the *top* of the photograph is the siphon, used by these largely nocturnal snails use to locate prey. Photograph by Kerry Matz.

to generate hypervariability in the loop regions between Cys residues of the standard frameworks (see Fig. 4) (1, 2). This may explain why conotoxins have highly conserved arrangements of cysteine residues; *Conus* peptides with new pharmacologic specificity and biological roles are most likely to evolve with one of the standard conotoxin frameworks because of the hypervariability-generating mechanism. In the *Conus* peptide system, extreme sequence hypervariability is observed between functionally homologous conotoxins. The same mechanism that gives rise to the wide variety of pharmacologically different conotoxins may also be responsible for sequence hypervariability within each pharmacological class. A set of peptide ligands from a single genus, targeted to the same binding pocket of a particular receptor, would normally be expected to be highly conserved in primary sequence; instead, a remarkable divergence is found.

Homologous peptides from two venoms that both target to presynaptic Ca^{2+} channels are shown in Table IIIA (16). If the sequences of these peptides, ω-conotoxins GVIA and MVIIA, are aligned, less than one-third of the non-cysteine amino acids are identical. Furthermore, the amino acids in corresponding loops are strikingly different; for example, ω-conotoxin GVIA has 3 residues of hydroxyproline, whereas ω-conotoxin MVIIA has none. Despite these substantial sequence differences, both ω-conotoxins target the same subset of calcium channels and elicit identical biological effects in most phylogenetic systems (17). Presumably, these peptides were both evolved to cause paralysis by inhibiting presynaptic calcium channels at fish neuromuscular junctions, since both come from fish-hunting *Conus*.

It is noteworthy that most of the amino acids conserved between ω-conotoxins GVIA and MVIIA are also conserved in a peptide with entirely different pharmacological specificity, the King-Kong peptide from *Conus textile* (2, 18). If all ω-conotoxin sequences are aligned, only the 6 cysteine residues and one glycine moiety are conserved (see Table III); these are all present in the King-Kong peptide. Thus, the high binding specificity of ω-conotoxins for calcium channels at vertebrate presynaptic termini must be due to the variable loop sequences and not the amino acid residues conserved in both ω-conotoxins and the King-Kong peptide (such as the Cys framework).

Considerable polymorphism occurs even in the same venom. Two α-conotoxins from *C. striatus* (19, 20) are shown in Table IIIB. The Cys residues are conserved, but fully two-thirds of the other amino acids differ in the two peptides. The hypervariability of conotoxin homologs extends to peptides which are not disulfide-bonded. Thus, as shown in Table IIIC, conantokins from two different *Conus* species, which are essentially identical in biological activity, are highly divergent in primary sequence (21).

TABLE I
Identified receptor targets of conotoxins
The detailed sequences of all peptides are shown in Ref. 16.

Targets identified	Conotoxin example from *C. geographus*	Size of conotoxin[a]	No. of disulfide bonds
Voltage-sensitive Ca^{2+} channel	ω-Conotoxin GVIA	27	3
Acetylcholine receptor	α-Conotoxin GI	13	2
Voltage-sensitive Na^{+} channel	μ-Conotoxin GIII	22	3
Vasopressin receptor	Conopressin-G	9	1
NMDA receptor	Conantokin-G	17	0

[a] No. of amino acids.

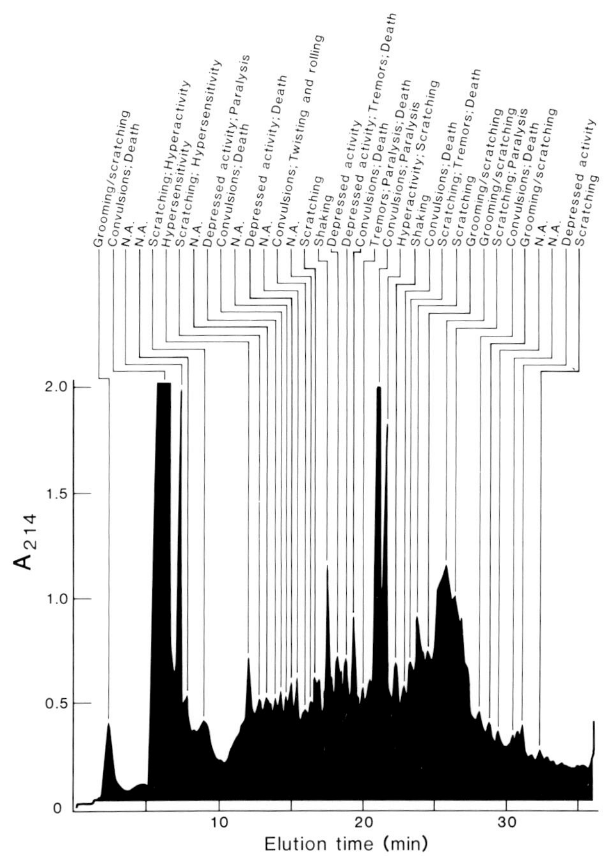

FIG. 3. **An HPLC analysis of a peptide fraction from *Conus magus* venom.** A peptide fraction from crude *C. magus* venom was obtained after size fractionation on Sephadex G-25 and reverse phase HPLC carried out as previously described. Each peak was assayed by intracranial injection of 0.5–2 nmol into mice (assuming average absorbance and molecular weight). Symptoms obtained are indicated above each peak. The two peaks that induce shaking are ω-conotoxins; *N.A.* indicates no biological activity observed.

TABLE II

Major conotoxin frameworks

Framework	Examples[a]
"4-loop" framework: 1 2 3 4 C– – –C– – –CC– – –C– – –C	ω-Conotoxins (*C. geographus, C. magus*); "King-Kong" peptides (*C. textile*)
"3-loop" framework: 1 2 3 CC– – –C– – –C– – –CC	μ-Conotoxins (*C. geographus*); "scratcher" peptide (*C. textile*)
"2-loop" framework: 1 2 CC– – –C– – –C	α-Conotoxins (*C. geographus, C. striatus*)

[a] Detailed sequences of all examples are in Ref. 1.

Conotoxin Sequence Degeneracy and Receptor-Ligand Interactions

Why is it possible for peptides with strikingly different primary sequences (such as ω-conotoxins GVIA and MVIIA) to target the same binding sites? Except for the conserved disulfide frameworks, which are demonstrably not the primary determinants of binding specificity, conotoxin homologs are surprisingly diverse in primary sequence. One explanation is that the conotoxin surfaces that interact with the receptor target (the "pharmacophore" in the language of pharmaceutical chemistry) have the same conformation despite divergent primary sequences. In this view, there are degenerate ways to get congruent conformations, and amino acid identity in specific positions is not obligatory.

Alternatively, ligands the size of conotoxins may interact with a "macrosite" on the receptor target that contains a number of "microsites." Each microsite could contribute to binding affinity upon contact with the ligand. The essence of this hypothesis is that only a fraction of all potential microsites actually make focal

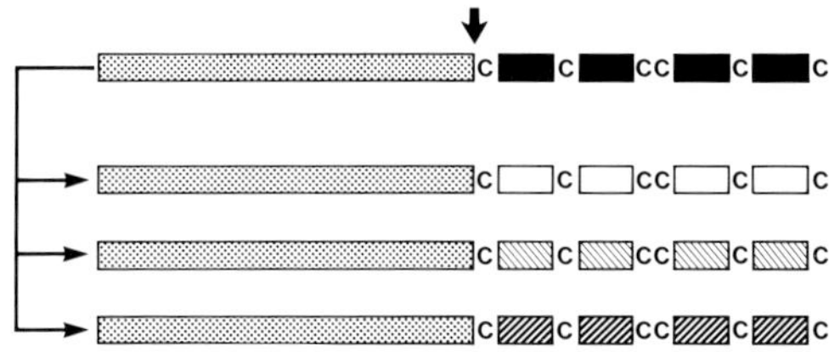

FIG 4. **Evolution of new conotoxins.** cDNA cloning has indicated that although the N-terminal end of the conotoxin precursors is highly conserved, the cone snails have a genetic mechanism for introducing rapid sequence changes specifically in loops (represented as *black bars* in the original peptide) between cysteine residues . The *arrow* represents the site of proteolytic cleavage to release the mature Cys-rich conotoxin from a prepropeptide precursor. By switching loops between cysteine residues at the gene level (perhaps by a cassette switching mechanism), three new peptides could be generated. The conservation of both the excised N-terminal preproregion and the Cys residues in the mature toxin probably guarantees that specific disulfide bonding is conserved. However, the new peptides may either have the same pharmacological specificity or entirely different pharmacological specificity from the original peptide and from each other.

TABLE III

Hypervariability of conotoxins

Sequences given are from Refs. 1, 17, and 19–21.

A. ω-Conotoxins from *C. geographus* and *C. magus*	
ω-Conotoxin GVIA	CKSPGSSCSPTSYNCCRS-CNPYTKRCY*[a]
ω-Conotoxin MVIIA	CKGKGAKCSRLMYDCCTGSCRSGK--C*
AA identities in 12 ω-conotoxins sequenced	C G C CC C C
King-Kong peptide (not an ω-conotoxin)	WCKQSGEMCNLLDQNCCDGYCIVLV--CT
B. α-conotoxins from *C. striatus*	
α-Conotoxin SI	ICCNPACGPKYSC*
α-Conotoxin SIA	YCCHPACGKNFDC*
C. Conantokins from *C. geographus* and *C. tulipa*	
Conantokin-G	GEγγLQγNQγLIRγKSN*
Conantokin-T	GEγγYQKMLγNLRγAEVKKNA*

[a] *, C-terminal amidation; P, hydroxyproline; γ, γ-carboxyglutamate.

contact with determinants on the conotoxin. Thus, two different conotoxins with the same pharmacological specificity could contact a different subset of microsites within the same macrosite. Therefore, a large number of diverse peptide structures could potentially bind a macrosite. An important prediction of this hypothesis is that pharmacologically homologous conotoxins with divergent primary sequences would not be conformationally identical, even at the contact surface with the receptor (Fig. 5).

Conotoxins have great utility for studying cell-surface receptors, particularly in the nervous sytem. In general, the receptor system under study is not the natural physiological target but one that is evolutionarily related. For example, ω-conotoxin GVIA is widely used to study mammalian central nervous system calcium channels, but not fish calcium channels, the natural target. In the macrosite model above, a receptor in the same class as the natural target could have many microsites conserved but a subset that may have diverged. Only the subset of conotoxin homologs with direct focal contact would have altered receptor affinity when a particular microsite is altered. Thus, a set of conotoxin homologs should all bind the natural target with high affinity; if tested on an evolutionarily related receptor, the set would not behave uniformly but in an eclectic fashion, some with high affinity and some not binding at all. There are experimental observations consistent with this prediction, *i.e.* the dramatically different behavior of α-conotoxins GI and SI on mammalian neuromuscular synapses, and of ω-conotoxin GVIA and MVIIA on amphibian neurotransmission. Such results lend some credence to the macrosite model in Fig. 5. However, the two models above are not mutually exclusive; both pharmacophore conformation degeneracy on the ligand and alternative microsite contacts on the receptor could conceivably contribute to the sequence divergence observed between any two conotoxin homologs.

Perspectives and Future Directions

At the present time, only a few *Conus* venoms have been surveyed, and even in the best characterized venom, only a minor

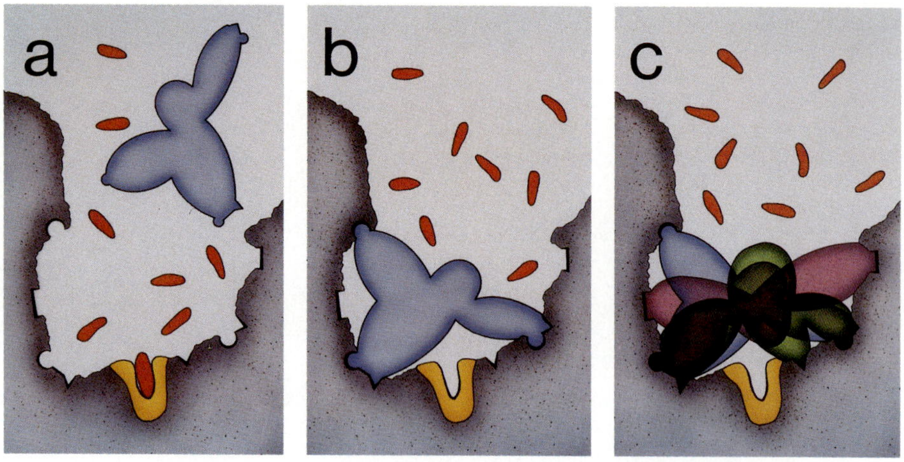

FIG 5. **The *Conus* toxin macrosite model.** A representation of a receptor binding pocket with a number of microsites which can potentially make focal contacts with conotoxin ligands is shown. The diagram illustrates a receptor with an endogenous ligand agonist (an example is the acetylcholine receptor); the endogenous ligands are the *orange* blobs, and the *yellow* region is the agonist binding site. The *middle panel* illustrates a conotoxin-blocking endogenous ligand binding, making three microsite focal contacts. As shown in the *right panel,* the macrosite can be alternatively occupied by conotoxins with different primary sequences, each making a different subset of focal contacts. In this way, even peptides with highly divergent sequences compete for binding to the same receptor pocket. In the example shown *blue* and *green* conotoxins share two focal contacts, while each shares one with the *purple* conotoxin. All three would serve as antagonists of this receptor.

fraction of peptides has been biochemically or pharmacologically characterized. However, this data base still permits a number of generalizations. First, there is remarkable pharmacological and biochemical diversity of small constrained peptides in each *Conus* venom. In *Conus geographus* venom, small peptide ligands target calcium channels, sodium channels, acetylcholine receptors, and NMDA receptors. Even more intriguing are the much larger number of biologically active peptides in the same venom for which receptors have not yet been identified.

In addition to the peptide diversity in an individual venom, an amazing sequence hypervariability between venoms is observed. No two *Conus* species have yet been found with the same conotoxin sequence. The present data base is best for the paralytic conotoxins; it seems reasonable to expect that every *Conus* venom will contain conotoxins directly paralytic to the prey. One obvious class of paralytics is conotoxins which inhibit acetylcholine receptors at neuromuscular junctions. Such agents have been described in all fish-hunting species examined and are very likely found in worm-hunting and mollusc-hunting *Conus* venoms as well. However, the toxins in fish-hunting species are presumably selected to inhibit fish acetylcholine receptors, while the corresponding toxins in the venoms of vermivorous *Conus* species would interact optimally with worm receptors. We can extrapolate from the data already collected that an acetylcholine receptor-targeted conotoxin in one species will have a significantly different sequence from that in any another *Conus* species. Thus, for the genus as a whole, there should be literally over a thousand different small peptides targeted to acetylcholine receptors. It seems likely that large sets of *Conus* peptides will be similarly targeted to many other receptors and ion channels.

The pharmacological potential of such substantial collections of small constrained peptides targeting to one class of receptors is immense. The acetylcholine receptor-targeted peptides can be tested on various acetylcholine receptors in different phylogenetic systems, such as the set of neuronal receptors in mammalian brain. While we cannot predict which peptides in the collection will have high affinity for a particular mammalian central nervous system acetylcholine receptor subtype, the natural repertoire of conotoxins should provide a rich source of ligands for discriminating between different receptor target subtypes.

As more information is collected about conotoxin design and synthesis in the natural system, it becomes increasingly feasible to apply similar strategies to generate conotoxin-like molecules *in vitro.* One particularly promising approach is to combine conotoxin biochemistry with newly developed peptide screening methods, such as the fUSE phage-peptide library technique of Scott and Smith (22). Recently, it was shown that the addition of a conotoxin module as a fusion to a phage coat protein did not affect phage viability (23). Thus, it is feasible to clone billions of conotoxin-like sequences onto vectors like fUSE phage and screen the library of conotoxin-like modules for interaction with cloned receptor targets. The phage that do exhibit affinity for the receptor target can then be amplified, the DNA sequenced, and the predicted disulfide-rich conotoxin-like peptide synthesized and tested for binding to the receptor target. Modification experiments to determine the focal contact points with the receptor are also feasible. Screens for conotoxins with any pharmacological specificity desired can be easily designed, *i.e.* small peptides that have very high affinity for one receptor subtype but which do not bind other closely related subtypes at all. Since these peptides will be constrained conformationally because of multiple disulfide bonds, their three-dimensional conformation could then be analyzed. This opens the door to pure chemical applications; for example, appropriate peptidomimetic derivatives can then be designed. The combination of a molecular genetic approach to allow screening of billions of sequences with the insight provided by the conotoxin system bridges molecular genetics and chemistry in a way that should permit exciting future applications in the pharmaceutical industry and many other areas of biotechnology.

REFERENCES

1. Olivera, B. M., Rivier, J., Clark, C., Ramilo, C. A., Corpuz, G. P., Abogadie, F. C., Mena, E. E., Woodward, S. R., Hillyard, D. R., and Cruz, L. J. (1990) *Science* **249,** 257–263
2. Woodward, S. R., Cruz, L. J., Olivera, B. M., and Hillyard, D. R. (1990) *EMBO J.* **1,** 1015–1020
3. Kohn, A. J. (1990) *Malacologia* **32,** 57–67
4. Kohn, A. J., and Nybakken, J. W. (1975) *Mar. Biol. (NY)* **29,** 211–274
5. Gray, W. R., Olivera, B. M., and Cruz, L. J. (1988) *Annu. Rev. Biochem.* **57,** 665–700
6. Rivier, J., Galyean, R., Simon, L., Cruz, L. J., Olivera, B. M., and Gray, W. R. (1987) *Biochemistry* **26,** 8508–8512
7. Mena, E. E., Gullak, M. F., Pagnozzi, M. J., Richter, K. E., Rivier, J., Cruz, L. J., and Olivera, B. M. (1990) *Neurosci. Lett.* **118,** 241–244
8. Czerwiec, E., DePotter, W., Convents, A., and Vauquelin, G. (1989) *Neurochem. Int.* **14,** 413–417
9. Elliot, E. J., and Raftery, M. A. (1979) *Toxicon* **17,** 259–268
10. Kobayashi, J., Nakamura, H., Hirata, Y., and Ohizumi, Y. (1982) *Biochem. Biophys. Res. Commun.* **105,** 1389–1395
11. Kobayashi, J., Nakamura, H., Hirata, Y., and Ohizumi, Y. (1982) *Life Sci.* **31,** 1085–1091
12. Schweitz, H., Renaud, J.-F., Randimbivololona, N., Preau, C., Schmid, A., Romey, G., Rakotovao, L., and Lazdunski, M. (1986) *Eur. J. Biochem.* **161,** 787–792
13. Gonoi, T., Ohizumi, Y., Kobayashi, J., Nakamura, H., and Catterall, W. A. (1987) *Mol. Pharmacol.* **32,** 691–698
14. Selsted, M. E., Brown, D. M., Delange, R. J., Harwig, S. S. L., and Lehrer, R. I. (1985) *J. Biol. Chem.* **260,** 4579–4584
15. Myers, R. A., McIntosh, J. M., Imperial, J., Williams, R. W., Oas, T., Haack, J. A., Hernandez, J. F., Rivier, J., Cruz, L. J., and Olivera, B. M. (1990) *J. Toxin-Toxin Revs.* **9,** 179–202
16. Olivera, B. M., Gray, W. R., Zeikus, R., McIntosh, J. M., Varga, J., Rivier, J., de Santos, V., and Cruz, L. J. (1985) *Science* **230,** 1338–1343
17. Cruz, L. J., Johnson, D. S., Imperial, J. S., Griffin, D., LeCheminant, G. W., Miljanich, G. P., and Olivera, B. M. (1988) *Curr. Top. Membr. Transp.* **33,** 417–429
18. Hillyard, D. R., Olivera, B. M., Woodward, S., Corpuz, G. P., Gray, W. R., Ramilo, C. A., and Cruz, L. J. (1989) *Biochemistry* **28,** 358–361
19. Zafaralla, G. C., Ramilo, C., Gray, W. R., Karlstrom, R., Olivera, B. M., and Cruz, L. J. (1988) *Biochemistry* **27,** 7102–7105
20. Myers, R. A., Zafaralla, G. C., Gray, W. R., Abbott, J., Cruz, L. J., and Olivera, B. M. (1991) *Biochemistry* **30,** 9370–9377
21. Haack, J. A., Rivier, J., Parks, T. N., Mena, E. E., Cruz, L. J., and Olivera, B. M. (1990) *J. Biol. Chem.* **265,** 6025–6029
22. Scott, J. K., and Smith, G. P. (1990) *Science* **228,** 1315–1317
23. Olivera, B. M., Cruz, L. J., Myers, R. A., Hillyard, D. R., Rivier, J., and Scott, J. K. (1992) in *Molecular Basis of Drug and Pesticide Action* (Duce, I., ed) Elsevier Science Publishing Co., New York, in press

THE JOURNAL OF BIOLOGICAL CHEMISTRY
Vol. 266, No. 34, Issue of December 5, pp. 22777–22780, 1991

Printed in U.S.A.

von Willebrand Factor

J. Evan Sadler

From the Howard Hughes Medical Institute, The Jewish Hospital of St. Louis, and the Departments of Medicine and of Biochemistry and Molecular Biophysics, Washington University School of Medicine, St. Louis, Missouri 63110

The existence of a "von Willebrand factor" (vWF)[1] was first suggested by Erik A. von Willebrand's discovery, reported in 1926 (1), of a serious hereditary bleeding disorder among residents of the Åland Islands in the Gulf of Bothnia, Finland. The disorder was clearly different from classical hemophilia because it was inherited as an autosomal dominant trait. von Willebrand named the disease "pseudohemophilia," but later authors have referred to it as von Willebrand disease (vWD). The nature of the defect in vWD was controversial for over 30 years. By 1957, however, transfusion studies indicated that the disease was caused by lack of a blood plasma factor (2). Within only the last decade the structures of the vWF protein and gene were determined, and many vWF structure-function relationships were elucidated. vWD is now known to be the most common inherited human bleeding disorder, and mutations causing several variants of vWD have been identified.

Biosynthesis and Structure of vWF

vWF is synthesized by vascular endothelial cells (3) and megakaryocytes (4) and is a useful immunohistochemical marker for these cell lineages. The ~360-kDa (2813-amino acid) primary translation product includes a 22-amino acid signal peptide, a very large ~95-kDa (741-amino acid) propeptide also known as von Willebrand antigen II (5), and a ~260-kDa (2050-amino acid) mature subunit (vWF) (Fig. 1). The vWF precursor contains four types of repeated domains (A–D) present in 2–5 copies each. Together these comprise more than 90% of the protein. The sequence of vWF is remarkable for the clustering of cysteine residues in amino-terminal and carboxyl-terminal regions; in fact, cysteine is the most abundant amino acid in the protein and accounts for 8.3% of the total (reviewed in Ref. 6). The pairing of 52 half-cystine residues has been determined (7, 8).

After translocation into the endoplasmic reticulum, pro-vWF subunits form dimers that are held together by disulfide bonds near the carboxyl termini (7, 9). In the Golgi apparatus, or possibly in a post-Golgi compartment, the pro-vWF dimers form multimers through one or more disulfide bonds involving cysteines between amino acids 459 and 464 of the mature subunit sequence (7, 10–12), and the propeptide is cleaved from most of the subunits. The resultant mature multimers range in size from dimers of approximately 500 kDa to species of more than 20 million daltons (Fig. 2) (reviewed in Ref. 14).

The vWF propeptide may be required for normal multimer formation. In transfected cells, propeptide deletions prevent multimerization (15, 16), while missense mutations that abolish propeptide cleavage do not (16). Purified pro-vWF dimers, but not mature vWF dimers, can be induced to form multimers *in vitro* in the presence of calcium ions and at low pH (17). These observations suggest that the vWF propeptide may directly catalyze the formation of disulfide bonds in vWF multimers (17).

Besides multimer formation and proteolytic processing, post-translational modifications of vWF include extensive Asn-linked and Thr/Ser-linked glycosylation (18) and sulfation of a restricted number of Asn-linked oligosaccharides (Fig. 1). The biosynthesis and structure of vWF apparently are not disrupted by inhibitors of oligosaccharide sulfation (19).

In endothelial cells, a fraction of vWF multimers is secreted constitutively (20). The remainder is stored in specialized organelles called Weibel-Palade bodies (21). These membrane-bound, 0.1 × 2–3-μm organelles contain longitudinally oriented tubular structures that represent tightly packed vWF multimers. The vWF propeptide appears to be necessary for the targeting of vWF to these storage granules (12). Cleaved propeptide is packaged together with vWF multimers, with a stoichiometry of one propeptide molecule per mature subunit (22, 23). vWF is found in similar tubular structures within the α-granules of platelets (24).

The contents of Weibel-Palade bodies and platelet α-granules are secreted in response to a variety of physiological stimuli. After secretion the cleaved propeptide circulates in the blood as a noncovalently associated dimer, independent of vWF (22). Whether the propeptide has a biological function, aside from its role in the multimerization of vWF, is not known.

Biological Function of vWF

vWF performs two important hemostatic functions. It is required for the adhesion of platelets to sites of vascular damage, linking specific platelet membrane receptors to constituents of subendothelial connective tissue (reviewed in Ref. 25). It also binds to and stabilizes blood coagulation factor VIII (antihemophilic factor) in the circulation (26). As a consequence of these activities, severe vWF deficiency can cause bleeding with two characteristic clinical patterns. 1) The defect in platelet adhesion is associated with bleeding from skin and mucosal surfaces like that caused by platelet disorders, and a prolonged skin bleeding time is a diagnostically useful feature of the disease. 2) The failure of vWF to stabilize factor VIII results in secondary factor VIII deficiency; this can cause spontaneous bleeding into joints and soft tissues, similar to that of classical hemophilia A.

Two platelet membrane glycoprotein receptors for vWF have been identified. Unactivated platelets bind vWF through the platelet GPIb-IX complex (27); inherited deficiency of GPIb-IX (Bernard-Soulier syndrome) (28) causes bleeding similar to that of vWD. After activation by thrombin or other agonists, platelets express a second binding site for vWF, the GPIIb-IIIa complex (29). GPIIb-IIIa belongs to a family of related heterodimeric cell surface proteins, called "integrins," that mediate a wide variety of cellular adhesive interactions (30). The GPIIb-IIIa complex is defective in Glanzmann thrombasthenia (31, 32), which also is associated with bleeding.

The constituent in the vascular wall to which vWF binds has not been identified conclusively, and more than one ligand may be physiologically important. Several macromolecules present in subendothelium have been shown to bind specifically to vWF, including various fibrillar collagens (33), collagen type VI (34), heparin (35), and sulfated glycolipids (sulfatides) (36, 37). The significance of binding to fibrillar collagens is uncertain, since binding of vWF to extracellular matrix is not inhibited upon elimination of fibrillar collagens either by collagenase digestion or by elaboration of matrix in the presence of α,α′-dipyridyl, a collagen synthesis inhibitor (38). Furthermore, monoclonal antibodies to vWF that inhibit binding to subendothelial matrix but not to fibrillar collagens, and *vice versa*, have been identified (39).

In contrast to the fibrillar collagens, collagen type VI is abundant in the extracellular matrix (40) and is resistant to both collagenase (41) and α,α′-dipyridyl (42). This collagen is an interesting candidate ligand for vWF. As discussed in a later section, it contains repeated noncollagenous domains that are homologous to the vWF A-domains, and it also contains a separate region near the carboxyl terminus that has limited sequence

[1] The abbreviations used are: vWF, von Willebrand factor; vWD, von Willebrand disease; GP, glycoprotein.

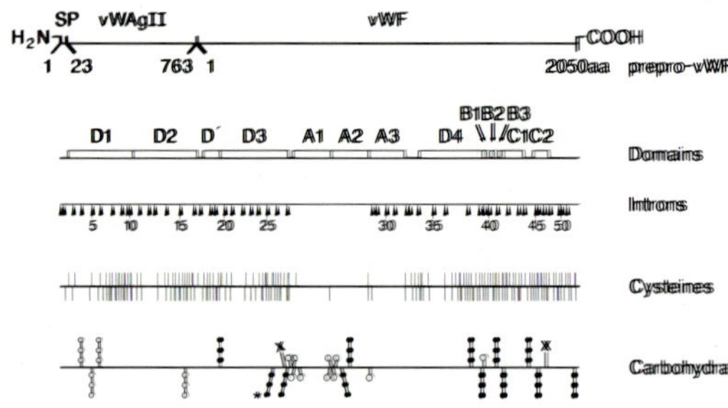

FIG. 1. **Structural features of cDNA and amino acid sequences for the human vWF precursor, prepro-vWF.** Prepro-vWF: the locations of the signal peptide (*SP*) or prepeptide, von Willebrand antigen II propeptide (*vWAgII*), and mature subunit (*vWF*) are indicated. Amino acid residues in the prepropeptide are numbered consecutively 1–763; residues 1–22 are the signal peptide, and residues 23–763 are the propeptide. Amino acid (*aa*) residues in the mature subunit are separately numbered 1–2050. Domains: the repeated domains are labeled *D1, D2, D′, D3, A1, A2, A3, D4, B1, B2, B3, C1,* and *C2.* Introns: the locations in the amino acid sequence of the 51 introns of the vWF gene are indicated by *arrowheads.* Every fifth intron is *numbered.* Cysteines: in regions of especially high cysteine content, one *mark* may represent more than 1 cysteine residue. Carbohydrate: potential sites of *N*-glycosylation are indicated by *open symbols* (○-○-○). *N*-Glycosylation sites shown to be utilized are indicated by *filled symbols* (●-●-●), and two potential sites shown not to be glycosylated are indicated by ×. One site of *N*-glycosylation labeled with an *asterisk* (*) occurs in the sequence Asn-Ser-Cys. Sites of *O*-glycosylation are indicated by *single open circles* (○). Adapted from Ref. 6.

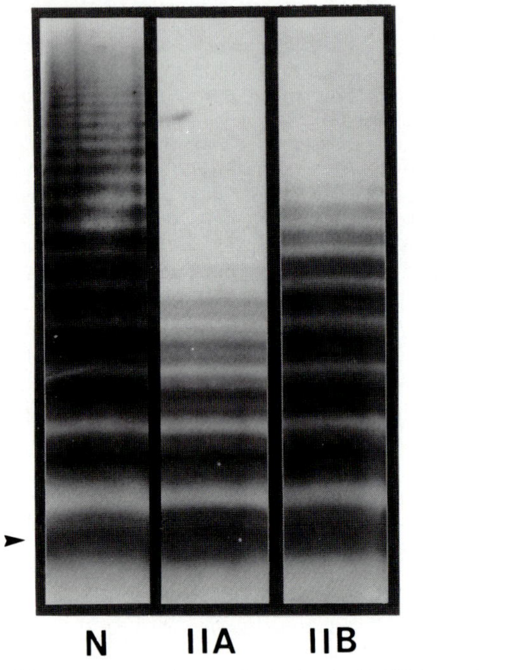

FIG. 2. **Multimer patterns of normal vWF and of vWF in variants of human vWD.** Samples of plasma were electrophoresed through a 1.4% agarose gel in the presence of sodium dodecyl sulfate and detected by Western blotting (13). *N,* normal plasma; *IIA,* plasma from a patient with vWD type IIA; *IIB,* plasma from a patient with vWD type IIB. The smallest visible multimer (*arrowhead*) is a dimer of ~500 kDa, and the increment of mass between adjacent multimers is ~500 kDa. In addition to the major multimer bands, there are often two minor "satellite" bands per multimer. The intensity and spacing of satellite bands have diagnostic significance for certain vWD variants.

similarity to platelet GPIb (43). Whether these domains of collagen type VI are involved in physiological binding to vWF is not yet known.

Structure-Function Relationships of vWF

The larger multimers of vWF are required for normal biological function, but the reasons for this are not completely understood. One factor is certainly that multimerization increases the number of ligand binding sites per vWF molecule. In addition, platelet adhesion depends upon vWF only at high wall shear rates that are characteristic of blood flow in the microcirculation (25). Thus, the physical properties of extended polymers under conditions of high shear may be critical for the function of vWF.

Many of the binding functions of vWF have been localized to discrete regions of the subunit (Fig. 3). The propeptide has at least one binding site for collagen (44); the physiological significance of this interaction is not known. The binding site for blood coagulation factor VIII is within the amino-terminal 272 amino acids of the mature subunit (45). A proteolytic fragment that contains this sequence also binds to heparin (46), but the major heparin binding site on vWF appears to be elsewhere, within domain A1.

Domain A1 contains binding sites for several macromolecules (Fig. 3), including platelet GPIb. In solution, human vWF does not bind spontaneously to GPIb. This interaction is induced physiologically by the binding of vWF to subendothelial connective tissue (25). *In vitro,* binding of vWF to GPIb can be induced by the antibiotic ristocetin (47), and the ristocetin-induced aggregation of platelets is the basis of several clinical assays of vWF function (6). Ristocetin binds to platelets (48, 49) and probably to vWF (49), but how this induces the binding of vWF to GPIb is not known. Discrete segments of domain A1 are proposed to mediate ristocetin-induced binding of vWF to platelet GPIb (50).

The venom of certain pit vipers of the *Bothrops* genus, particularly *Bothrops jararaca,* contains a protein called "botrocetin" that also induces vWF binding to GPIb (51, 52). Botrocetin binds directly to domain A1, within a disulfide loop bounded by Cys^{509} and Cys^{695} (8, 52, 53). This same region contains distinct binding sites for heparin (54), sulfatides (36, 37), and collagen (10). Despite some structural similarities, heparin and sulfatides do not bind competitively to vWF (36). A second collagen-binding site has been localized within domain A3 (55).

The binding site on vWF for the GPIIb-IIIa complex of activated platelets has been localized to the tetrapeptide sequence, Arg-Gly-Asp-Ser(1747), near the carboxyl-terminal end of domain C1 (Fig. 3) (56). Arg-Gly-Asp sequences occur in fibrinogen, fibronectin, and vitronectin; these proteins also bind to GPIIb-IIIa. Several other integrins besides GPIIb-IIIa, including receptors for fibronectin and vitronectin, recognize Arg-Gly-Asp sequences within their ligands but do not bind vWF. These interactions are required for the cell attachment activities of fibronectin, vitronectin, and other proteins (reviewed in Ref. 57).

Gene Structure and Evolution of vWF

The vWF gene is located at the tip of the short arm of chromosome 12 (58, 59). It spans ~180 kilobases and contains 52 exons (60–63). Except for the A domains, gene segments encoding homologous vWF repeated domains tend to have similarly placed intron-exon boundaries. A partial, unprocessed vWF pseudogene is located on chromosome 22q11-13. The pseudogene spans 21–29 kilobases and corresponds to vWF exons 23–34, which encode ~34% of the vWF precursor sequence including the triplicated A domains. The vWF gene and pseudogene have diverged only ~3.1%, suggesting that the pseudogene arose recently, perhaps <20–30-million years ago (64).

Since the A domains were identified in vWF (65), homologous A-like domains have been found in several other proteins: the serine protease complement factor B and the related complement component C2; two subgroups of integrin α-subunits, the three

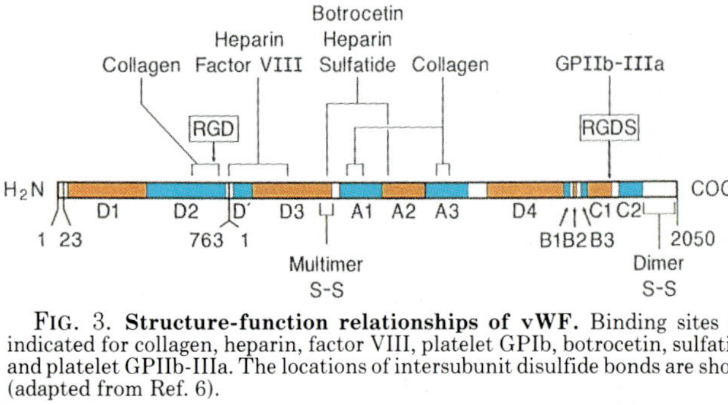

FIG. 3. **Structure-function relationships of vWF.** Binding sites are indicated for collagen, heparin, factor VIII, platelet GPIb, botrocetin, sulfatide, and platelet GPIIb-IIIa. The locations of intersubunit disulfide bonds are shown (adapted from Ref. 6).

leucocyte adhesion receptors (Mac-1, LFA-1, and p150,95) and two collagen receptors ($\alpha1\beta1$ and $\alpha2\beta1$); cartilage matrix protein; and all three chains of collagen type VI (reviewed in Ref. 66). To date there are 12 known genes that together encode at least 29 vWF A-like domains, often in association with distinct, unrelated, structural motifs. The vWF C domains share limited sequence similarity with and may be homologous to portions of thrombospondin, noncollagenous domains of procollagen types α-1(I) and α-1(III) (67), and a *Xenopus laevis* integumentary mucin (68). Its highly repeated structure and the existence of A-domain and possibly C-domain homologues throughout the human genome indicate that vWF has a complex evolutionary history marked by repeated gene segment duplications and exon shuffling.

Molecular Defects Causing vWD

Inherited defects of vWF function can be detected by laboratory testing in nearly 1% of the population. Fortunately, clinically significant vWD is much less common and affects ~125 per million population (6). vWD is classified into three major categories, each of which is heterogeneous. vWD type I refers to a partial quantitative deficiency of vWF, is inherited as an autosomal dominant trait, and accounts for ~80% of all vWD. vWD type II refers to qualitative abnormalities of vWF and accounts for 15–20% of vWD. Most type II variants have an abnormality of vWF multimer structure, usually a deficiency of the largest multimers (Fig. 2). vWD type III refers to a severe, autosomal recessive disease that is characterized by virtual absence of vWF. These patients are uncommon, with a prevalence of 0.5–5 per million population, and appear to be homozygous or compound heterozygous for defects at the vWF locus (6). A subset of patients with vWD type III have total (69, 70) or partial (71) deletions of the vWF gene. Such deletions predispose to the development of alloantibody inhibitors of vWF, a serious complication of transfusion therapy. The parents of patients with vWD type III are obligate heterozygotes, who often have moderate decreases in plasma vWF activity and are usually asymptomatic. One of the remaining mysteries in vWD is that patients with vWD type I and parents of patients with vWD type III can have similarly decreased laboratory indices of vWF function, and yet only patients with vWD type I have a bleeding diathesis.

Molecular defects have been characterized in several variants of vWD type II, and in many cases these defects have provided information concerning vWF structure-function relationships. In vWD type IIA, the vWF protein lacks intermediate and high molecular weight multimers (Fig. 2) and has decreased affinity for platelet GPIb. Seven candidate missense mutations causing vWD type IIA have been reported, and all but one cluster within a restricted segment of domain A2 (72–76), near a site in normal vWF (Tyr^{842}-Met^{843}) that is sensitive to proteolysis *in vivo* (Fig. 4) (77). The vWD type IIA phenotype may be caused by two distinct mechanisms in specific subgroups of these mutations. In one group, typified by the $Arg^{834} \rightarrow Trp$ mutation, a full range of vWF multimers is produced, but they are abnormally sensitive to proteolytic degradation in the circulation. This results in a type II multimer pattern and decreased vWF function. In the second group, which includes the $Val^{844} \rightarrow Asp$ mutation, there is defective intracellular processing, with secretion of vWF having a type II multimer pattern and decreased function (72, 78).

vWD type IIB is a less common variant in which vWF multimers of normal size are secreted, but these multimers have exaggerated affinity for platelet GPIb. This gain of function is associated paradoxically with bleeding, possibly because the largest vWF multimers spontaneously associate with platelets and are cleared from the circulation. The remaining vWF in the circulation is relatively deficient in large multimers (Fig. 2) and has decreased function. The seven reported mutations in vWD type IIB cluster in a single disulfide loop of domain A1 (Fig. 4) (79–83), and they lie within segments of this loop that appear to interact directly with botrocetin (53). These mutations may induce a conformational transition in vWF, similar to the effect of botrocetin binding, that promotes the association of vWF with platelet GPIb.

Recently, a variant of vWD was described in which the binding of vWF to factor VIII is defective, resulting in accelerated clearance of factor VIII (84, 85). Patients with this disorder, tentatively referred to as vWD "Normandy" (85), have factor VIII deficiency that mimics hemophilia A except that the inheritance is autosomal rather than X-linked recessive. Accurate discrimination between hemophilia A and vWD "Normandy" is important for proper treatment of bleeding episodes and for genetic counseling. Three different mutations have been characterized that cause this phenotype: $Thr^{28} \rightarrow Met$ in exon 18, $Arg^{53} \rightarrow Gln$ in exon 19, and $Arg^{91} \rightarrow Gln$ in exon 20 (60, 86–90). These mutations lie within the amino-terminal 100 amino acids of the mature vWF subunit, consistent with the localization of the factor VIII binding site on vWF in the same region.

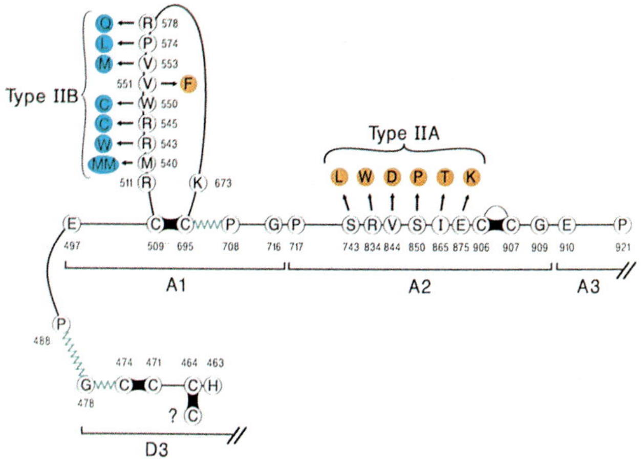

FIG. 4. **Mutations in vWF exon 28 causing vWD type IIA and type IIB.** The segment of mature vWF shown is encoded by exon 28 and includes amino acid residues 463–921. The positions of repeated domains D3, A1, A2, and A3 are indicated. The *green zigzag* segments from Cys^{474}-Pro^{488} and Cys^{695}-Pro^{708} indicate regions proposed to interact directly with platelet GPIb (51). Mutations reported to cause vWD type IIA (*orange circles*) and type IIB (*blue circles*) are indicated by *brackets*; one proposed type IIA mutation ($Val^{551} \rightarrow Phe$, *orange circle*) occurs in the region of the type IIB mutations (adapted from Ref. 79).

Concluding Remarks

Since von Willebrand's discovery, 65 years ago, of the disease that now bears his name, substantial progress has been made toward understanding vWF structure, biological function, molecular biology, and pathophysiology. Further characterization of mutations in vWD, combined with *in vitro* mutagenesis, will continue to provide insight into the biological function and structure-function relationships of vWF. These studies will also illuminate the physiological significance of the many specific vWF-ligand interactions that have been described *in vitro* and eventually will result in improved therapeutic interventions for patients with vWD.

Acknowledgments—I thank Anna Randi for performing the multimer gel electrophoresis of Fig. 2 and Zaverio Ruggeri for providing information on the interaction between vWF and botrocetin prior to publication and for supplying the plasma sample for vWD type IIA. I would also like to thank David Ginsburg and the members of my laboratory for helpful review of this manuscript.

REFERENCES

1. von Willebrand, E. A. (1926) *Finska Lakarsallskapets Handlingar* **68,** 87–112
2. Nilsson, I. M., Blombäck, M., Jorpes, E., Blombäck, B., and Johansson, S.-A. (1957) *Acta Med. Scand.* **159,** 179–188
3. Jaffe, E. A., Hoyer, L. W., and Nachman, R. L. (1973) *J. Clin. Invest.* **52,** 2757–2764
4. Nachman, R., Levine, R., and Jaffe, E. A. (1977) *J. Clin. Invest.* **60,** 914–921
5. Montgomery, R. R., and Zimmerman, T. S. (1978) *J. Clin. Invest.* **62,** 1498–1507
6. Sadler, J. E. (1989) in *The Metabolic Basis of Inherited Disease* (Scriver, C. R., Beaudet, A. L., Sly, W. S., and Valle, D., eds) 6th Ed., pp. 2171–2188, McGraw-Hill Book Co., New York
7. Marti, T., Rösselet, S. J., Titani, K., and Walsh, K. A. (1987) *Biochemistry* **26,** 8099–8109
8. Andrews, R. K., Gorman, J. J., Booth, W. J., Corino, G. L., Castaldi, P. A., and Berndt, M. C. (1989) *Biochemistry* **28,** 8326–8336
9. Voorberg, J., Fontijn, R., Calafat, J., Janssen, H., van Mourik, J. A., and Pannekoek, H. (1991) *EMBO J.* **113,** 195–205

10. Mohri, H., Yoshioka, A., Zimmerman, T. S., and Ruggeri, Z. M. (1989) *J. Biol. Chem.* **264,** 17361-17367
11. Voorberg, J., Fontijn, R., van Mourik, J. A., and Pannekoek, H. (1990) *EMBO J.* **9,** 797-803
12. Wagner, D. D., Saffaripour, S., Bonfanti, R., Sadler, J. E., Cramer, E. M., Chapman, B., and Mayadas, T. N. (1991) *Cell* **64,** 403-413
13. Raines, G., Aumann, G., Sykes, S., and Street, A. (1990) *Thromb. Res.* **60,** 201-212
14. Mayadas, T. N., and Wagner, D. D. (1991) *Ann. N. Y. Acad. Sci.* **614,** 153-166
15. Verweij, C. L., Hart, M., and Pannekoek, H. (1987) *EMBO J.* **6,** 2885-2890
16. Wise, R. J., Pittman, D. D., Handin, R. I., Kaufman, R. J., and Orkin, S. H. (1988) *Cell* **52,** 229-236
17. Mayadas, T. N., and Wagner, D. D. (1989) *J. Biol. Chem.* **264,** 13497-13503
18. Titani, K., Kumar, S., Takio, K., Ericsson, L. H., Wade, R. D., Ashida, K., Walsh, K. A., Chopek, M. W., Sadler, J. E., and Fujikawa, K. (1986) *Biochemistry* **25,** 3171-3184
19. Carew, J. A., Browning, P. J., and Lynch, D. C. (1990) *Blood* **76,** 2530-2539
20. Sporn, L. A., Marder, V. J., and Wagner, D. D. (1986) *Cell* **46,** 185-190
21. Wagner, D. D., Olmsted, J. B., and Marder, V. J. (1982) *J. Cell Biol.* **95,** 355-360
22. Wagner, D. D., Fay, P. J., Sporn, L. A., Sinha, S., Lawrence, S. O., and Marder, V. J. (1987) *Proc. Natl. Acad. Sci. U. S. A.* **84,** 1955-1959
23. Ewenstein, B. M., Warhol, M. J., Handin, R. I., and Pober, J. (1987) *J. Cell Biol.* **104,** 1423-1433
24. Cramer, E. M., Meyer, D., le Menn, R., and Breton-Gorius, J. (1985) *Blood* **66,** 710-713
25. Weiss, H. J. (1991) *Ann. N. Y. Acad. Sci.* **614,** 125-137
26. Tuddenham, E. G. D., Lane, R. S., Rotblat, F., Johnson, A. J., Snape, T. J., Middleton, S., and Kernoff, P. B. A. (1982) *Br. J. Haematol.* **52,** 259-267
27. Sakariassen, K. S., Nievelstein, P. F. E. M., Coller, B. S., and Sixma, J. J. (1986) *Br. J. Haematol.* **63,** 681-691
28. Caen, J. P., Nurden, A. T., Jeanneau, C., Michel, H., Tobelem, G., Levy-Toledano, S., Sultan, Y., Valensi, F., and Bernard, J. (1976) *J. Lab. Clin. Med.* **87,** 586-596
29. Fujimoto, T., Ohara, S., and Hawiger, J. (1982) *J. Clin. Invest.* **69,** 1212-1222
30. Hynes, R. O. (1987) *Cell* **48,** 549-550
31. Nurden, A. T., and Caen, J. P. (1974) *Br. J. Haematol.* **28,** 253-260
32. Phillips, D. R., and Poh Agin, P. (1977) *J. Clin. Invest.* **60,** 535-545
33. Santoro, S. A. (1981) *Thromb. Res.* **21,** 689-693
34. Rand, J. H., Patel, N. D., Schwartz, E., Zhou, S.-L., and Potter, B. J. (1991) *J. Clin. Invest.* **88,** 253-259
35. Madaras, F., Bell, W. R., and Castaldi, P. A. (1978) *Haemostasis* **7,** 321-331
36. Roberts, D. D., Williams, S. B., Gralnick, H. R., and Ginsburg, V. (1986) *J. Biol. Chem.* **261,** 3306-3309
37. Christophe, O., Obert, B., Meyer, D., and Girma, J.-P. (1991) *Thromb. Haemostasis* **65,** 797
38. Wagner, D. D., Urban-Pickering, M., and Marder, V. J. (1984) *Proc. Natl. Acad. Sci. U. S. A.* **81,** 471-475
39. de Groot, P. G., Ottenhof-Rovers, M., van Mourik, J. A., and Sixma, J. J. (1988) *J. Clin. Invest.* **82,** 65-73
40. Trüeb, B., Schreier, T., Bruckner, P., and Winterhalter, K. H. (1987) *Eur. J. Biochem.* **166,** 699-703
41. von der Mark, H., Sumailley, M., Wick, G., Fleischmajer, R., and Timpl, R. (1984) *Eur. J. Biochem.* **142,** 493-502
42. Colombatti, A., and Bonaldo, P. (1987) *J. Biol. Chem.* **262,** 14461-14466
43. Bonaldo, P., and Colombatti, A. (1989) *J. Biol. Chem.* **264,** 20235-20239
44. Takagi, J., Fujisawa, T., Sekiya, F., and Saito, Y. (1991) *J. Biol. Chem.* **266,** 5575-5579
45. Foster, P. A., Fulcher, C. A., Marti, T., Titani, K., and Zimmerman, T. S. (1987) *J. Biol. Chem.* **262,** 8443-8446
46. Fretto, L. J., Fowler, W. E., McCaslin, D. R., Erickson, H. P., and McKee, P. A. (1986) *J. Biol. Chem.* **261,** 15679-15689
47. Nachman, R. L., Jaffe, E. A., and Weksler, B. W. (1977) *J. Clin. Invest.* **59,** 143-148
48. Coller, B. S. (1978) *J. Clin. Invest.* **61,** 1168-1175
49. Scott, J. P., Montgomery, R. R., and Retzinger, G. S. (1991) *J. Biol. Chem.* **266,** 8149-8155
50. Mohri, H., Fujimura, Y., Shima, M., Yoshioka, A., Houghten, R. A., Ruggeri, Z. M., and Zimmerman, T. S. (1988) *J. Biol. Chem.* **263,** 17901-17904
51. Read, M. S., Shermer, R. W., and Brinkhous, K. M., (1978) *Proc. Natl. Acad. Sci. U. S. A.* **75,** 4514-4518
52. Fujimura, Y., Holland, L. Z., Ruggeri, Z. M., and Zimmerman, T. S. (1987) *Blood* **70,** 985-988
53. Sugimoto, M., Mohri, H., McClintock, R. A., and Ruggeri, Z. M. (1991) *J. Biol. Chem.* **266,** 18172-18178
54. Fujimura, Y., Titani, K., Holland, L. Z., Roberts, J. R., Kostel, P., Ruggeri, Z. M., and Zimmerman, T. S. (1987) *J. Biol. Chem.* **262,** 1734-1739
55. Roth, G. J., Titani, K., Hoyer, L. W., and Hickey, M. J. (1986) *Biochemistry* **25,** 8357-8361
56. Berliner, S. K., Niiya, K., Roberts, J. R., Houghten, R. A., and Ruggeri, Z. M. (1988) *J. Biol. Chem.* **263,** 7500-7505
57. D'Souza, S. E., Ginsberg, M. H., and Plow, E. F. (1991) *Trends Biochem. Sci.* **16,** 246-250
58. Ginsburg, D., Handin, R. I., Bonthron, D. T., Donlon, T. A., Bruns, G. A. P., Latt, S. A., and Orkin, S. H. (1985) *Science* **228,** 1401-1406
59. Verweij, C. L., de Vries, C. J. M., Distel, B., van Zonneveld, A.-J., van Kessel, A. G., van Mourik, J. A., and Pannekoek, H. (1985) *Nucleic Acids Res.* **13,** 4699-4717
60. Mancuso, D. J., Tuley, E. A., Westfield, L. A., Worrall, N. K., Shelton-Inloes, B. B., Sorace, J. M., Alevy, Y. G., and Sadler, J. E. (1989) *J. Biol. Chem.* **264,** 19514-19527
61. Collins, C. J., Underdahl, J. P., Levene, R. B., Ravera, C. P., Morin, M. J., Dombalagian, M. J., Ricca, G., Livingston, D. M., and Lynch, D. C. (1987) *Proc. Natl. Acad. Sci. U. S. A.* **84,** 4393-4397
62. Assouline, Z., Kerbiriou-Nabias, D. M., Piétu, G., Thomas, N., Bahnak, B. R., and Meyer, D. (1988) *Biochem. Biophys. Res. Commun.* **153,** 1159-1166
63. Bonthron, D., and Orkin, S. H. (1988) *Eur. J. Biochem.* **171,** 51-57
64. Mancuso, D. J., Tuley, E. A., Westfield, L. A., Lester-Mancuso, T. L., Le Beau, M. M., Sorace, J. M., and Sadler, J. E. (1991) *Biochemistry* **30,** 253-269
65. Sadler, J. E., Shelton-Inloes, B. B., Sorace, J. M., Harlan, J. M., Titani, K., and Davie, E. W. (1985) *Proc. Natl. Acad. Sci. U. S. A.* **82,** 6394-6398
66. Colombatti, A., and Bonaldo, P. (1991) *Blood* **77,** 2305-2315
67. Hunt, L. T., and Barker, W. C. (1987) *Biochem. Biophys. Res. Commun.* **144,** 876-882
68. Probst, J. C., Gertzen, E.-M., and Hoffman, W. (1990) *Biochemistry* **29,** 6240-6244
69. Shelton-Inloes, B. B., Chehab, F. F., Mannucci, P. M., Federici, A. B., and Sadler, J. E. (1987) *J. Clin. Invest.* **79,** 1459-1465
70. Ngo, K. Y., Glotz, V. T., Koziol, J. A., Lynch, D. C., Gitschier, J., Ranieri, P., Ciavarella, N., Ruggeri, Z. M., and Zimmerman, T. S. (1988) *Proc. Natl. Acad. Sci. U. S. A.* **85,** 2753-2757
71. Peake, I. R., Liddell, M. B., Moodie, P., Standen, G., Mancuso, D. J., Tuley, E. A., Westfield, L. A., Sorace, J. M., Sadler, J. E., Verweij, C. L., and Bloom, A. L. (1990) *Blood* **75,** 654-661
72. Ginsburg, D., Konkle, B. A., Gill, J. C., Montgomery, R. R., Bockenstedt, P. L., Johnson, T. A., and Yang, A. Y. (1989) *Proc. Natl. Acad. Sci. U. S. A.* **86,** 3723-3727
73. Chang, H.-Y., Chen, Y.-P., Chediak, J. R., Levene, R. B., and Lynch, D. C. (1989) *Blood* **74,** Suppl. 1, 131a
74. Iannuzzi, M. C., Hidaka, N., Boehnke, M., Bruck, M. E., Hanna, W. T., Collins, F. S., and Ginsburg, D. (1991) *Am. J. Hum. Genet.* **48,** 757-763
75. Lavergne, J. M., Ribba, A. S., Bahnak, B. R., de Paillette, L., Derlon, A., Meyer, D., and Pietu, G. (1991) *Thromb. Haemostasis* **65,** 738
76. Sugiura, I., Matsuchita, T., Tanimoto, M., Takamatsu, J., Kamiya, T., Saito, H., Furuya, H., and Kato, Y. (1991) *Thromb. Haemostasis* **65,** 763
77. Dent, J. A., Berkowitz, S. D., Ware, J., Kasper, C. K., and Ruggeri, Z. M. (1990) *Proc. Natl. Acad. Sci. U. S. A.* **87,** 6306-6310
78. Lyons, S. E., and Ginsburg, D. (1990) *J. Cell Biol.* **111,** 464a
79. Randi, A. M., Rabinowitz, I., Mancuso, D. J., Mannucci, P. M., and Sadler, J. E. (1991) *J. Clin. Invest.* **87,** 1220-1226
80. Cooney, K. A., Nichols, W. C., Bruck, M. E., Bahou, W. F., Shapiro, A. D., Bowie, E. J. W., Gralnick, H. R., and Ginsburg, D. (1991) *J. Clin. Invest.* **87,** 1227-1233
81. Ware, J., Dent, J. A., Azuma, H., Sugimoto, M., Kyrle, P. A., Yoshioka, A., and Ruggeri, Z. M. (1991) *Proc. Natl. Acad. Sci. U. S. A.* **88,** 2946-2950
82. Kroner, P. A., Kluessndorf, M. L., Trieu, H., and Montgomery, R. R. (1991) *Clin. Res.* **39,** 326a
83. Ribba, A. S., Lavergne, J. M., Bahnak, B. R., Derlon, A., Piétu, G., and Meyer, D. (1991) *Blood* **78,** 1738-1743
84. Nishino, M., Girma, J.-P., Rothschild, C., Fressinaud, E., and Meyer, D. (1989) *Blood* **74,** 1591-1599
85. Mazurier, C., Dieval, J., Jorieux, S., Delobel, J., and Goudemand, M. (1990) *Blood* **75,** 20-26
86. Gaucher, C., Jorieux, S., Mercier, B., Oufkir, D., and Mazurier, C. (1991) *Blood* **77,** 1937-1941
87. Gaucher, C., Mercier, B., Jorieux, S., Oufkir, D., and Mazurier, C. (1991) *Br. J. Haematol.* **78,** 506-514
88. Tuley, E. A., Gaucher, C., Jorieux, S., Worrall, N. K., Sadler, J. E., and Mazurier, C. (1991) *Proc. Natl. Acad. Sci. U. S. A.* **88,** 6377-6381
89. Kroner, P. A., Friedman, K. D., Fahs, S., Scott, J. P., and Montgomery, R. R. (1991) *J. Biol. Chem.* **266,** 19146-19149
90. Cacheris, P. M., Nichols, W. C., and Ginsburg, D. (1991) *J. Biol. Chem.* **266,** 13499-13502

THE JOURNAL OF BIOLOGICAL CHEMISTRY
Vol. 266, No. 35, Issue of December 15, pp. 23517–23520, 1991

Printed in U.S.A.

CD45

A PROTOTYPE FOR TRANSMEMBRANE PROTEIN TYROSINE PHOSPHATASES*

Ian S. Trowbridge

From the Department of Cancer Biology, The Salk Institute, San Diego, California 92186-5800

Historical Perspective

The identification of CD45 (T200 or leukocyte common antigen) as a protein tyrosine phosphatase (PTP)[1] resulted from the convergence of two independent lines of investigation. Immunological characterization of lymphocyte cell surface molecules which began almost 30 years ago (1, 2) has reaped a rich harvest recently as the functional roles of many molecules initially detected on the basis of their antigenicity have been elucidated. CD45 is a product of this experimental approach, attracting early attention because of its abundance, distinctive tissue distribution, and novel structural features (reviewed in Ref. 3). It was characterized in the mid-1970s as a major transmembrane glycoprotein found only on hematopoietic cells. As a consequence of its tissue distribution, CD45 has found clinical utility in the diagnosis of undifferentiated lymphoma (4). CD45 is structurally heterogeneous consisting of a family of isoforms ranging in M_r from ~180,000 to 220,000 that are distributed in characteristic patterns within the hematopoietic system: B cells expressing predominantly the 220,000 isoform; thymocytes the 180,000 isoform; and T cells different patterns of isoforms that appear to correlate with function and prior antigenic exposure (3). The primary structures of rat, mouse, and human CD45 deduced from cDNA sequencing in the mid-1980s revealed that the cytoplasmic domain of CD45 was unusually large (3), confirming an inference from earlier studies of a mutant CD45 molecule (5). This work assumed broader significance when it was found that the two imperfect tandem repeats within the cytoplasmic domain of CD45 had highly significant sequence similarity with human placental PTP1B (6). CD45 was quickly shown to have PTP activity (7), raising the possibility that transmembrane PTPs may represent a novel class of receptors that play an active role in the regulation of cell growth. This review will focus upon CD45 as a prototype for transmembrane PTPs and describe our current understanding of the structural and functional properties of the now diverse and rapidly growing family of intracellular and transmembrane PTPs. For other recent reviews see Refs. 8 and 9.

Structural Features of CD45: Requirements for Activity

The structure of CD45 is consistent with a role as a receptor. It consists of a large, heavily glycosylated, amino-terminal external domain of ~400–500 residues, a single transmembrane region, and a large highly conserved carboxyl-terminal cytoplasmic domain of ~700 residues (Fig. 1). Isoforms of CD45 are generated by the alternative splicing of at least three exons (4, 5, and 6), each encoding ~50 amino acids that are inserted close to the amino terminus. These segments are rich in small aliphatic amino acids and contain multiple sites for *O*-linked glycosylation. Thus, higher M_r isoforms are generated by modification of the polypeptide structure of the external domain and addition of carbohydrate. All eight possible CD45 isoforms generated by the alternative splicing of these three exons have been detected at the level of mRNA (10), and alternative splicing of exon 7 may also occur (11). Different isoforms of CD45 have similar intrinsic PTP activity (12). As the pattern of expression of different isoforms is tightly regulated during lymphoid development and is conserved between species, the structural variation in the external domain is likely to be of functional significance, perhaps by modulating the specificity of ligand interaction. Visualization of CD45 by low-angle shadowing suggests that the external domain is an extended rod (13) and thus could interact with surface molecules on other cells. The external domain of CD45 has been produced as a soluble recombinant protein in Chinese hamster ovary cells, but efforts to identify a ligand or demonstrate specific binding to cells using this material have so far been unsuccessful (14). However, indirect evidence has recently been reported that, in the human, the B cell adhesion molecule, CD22, interacts with T cells by binding to the smallest isoform of CD45, CD45R0 (15). Two forms of CD22 have been identified, and only the larger form, CD22β, binds to T cells, whereas the smaller form, CD22α, binds to erythrocytes and monocytes. The binding of a soluble recombinant CD22β immunoglobulin hybrid protein to T cells correlates with the expression of CD45R0 and is inhibited by UCHL-1, a monoclonal antibody that reacts with an antigenic determinant restricted to the CD45R0 isoform of human CD45 (15). However, specific binding of CD22β either to cells transfected with CD45R0 or the purified glycoprotein has not been demonstrated. Moreover, if CD22 is a physiological ligand for CD45, it might be expected that its interaction with CD45R0 would trigger a change in PTP activity. To further complicate the picture, CD22β also reacts with tonsillar B cells and some B cell lines by binding to CD75, a cell surface α2–6 sialyltransferase (15).

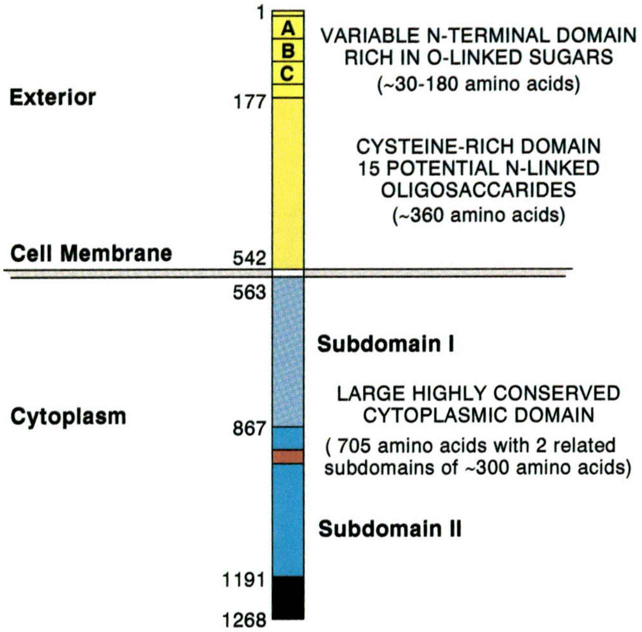

FIG. 1. **Schematic representation of the structure of CD45.** The numbering system used is for the mouse CD45R (ABC) or B cell isoform of CD45. Regions *A*, *B*, and *C* are the extra segments (residues 8–50, 51–99, and 100–146, respectively) encoded by the alternatively spliced exons 4, 5, and 6. The external domain (residues 1–541), transmembrane region (residues 542–563), and cytoplasmic domain (residues 564–1268) are shown. The acidic insert region in the second tandem repeat (residues 958–978) is also shown colored *magenta*.

The major structural features of the cytoplasmic domain of mouse CD45 are typical of most transmembrane PTPs: a mem-

* Work from this laboratory was supported by National Institutes of Health Grant CA 17733.

[1] The abbreviations used are: PTP, protein tyrosine phosphatase; PTK, protein tyrosine kinase.

```
CD45 PTP Domain I       641  NQNKNRYVDILPYDYNRVELSEI
CD45 PTP Domain II      932  NKKKNRNSNVVPYDFNRVPLKHE
Consensus Sequence           N--kNR------yd--RV-L---

664  NGDAG                   STYINASYIDGFKEPRKYIAAQGPRD
955  LEMSKESEPESDESSDDDSDSEETSKYINASFVMSYWKPEMMIAAQGPLK
     ------------------------dYiNAs---gy------Ia-QGPl-

695  ETVDDFWRMIWEQKATVIVMVTRCEEGNRNKCAEYWPSMEEGTRAFKDIV
1005 ETIGDFWQMIFQRKVKVIVMLTELVNGDQEVCAQYWG   EGKQTYGDME
     -T--DFWrM-We------VM-t---e----KC-qYWP-------------

745  VTINDHKRCPDYIIQKLNVAHKKEKATGREVTHIQFTSWPDHGVPEDPHL
1052 VEMKDTNRASAYTLRTFELRHSKRKEP RTVYQYQCTTWKGEELPAEPKD
     v----------y--r-f-----------R---q-----WP--gvP-----

795  LL         KL RRRVNAFSNFFSGPIVVHCSAGVGRTGTYIGIDAMLE
1101 LVSMIQDLKQKLPKASPEGMKYHKHASILVHCRDGSQQTGLFCALFNLLE
     ------------------------gP--vHCsaG-grtg--------l-

836  GLEAEGKVDVYGYVVKLRRQRCLMVQVEAQYILIHQALVE
1151 SAETEDVVDVFQVVKSLRKARPGVVCSYEQYQFLYDIIAS
     ----e--vd----v---R-qR---vqt--Qy-f---a--e
```

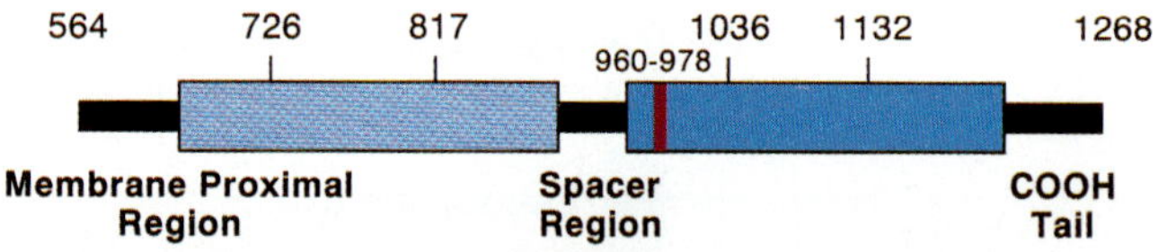

FIG. 2. **Sequence and structure of murine CD45 cytoplasmic domain.** Schematic representation of murine CD45 cytoplasmic domain showing the tandem PTP domains, domain I (235 residues in *light blue*) and domain II (253 residues in *dark blue*), including the acidic insert region (residues 960–978 in *magenta*, the membrane-proximal region (77 residues), spacer region (56 residues), and carboxyl-terminal tail (78 residues). The positions of the four conserved cysteine residues are shown (726, 817, 1036, 1132). The sequences of the two aligned PTP domains are given. The consensus sequence of the PTP domains from the PTPs of higher eukaryotes shown in Fig. 3 and HPTPs γ, δ, and ζ (Ref. 35) is also presented. Conserved residues found in all 22 PTP domains are shown in *red*. In the consensus sequence, residues conserved in more than 70% of PTP domains are *capitalized* in *black* and residues conserved in more than 50% of the PTP domains are in *lower case*. The conserved region around the cysteine residue required for catalytic activity (shown in *white* on *red* background) is in the *pale blue box*; the acidic insert is in the *yellow box*.

brane proximal region, followed by tandem catalytic domains of ~240 residues separated by a short spacer region, and a carboxyl-terminal tail (Fig. 2). Within the second PTP domain of CD45, there is a unique acidic region of 19 residues that contains multiple potential sites for serine phosphorylation. Recombinant CD45 cytoplasmic domain has intrinsic PTP activity independent of the external domain (12, 16). Mutation of a cysteine residue in the carboxyl-terminal region of the first PTP domain abolishes activity suggesting that the second PTP domain is inactive against the substrates commonly used *in vitro* (16). However, deletion of PTP domain II also abolishes activity, indicating that it is required for PTP domain I to be active. Whether the second PTP domain is required to maintain the structural integrity of the first PTP domain or whether it plays a regulatory role is unclear. The membrane proximal region, but not the carboxyl-terminal tail of CD45, is also required for activity[2] (9).

Comparison of the sequences of transmembrane and intracellular PTPs reveals that their catalytic domains contain several short conserved segments (Fig. 2). The most highly conserved of these is a region of 11 amino acids that includes the essential cysteine residue and a GXGXXG motif resembling the glycine-rich loop associated with the nucleotide binding site of certain dehydrogenases, G-proteins, and protein kinases (17, 18). Mutagenesis of CD45 and the related transmembrane PTP, LAR, indicate that many residues within this region are essential for activity and that alteration of the others reduces activity (9, 19). PTP domain II of CD45 is unusual in that 5 residues within this region differ from the PTP consensus sequence (Fig. 2). Single substitutions of conserved residues in the first PTP domain of CD45 with the nonconserved residue at the equivalent position in the second PTP domain abolished activity indicating that if the second PTP domain has activity its substrate specificity is likely to be very different from that of domain I (9). Recently, it has been shown for a rat brain PTP that the essential cysteine forms a covalent thiol phosphate intermediate during catalysis (20), and it is likely, therefore, that other residues within this conserved region contribute to the enzyme active site. The other conserved segments in PTP catalytic domains contain invariant aromatic residues that may play structural roles. Alteration of Tyr to Phe in the YINAS sequence (Fig. 2) has no effect on the activity of CD45 but is associated with a temperature-sensitive phenotype in LAR (9, 21). The tyrosine residue in the conserved QYWP sequence is essential for the activity of CD45 (9).

Role of CD45 in Lymphocyte Signal Transduction

Our current understanding of the role of CD45 in lymphocyte signal transduction rests heavily upon the use of CD45⁻ mutant cell lines, a somatic genetic approach pioneered by Hyman (22). The most important fact to emerge is that CD45 is essential for antigen-induced proliferative responses of T cells. CD45⁻ mutants of helper and cytotoxic mouse T cell lines have been derived by immunoselection and are markedly impaired in their ability to respond to antigen (23, 24); antigen responsiveness is restored in CD45⁺ revertants. The most simple interpretation of these data is that CD45 provides a positive signal necessary for proliferation in response to antigen, possibly by acting synergistically with protein tyrosine kinases (PTKs) to produce a hysteretic system as recently proposed (8). It is of interest that a detectable phenotype is associated with the loss of CD45 as this indicates that other PTPs expressed in T cells, such as LRP and T cell PTP (see Fig. 3), cannot fully compensate for the deficiency and suggests each PTP will play a distinct role. Other studies of CD45⁻ mutant human T cell lines show that CD45 is required for the coupling of T cell antigen receptor signals to the phosphoinositol second messenger system (25) and for the rapid increase in tyrosine phosphorylation induced by activation of the T cell receptor (26). CD45 also appears to be necessary for signaling via the B cell antigen receptor (27).

The mechanism by which CD45 presumably regulates antigen responses is by the dephosphorylation of tyrosine residues on one or more substrates. The search for physiological substrates of PTPs is just beginning, and currently little is known. In the case of CD45, a comparison of the phosphotyrosine levels of the cellular proteins in three independent pairs of CD45⁻ mutant and CD45⁺ parental murine T lymphomas demonstrated that the

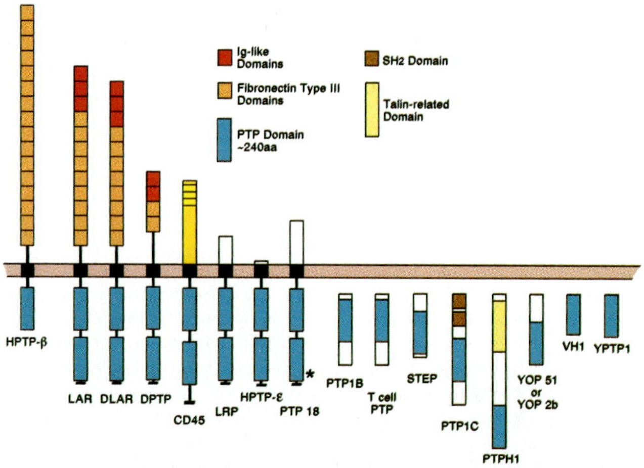

FIG. 3. **Schematic representation of the PTP family.** PTPs are oriented with their NH_2 termini toward the top of the diagram. Transmembrane PTPs are type I membrane proteins with an amino-terminal extracellular domain (shown as extending above the cell membrane), a single transmembrane region, and a carboxyl-terminal cytoplasmic domain. HPTP-α, -β, and -ϵ, LAR, CD45, PTP 1B, T cell PTP, STEP, PTP1C, and PTPH1 have been identified in the human and, in some cases, other mammalian species, PTP 18 in the rat, DLAR and DPTP in *Drosophila*, YOP 51 and YOP 26 in *Yersinia*, VH1 in vaccinia virus, and YPTP in yeast. LRP (33) has been cloned by several groups, and its various synonyms are HLPR (34), HPTP-α (35), and RPTP-α (36). The *asterisk* by PTP 18 indicates that an aspartate residue has replaced the conserved cysteine residue found in the carboxyl-terminal region of the PTP II subdomains of other transmembrane PTPs. See text for further details.

[2] P. Johnson, H. L. Ostergaard, C. Wasden, and I. S. Trowbridge, unpublished results.

expression of CD45 correlated with a specific decrease in the level of phosphorylation of Tyr-505 in $p56^{lck}$, the *src*-related PTK found associated with CD4 and CD8 surface glycoproteins in T cells (12). Antibody-mediated co-clustering of CD45 and CD4 induces dephosphorylation of $p56^{lck}$ providing further evidence that CD45 can act on $p56^{lck}$ *in vivo* (28). As dephosphorylation of Tyr-505 would be expected to activate $p56^{lck}$ this might be one mechanism by which CD45 could influence antigen responses in a positive manner. However, confirmation of this notion awaits a clearer understanding of the roles of $p56^{lck}$ and other *src*-related PTKs such as $p59^{fyn}$ in T cell signaling. Substrates of CD45 in other leukocytes have not been identified.

Regulation of CD45 PTP Activity

The abundance and high intrinsic specific activity of CD45 against artificial substrates *in vitro* (29) suggest that its activity is tightly regulated in order for PTKs to deliver signals via tyrosine phosphorylation and that in some circumstances it would be necessary to suppress activity. Little change in the level of expression of CD45 takes place during lymphoid development although, as noted earlier, programmed changes in isoform expression occur that may modify interactions with specific ligands. There are several mechanisms by which CD45 activity might be modified, including ligand binding, phosphorylation or other covalent modification, or redistribution in the cell membrane, but as for other PTPs little definitive data exists. There is now, however, strong evidence that CD45 activity can be negatively regulated *in vivo*. Treatment of a variety of functional mouse T cell lines or thymocytes with 1–2 μM ionomycin reduced CD45 activity by 50–90% (30). The inactivation of CD45 was shown to be due to increased intracellular calcium, but the kinetics of inactivation indicated that this was an indirect effect involving secondary events occurring during the first 30–40 min after the induction of a Ca^{2+} flux (30). The decrease in PTP activity correlated with a selective decrease in serine phosphorylation at specific sites in the molecule. As a transient Ca^{2+} flux is induced when T cells respond to antigen, it is possible that inactivation of CD45 by an increase in intracellular free Ca^{2+} occurs under physiological conditions. Demonstration that this effect is mediated by changes in serine phosphorylation will require identification of the enzymes and specific sites involved. To date, changes in CD45 activity following phosphorylation with a variety of Ser/Thr kinases *in vitro* have not been observed (29). Transient phosphorylation of CD45 on tyrosine has also been reported after stimulation of a human T cell leukemia line Jurkat with phytohemagglutinin or anti-CD3 antibodies (31). However, the demonstration of tyrosine phosphorylation required pretreatment of cells with phenylarsine oxide, a PTP inhibitor, precluding determination of whether tyrosine phosphorylation changed the activity of CD45. A potential difficulty in studying the regulation of PTPs is that, as for PTKs, the *in vitro* assays currently employed may not accurately reflect the activity of the enzymes against physiological substrates *in vivo*.

Other Members of the PTP Family

The PTP family continues to grow in size and diversity. Currently, the complete primary structures of at least eight putative transmembrane PTPs and eight intracellular PTPs have been deduced from cDNA nucleotide sequences (Fig. 3) (3, 6, 16, 32–48). In addition, partial cDNA clones of other apparently novel PTPs have been obtained from higher eukaryotes (35, 37), and 27 distinct PTPs have been identified in the prochordate, *Styela plicata*, by the polymerase chain reaction technique (49). With the exception of HPTP-β, the cytoplasmic domains of all the transmembrane PTPs are similar in structure and have tandem catalytic domains. The cytoplasmic domains of HPTP-β, LAR, CD45, and HPTP-α have been expressed as recombinant proteins and formally shown to have intrinsic PTP activity *in vitro* (12, 16, 19, 35). The significance of the tandem catalytic domains of the transmembrane PTPs is unclear. The second catalytic domains of LAR and CD45 are not active against the artificial substrates commonly used to assay PTP activity *in vitro* (9, 19), and the second domain of rat PTP18 apparently lacks the essential cysteine residue (44). In contrast to the situation with CD45, in which the second catalytic domain is required for activity of the first domain (9, 19), the single catalytic domain of HPTP-β and the independently expressed first domain of LAR are both active (35, 50). The external domains of the transmembrane PTPs vary in structure, but it seems too early to classify them on this basis particularly as most structures have been deduced from cDNA sequences and the encoded proteins may be subject to proteolytic cleavage or be associated with other subunits. The external domain of CD45 is unique, whereas those of LAR, DLAR, DPTP, and HPTP-β are related, being composed of repeating units of immunoglobulin and fibronectin type III repeats (Fig. 3). This has led to the suggestion that this latter group of PTPs is involved in cell-cell interactions (35). The external domain of LRP is 123 residues (33) and that of HPTP-ϵ is 27 residues (35); both are likely to be heavily glycosylated and, in view of the size of their external domains, are the most likely of the transmembrane PTPs to be associated with other subunits. CD45 is the only receptor-like PTP known to have a highly restricted tissue distribution; LAR and LRP appear to be expressed on a wide variety of tissues based on Northern blot analysis of mRNA levels (32–34, 36), and the tissue distributions of the other receptor-like PTPs have not been reported. Antibodies specific for the external domains of each of the transmembrane PTPs will be required for further analysis of their tissue distribution and function.

Five intracellular mammalian PTPs have been isolated; PTP 1B (6, 38) and T cell PTP (39) have a broad tissue distribution and contain ~430 amino acids with a single PTP domain located in their amino-terminal region. The carboxyl-terminal regions of these two PTPs appear to be important in determining their intracellular localization and regulation of their enzymic activity (8, 39). The full-length intracellular PTPs are associated with the particulate fraction of cell homogenates and have hydrophobic carboxyl termini that may serve as membrane anchors. PTP 1B is readily cleaved to give a soluble proteolytic fragment of 321 amino acids, and this was the molecular species originally isolated from human placenta (6, 51). A cDNA encoding a putative brain-specific intracellular PTP of ~369 amino acids was isolated from a rat striatal cDNA library (41). Within the brain, by Northern blot analysis, the PTP appears highly enriched within the striatum relative to other areas, hence the name STEP (striatum-enriched phosphatase). PTPH1 is an intracellular PTP of 913 amino acids and has a region of similarity in the amino-terminal region (residues 30–357) with the amino-terminal regions of the cytoskeletal associated proteins band 4.1, ezrin, and talin (42). Most recently, a fifth human intracellular PTP, designated PTP1C, has been identified that, interestingly, has in its amino-terminal region two SH2 domains (43), structural motifs found in various nonreceptor PTKs and other cytoplasmic signaling proteins (52). As with other SH2 domains, the SH2 sequences of PTP1C could form high affinity complexes with the activated epidermal growth factor receptor and other phosphotyrosine-containing proteins (43).

Two other novel PTPs have been identified in pathogenic organisms (44–46). One was identified by a computer search for sequences related to conserved PTP sequences in the bacterial genus *Yersinia*, which is comprised of three species of bacteria that are causative agents in human disease. Each *Yersinia* species contains a plasmid encoding proteins associated with virulence. In the case of *Yersinia pseudotuberculosis*, it has been shown that the Yop H gene on such a plasmid encodes a PTP that is an essential virulence determinant and that can dephosphorylate tyrosine residues of proteins in host macrophages (45). The other PTP was identified in vaccinia virus, and interestingly, although this protein, VH1, is clearly a member of the PTP family and can dephosphorylate tyrosine residues, it also has significant activity against phosphoserine (46). It is not apparent, however, from a comparison of the sequence of the vaccinia PTP with other PTPs and Ser/Thr phosphatases what structural features account for this dual enzymatic activity. An intracellular PTP lacking a carboxyl-terminal regulatory region has been identified in the budding yeast *Saccharomyces cerevisiae* (47). Another putative 555-amino acid PTP has also been identified in the fission yeast, *Schizosaccharomyces pombe*, but the recombinant protein expressed in bacteria lacked detectable enzymatic activity (48).

Concluding Remarks

Remarkable progress has been made over the last two years in understanding the role of CD45 in lymphocyte activation and identifying other members of the PTP family. The most important outstanding question is whether CD45 and the other transmembrane PTPs are receptors and, if so, what are their respective ligands. The identification of physiological substrates for individual PTPs is also a key issue. Overexpression of each of the PTPs would likely aid in the search for substrates and to further define their functions, but this has proved difficult for CD45. It seems likely that many of the intracellular PTPs will be found to be localized in specific regions of the cell, and there is evidence that some are membrane-associated. Identification of an intracellular PTP with SH2 domains implies that specific protein-protein interactions may regulate PTP activity by promoting physical association with phosphotyrosine-containing substrates as observed for nonreceptor PTKs that have SH2 motifs. As antagonists of PTK action, PTPs might be expected to function as anti-oncogenes; however, there is currently little data to support this idea. Loss of CD45 is not usually associated with malignant transformation of hematopoietic cells, but careful studies to determine whether there are changes in CD45 PTP activity associated with specific types of leukemia or lymphoma have not been performed. With one possible exception (53), loss or inactivation of other PTPs has not yet been associated with malignant transformation of other cell types. Overexpression of the truncated form of T cell PTP lacking the carboxyl-terminal regulatory domain reduces the growth rate of baby hamster kidney cells 2-fold (54). Further, there is almost a 10-fold elevation in the activity of membrane-associated PTP activity in density-dependent growth-arrested fibroblasts implying that increased PTP activity may be involved in the inhibition of cell growth in high density cultures (55). The role of PTPs in the regulation of the cell cycle is also a potentially fertile area that is only just beginning to be explored (8). The next few years are likely to produce answers to many of these questions as PTPs take their place alongside PTKs as crucial elements in the regulation of cell growth and differentiation.

Acknowledgments—I thank R. Hyman and J. Lesley for their comments on the manuscript and the many colleagues, particularly Hanne Ostergaard and Pauline Johnson, who contributed to the work cited from my own laboratory.

REFERENCES

1. Boyse, E. A., and Old, L. J. (1969) *Annu. Rev. Genet.* **3,** 269–290
2. Reif, A. E., and Allen, J. M. V. (1963) *Nature* **200,** 1332–1333
3. Thomas, M. L. (1989) *Annu. Rev. Immunol.* **7,** 339–369
4. Battifora, H., and Trowbridge, I. S. (1983) *Cancer* **51,** 816–821
5. Hyman, R., Trowbridge, I., Stallings, V., and Trotter, J. (1982) *Immunogenetics* **15,** 413–420
6. Charbonneau, H., Tonks, N. K., Walsh, K. A., and Fischer, E. H. (1988) *Proc. Natl. Acad. Sci. U. S. A.* **85,** 7182–7186
7. Tonks, N. K., Charbonneau, H., Diltz, C. D., Fischer, E. H., and Walsh, K. A. (1988) *Biochemistry* **27,** 8696–8700
8. Fischer, E. H., Charbonneau, H., and Tonks, N. K. (1991) *Science* **253,** 401–406
9. Trowbridge, I. S., Ostergaard, H., and Johnson, P. (1991) *Biochim. Biophys. Acta* **1095,** 46–56
10. Saga, Y., Furukawa, K., Rogers, P., Tung, J.-S., Parker, D., and Boyse, E. A. (1990) *Immunogenetics* **31,** 296–306
11. Chang, H.-L., Lefrancois, L., Zaroukian, M. H., and Esselman, W. J. (1991) *J. Immunol.* **147,** 1687–1693
12. Ostergaard, H. L., Shackelford, D. A., Hurley, T. R., Johnson, P., Hyman, R., Sefton, B. M., and Trowbridge, I. S. (1989) *Proc. Natl. Acad. Sci. U. S. A.* **86,** 8959–8963
13. Woollett, G. R., Williams, A. F., and Shotton, D. M. (1985) *EMBO J.* **4,** 2827–2830
14. Trowbridge, I. S., Ostergaard, H., Shackelford, D., Hole, N., and Johnson, P. (1991) *Adv. Protein Phosphatases* **6,** 227–250
15. Stamenkovic, I., Sgroi, D., Aruffo, A., Sy, M. S., and Anderson, T. (1991) *Cell* **66,** 1133–1144
16. Streuli, M., Krueger, N. X., Tsai, A. Y. M., and Saito, H. (1989) *Proc. Natl. Acad. Sci. U. S. A.* **86,** 8698–8702
17. Taylor, S. S., Buechler, J. A., and Yonemoto, W. (1990) *Annu. Rev. Biochem.* **59,** 971–1005
18. Pai, E. F., Kabsch, W., Krengel, U., Holmes, K. C., John, J., and Wittinghofer, A. (1989) *Nature* **341,** 209–214
19. Streuli, M., Krueger, N. X., Thai, T., Tang, M., and Saito, H. (1990) *EMBO J.* **9,** 2399–2407
20. Guan, K. L., and Dixon, J. E. (1991) *J. Biol. Chem.* **266,** 17026–17030
21. Tsai, A. Y. M., Itoh, M., Streuli, M., Thai, T., and Saito, H. (1991) *J. Biol. Chem.* **266,** 10534–10543
22. Hyman, R. (1985) *Biochem. J.* **225,** 27–40
23. Pingel, J. T., and Thomas, M. L. (1989) *Cell* **58,** 1055–1065
24. Weaver, C. T., Pingel, J. T., Nelson, J. O., and Thomas, M. L. (1991) *Mol. Cell. Biol.* **11,** 4415–4422
25. Koretzky, G. A., Picus, J., Thomas, M. L., and Weiss, A. (1990) *Nature* **346,** 66–68
26. Koretzky, G. A., Picus, J., Schultz, T., and Weiss, A. (1991) *Proc. Natl. Acad. Sci. U. S. A.* **88,** 2037–2041
27. Justement, L. B., Campbell, K. S., Chien, N. C., and Cambier, J. C. (1991) *Science* **252,** 1839–1842
28. Ostergaard, H. L., and Trowbridge, I. S. (1990) *J. Exp. Med.* **172,** 347–350
29. Tonks, N. K., Diltz, C. D., and Fischer, E. H. (1990) *J. Biol. Chem.* **265,** 10674–10680
30. Ostergaard, H. L., and Trowbridge, I. S. (1991) *Science* **253,** 1423–1425
31. Stover, D. R., Charbonneau, H., Tonks, N. K., and Walsh, K. A. (1991) *Proc. Natl. Acad. Sci. U. S. A.* **88,** 7704–7707
32. Streuli, M., Krueger, N. X., Hall, L. R., Schlossman, S. F., and Saito, H. (1988) *J. Exp. Med.* **168,** 1553–1562
33. Matthews, R. J., Cahir, E. D., and Thomas, M. L. (1990) *Proc. Natl. Acad. Sci. U. S. A.* **87,** 4444–4448
34. Jirik, F. R., Janzen, N. M., Melhado, I. G., and Harder, K. W. (1990) *FEBS Lett.* **273,** 239–242
35. Krueger, N. X., Streuli, M., and Saito, H. (1990) *EMBO J.* **9,** 3241–3252
36. Sap, J., D'Eustachio, P. D., Givol, D., and Schlessinger, J. (1990) *Proc. Natl. Acad. Sci. U. S. A.* **87,** 6112–6116
37. Kaplan, R., Morse, B., Huebner, K., Croce, C., Howk, R., Ravera, M., Ricca, G., Jaye, M., and Schlessinger, J. (1990) *Proc. Natl. Acad. Sci. U. S. A.* **87,** 7000–7004
38. Chernoff, J., Schievella, A. R., Jost, C. A., Erikson, R. L., and Neel, B. G. (1990) *Proc. Natl. Acad. Sci. U. S. A.* **87,** 2735–2739
39. Cool, D. E., Tonks, N. K., Charbonneau, H., Walsh, K. A., Fischer, E. H., and Krebs, E. G. (1989) *Proc. Natl. Acad. Sci. U. S. A.* **86,** 5257–5261
40. Guan, K., Haun, R. S., Watson, S. J., Geahlen, R. L., and Dixon, J. E. (1990) *Proc. Natl. Acad. Sci. U. S. A.* **87,** 1501–1505
41. Lombroso, P. J., Murdoch, G., and Lerner, M. (1991) *Proc. Natl. Acad. Sci. U. S. A.* **88,** 7242–7246
42. Yang, Q., and Tonks, N. K. (1991) *Proc. Natl. Acad. Sci. U. S. A.* **88,** 5949–5953
43. Shen, S. H., Bastien, L., Posner, B. I., and Chretien, P. (1991) *Nature* **352,** 736–739
44. Guan, K., and Dixon, J. E. (1990) *Science* **249,** 553–556
45. Bliska, J. B., Guan, K., Dixon, J. E., and Falkow, S. (1991) *Proc. Natl. Acad. Sci. U. S. A.* **88,** 1187–1191
46. Guan, K., Broyles, S. S., and Dixon, J. E. (1991) *Nature* **350,** 359–362
47. Guan, K., Deschenes, R. J., Qiu, H., and Dixon, J. E. (1991) *J. Biol. Chem.* **266,** 12964–12970
48. Ottilie, S., Chernoff, J., Hannig, G., Hoffman, C. S., and Erikson, R. L. (1991) *Proc. Natl. Acad. Sci. U. S. A.* **88,** 3455–3459
49. Matthews, R. J., Flores, E., and Thomas, M. L. (1991) *Immunogenetics* **33,** 33–41
50. Cho, H., Ramer, S. E., Itoh, M., Winkler, D. G., Kitas, E., Bannwarth, W., Burn, P., Saito, H., and Walsh, C. T. (1991) *Biochemistry* **30,** 6210–6216
51. Tonks, N. K., Diltz, C. D., and Fischer, E. H. (1988) *J. Biol. Chem.* **263,** 6731–6737
52. Koch, C. A., Anderson, D., Moran, M. F., Ellis, C., and Pawson, T. (1991) *Science* **252,** 668–674
53. LaForgia, S., Morse, B., Levy, J., Barnea, G., Cannizzaro, L. A., Li, F., Nowell, P. C., Boghosian-Sell, L., Glick, J., Weston, A., Harris, C. C., Drabkin, H., Patterson, D., Croce, C. M., Schlessinger, J., and Huebner, K. (1991) *Proc. Natl. Acad. Sci. U. S. A.* **88,** 5036–5040
54. Cool, D. E., Tonks, N. K., Charbonneau, H., Fischer, E. H., and Krebs, E. G. (1990) *Proc. Natl. Acad. Sci. U. S. A.* **87,** 7280–7284
55. Pallen, C. J., and Tong, P. H. (1991) *Proc. Natl. Acad. Sci. U. S. A.* **88,** 6996–7000

Minireview

THE JOURNAL OF BIOLOGICAL CHEMISTRY
Vol. 266, No. 36, Issue of December 25, pp. 24233-24236, 1991

Printed in U.S.A.

Biological Role and Regulation of the Universally Conserved Heat Shock Proteins

Debbie Ang‡§, Krzysztof Liberek‡¶, Dorota Skowyra¶, Maciej Zylicz¶, and Costa Georgopoulos‡§

From the ‡Department of Cellular, Viral, and Molecular Biology, University of Utah Medical Center, Salt Lake City, Utah 84132 and the ¶Division of Biophysics, Department of Molecular Biology, University of Gdansk, Kladki 24, Gdansk 80-822, Poland

The Heat Shock Response

The heat shock (HS)[1] or stress response has been conserved in evolution, from bacteria to man. Although the original discovery by Ritossa (1) in *Drosophila* passed almost unnoticed, the HS response has been vigorously investigated by numerous laboratories around the world during the past 10 years. The continuous interest in the HS response rests on (*a*) its universality, (*b*) the tremendous conservation of the structure and function of many of the HS proteins (HSPs), (*c*) the regulation of the HS response as a paradigm of gene expression, and (*d*) the insights that it continues to provide in the pathways of protein folding, oligomerization, secretion, and degradation. Over the last 2 years many excellent reviews have been written (2–9).

In prokaryotes, as exemplified by *Escherichia coli*, the major HSPs are coded by single genes expressed constitutively at all temperatures. Following a temperature shift or treatment with protein-damaging agents, such as ethanol, the rate of expression of these genes abruptly accelerates. Within a few minutes the rate of expression subsides, reaching a new steady state level, characteristic of a given temperature. The HS response is positively regulated at the transcriptional level by the σ^{32} polypeptide, the product of the *rpoH* (*htpR*) gene (reviewed in Refs. 2, 4, and 5). A new wrinkle has been recently added to *E. coli*'s HS response, namely the discovery of a new set of HS genes, positively regulated at the transcriptional level by the σ^{24} (σ^{E}) polypeptide (reviewed in Ref. 4). Interestingly, one of the promoters of the *rpoH* (*htpR*) gene is transcribed by the $E\sigma^{24}$ RNA polymerase holoenzyme. This result suggests a network of expression control between the two HS pathways. The gene coding for σ^{24} has not been discovered as yet.

In eukaryotes, an analogous situation exists. The HS response is controlled mostly at the transcriptional level by a positively acting heat shock factor (HSF), which binds to specific regions of DNA called heat shock elements, located upstream of HS gene promoters (reviewed in Ref. 10). Recent reports point to the existence of multiple HSFs, some of which are themselves HS-inducible (11) (reviewed in Ref. 7). Certain mutations in the yeast HSF gene result in failure to induce HSPs at the nonpermissive temperature coupled with defects in mitochondrial protein import and cell cycle progression, suggesting that HSF may help coordinate these processes under all conditions (12).

Very often in eukaryotes each family of HSPs includes several subsets: HSCs (heat shock cognates), which are expressed constitutively, HSPs (heat shock proteins) which are largely expressed under conditions of stress, and sometimes a third class of proteins which are expressed constitutively but whose rate of synthesis is significantly augmented following stress. The synthesis of some members of this last category is increased following glucose starvation, and they have been named GRPs (glucose-regulated proteins). As discussed below, all of these protein members appear to play similar roles, although some specialization of function may arise, partly dictated by the differential location of these proteins (*e.g.* cytosol, endoplasmic reticulum, mitochondria, etc.).

Need for Heat Shock Proteins

The foresight of Hightower (13) and Pelham (14) has played a decisive role in our understanding of HSP function. Under normal conditions, a polypeptide chain must be correctly folded, processed, localized, and, in some cases, complexed with other polypeptides to properly perform its biological function. Obviously this is the culmination of a complex biological process, with many pitfalls along the way. For example, as the growing polypeptide emerges from the ribosome, it is subject to premature contact with other protein domains, either intra- or interspecific, because of the high cytosolic protein concentration. Such premature interaction among protein domains must be avoided to prevent misfolding (reviewed in Refs. 6 and 15–17). In the case of proteins crossing membranes, it is clear that only unfolded proteins can properly traverse biological membranes (reviewed in Ref. 18), again creating the opportunity for inappropriate interactions among protein domains. In both cases, the need for some sort of "chaperoning" activity to maintain proteins in an unfolded state, yet prevent undesirable interactions, is clear. While the elegant work of Anfinsen (19), resulting in the successful *in vitro* refolding of purified ribonuclease, left us with the generally correct conclusion that the primary sequence of a polypeptide chain dictates its final, three-dimensional structure, under physiological conditions (25–37 °C and high protein concentrations), most unfolded polypeptide chains (kept in 8 M urea or guanidine), in the absence of any auxiliary proteins, misfold upon rapid dilution into refolding mixtures. This is mainly due to intermolecular aggregation between reactive (hydrophobic or hydrophilic) groups. A high level of accurately folded proteins can be achieved when both the protein concentration and the reaction temperature are kept very low (20). However, conditions in the cytosol are quite the opposite, and help is needed to prevent misfolding or aggregation of nascent polypeptides. Attention has recently been focused on a set of proteins called "chaperones" (21, 22), some of which were originally discovered as HSPs, whose biological role is to maintain and shield the unfolded state of newly synthesized proteins, thus (*a*) preventing them from misfolding or aggregating, (*b*) allowing them to traverse biological membranes, and (*c*) allowing them to fold properly, thus leading to oligomerization. Under conditions of stress, chaperones protect other proteins from heat denaturation or, once damage has occurred, disaggregate and allow them to refold back to an active form (see below). Thus chaperones, a large subset of which are HSPs, have also evolved as part of a cellular safety or rescue mechanism.

The hsp70/dnaK Family

This class of proteins has been universally conserved, its members being at least 50% identical to each other at the amino acid sequence level. In *E. coli* this class is represented by the single copy *dnaK* gene, whereas yeast possesses at least eight gene copies (23). Part of the reason for the abundance of gene copies in eukaryotes is the fact that some of their gene products are found exclusively in specialized organelles such as the endoplasmic reticulum, mitochondria, and chloroplasts (22, 23). The hsp70 proteins are the "workhorse" of the chaperones, not only because of their promiscuity in binding to other unfolded polypeptides (6,

§ Present address: Dept. of Medical Biochemistry, University of Geneva, 1, rue Michel-Servet, 1211 Geneva 4, Switzerland.

[1] The abbreviations used are: HS, heat shock; HSC, heat shock cognate; HSP, heat shock protein; HSF, heat shock factor.

15, 24–26) but also due to their relative abundance in the various cellular compartments.

The *E. coli dnaK* gene was originally discovered because mutations in it block λ DNA replication (5). Subsequently it was shown that dnaK performs almost indispensable bacterial functions, since deletion of the gene can be tolerated only within a narrow temperature range, and even then, extragenic suppressors accumulate very rapidly (27).[2] In yeast, deletion of some HSP70 gene members can be tolerated if the product of another family member is also present in the same cellular compartment to replace the missing function (23). However, the gene becomes essential if its product is the only member found in the compartment (18, 28). In some instances, the polypeptide to be imported is maintained unfolded in the cytosol and "pulled" into the organelle by the various hsp70 resident members (18, 28, 29).

The biochemical properties of this family of proteins are described in the following paragraphs.

An Extremely Weak ATPase Activity—The ATPase domain lies in the amino-terminal portion of the protein. The protease-resistant, ~44-kDa amino-terminal fragment of the bovine hsc70 polypeptide has been crystallized and its structure solved (30). A remarkable similarity between this structure and that of the rabbit skeletal muscle actin has been observed, despite the low sequence identity between them (7, 31).

Promiscuous Binding to Various "Unstructured" or Unfolded Polypeptides (Reviewed in Refs. 15 and 24), Which Is Inhibited by ATP, but in Some Instances Accelerated by ADP (32)—Binding of unfolded polypeptide stimulates hsp70's ATPase activity, followed by release of the bound polypeptide. Using a set of peptides of random sequence, Flynn *et al.*[3] showed that peptides of 7 amino acid residues maximally stimulate hsc70's ATPase activity and that cycles of binding and release from hsc70 result in the enrichment of peptides with aliphatic amino acid residues at all 7 positions of the peptide ligand. These results suggest that the peptide binding site in hsc70 accommodates the peptide ligand in an environment energetically equivalent to the interior of a folded protein.[3] In another example, self-peptides have been identified that bind to a purified human class I major histocompatibility complex molecule, HLA-B27, a protein also capable of promiscuous binding to peptides (33, 34). These authors found that the nonamer polypeptides are bound in an extended conformation, with an arginine residue at position 2 and more random residues at other positions. Interestingly, the putative consensus binding domains of HLA and hsp70 appear to share important structural features (7, 35, 36).

Other Biochemical Properties—These include a perfect correlation between ATP hydrolysis, the release of the bound polypeptide, and a dramatic conformational change in hsp70 proteins, as exemplified by dnaK (26); the protection of other enzymes from thermal inactivation (37); and disaggregation of some protein aggregates (37, 38), including the preprimosomal protein complex assembled at *ori*λ (39, 40), and the phage P1 repA dimer (41), as well as restoration of activity to purified mutant polypeptides (42). In addition, dnaK in *E. coli* has been shown to functionally interact with two other *E. coli* HSPs, dnaJ and grpE, which stimulate dnaK's ATPase activity (43) (see below).

The dnaJ and grpE Proteins

The *dnaJ* and *grpE* genes of *E. coli* were originally discovered because mutations in them block bacteriophage λ growth. The *dnaJ* gene forms an operon with the *dnaK* gene, the order being promoter-*dnaK-dnaJ*, whereas *grpE* maps elsewhere and is monocistronic. Bacterial homologues to the *dnaJ* gene have been discovered (5).

The dnaJ and grpE proteins are absolutely essential for bacteriophage λ DNA replication as well as bacteriophage P1 plasmid replication *in vivo* and *in vitro* (39–41). The presence of both grpE and dnaJ stimulates dnaK's ATPase activity manyfold, dnaJ specifically accelerating the hydrolysis step, and grpE specifically stimulating the nucleotide release step. These results suggest that one role of dnaJ and grpE is to facilitate the intracellular recycling of dnaK (43) (analogous to the groES role shown below in Fig. 1). In this respect, it is interesting that in some organisms the three genes form an operon, *grpE-dnaK-dnaJ* (44).

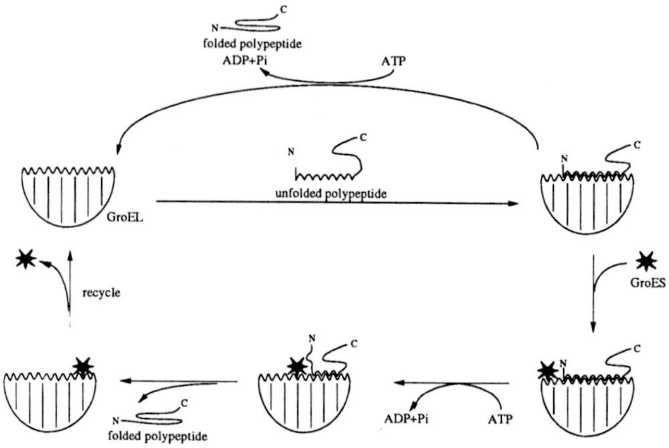

FIG. 1. **A model for the action of the groEL/groES chaperonins.** The groEL chaperonin binds to many, but perhaps not all, unfolded polypeptides, some of which may still be nascent. The hydrolysis of ATP allows the release of some polypeptides. ATP hydrolysis coupled with the "cogwheel" action of groES results in release of the rest. For details, see text and Refs. 16, 48, 49, 52, and 56.

dnaJ and grpE may also serve to "target" other proteins for action by dnaK, *e.g.* in the monomerization of P1 repA dimers, formation of a dnaJ-repA complex is a necessary first step. dnaK is attracted to this complex, either through an affinity for dnaJ or because the dnaJ-repA interaction makes repA a better substrate for dnaK binding (41). dnaJ also stabilizes the complex formed between dnaK and λP or unfolded bovine pancreatic trypsin inhibitor.[2] In contrast, the ability of dnaK to restore activity to the mutant dnaA5 and dnaA46 polypeptides requires grpE but is inhibited by dnaJ.[4] Thus, dnaJ and grpE may also participate directly in some of dnaK's functions, besides stimulating its ATPase activity.

A recent flurry of publications has revealed the presence of at least four genes in yeast with partial or total homology to dnaJ (45–47). The conservation of function is such that the cell growth phenotypes of one of these genes, YDJ1, are suppressed by plasmids carrying either the SIS1 yeast homologue (46) or even the *E. coli dnaJ* gene.[5] The function of some of these gene products appears to be necessary in protein sorting or proper nuclear structure and function. Perhaps these dnaJ-like proteins perform their functions by interacting with members of the hsp70 family located in the same cellular compartment in a manner similar to *E. coli*'s dnaJ and dnaK proteins.

The hsp60/groEL Family

The hsp60/groEL family of proteins has been widely conserved across evolution, although, to date, its members have been found only in bacteria, chloroplasts, and mitochondria. Thus, they are designated by the specialized term, "chaperonins" (22).

The *groES* and *groEL* genes of *E. coli* were originally discovered because mutations in them block bacteriophage growth at the level of assembly of the dodecameric head-tail connector structure (48). Subsequent studies have established the following facts.

(*a*) The amino acid sequence and function (see below) of the groEL protein has been widely conserved, being approximately 50% identical at the amino acid sequence level to the hsp60 protein of eukaryotes, present in all mitochondria, and the ribulose-bisphosphate carboxylase/oxygenase-binding proteins, present in all chloroplasts (22). Similarly, the groES protein also appears to be universally conserved since its homologue has been recently discovered in mitochondria (49).

(*b*) The two genes are completely indispensable for *E. coli*

[2] D. Ang, K. Liberek, D. Skowyra, M. Zylicz, and C. Georgopoulos, unpublished data.

[3] Flynn, G. C., Pohl, J., Flocco, M. T., and Rothman, J. E. (1991) *Nature* **353,** 726–730.

[4] J. Kaguni, unpublished data.

[5] A. Caplan, personal communication.

growth at all temperatures and under all conditions tested (5). The *groES* and *groEL* genes form an operon expressed from both a σ^{32}- and a σ^{70}-dependent promoter, resulting in a substantial rate of transcription of the *groE* operon, even in the absence of σ^{32} (in *rpoH*Δ strains) (4). Although wild type *E. coli* deleted for the *rpoH* gene is unable to form colonies above 20 °C (4), the overexpression of the *groES groEL* operon allows it to grow up to 39 °C, suggesting that the groE proteins have important housekeeping functions (50). The groES and groEL proteins functionally interact, leading to an inhibition of groEL's ATPase activity (48). The interaction is also absolutely necessary for the release of some unfolded polypeptides bound to groEL (see below). The groEL/groES interaction may be promoted by the presence of potential common structural features (51).

(*c*) hsp60/groEL proteins can bind various unfolded polypeptides *in vitro*. While it is unclear what exact feature(s) of unfolded polypeptides are recognized by hsp60, the binding appears to be both promiscuous and amino acid sequence-independent (16). Following denaturation, approximately half of total *E. coli* proteins bind to groEL upon dilution from the denaturant.[6] Recent reports suggest that hsp60 may recognize unfolded polypeptides in a "molten globule," collapsed structure (52) and that, upon binding, an oligopeptide in the polypeptide assumes an α-helical structure (53).

(*d*) The binding of unfolded (perhaps nascent?) polypeptides by hsp60 promotes their correct assembly by preventing premature inter- or intramolecular interactions that can lead to aggregation. Some polypeptides, like pre-β-lactamase, are readily released from groEL following ATP hydrolysis (54), whereas the release of other bound polypeptides, such as prokaryotic ribulose-bisphosphate carboxylase/oxygenase and rhodanese, necessitates the simultaneous action of the groES protein (52, 55, 56). The groES/groEL interaction has been shown to diminish the number of hydrophobic patches exposed on groEL's surface, suggesting that the complex is stabilized by hydrophobic interactions (56). Why is groES absolutely necessary for the release of some groEL-bound polypeptides? Fig. 1 depicts a naive model of groES protein action in which it accelerates the release of bound polypeptides by simply displacing them from groEL. Another possibility is that the binding and release of polypeptides by groEL is independent of groES, but perhaps some polypeptides can be quickly "recaptured" by groEL before achieving a properly folded state. groES may be required to shift the equilibrium and prevent such a reassociation with groEL.

(*e*) Although the formation of many oligomeric structures requires groEL/hsp60 proteins, there is no evidence whatsoever that groEL/hsp60 participates in the oligomerization process itself. Rather, it appears that groEL/hsp60 helps to correctly fold the monomers and that the monomers spontaneously oligomerize to form their correct oligomeric structures. In this respect, it is perhaps amusing that the efficient assembly of groEL and hsp60 themselves from monomers is catalyzed by the presence of their corresponding functional, decatetrameric structures (57, 58).

The hsp90 Family

Again, in prokaryotes, this family is represented by a single gene, called *htpG,* which is 40% identical to its eukaryotic counterparts. The htpG protein is a shortened version of its eukaryotic homologue, missing a stretch of 50 hydrophilic amino acid residues near its amino terminus and the last 35 amino acid residues from its carboxyl terminus with respect to hsp90 (59). The *htpG* gene can be easily deleted in *E. coli,* the deleted strain exhibiting no detectable phenotypes, except slow growth at very elevated temperatures (60). Perhaps other *E. coli* proteins can partially substitute for the missing htpG function.

In yeast, there are two versions of the HSP90 gene, one HS-inducible and one expressed constitutively. Whereas the deletion of either of these two genes can be tolerated, the deletion of both genes is lethal. Interestingly, expression of the mammalian homologue allows the deletion of both yeast gene copies (61), demonstrating functional conservation.

Both the prokaryotic and eukaryotic members of this family are dimeric, with phosphorylated isoforms (9, 62). The eukaryotic homologues constitute a large percentage of the total cellular protein, suggesting that they play important cellular roles. The eukaryotic protein associates with a number of interesting proteins, including pp60^{src}, members of the steroid receptor family, and a kinase that phosphorylates the α subunit of the eukaryotic initiation factor 2 (9).

Although the function of the hsp90 proteins is unclear, it is possible that they have a role similar to the hsp60 and hsp70 proteins, *i.e.* binding to nascent polypeptides, and either escorting these polypeptides to their proper cellular location, as is the case with pp60^{src}, or maintaining them in a form competent for correct folding. In the case of the steroid hormone receptors, it is known that the hsp90-steroid hormone receptor complex also includes hsp70, a minor hsp56 heat shock protein as well as 50- and 25-kDa proteins (63, 64). In the absence of steroid hormone, the receptor is kept inactive in this large complex and is activated upon hormone addition. A suggested sequence of events includes the release of the hsp90, hsp56, and 50- and 25-kDa polypeptides upon steroid hormone addition. The remaining hsp70-steroid receptor complex dissociates upon ATP hydrolysis, allowing the formation of functional steroid hormone receptor dimer (63). Recent experiments suggest that complex formation between steroid hormone receptor polypeptide and hsp90 is a prerequisite to the eventual activation of the receptor, *i.e.* in the absence of high levels of hsp90, very little active steroid hormone receptor is made (61). Thus, the initial capture and silencing of the steroid hormone receptor is an obligatory prerequisite to its eventual correct folding and assembly.

Other Selected Heat Shock Proteins

Two additional functions of HSPs may be to help proteolyze irreversibly damaged polypeptides and to provide thermotolerance to the organism. Some of the HSPs have been shown to be either proteases themselves, such as the *E. coli* ATP-dependent lon and clpP proteases (4, 65), or contribute to proteolysis indirectly, by helping "present" to proteases those polypeptides destined to be degraded, *e.g.* the eukaryotic ubiquitin protein, whose conjugation to polypeptides "tags" them for eventual proteolysis (8). Mutations in the major chaperone proteins in *E. coli* lead to hypodegradation of abnormal proteins (4), suggesting that enhancement of proteolytic activity during a heat shock is mediated through a "presentation" of chaperone-bound polypeptides to the various proteases (4). An exciting recent discovery is the characterization of the *E. coli clpB* gene, whose expression is under HS regulation. Homologues to clpB are present in all organisms, including *E. coli* itself, which also expresses the clpA protein (66, 67). It is likely, but not yet shown, that clpB complexes with clpP in *E. coli* to form an active protease similar to the clpA-clpP complex. However, some evidence exists to suggest that clpB and its shorter in-frame translation product may also act as chaperones (discussed in Ref. 67). Inactivation of *clpB* in *E. coli* leads to no detectable phenotype other than a slowing of cell growth at 44 °C (67).

It has been known for some time that when organisms are exposed to mild stress conditions, they are better able to tolerate subsequent exposures to even harsher stress conditions. The various studies in this broad field of acquired thermotolerance have been reviewed by Hahn and Li (68) and by Hightower (7). Depending on the experimental system under study, various HSPs have been implicated as contributing to thermotolerance, including hsp70, hsp27, and recently, hsp104 (7, 69). While the role of the hsp70 proteins in acquired thermotolerance can be easily understood in terms of both their protective and disaggregating abilities (37), the mechanism by which other HSPs contribute to this phenomenon is not clear. An interesting recent twist to the story of thermotolerance is the finding that the hspl04 protein of yeast (whose function is required for thermotolerance (69)) and the clpB protein of *E. coli* are highly homologous (67) and that the HSP104 gene has been universally conserved (70).

[6] G. Lorimer, personal communication.

The Regulation of the Heat Shock Response

Early studies in *E. coli* established dnaK as a key player in the negative autoregulation of the heat shock response. Mutations in *dnaK* result in the overproduction of HSPs even in the absence of a heat shock stimulus. Subsequent studies showed that the σ^{32} heat shock transcription factor, a subunit of *E. coli* RNA polymerase, is extremely unstable (half-life, <1 min!) and that mutations in *dnaK, dnaJ*, or *grpE* lead to a dramatic stabilization of its half-life (4, 5, 17). Recent experiments have shown that purified wild type dnaK, but not mutant dnaK756, can bind to purified σ^{32} polypeptide, which is released following ATP hydrolysis.[7] Thus, autoregulation of the heat shock response in *E. coli* can be explained in the following way. Under non-heat shock conditions, a large pool of free dnaK protein accumulates. Free dnaK protein can bind σ^{32}, sequestering it from RNA polymerase and/or "presenting" it to proteases, leading to its rapid degradation. Under heat shock conditions, *i.e.* in the presence of excess unfolded, misfolded, or aggregated polypeptides, dnaK associates with these proteins, allowing σ^{32} to associate with RNA polymerase core, leading to transcription from heat shock promoters and thus higher levels of heat shock gene expression. As the level of damaged proteins decreases, thus no longer occupying dnaK, free dnaK levels increase, resulting in "recapture" of σ^{32} and dampening of the heat shock response. dnaJ and grpE may directly aid the formation or stabilization of the dnaK-σ^{32} complex. Alternatively, they may regulate the heat shock response indirectly, by accelerating dnaK's ATPase activity, thus allowing it to recycle more efficiently. Preliminary evidence suggests that a similar mechanism of autoregulation operates in eukaryotes between the heat shock factor, HSF, and one or more of the major HSPs (7).[8]

Thus, the study of the heat shock response has given us insights not only into how the cell deals with rapid, unfavorable changes in its environment but also into fundamental cellular processes. The next 10 years are certain to bring a wealth of new information and perhaps development of some practical applications using heat shock proteins.

REFERENCES

1. Ritossa, F. (1962) *Experientia (Basel)* **18,** 571–573
2. Neidhardt, F. C., and VanBogelen, R. A. (1987) in *Escherichia coli and Salmonella typhimurium: Cellular and Molecular Biology* (Neidhardt, F. C., Ingraham, J. L., Low, K. B., Magasanik, B., Schaechter, M., and Umbarger, H. E., eds) pp. 1334–1345, American Society for Microbiology, Washington, D. C.
3. Lindquist, S., and Craig, E. A. (1988) *Annu. Rev. Genet.* **22,** 631–677
4. Gross, C. A., Straus, D. B., Erickson, J. W., and Yura, T. (1990) in *Stress Proteins in Biology and Medicine* (Morimoto, R., Tissieres, A., and Georgopoulos, C., eds) pp. 166–190, Cold Spring Harbor Laboratory, Cold Spring Harbor, NY
5. Georgopoulos, C., Ang, D., Liberek, K., and Zylicz, M. (1990) in *Stress Proteins in Biology and Medicine* (Morimoto, R., Tissieres, A., and Georgopoulos, C., eds) pp. 191–221, Cold Spring Harbor Laboratory, Cold Spring Harbor, NY
6. Morimoto, R., Tissieres, A., and Georgopoulos, C. (1990) in *Stress Proteins in Biology and Medicine* (Morimoto, R., Tissieres, A., and Georgopoulos, C., eds) pp. 1–36, Cold Spring Harbor Laboratory, Cold Spring Harbor, NY
7. Hightower, L. E. (1991) *Cell* **66,** 191–197
8. Schlesinger, M. J. (1990) *J. Biol. Chem.* **265,** 12111–12114
9. Welch, W. J. (1990) in *Stress Proteins in Biology and Medicine* (Morimoto, R., Tissieres, A., and Georgopoulos, C., eds) pp. 223–278, Cold Spring Harbor Laboratory, Cold Spring Harbor, NY
10. Sorger, P. K. (1991) *Cell* **65,** 363–366
11. Schuetz, T. J., Gallo, G. J., Sheldon, L., Tempst, P., and Kingston, R. E. (1991) *Proc. Natl. Acad. Sci. U. S. A.* **88,** 6911–6915
12. Smith, B. J., and Yaffe, M. B. (1991) *Mol. Cell. Biol.* **11,** 2647–2655
13. Hightower, L. E. (1980) *J. Cell. Physiol.* **102,** 407–427
14. Pelham, H. R. B. (1986) *Cell* **46,** 959–961
15. Rothman, J. E. (1989) *Cell* **59,** 591–601
16. Landry, S. J., and Gierasch, L. M. (1991) *Trends Biochem. Sci.* **16,** 159–164

[7] K. Liberek, T. P. Galitski, M. Zylicz, and C. Georgopoulos, submitted for publication.

[8] R. Morimoto, personal communication.

17. Craig, E. A., and Gross, C. A. (1991) *Trends Biochem. Sci.* **16,** 135–139
18. Neupert, W., Hartl, F.-U., Craig, E. A., and Pfanner, N. (1990) *Cell* **63,** 447–450
19. Anfinsen, C. B. (1973) *Science* **181,** 223–230
20. Buchner, J., Schmidt, M., Fuchs, M., Jaenicke, R., Rudolph, R., Schmid, F. X., and Kiefhaber, T. (1991) *Biochemistry* **30,** 1586–1591
21. Ellis, J. (1987) *Nature* **328,** 378–379
22. Ellis, R. J., and van der Vies, S. M. (1991) *Annu. Rev. Biochem.* **60,** 321–347
23. Craig, E. A. (1990) in *Stress Proteins in Biology and Medicine* (Morimoto, R., Tissieres, A., and Georgopoulos, C., eds) pp. 301–321, Cold Spring Harbor Laboratory, Cold Spring Harbor, NY
24. Beckmann, R. P., Mizzen, L. A., and Welch, W. J. (1990) *Science* **248,** 850–854
25. Pelham, H. R. B. (1990) in *Stress Proteins in Biology and Medicine* (Morimoto, R., Tissieres, A., and Georgopoulos, C., eds) pp. 287–299, Cold Spring Harbor Laboratory, Cold Spring Harbor, NY
26. Liberek, K., Skowyra, D., Zylicz, M., Johnson, C., and Georgopoulos, C. (1991) *J. Biol. Chem.* **266,** 14491–14496
27. Bukau, B., and Walker, G. C. (1990) *EMBO J.* **9,** 4027–4036
28. Kang, P.-J., Ostermann, J., Shilling, J., Neupert, W., Craig, E. A., and Pfanner, N. (1990) *Nature* **348,** 137–142
29. Scherer, P. E., Krieg, U. C., Hwang, S. T., Vestweber, D., and Schatz, G. (1991) *EMBO J.* **8,** 4315–4322
30. Flaherty, K. M., DeLuca-Flaherty, C., and McKay, D. B. (1990) *Nature* **346,** 623–628
31. Flaherty, K. M., McKay, D. B., Kabsch, W., and Holmes, K. C. (1991) *Proc. Natl. Acad. Sci. U. S. A.* **88,** 5041–5045
32. Palleros, D. R., Welch, W. J., and Fink, A. L. (1991) *Proc. Natl. Acad. Sci. U. S. A.* **88,** 5719–5723
33. Madden, D. R., Gorga, J. C., Strominger, J. L., and Wiley, D. C. (1991) *Nature* **353,** 321–325
34. Jardetsky, T. S., Lane, W. S., Robinson, R. A., Madden D. R., and Wiley, D. C. (1991) *Nature*, in press
35. Sadis, S., Raghavendra, K., and Hightower, L. E. (1990) *Biochemistry* **29,** 8199–8206
36. Rippmann, F., Taylor, W. R., Rothbard, J. B., and Green, N. M. (1991) *EMBO J.* **10,** 1053–1059
37. Skowyra, D., Georgopoulos, C., and Zylicz, M. (1990) *Cell* **62,** 939–944
38. Hwang, D. S., and Kornberg, A. (1990) *Cell* **63,** 325–331
39. Zylicz, M., Ang, D., Liberek, K., and Georgopoulos, C. (1989) *EMBO J.* **8,** 1601–1608
40. Alfano, C., and McMacken, R. (1989) *J. Biol. Chem.* **264,** 10709–10718
41. Wickner, S., Hoskins, J., and McKenney, K. (1991) *Proc. Natl. Acad. Sci. U. S. A.* **88,** 7903–7907
42. Hwang, D. S., and Kaguni, J. M. (1991) *J. Biol. Chem.* **266,** 7537–7541
43. Liberek, K., Marszalek, J., Ang, D., Georgopoulos, C., and Zylicz, M. (1991) *Proc. Natl. Acad. Sci. U. S. A.* **88,** 2874–2878
44. Wetzstein, M., and Schumann, W. (1990) *Nucleic Acids Res.* **18,** 1289
45. Blumberg, H., and Silver, P. A. (1991) *Nature* **349,** 627–630
46. Caplan, A. J., and Douglas, M. G. (1991) *J. Cell Biol.* **114,** 609–621
47. Luke, M. M., Sutton, A., and Arndt, K. T. (1991) *J. Cell Biol.* **114,** 623–638
48. Zeilstra-Ryalls, J., Fayet, O., and Georgopoulos, C. (1991) *Annu. Rev. Microbiol.* **45,** 301–325
49. Lubben, T. H., Gatenby, A. A., Donaldson, G. K., Lorimer, G. H., and Viitanen, P. V. (1990) *Proc. Natl. Acad. Sci. U. S. A.* **87,** 7683–7687
50. Kusukawa, N., and Yura, T. (1988) *Genes & Dev.* **2,** 874–882
51. Martel, R., Cloney, L. P., Pelcher, L. E., and Hemmingsen, S. M. (1990) *Gene (Amst.)* **94,** 181–187
52. Martin, J., Langer, T., Boteva, R., Schramel, A., Horwich, A. L., and Hartl, F.-U. (1991) *Nature* **352,** 36–42
53. Landry, S. J., and Gierasch, L. M. (1991) *Biochemistry* **30,** 7359–7362
54. Laminet, A. A., Ziegelhoffer, T., Georgopoulos, C., and Pluckthün, A. (1990) *EMBO J.* **9,** 2315–2319
55. Goloubinoff, P., Christeller, J. T., Gatenby, A. A., and Lorimer, G. H. (1989) *Nature* **342,** 884–889
56. Mendoza, J. A., Rogers, E., Lorimer, G. H., and Horowitz, P. M. (1991) *J. Biol. Chem.* **266,** 13044–13049
57. Lissin, N. M., Venyaminov, S. Y., and Girshovich, A. S. (1990) *Nature* **348,** 339–342
58. Cheng, M. Y., Hartl, F.-U., and Horwich, A. L. (1990) *Nature* **348,** 455–458
59. Bardwell, J. C. A., and Craig, E. A. (1987) *Proc. Natl. Acad. Sci. U. S. A.* **84,** 5177–5181
60. Bardwell, J. C. A., and Craig, E. A. (1988) *J. Bacteriol.* **170,** 2977–2983
61. Picard, D., Khursheed, B., Garabedian, M. J., Fortin, M. G., Lindquist, S., and Yamamoto, K. R. (1990) *Nature* **348,** 166–168
62. Spence, J., and Georgopoulos, C. (1989) *J. Biol. Chem.* **264,** 4398–4403
63. Smith, D. F., Faber, L. E., and Toft, D. O. (1990) *J. Biol. Chem.* **265,** 3996–4003
64. Sanchez, E. R. (1990) *J. Biol. Chem.* **265,** 22067–22070
65. Kroh, H. E., and Simon, L. D. (1990) *J. Bacteriol.* **172,** 6026–6034
66. Kitagawa, M., Wada, C., Yoshioka, S., and Yura, T. (1991) *J. Bacteriol.* **173,** 4247–4253
67. Squires, C. L., Pedersen, S., Ross, B. M., and Squires, C. (1991) *J. Bacteriol.* **173,** 4254–4262
68. Hahn, G. M., and Li, G. (1990) in *Stress Proteins in Biology and Medicine* (Morimoto, R., Tissieres, A., and Georgopoulos, C., eds) pp. 79–100, Cold Spring Harbor Laboratory, Cold Spring Harbor, NY
69. Sanchez, Y., and Lindquist, S. L. (1990) *Science* **248,** 1112–1115
70. Parsell, D. A., Sanchez, Y., Stitzel, J. D., and Lindquist, S. (1991) *Nature* **353,** 270–273

The Journal of Biological Chemistry

1988 Minireview Compendium

CONTENTS*

* Page numbers refer to the original page numbers as printed in the 1988 issues.

Author Index

The Journal of Biological Chemistry

1989 Minireview Compendium

CONTENTS*

* Page numbers refer to the original page numbers as printed in the 1989 issues.

Author Index

The Journal of Biological Chemistry

1990 Minireview Compendium

CONTENTS*

* Page numbers refer to the original page numbers as printed in the 1990 issues.

The Tables of Contents for the 1988 and 1989 Minireview Compendiums are reprinted at the end of this volume.

Author Index